The Origin and Evolution of Mammals

The Origin and Evolution of Mammals

T. S. Kemp

Oxford University

OXFORD

UNIVERSITY PRESS

OXFORD
UNIVERSITY PRESS

Great Clarendon Street, Oxford OX2 6DP
Oxford University Press is a department of the University of Oxford.
It furthers the University's objective of excellence in research, scholarship,
and education by publishing worldwide in

Oxford New York

Auckland Cape Town Dar es Salaam Hong Kong Karachi
Kuala Lumpur Madrid Melbourne Mexico City Nairobi
New Delhi Shanghai Taipei Toronto

With offices in

Argentina Austria Brazil Chile Czech Republic France Greece
Guatemala Hungary Italy Japan South Korea Poland Portugal
Singapore Switzerland Thailand Turkey Ukraine Vietnam

Oxford is a registered trade mark of Oxford University Press
in the UK and in certain other countries

Published in the United States
by Oxford University Press Inc., New York

A catalogue record for this book is available from the British Library

Library of Congress Cataloging in Publication Data
Data available

ISBN 0 19 850760 7 (hbk)
ISBN 0 19 850761 5 (pbk)

10 9 8 7 6 5 4 3 2

Typeset by Newgen Imaging Systems (P) Ltd., Chennai, India
Printed in Great Britain
on acid-free paper by
Antony Rowe Ltd., Chippenham, Wiltshire

Preface

This book arose from twin aims: to update my 1982 book on the origin of mammals in relation to the steady stream of new fossils that have been found or re-described in the last two decades; and to elaborate at a similar level the story of what happened next – the radiation of mammals after the end of the Cretaceous. In the latter case even more than in the former, constraints of space and time have meant severely limiting this enormous field and even then I have been able to do little more than point to the most important fossils, outline the ideas, and give guidance to appropriate recent literature for the details. The whole corresponds to a twelve lecture option I give to Oxford Honour students, and I am very grateful to all those who have, over the years, attended in encouraging numbers and stimulated me with their interest, despite the alternative, fashionable attractions on offer in the biological sciences. I hope they will find this volume helpful.

I am particularly grateful to Professor Zofia Kielan-Jaworowska who read and very helpfully commented on Chapter 5, and was generous enough to allow me access to the manuscript of her joint new book *Mammals from the age of dinosaurs: origins, evolution, and structure*, a magnificent, comprehensive work which will be the standard account of Mesozoic mammals for the foreseeable future. I should also like to take the opportunity of thanking Fernando Abdala for reading Chapter 3, and to Sam Turvey for reading Chapters 6 and 7; both made extremely helpful suggestions for improving the manuscript. I am grateful to Kuba Kopycinski for his helpful advice on electronic preparation of figures.

Acknowledgements are due to the following publishers and journals for kindly giving permission to reproduce illustrations.

Acta Palaeontologica Polonica; American Journal of Science; Bernard Price Institute for Paleontological Research; C.N.R.S., Paris; California University Press; Cambridge University Press; Carnegie Museum of Natural History; Columbia University Press; Elsevier Ltd; *Evolution*; Geological Society of America; Indiana University Press; John Wiley and Sons; *Journal of Morphology*; Kluwer Academic Press; Monash University, Melbourne; Musée Nationale d'Histoire Naturelle, Paris; Museum of Comparative Zoology, Harvard; Museum of Northern Arizona Press; National Academy of Sciences, U.S.A.; National Geographic Society; National Museum of Bloemfontein; National Research Council of Canada; Natural History Museum, London; *Nature*; Oxford University Press; *Paleobiology; Paleontologicheskii Zhurnal*; Prentice Hall; Royal Society of Edinburgh; Royal Society of London; Royal Zoological Society of New South Wales; *Science*; Selbstverlag Fachbereich Geowissenschaften; Serviços Geológicos de Portugal; Smithsonian Institution Press; Society for the Study of Mammalian Evolution; Society of Vertebrate Paleontology; South African Museum; Springer-Verlag; University of Michigan; University of Wyoming Department of Geology and Geophysics; W. H. Freeman; *Zoological Journal of the Linnean Society*.

University Museum and St John's College
Oxford

For Małgosia
with love and thanks

Contents

1 Introduction

The definition of Mammalia   1

A sketch of the plot   3

Palaeobiological questions   4

2 Time and classification

The geological timescale   6

Classification   9

3 Evolution of the mammal-like reptiles

The vertebrate conquest of land: origin of the Amniota   14

Pelycosauria: the basal synapsid radiation   19

The origin and early radiation of the Therapsida   26

Biarmosuchia   31

Dinocephalia   33

Anomodontia   39

Gorgonopsia   52

Therocephalia   56

Cynodontia   60

Interrelationships of Cynodontia and the phylogenetic position of Mammalia   75

Overview of the interrelationships and evolution of the Therapsida   78

The palaeoecology and evolution of Synapsida   80

4 Evolution of mammalian biology

Feeding   90

Locomotion   101

Sense organs and brain   113

Growth and development   120

Temperature physiology   121

An integrated view of the origin of mammalian biology   129

The significance of miniaturisation   135

5 The Mesozoic mammals

The diversity of the Mesozoic mammals   138

An overview of the interrelationships and evolution   180

The general biology of the Mesozoic mammals   183

End of the era: the K/T mass extinction and its aftermath   186

6 Living and fossil marsupials

Living marsupials and their interrelationships 191

The fossil record of marsupials 196

An overview of marsupial evolution 216

7 Living and fossil placentals

Living placentals and their interrelationships 222

Cretaceous fossils 226

Palaeocene fossils: the archaic placentals 230

The origin and radiation of the modern orders 250

Overview of placental evolution 274

References 291

Index 325

CHAPTER 1

Introduction

There are about 4,600 species of animals today that are called mammals because, despite an astonishing diversity of form and habitat, they all share a long list of characters not found in any other organisms, such as the presence of mammary glands, the single bone in the lower jaw, and the neocortex of the forebrain. This makes them unambiguously distinct from their closest living relatives, and their unique characters together define a monophyletic taxon, the class Mammalia. Three subgroups are readily distinguished amongst the living mammals. The Monotremata are the egg-laying mammals of Australasia, consisting only of two species of echidna and a single platypus species; for all their primitive reproductive biology, monotremes are fully mammalian in their general structure and biology. The Marsupialia, or Metatheria are the pouched mammals, whose approximately 260 species dominate the mammalian fauna of Australia, and also occur as part of the indigenous fauna of the Americas. By far the largest group of living mammals are the Placentalia, or Eutheria with about 4,350 species divided into usually eighteen recent orders.

It is virtually unanimously accepted that the closest living relatives, the sister group, of mammals consists of the reptiles and the birds. The only serious dissent from this view in recent years was that of Gardiner (1982) who advocated that the birds alone and mammals were sister groups, the two constituting a taxon Haemothermia, defined among other characters by the endothermic ('warm-blooded') temperature physiology. Gardiner certainly drew attention to some remarkable similarities between birds and mammals, notably the details of the endothermic processes, the enlarged size and surface folding of the cerebellum, and a number of more superficial morphological features. There was also some molecular sequence data supporting the

Haemothermia concept, including the beta-globin gene and 18S rRNA. Gardiner's view briefly became a *cause célèbre* in part for its sheer heterodoxy, but all concerned have since rejected it on the grounds that a careful, comprehensive analysis of the characters supports the traditional view (Kemp 1988*b*), particularly if the characters of the relevant fossils are taken into account (Gauthier, Kluge, and Rowe 1988). Nor does it receive any significant support at all from the mass of nucleotide sequence data now available.

The definition of Mammalia

The formal definition of Mammalia is simple as far as the living mammals are concerned, because of the large number of unique characters they possess. However, the fossil record makes the situation a good deal less clear-cut. The first part of this book is about the extraordinarily good fossil record of animals that were, to varying degrees, intermediate in grade between the modern mammals and the last common ancestor between them and their living sister group. They are known informally as the 'mammal-like reptiles', and more formally as the non-mammalian Synapsida. By definition, a mammal-like reptile possesses some, but not all the characters that define living mammals. A semantic difficulty arises about defining a mammal because a decision has to be taken on which, if any of the mammal-like reptiles should be included. The earliest, most primitive ones have very few mammalian characters, just a small temporal fenestra in the skull and an enlarged canine tooth in the jaw. In contrast, there are later forms that possess almost all the modern mammalian skeletal characters, lacking only a few of the postcranial skeleton ones like a scapula spine, and fine details of the ankle structure. If the definition of a mammal is based rigorously upon possession of *all* the characters of living mammals, then some fossil forms that are extremely mammalian in anatomy, and by inference in their general biology, are excluded. If, on the other hand, a mammal is defined as an animal that possesses *any* of the modern mammal characters, then some extremely non-mammalian forms, primitive, sprawling-limbed, and no doubt scaly, ectothermic creatures must be included. If a compromise is sought by using certain selected living mammal characters as the basis of the definition, then it becomes an arbitrary decision as to exactly which characters are to be given defining status.

After rather a lot of fruitless discussion during the last half-century about this issue generally, two alternative approaches emerged. Developing one of Willi Hennig's (1966) proposals in his original prescription for cladistics, Jefferies (1979) and Ax (1987) proposed distinguishing a 'crown' group for the living members of a taxon from a 'stem' group for all the more basal fossil forms possessing some but not all the characters of the crown group. Rowe (1988, 1993) adopted this solution and formally defined the taxon Mammalia as those organisms possessing, or presumed to have possessed, all the characters of living mammals. All those fossil groups that possess at least one, but less than all of the characters of living mammals can be referred to as stem-group Mammalia, a paraphyletic, but informally recognisable taxon.

Developments of both technology and fashion in systematics have since overtaken the simple concept of crown versus stem groups based respectively on all, or less than all the defining characters (Benton 2000). The ability of computer programs such as PAUP to handle vastly more amounts of information has led to computer-generated cladograms involving large numbers of taxa and characters. Even in the maximally parsimonious cladogram, many of the characters are inevitably homoplasic, occurring independently in different groups. Therefore, it is impossible to read from the cladogram simple lists of characters that define the various contained groups. How much more so is this true of phylogenetic trees based on molecular sequence data, which can involve computation of several thousand nucleotides? The formal creation

of taxon definitions has therefore shifted from definitions based on lists of characters, to labelling nodes in the cladogram: a taxon is defined as the clade that includes the common ancestor of subgroups X and Y. The evidence for the existence of a clade so defined is that it occurs in the best-supported cladogram. In this rarified sense, Mammalia is the clade that includes the common ancestor of monotremes, marsupials, and placentals. The shift in fashion that has simultaneously affected the naming of groups, whether character- or node-based, is a reluctance to accept the vagueness of paraphyletic stem groups, but as far as reasonable to name every monophyletic group in it. Thus Rowe (1988, 1993) introduced the formal names Mammaliamorpha and Mammaliaformes at two of the nodes on the stem lineage below Mammalia, and two more above it, Theriomorpha, Theriiformes, before the node Theria that represents the common ancestor of marsupials and placentals.

An altogether different perspective on defining Mammalia is based on traditional palaeobiological practice (e.g. Simpson 1960; Kemp 1982). An arbitrary decision is made about which characters to select as defining characters, and therefore which particular node on the stem lineage to label Mammalia. Characters deemed appropriate are those reflecting the evolution of the fundamental mammalian biology. The essence of mammalian life is to be found in their endothermic temperature physiology, greatly enlarged brain, dentition capable of chewing food, highly agile, energetic locomotion, and so on. The organisms that achieved this grade of overall organisation are deemed to be Mammalia, and consequently those characters that they possess are the defining characters of the group. In this view, missing an odd few refinements such as, free ear ossicles or the details of the ankle joint is insufficient justification for denying mammalian status to a fossil that is otherwise mammalian in structure. Around the end of the Triassic Period, about 205 Ma, a number of fossils are found of very small animals that have the great majority of the skeletal characters of modern mammals. The brain is enlarged and the postcranial skeleton differs from that of a modern mammal only in a few minor details. Their novel feature is the jaw mechanism. The dentition is fully differentiated with transversely occluding molar teeth that functioned in the unique manner of basic living mammals, and there is a new jaw hinge between the dentary bone of the lower jaw and the squamosal of the skull that permitted a much stronger bite. Such animals were undoubtedly mammalian in the biological sense although not strictly members of the crown-group Mammalia. The most primitive of these forms is called *Sinoconodon*, and most palaeobiologists believe that they should formally be members of the clade designated Mammalia (e.g. Crompton and Sun 1985; Kielan-Jaworowska *et al.* 2004). The definition of Mammalia thus becomes: synapsids that possess a dentary-squamosal jaw articulation and occlusion between lower and upper molars with a transverse component to the movement. This has exactly the same membership as the clade that includes the common ancestor of *Sinoconodon*, living mammals, and all its descendants.

This is the concept of a mammal that is used in the chapters that follow, in the belief that conceding a degree of subjectivity in the choice of what is a mammal is a small price to pay for allowing the focus of the work to be on the origin and evolution of the quality of 'mammalness' just as much as on the origin and evolution of the taxon Mammalia.

A sketch of the plot

The story of the origin and evolution of mammals as told by the fossil record falls into three distinct phases, the first of which led to the origin of mammals as such. The lineage of amniotes that culminated in the mammals made its first appearance in the fossil record of the Pennsylvanian (Upper Carboniferous) about 305 Ma. The best-known form from this time is *Archaeothyris*, which existed very soon after the initial appearance of the terrestrially adapted amniote animals. From a hypothetical ancestor comparable to *Archaeothyris*, the radiation of mammal-like reptiles commenced and occupied the succeeding 100 Ma. Taxa appeared, flourished, and went extinct in a complex kaleidoscopic pattern of successive groups of small-, medium-, and large-sized carnivores and herbivorous groups, which over time exhibited increasing numbers of mammalian characters superimposed on their particular specialisations. In the latest

Triassic Period, 210 Ma, one particular lineage of small carnivores culminated in the first mammal.

The second phase of the unfolding story consists of what are referred to as the Mesozoic mammals. These were the numerous subgroups that radiated from the mammalian ancestor during the 140 Ma duration of the Jurassic and Cretaceous Periods, the time during which the world's terrestrial fauna was dominated by the dinosaurs. A good deal of evolution, especially dental evolution occurred, but none of them ever evolved the medium or large body size found in so many modern mammals. For two-thirds of the whole of their history, mammals remained small animals with the largest being barely larger than a cat, and the vast majority of the size of living shrews, mice, and rats. With hindsight, the most important evolutionary event was the origin of the modern mammalian kind of molar tooth, known as the tribosphenic tooth, and with it the roots of the two major modern taxa, marsupials and placentals.

Sixty-five million years ago, the mass extinction marking the close of the Mesozoic Era saw the end of the dinosaurs along with the extinction or severe depletion of many other taxa. Several mammal lineages survived this event, and within a mere 2–3 Ma had radiated explosively to produce a plethora of new small forms but also, for the first time, mammals of middle to large body size. This was the commencement of the great Tertiary radiation of placental and, to a lesser but equally interesting degree, marsupial mammals during which many unfamiliar, often quite bizarre kinds evolved, flourished, and disappeared. The Tertiary world has been a period of dramatic biogeographical and climatic change, against which is set this extraordinary evolutionary pageant. The old, single supercontinent of Pangaea occupied by the mammal-like reptiles had long since broken into Laurasia in the north and Gondwana in the south, and by the early Tertiary, Gondwana itself was breaking up. Africa and India for a short while, and South America and Australia for most of the era were island continents with extremely limited biotic contact with each other and with the northern land masses. Even amongst the Laurasian continents of North America, Europe, and Asia, interconnections formed and broke from time to time as land masses shifted, sea levels varied, and ice ages alternated with warm times. Thus, for much of the Tertiary, independent evolutionary radiations of mammals were occurring simultaneously in different areas, with secondary contacts between hitherto isolated faunas adding complex patterns of dispersal and competition. As time passed the mammalian fauna contained more and more of the familiar groups of today. The final great event in the evolution of mammals, and the cause of the last touches to the shaping of the modern fauna, occurred a mere 10–20 thousand years ago, when most of the species of larger mammals disappeared from the fossil record. Whether this was due to environmental change or over-hunting by humans is still vigorously debated.

Palaeobiological questions

The quality of the fossil record documenting this series of events is remarkably good. Vertebrate skeletons generally have a high capacity for being preserved, and because they are so anatomically complex, many characters are potentially available for taxonomic analysis. A further consequence of its complex, multifunctional nature is that a more or less complete, well-preserved vertebrate skeleton reveals a great deal of biological information about the life of its one-time possessor. Unfortunately, set against this high information content is the fact that good specimens do tend to be rare, because vertebrates are relatively large animals that therefore live in relatively small populations compared to many other kinds of organisms. In the case of terrestrial vertebrates the rarity is exaggerated because the only usual way that a fossil forms is if the dead organism suffers the unlikely fate of ending up buried in sediments forming on the bed of a lake, delta, or sea. There are wonderful exceptions to this rule of improbability, and a handful of freakish conditions were responsible for some Lagerstätten in which the preservation of mammals and other taxa is superb, and which open unique windows onto the faunas of the time. The Early Cretaceous Yixian Formation of Liaoning Province of China is one such (Luo 1999; Zhou *et al.* 2003), with complete skeletons of a variety of little mammals, some even with impressions of the pelt. The Grube Messel Eocene locality, an old opencast oil-shale mine near

Frankfurt that is early Eocene in age contains a sample of forest-dwelling mammals of the time that even includes gut contents and muscle and skin impressions (Schaal and Zeigler 1992). At the other end of the temporal scale, the famous Rancha La Brea asphalt pits, within the city boundary of Los Angeles, were trapping mammals in seeping oil from the Late Pleistocene 40,000 years ago to about 4,500 years ago (Marcus and Berger 1984). Even such fossiliferous localities of these do not break the general rule that fossil tetrapod vertebrates are individually highly informative, but collectively very uncommon.

The palaeobiological questions that can be addressed with reference to the origin and evolution of mammals are circumscribed by these qualities of the synapsid fossil record. Microevolutionary issues such as intraspecific changes in response to selection pressures, or the process of speciation cannot usually be approached using these particular fossils, although there are a number of cases where Tertiary mammal species have been used to demonstrate microevolutionary trends, particularly, though not exclusively from the Pleistocene, as any palaeobiological textbook will attest. However, there are several general macroevolutionary questions concerning events whose time course is a matter of millions to tens and hundreds of millions of years where the synapsid fossil record has been and continues to be extremely important.

Concerning the first phase of synapsid radiation, the big question is that of how a major new kind of organism evolves, and the case of the origin of mammals is far and away the example with the most revealing fossil record. Very many characters changed in the course of the transition from ancestral amniote to mammal, and yet biological integration between all the structures and processes of each and every intermediate organism must have been maintained. From the pattern and sequence of acquisition of mammalian characters inferred from the cladistic analysis of mammal-like reptiles, much light is thrown on the process of the origin of a new higher taxon. From palaeoecological analysis, something of the environmental backdrop to the process is revealed.

The most compelling macroevolutionary question concerning the second phase of synapsid evolution, the Mesozoic mammals, is to do with evolutionary constraints and potentials. Why did none of them evolve large or even medium body size for 140 Ma, and then suddenly several of them did so independently of one another? Was it due to competitive exclusion by the contemporary dinosaurs, which would indicate that this is an extraordinarily powerful evolutionary force, or was it a consequence of some physiological or environmental constraint acting on these early mammals, in which case what kind of environmental trigger was necessary to release it?

Finally, the third phase of the story, the great Tertiary mammal fossil record lends itself extraordinarily well to an analysis of the nature of adaptive radiation. Much is revealed about the way in which a simply expressed phylogeny is actually the outcome of a complex interplay between the evolving organisms and the dramatically changing climates and palaeogeography that characterise the terrestrial sphere. It allows such questions to be posed as, why is convergent evolution of many ecotypes such as small omnivores, ungulates, carnivores, or anteaters so common? Why do some groups successfully invade new areas while others completely fail to do so? Why do older taxa tend to go extinct in the face of younger ones? Why have some very old taxa nevertheless survived? A powerful new technique has arrived in the last decade that is recasting many questions about the Tertiary radiation of mammals. Viewed not long ago with suspicion verging on scepticism by the palaeontological community, molecular taxonomy has emerged as a major player. Inferring phylogenetic relationships and times of divergence from gene sequences tens of kilobases long, and applying highly sophisticated analytical methods, a complete and well-supported phylogeny of living mammals is close to realisation. Initially there were profound conflicts between fossils and molecules, but the two are coming to be seen to be telling the same stories more and more. Some of these are most surprising. Whales are indeed a specialised branch of even-toed ungulates, and there are now fossils as well as molecules to show it. A morphologically highly disparate group of orders based in Africa are beyond question a monophyletic superorder, Afrotheria, as far as molecular evidence is concerned; now it actually makes biogeographic sense as well. These are exciting times to be a palaeomammalogist!

CHAPTER 2

Time and classification

The age and the classification of a particular fossil are the two fundamental properties necessary to begun understanding how it fits into the evolutionary patterns revealed by the fossil record. There are often misunderstandings of one or other of these by specialists. Evolutionary biologists on occasion express far too optimistic a view of how accurately fossils can actually be dated, both absolutely and relative to one another. Geologists have been known to have a rather limited view of how modern systematic methods are used to infer relationships from large amounts of information, be it morphological or molecular. In this chapter, a brief outline of the principles underlying the construction of the geological timescale, and of a classification are given, along with reference timescales and classifications for use throughout the following chapters.

The geological timescale

The creation of a timescale for dating the events recorded in the rocks since the origin of the Earth is one of the greatest achievements of science, unspectacular and taken for granted as it may often be. It is also unfinished business insofar as there are varying degrees of uncertainty and inaccuracy about the dates of many rock exposures, none more so than among the mostly continental, rather than marine sediments containing the fossils with which this work is concerned. A geological timescale is actually a compilation of the results of two kinds of study. One is recognising the temporal sequence of the rocks, and agreeing arbitrarily defined boundaries between the named rock units, the result of which is a chronostratic timescale. The other is calibration of the sequence and its divisions in absolute time units of years before present, a chronometric timescale.

Chronostratic timescale

It is simple in principle to list the relative temporal order of events, such as the occurrence of fossils, in a single rock unit, although even here the possibility of missing segments, known as hiatuses, in local parts of the unit, or of complex folding movements of the strata disturbing the order must not be forgotten. The biggest problem is correlating relative dates between different units in different parts of the world. The potential markers available for correlation are the numerous kinds of signals in the rocks of particular events that had a widespread, ideally global effect over a geologically brief period of time. Historically, the occurrence of particular fossil species was the most important, followed by evidence of climatic change such as tillites indicating glaciation, and evaporates associated with aridity. Changes in sea levels are indicated by shifting coastline sediments, and periods of intense volcanic activity by igneous rock extrusions. More modern techniques reveal characteristic sequences of reversals of the magnetic field. Ratios of stable isotopes have become particularly important geochemical signals of several kinds of physical and biotic events. If the ideal marker is a clear geochemical signal of a brief, but globally manifested event that affected all environments, then the best are the effects of a massive bolide impact, such as the enhanced iridium levels marking the Cretaceous–Cenozoic boundary. Unfortunately, such ideal examples are extremely infrequent.

After global correlation has been completed, the boundaries between the named chronostratigraphic units have to be arbitrarily defined by selecting a single point in a particular exposure, known as a GSSP (Global Stratotype Section and Point) or a 'Golden Spike'. To gain maximum international agreement, a hierarchy of committees under the aegis of the IUGS (International Union of Geological Sciences) ratifies these decisions. The units themselves are organised hierarchically, and within the Phanerozoic Eon this is first into three Eras, Palaeozoic, Mesozoic, and Cenozoic. These are then divided successively into Periods, Epochs, and Stages as shown on the chart (Fig. 2.1). The final division is usually into Zones, which are defined by the presence of one, or a few characteristic fossil species. Zones are therefore primarily local rather than global, and typically less than 1 Ma in duration.

Correlation is most problematic when comparing very different environments, for there may be virtually no geological markers shared between them. Because most sedimentary rocks are marine in origin, most of the global chronostratic timescale is based on these. There is particular difficulty in correlating marine and continental rocks, and therefore in relatively dating the latter. In a number of cases, the position in practice is that a local chronostratic timescale exists that has not yet been accurately correlated with the global scale. The new techniques, particularly stable isotope ratios are, however, rapidly improving the situation, a notable relevant example being the mammal-bearing Cenozoic rocks of North America. Traditionally a succession of North American Land Mammal Ages (NALMA) was recognised, whose relationship to the global Cenozoic timescale was poorly understood, but the correlation between the two is now virtually complete (Prothero 1997). The equivalent LMA's of other parts of the world are not yet completely correlated with those of North America, but undoubtedly this work will be completed quite soon.

Earlier continental deposits relevant here are less well correlated with the global timescale, especially the Permo-Carboniferous of North America, the Late Permian of South Africa and Russia, and the Triassic of South Africa, South America, and Asia. They are shown in Fig. 2.2.

Chronometric timescale

The only way of absolutely dating points in the chronostratigraphic timescale is radiometry, based on isotope decay rates. The principle is that there is a constant rate of spontaneous decay of a specified radioactive isotope into its daughter product. By knowing the ratio of the isotopes at the start and at the end of a time period, and the rate of decay, the absolute elapsed time of the period can be calculated. However, since the ratio of isotopes at the start of the period cannot be measured, the technique depends on the assumption that at the time of formation of the sample being dated, it contained 100% of the mother isotope and none of the daughter product, which can only be true for a sample that was formed by crystallising out of molten rock. Therefore only igneous rocks can be radiometrically dated from their time of formation. Sedimentary rocks, including virtually all the fossil-bearing rocks, are a random mixture of particles of rocks of differing ages, derived by weathering processes going on somewhere quite other than where they finally settled. Fossils cannot be directly dated! The one exception to this assertion is radiocarbon dating which is somewhat different. It depends on the fact that while alive and metabolising, an organism's tissues have the same ratio of the radioactive ^{14}C to the stable ^{12}C as in the atmosphere, about 1%. After death, the ^{14}C decays to ^{12}C and so, assuming the ratio in the atmosphere was the same at the time of death as it is today, the ratio in the organic sample depends only on the rate of decay. However, the half-life of this particular process is a mere 5,730 years, which is too short to date accurately organic samples much more than 50,000 years old. It is primarily an archaeological and not a palaeontological tool.

The most widely used isotope pair for construction of the global chronometric timescale is potassium–argon, in which the isotope ^{40}K decays to ^{40}Ar, with a half-life of about 1,500 million years. Uranium–lead ($^{235}Ur/^{207}Pb$: half-life about 700 Ma), and Rubidium–Strontium ($^{87}Rb/^{87}Sr$: half-life 50,000 Ma) are also commonly used. The accuracy of radiometric dating can be compromised by several factors. Secondary loss or addition of isotopes may have occurred during the lifetime of the rock sample, particularly if it had been exposed to high temperatures. The half-life is difficult to measure with complete precision for such a slow process. Laboratory techniques for measuring what are often extremely small percentages introduce experimental errors. The outcome is that radiometric dates are generally only accurate to about 2%. While for many purposes this is adequate,

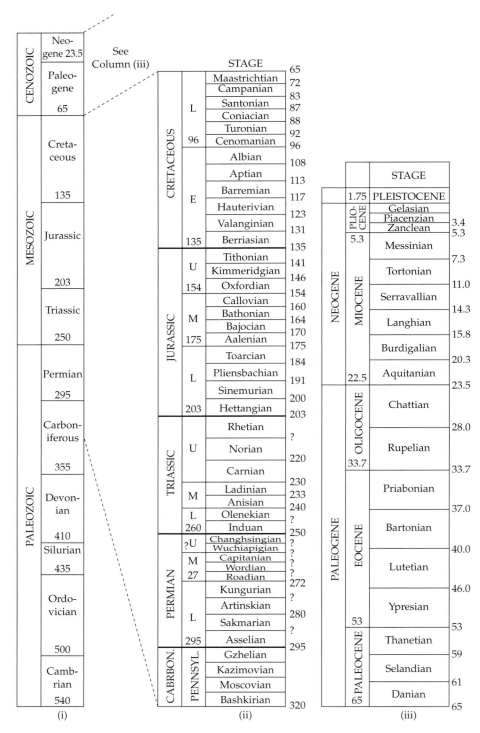

Figure 2.1 The International Stratigraphic Chart for the Phanerozoic (2000 version) modified from Briggs and Crowther (2001).

it means that a 200 Ma igneous rock of, say, Upper Triassic when the first mammals appeared may be dated at plus or minus 4 Ma, a significant time in evolutionary terms.

Absolute dating of an actual fossil requires that the age of the sediment containing it is assessed relative to available absolutely dated rocks that are close to it in time and space. This often demands such further error-prone procedures as measuring the thickness, and estimating the rate of deposition of the sediments separating the fossil layer and the igneous rock exposure.

The International Stratigraphic Chart
The relevant parts of the 2000 version of the International Stratigraphic Chart for Phanerozoic time are presented in Fig. 2.1 for reference. It is not always completely consistent with the text that follows because a lot of the literature on synapsids has not been based upon this particular timescale. One of the most prominent inconsistencies is the frequent use by synapsid palaeontologists of a twofold division of the Permian into Early and Late Epochs, compared to the threefold division now adopted, of Lower, Middle, and Upper. Another is the difference between the European-dominated use of Lower, Middle, and Upper Epochs for the Carboniferous, in contrast to the North American twofold arrangement into Mississippian and Pennsylvanian that has been adopted by the international scheme.

Classification

For all organisms, fossil and living, the first requirement for practically any evolutionary interpretation is a cladistic analysis of their relationships, whether expressed as a cladogram, or as a formal classification of monophyletic groups. A hypothesis of the branching pattern that connects species and taxa allows objective inferences to be made of the pattern and sequence of evolution of characters, and of the biogeographic history of the group. The battles of the 1970s and 1980s over the appropriateness and significance of cladism have long since been won and lost, and the focus has turned from the question of principle to the question of how actually to analyse the huge amounts of information about characters and taxa now available, and requiring ever-more powerful computer and statistical techniques.

Morphological characters

Dealing with the morphological characters of fossils still relies primarily on parsimony, that is to say, the assumption of a model of evolution based on minimum evolutionary change. No other model, for instance one that might make assumptions about different probabilities of change in different sorts of characters, is defensible in the absence of any independent method of measuring such probabilities. Therefore, minimising the total number of inferred evolutionary changes in characters generates the simplest hypothesis of the overall pattern of evolution. Apart from the obvious incorrectness of the assumption of equal probability of change in all characters, morphological analysis suffers from the intractable problem of what constitutes a unit character. That different authors can study exactly the same fossils but then produce different cladograms, sometimes markedly different ones, is mainly due to the unit character issue. As an extreme example, a single mammalian molar tooth could be coded as a single character state such as 'hypsodont', or by maybe a dozen characters, one for each cusp or loph. Searching for whether proposed multiple characters are always correlated with one another, and assuming in such a case that they should be treated as a single one, potentially could remove highly informative evidence of relationships, if the characters were in fact independently evolved. This and other problems such as how to assess which characters are homologous with which, how to treat multistate characters, what to do about characters not preserved in some of the specimens, and so on are discussed at length in the literature, for example, by Kitching *et al.* (1998) and will not be pursued here. But the fact remains that for all the confidence some authors place in their long lists of morphological characters, their application of such programmes as PAUP, and the associated statistical tests for how well the data fits and how robust the cladogram is, there remain major outstanding disagreements between them over several critical parts of the synapsid cladogram.

Molecular taxonomy

The rise of molecular taxonomy based on sequences of nucleotides in genes has revolutionised the field of classification of living taxa, and nowhere has there

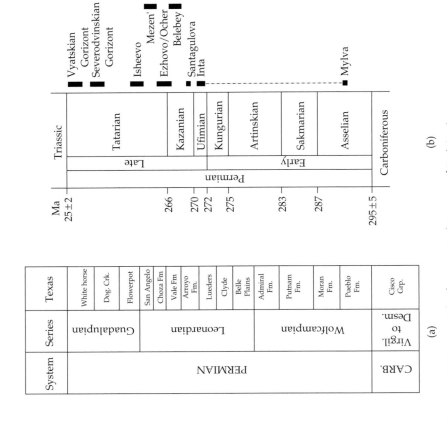

Figure 2.2 (a) The Permo-Carboniferous stratigraphic sequence of North America. (b) The Permian stratigraphic sequence of Russia (after Modesto and Rybczynski 2000). (c) The Permo-Triassic sequence of Assemblage Zones of South Africa.

been a more fundamental impact than on mammalian systematics. It began with the now very primitive-looking immunological techniques generating such radical proposals as the close relationship of humans to the African apes (Sarich and Wilson 1967), and the monophyly of the Australian marsupials (Kirsch 1977). With the increased availability of comparative sequence, initially of proteins and subsequently nucleic acids, unfamiliar new hypotheses arrived, like the monophyly of Afrotheria, a morphologically very mixed bag of placental orders with no evident morphological characters in common at all, and the improbable-looking sister-group relationship of whales to hippos. However, both these are now widely, if not universally agreed upon by palaeontologists, who indeed have found support from fossils. In fairness of course, other early molecular proposals have disappeared without trace, such as guinea pigs not being rodents. In equal fairness, though, rejection of the latter was due primarily to further molecular evidence, rather than morphological evidence. The present position is that relatively enormous volumes of sequence data are already available across virtually the whole of living mammal diversity at the level of orders, and increasingly of families and lower taxa. This includes a mixture of sources. Many, and increasing numbers of complete mitochondrial genomes have been sequenced, along with a variety of nuclear genes. Databases for analysis are already ten to twenty thousand nucleotides long (10–20 kb), with presumably vastly increased potential resolving power. Nucleotide sequences have several advantages over morphological descriptions besides the more or less limitless amount of information available. It is easy in principle to recognise a single nucleotide as a unit character. It is also possible to have estimates of differential probabilities of different kinds of nucleotide substitutions, depending, for example, on whether it is a transition or a transversion, or first, second, or third placed position in the codon. Therefore, more realistic models of the evolutionary process are possible for inferring relationships from the data, with the help of suitably sophisticated methods such as maximum likelihood, and Bayesian inference. Again this subject is extensively discussed in several texts such as Salemi and Vandamme (2003).

A few years ago, it was legitimate to question whether molecular phylogeny was more reliable than morphological, and consequently whether it should be allowed to overrule the latter in cases of conflicting hypotheses of relationships. A common reaction was that molecular and morphological data should be combined to produce a compromise phylogeny. This was always a conceptually difficult thing to achieve, and all too often simply resulted in a very poorly resolved cladogram instead. However, the increasing number of occasions on which newer palaeontological evidence, both morphological and biogeographical, has tended to support, or at least be compatible with the molecular evidence suggests that with a careful enough analysis, molecular data is perfectly capable of resolving the interrelationships of mammalian groups that have living members. Indeed, it is heralding an exciting new phase in mammalian palaeobiology. With much more strongly supported and widely agreed phylogenies to hand, the fossil record is open to a whole new set of improved interpretations about the complex interplay between phylogenesis, biogeography, and palaeoecology that constitutes the history of mammals.

A classification of Synapsida

The classifications in Tables 2.1–2.4 are designed for general reference. All the formally expressed groups are monophyletic clades, but not all relationships between them are fully resolved to dichotomies. In the classifications of the Synapsida and the Mammalia, certain ancestral, or paraphyletic group names are included but in quotation marks, either because they are so familiar from the literature that they remain useful nomenclatural handles, or because the relationships of their constituent subgroups are inadequately known. Monophyletic groups whose cladistic position is relatively uncertain are indicated by a query.

These classifications are designed for the relatively uninitiated in the field of synapsid systematics. They are therefore a compromise between a fully cladistic classification that would be overly name-bound, and would tend to convey overconfidence in the expressed relationships, and an 'evolutionary' classification with undeclared paraphyletic groups and the consequent loss of genealogical information. For more detail and discussion, the appropriate chapters should be consulted.

Table 2.1 Classification of the Synapsida

	SYNAPSIDA
Subclass	"PELYCOSAURIA"
	Caseasauria
	Eothyrididae
	Caseidae
	"Eupelycosauria"
	Varanopseidae
	Ophiacodontidae
	Edaphosauridae
	Sphenacodontia
	Haptodontidae
	Sphenacodontidae
Subclass	THERAPSIDA
	Nikkasauridae (?)
	Niaftosuchus (?)
	Estemmenosuchidae (?)
	Biarmosuchia
	Biarmosuchidae
	Burnetiidae
	Dinocephalia
	Brithopia
	Titanosuchia
	Titanosuchidae
	Tapinocephalidae
	Styracocephalidae
	Anomodontia
	Anomocephalus
	Patronomodon
	Venyukovioidea
	Dromasauroidea
	Dicynodontia
	Eodicynodontidae
	Eudicynodontia
	Theriodontia
	Gorgonopsia (?)
	Eutheriodontia
	Therocephalia
	Cynodontia
	"Procynosuchia"
	Procynosuchidae
	Dviniidae
	Epicynodontia
	Galesauridae
	Thrinaxodontidae
	Eucynodontia

For further details see Chapter 3. Modified after Reisz (1986), Kemp (1982, 1988c), Hopson (1991), and Hopson and Kitching (2001).

Table 2.2 Classification of the Mesozoic mammals

	CLASS MAMMALIA
	Adelobasileus (?)
	Sinocondon
	Kuehneotheriidae (?)
	Morganucodontia
	Morganucodontidae
	Megazostrodontidae
	Docodonta
	Hadroconium
	"Eutriconodonta"
	Triconodontidae
	Gobiconodontidae
	Amphilestidae
Subclass	ALLOTHERIA
	Haramiyida (?)
	Multituberculata
	"Plagiaulacida"
	Cimolodonta
	TRECHNOTHERIA
	"Symmetrodonta"
	Amphidontidae
	Tinodontidae
	Spalacotheriidae
	"Eupantotheria"
	Dryolestidae
	Paurodontidae
	Vincelestidae
	Amphitheriidae
	Shuotheriidae (?)
Subclass	TRIBOSPHENIDA (=BOREOSPHENIDA)
	Aegialodontidae
	Metatheria
	Deltatheriidae (?)
	Marsupialia
	Placentalia (=Eutheria)
Subclass	AUSTRALOSPHENIDA (?)
	Ausktribosphenida
	Monotremata

Simplified from Kielan-Jaworowska *et al.* (2004). The arrangement is cladistic except in the case of groups in quotation marks, which are paraphyletic, groups with query marks whose cladistic position is in doubt, and listed families which are left unresolved. Taxon names in parentheses are those used by Kielan-Jaworowska *et al.* (2004). For further details see Chapter 5.

Table 2.3 Classification of marsupials

MARSUPIALIA

AMERIDELPHIA
  †*Kokopellia*
  †Stagonodontidae
  †Pediomyidae
  †Glasbiidae
  Didelphimorphia
    Didelphidae
    †Sparassocynidae
  Sparassodontia
    †Mayulestidae
    †Borhyaenidae
  Paucituberculata
    †Caroloameghinoidea
    †Polydolopoidea
    Caenolestoidea
    †Argyrolagoidea
AUSTRALIDELPHIA
  Microbiotheria
  †Yalkaparidontia
  Peramelemorphia
  Dasyuromorphia
  Notoryctemorphia
  Diprotodontia

† extinct taxa

Table 2.4 Classification of the orders of living mammals
Based on the molecular analysis of Murphy et al. (2001*b*)

PLACENTALIA

AFROTHERIA
      Macroscelidea
      Tubulidentata
    Paenungulata
      Proboscidea
      Hyracoidea
      Sirenia
    Afrosoricida
      Tenreca
      Chrysochlorida
XENARTHRA
      Xenarthra
BOREOEUTHERIA
  LAURASIATHERIA
      Cetartiodactyla
      Perissodactyla
      Carnivora
      Chiroptera
      Eulipotyphla
  EUARCHONTOGLIRES
    Glires
      Rodentia
      Lagomorpha
    Euarchonta
      Primates
      Dermoptera
      Scandentia

CHAPTER 3

Evolution of the mammal-like reptiles

Mammals, along with the biologically remarkably similar birds, are the vertebrates that are most completely adapted to the physiological rigours of the terrestrial environment. Whilst all the terrestrial dwelling tetrapods can operate in the absence of the buoyancy effect of water, and can use the gaseous oxygen available, mammals have in addition evolved a highly sophisticated ability to regulate precisely the internal temperature and chemical composition of their bodies in the face of the extremes of fluctuating temperature and the dehydrating conditions of dry land. From this perspective, the origin of mammalian biology may be said to have commenced with the emergence of primitive tetrapods onto land around 370 Ma, in the Upper Devonian.

The vertebrate conquest of land: origin of the Amniota

Until the 1990s, the only Devonian tetrapod at all well known was *Ichthyostega* from east Greenland (Fig. 3.1(c)), as described by Jarvik (e.g. Jarvik 1980, 1996). Famous for its combination of primitive fish-like characters such as the lateral line canals, bony rays supporting a tail fin, and remnants of the opercular bones, with fully tetrapod characters such as the loss of the gills and opercular cover, robust ribcage, and of course large feet with digits, *Ichthyostega* provided more or less all the fossil information there was relating to the transition from a hypothetical rhipidistian fish to a tetrapod. Subsequently, however, an ever-increasing number of other Upper Devonian tetrapods have been described, and the emerging picture of the origin of vertebrate terrestriality has become more complicated and surprising (Ahlberg and Milner 1994; Clack 2002). The earliest forms are Upper Frasnian in age, and include *Elginerpeton* from the Scottish

locality of Scat Craig (Ahlberg 1995, 1998). So far known only from a few bones of the limbs and jaws, *Elginerpeton* adds little detail to the understanding of the evolution of tetrapods except to demonstrate that the process had commenced at least 10 million years prior to the existence of *Ichthyostega*.

The next oldest tetrapods are Fammenian in age and include *Ichthyostega*, and a second east Greenland form, *Acanthostega* (Fig. 3.1(a and b)), which has been described in great detail (Coates and Clack 1990, 1991; Coates 1996). This genus has proved to be very surprising because, despite being of the same age and apparently living in the same habitat as *Ichthyostega*, it was evidently not adapted for actual terrestrial life. In common with *Ichthyostega*, enclosed lateral line canals and a fish-like tail were present, but furthermore, a full set of ossified gill bars covered by a bony operculum was still present. The limbs, while certainly tetrapodal in lacking fin rays, were relatively short, stubby, and bore eight digits, and could not possibly have supported the weight of the animal out of water. Given its more basal cladistic position compared to *Ichthyostega* and all later tetrapods, the anatomy of both the limbs and the gills of *Acanthostega* led to the proposal that the tetrapod limb originally evolved as a specialised organ for aquatic locomotion. Indeed, Coates and Clack (1995) suggested that the whole Devonian tetrapod radiation including not only *Acanthostega*, but also *Ichthyostega* and the possible basal amniote *Tulerpeton* (Lebedev and Coates 1995) were entirely aquatic animals. Support for this idea has recently come from the discovery that *Ichthyostega* does in fact also possess gill bars, and has a unique ear structure designed for hearing under water rather than in air (Clack *et al.* 2003). This view is consistent with the argument, put forward by Janis and Farmer (1999), that a primarily

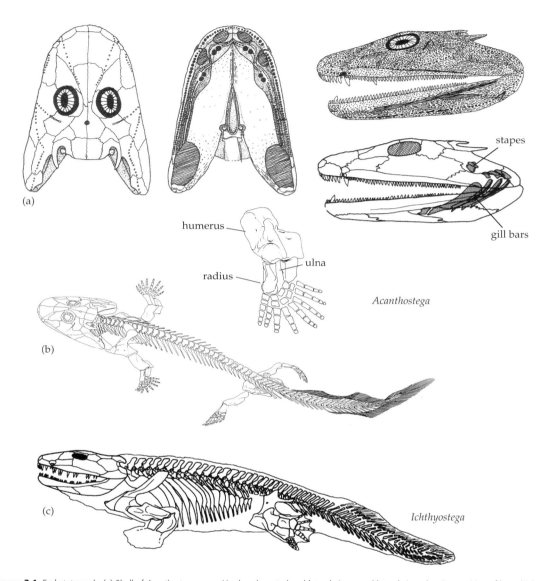

stapes

humerus

ulna

radius

gill bars

Acanthostega

(a)

(b)

(c)

Ichthyostega

Figure 3.1 Early tetrapods. (a) Skull of *Acanthostega gunnari* in dorsal, ventral and lateral views, and lateral view showing position of branchial arches. Skull length up to 20 cm (Clack 2002) (b) Reconstruction of the skeleton, and the bones of the forelimb of *Acanthostega gunnari* (Coates 1996). (c) Skeleton of *Ichthyostega*. Length approx. 1 m (Coates and Clack 1995).

aquatic tetrapod would not be expected to lose its gills even if living in low-oxygen waters, because of the effectiveness of gills in nitrogen and carbon dioxide excretion. On this argument, only gill-less tetrapods would be expected to have been in the habit of spending at least periods of their life completely out of the water.

Once into the Carboniferous, a substantial radiation of tetrapods commenced and there were lineages showing a wide range of respective degrees of terrestrial adaptation (Clack 2002). Many had secondarily reduced or even lost the limbs and developed streamlined bodies, while the great majority retained lateral lines. Although only demonstrable in the few cases where larval forms are known, presumably they mostly retained the anamniotic egg laid in water and hatching into a gilled, fully aquatic larval form.

Evolutionary speculations concerning the origin of tetrapods, as for other major new kinds of organisms, have been dominated by the idea of preadaptation. This is defined as the process whereby a structure of an organism that is adapted for a particular function in a particular habitat, becomes co-opted for a different function in a different habit. In fact, this is not a very helpful concept, because characters such as limbs and lungs in pre-tetrapods evidently did not change their function on attaining land. Rather, they were adaptations for aspects of the environment that were common to both the shallow water and to the muddy bank alongside. It is more plausible to view the origin of tetrapods as a transition along an ecological gradient in which there are no abrupt barriers. To be adapted to shallow, low-oxygen water a vertebrate requires the capacity for substrate locomotion and air breathing, a feeding mechanism suitable for small terrestrial organisms falling in, and aerial sense organs. These same adaptations must exist in a vertebrate adapted to life on the muddy bank adjacent to the water. Even such apparently radical differences between the two habitats as loss of buoyancy, increased dryness, and exposure to heat are in effect gradual across the boundary insofar as an organism can ameliorate them by temporarily returning to the protection of the water. From this perspective, the radiation of early tetrapods consists of lineages adapted to various points along the water to land gradient.

Within the context of the Carboniferous radiation of tetrapods, the lineage that evolved the most extreme terrestrial adaptations, to the extent that its members were eventually completely independent of free-standing water were the Amniota, represented today by the reptiles, birds, and mammals. The modern reptiles, lizards, snakes, crocodiles, and turtles, possess internal fertilisation and the cleidoic egg, greatly reduced skin permeability, a water-reabsorbing cloaca, and use of solid uric acid for excretion. Such physiological characters as these cannot normally be demonstrated in fossils, but fortunately the modern forms also possess a number of osteological characters that do permit

a diagnosis of the group Amniota, so that exclusively fossil groups can be added. Laurin and Reisz (1995) list nine skeletal synapomorphies, including for instance the entry of the frontal bone of the skull into the orbital margin, a row of large teeth on the transverse flange of the pterygoid, absence of enamel infolding of the teeth, two coracoid bones in the shoulder girdle, and the presence of a single astragalus bone formed from fusion of three of the ankle bones. The earliest known possible amniote is *Casineria* (Paton *et al.* 1999), which comes from the Middle Carboniferous of Scotland. Unfortunately, only a partial postcranial skeleton is preserved, so its correct taxonomic attribution remains dubious. The earliest certain amniote fossils occur in the Upper Carboniferous (Middle Pennsylvanian) of South Joggins, Nova Scotia. *Hylonomus* and *Paleothyris* are found preserved in a remarkable way, entombed within approximately 315 Ma petrified lycopod tree stumps (Carroll 1969). They are relatively small forms with a body length of about 10 cm, and are members of a group Protorothyrididae (Fig. 3.2(b) and (e)), which is related to the diapsid and testudine reptiles (Carroll 1972; Laurin and Reisz 1995; Berman *et al.* 1997).

Surprisingly perhaps, a second distinct group of amniotes appears at about the same time, characterised most significantly by the presence of a single opening low down in the cheek region of the skull. This is the synapsid version of temporal fenestration and is the most prominent, unambiguous hallmark of the Synapsida, the mammal-like reptiles and mammals (Fig. 3.2(c)). Carroll (1964) attributed some fragmentary remains from the same locality as *Hylonomus* to the synapsids, naming them *Protoclepsydrops*, although in his monograph on the pelycosaurs, Reisz (1986) was unable positively to confirm this conclusion. Be that as it may, there are indisputable pelycosaur synapsids (Fig. 3.2(f)) only very slightly younger than *Hylonomus*, namely *Archaeothyris* (Reisz 1972).

There is no serious questioning of the monophyly of the amniotes, including both protorothyridids and synapsids, but the relationships of the group to other Carboniferous tetrapod taxa is presently a matter of intense debate and disagreement. Amongst proposed stem-group amniotes, the earliest and most basal is the poorly known Late Devonian

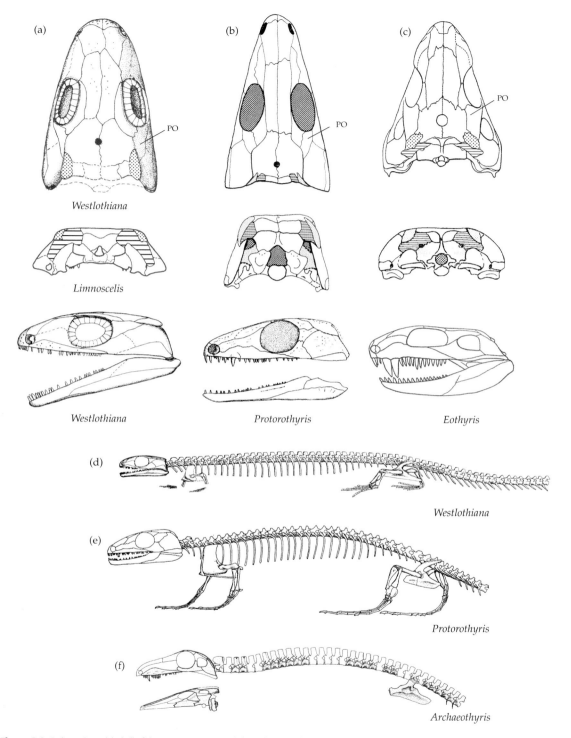

Figure 3.2 *Early amniotes.* (a) Skull of the stem amniote *Westlothiana lizziae* in dorsal and lateral views with occipital view of *Limnoscelis* (*Westlothiana* from Smithson *et al.* 1994; *Limnoscelis* from Laurin and Reisz 1995). (b) Dorsal, occipital, and lateral views of the skull of *Protorothyris* (Carroll 1964; Panchen 1972). (c) Dorsal, occipital, and lateral views of *Eothyris* (Romer and Price 1940; Panchen 1972). (d) Postcranial skeleton of *Westlothiana lizziae*. Presacral length approx. 11 cm (Smithson *et al.* 1994). (e) Postcranial skeleton of *Protorothyris*. Presacral length approx. 12 cm (Carroll 1964). (f) Postcranial skeleton of *Archaeothyris florensis*. Presacral length approx. 30 cm (Reisz 1972). PO, postorbital. Supratemporal stippled; Tabular cross-hatched.

Tulerpeton (Lebedev and Coates 1995), which, if correctly interpreted, indicates that the living Amniota must have separated from their closest living relatives, the Amphibia, at least 360 Ma. If *Tulerpeton* is rejected, the next candidate for earliest stem-amniote is the 333 Ma Lower Carboniferous *Westlothiana* (Fig. 3.2(a) and (d)), which is known from complete skeletons (Smithson *et al.* 1994). It was a small, short-limbed, superficially very reptile-like animal with a number of amniote characters such as the structure of the vertebral column and certain details of the skull bone pattern. There is a considerable diversity of later Carboniferous tetrapods that are widely, if not universally considered stem-group amniotes (Clack and Carroll 2000; Heatwole and Carroll 2000). The seymouriids, once believed to be the ancestral reptile group, include quite large highly terrestrially adapted forms. Probably the majority view at present is that the apparently herbivorous diadectids along with the large, crocodile-like limnoscelids together constitute the closest relatives of amniotes (Laurin and Reisz 1995, 1997; Berman *et al.* 1997; Gauthier *et al.* 1988). Other analyses place the diadectomorphs as the sister group of the Synapsida alone, and therefore regard them as actual members of the Amniota (Panchen and Smith 1988; Berman 2000).

Whatever the exact relationships might be, it is clear that Amniota was the one lineage within the complex radiation of early tetrapods that achieved a high level of terrestrial adaptation. Assuming that the protorothyridids and synapsids are indeed the two sister groups constituting a monophyletic Amniota, the structure of the hypothetical common ancestor of the two can be inferred from a comparison of the most primitive, Carboniferous members such as *Hylonomus* and *Archaeothyris*, respectively. It was a relatively small, superficially lizard-like animal, with a presacral body length of maybe 10 cm.

The classic view of the evolution of the skull in amniotes was that the synapsid skull such as that of *Archaeothyris* evolved from an anapsid skull such as that of a protorothyridid by little more than opening of a temporal fenestra low down behind the orbit (e.g. Carroll 1970*b*). However, Kemp (1980*b*, 1982) noted that the protorothyridid skull also possessed uniquely derived characters, absent from the synapsids. Notably, the postorbital, supratemporal, and

tabular bones were all reduced in size. The implication is that the common ancestral amniote skull must have combined the absence of a temporal fenestra with large postorbital, supratemporal, and tabular bones. This configuration is found in certain early tetrapods such as the limnoscelids and probably *Westlothiana* (Fig. 3.2(a)), both considered stem-group amniotes. Significantly, in both these latter cases the posterior part of the skull appears to have retained at least some degree of flexibility between the skull table and the postorbital cheek region. This in turn is probably part of a type of cranial kinetism found in the most primitive of tetrapods, which is associated with suction feeding in water. Thus the hypothetical ancestral amniote was by inference as much an aquatic feeder as a fully evolved terrestrial one. The strengthening of the hind part of the skull that is necessary for an actively biting animal feeding on land must have evolved subsequently, differently, and therefore independently in the two respective amniote groups, explaining the difference in cranial structure between them. In addition to the repatterning of the temporal bones, other features of the evolution of the amniote skull are related to strengthening of the skull. The back of the skull, the occiput, was braced firmly against the cheek to resist the increasing forces generated by the increased jaw closing musculature, whilst elaboration of the tongue and hyoid apparatus in the floor of the mouth may well have occurred to assist in the capture and oral manipulation of food (Lauder and Gillis 1997). To judge by the homodont dentition of sharp teeth rounded in cross-section, insects and other invertebrates constituted the principal diet.

There were new features of the postcranial skeleton of the ancestral amniote, although as Sumida (1997) shows, evolution of the characteristic amniote postcranial skeleton was quite gradual, with characters accumulating in several of the stem-amniote groups, and leading eventually to a highly terrestrially adapted form (Fig. 3.2(e)). Early on there was the evolution of the typical amniote vertebral structure in which the spool-shaped pleurocentral element and neural arch are firmly connected, and small ventral intercentra lie between adjacent pleurocentra. The limbs became relatively larger. The most characteristic postcranial feature acquired at the fully amniote level is the fusion of three of the ankle bones to form an astragalus, increasing the strength and precision of movements of the ankle.

Compared to the living amphibians, there are many physiological, soft tissue, and behavioural features characteristic of living amniotes, but of course very little is known about how, when, and in what sequence they evolved. Of all these, none is more significant than the amniote egg. However, understanding the evolution of this and the associated reproductive mode of amniotes is frustratingly difficult, not only because reproductive physiology is generally indecipherable from fossils, but also because the modern amniotes are very similar to one another in this respect, and very different from their closest living relatives, the Amphibia. (Stewart 1997). At best, occasional individual amphibian species demonstrate possible analogues of amniote features, such as internal fertilisation, or the laying of terrestrial eggs. A number of authors have stressed the primary importance of increased egg size in tetrapods that have direct terrestrial development compared to amphibians with the typical aquatic larval phase. Carroll (1970*a*) noted the relationship between body size and egg size in plethodontids, a group of amphibians that includes several that lay terrestrial eggs from which miniature adults emerge directly, without a larval phase. He inferred that the first amniotes to lay eggs on land that still lacked extraembryonic membranes must have been very small organisms with a precaudal body length of about 40 mm, because otherwise the eggs they laid would have been too large for adequate gas exchange. Packard and Seymour (1997) speculated that the initial evolutionary innovation towards the amniotic egg was the reduction of the jelly layers that surround a typical anamniotic egg and their replacement by a yolk sac, in order to increase the rate of gas diffusion from air to embryo. At the same time, the evolution of a proteinaceous egg membrane would have increased the mechanical strength of the egg, as well as further increasing its permeability. Given that these changes permitted some increase in egg size, further increase would have required more yolk provision, consequently imposing meroblastic rather than holoblastic development and requiring a vascular system for the transport of food from the yolk sac to the embryonic cells, an argument previously adduced by Elinson (1989). On this argument therefore, the extended yolk

sac was the first of the extraembryonic membranes to evolve. The allantoic sac was perhaps originally a small extension of the embryonic gut that functioned as storage of excretory products, and later the amniotic folds evolved to create a space surrounding the embryo so that the allantois could enlarge and become a functional lung. At any event, the final outcome of whatever sequence of evolutionary modifications occurred was the full amniotic egg, capable of developing on dry land and producing a miniature adult no longer dependent on free-standing water. The significance of this for the effective invasion of land, and ultimately the origin of mammals can hardly be exaggerated.

Other biological aspects of the origin of amniotes are equally obscure, such as the origin of amniote skin and scales, and their various protective and supportive functions (Frolich 1997). Virtually nothing is known of the evolution of the various water-conservation adaptations.

Pelycosauria: the basal synapsid radiation

The taxon Pelycosauria was introduced by E. D. Cope in the nineteenth century for the North American members of what subsequently became recognised as the most primitive members of the Synapsida. By the time of Romer and Price's (1940) *Review of the Pelycosauria*, which is one of the great classics of vertebrate palaeontological literature, the term was universally accepted. With the advent of cladistic classification, the group Pelycosauria was recognised as paraphyletic, but it continues to be widely used for 'basal', or 'non-therapsid' synapsids, in which informal, but extremely useful sense it has been adopted here.

Reisz (1986) noted the synapomorphies of the Synapsida, which are therefore the characters that distinguish pelycosaurs from non-synapsid amniotes:

- temporal fenestra primitively bounded above by the squamosal and postorbital bones
- occipital plate broad and tilted forwards
- reduced post-temporal fenestra
- postparietal bone single and medial
- septomaxilla with a broad base and massive dorsal process.

There are numerous characters in which the pelyco-saurs retain the primitive condition compared to the therapsids. These include the relatively small size of the temporal fenestra, the absence of distinctive canine teeth in the upper and the lower jaws, and the absence of a reflected lamina of the angular bone of the mandible. In the postcranial skeleton, the limbs are relatively short and heavily built and the girdles massive, indicating a lumbering, sprawling gait.

The modern cladistic framework for understanding the interrelationships of the pelycosaurs was established by Reisz (1986), who abandoned the largely paraphyletic groupings of Romer and Price (1940). This was slightly modified by Berman *et al.* (1995) as a result of re-study of some of the critical genera, and their cladogram is illustrated here (Fig. 3.3). The biggest difference from earlier understanding of the taxonomy is the recognition of a basal group of pelycosaurs which, although tending to be highly specialised, possess a primitive skull structure. These, the Caseasauria, have broad, low skulls, large postorbital and supratemporal bones, and a very limited exposure of the frontal bone in the margin of the orbit. The rest of the pelycosaurs form a group now named the Eupelycosauria.

Caseasauria: Eothyrididae

Eothyris (Fig. 3.2(c)) is known from a single specimen of a small skull, only about 6 cm in length, from the Early Permian Wichita Formation of Texas. The most striking feature of the skull is its exceptionally broad, flat shape and unique dentition, which consists of two very large upper caniniform teeth on each side. No comparably enlarged lower teeth are present. Set within a row of otherwise small, pointed teeth, the upper canines indicate a specialised, presumably carnivorous mode of life, but nothing at all is known of the postcranial skeleton, which might have thrown further light on its habits.

Langston (1965) described another eothyridid, *Oedaleops*, from a single skull (Fig. 3.4(a)), which has a similar size and proportions to *Eothyris* but in which the caniniform teeth are much less prominent. It dates from the contemporaneous Cutler Formation of New Mexico but is more primitive than *Eothyris* by virtue of its more generalised dentition.

Caseasauria: Caseidae

The caseids (Fig. 3.4(b) and (c)) share with the eothyridids an enlargement of the external nostrils

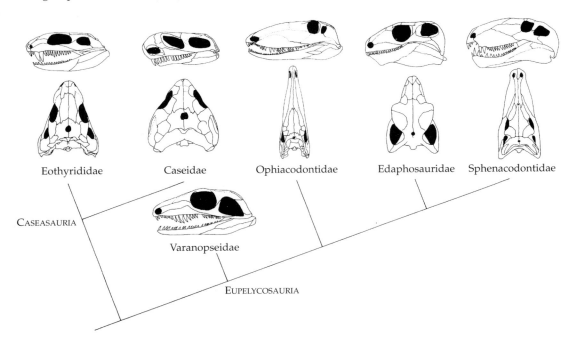

Figure 3.3 Phylogeny of the families of pelycosaurs (Kemp 1988).

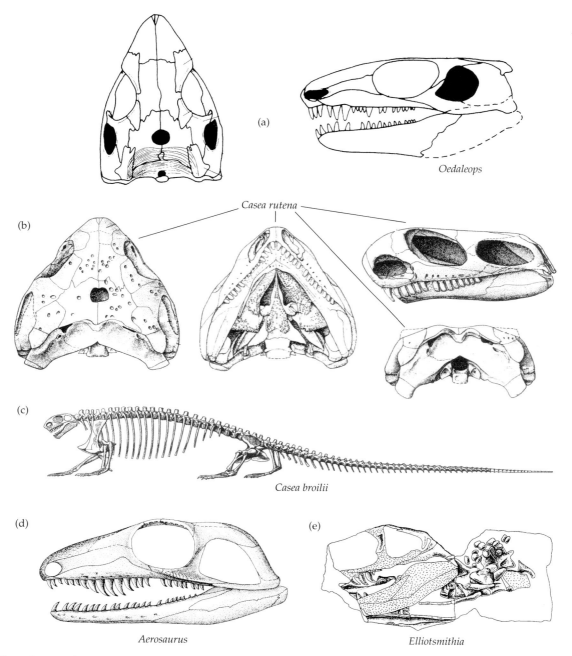

Figure 3.4 Caseid and varanopeid pelycosaurs. (a) Skull of *Oedaleops campi* in dorsal and lateral views (Kemp 1982 after Langston 1965). (b) Skull of *Casea rutena* in dorsal, ventral, lateral and posterior views (Sigogneau-Russell and Russell 1974). (c) Reconstruction of the skeleton of *Casea broilii* (Romer and Price 1940). (d) Skull of *Aerosaurus wellesi* in lateral view (Langston and Reisz 1981). (e) Skull of *Elliotsmithia* as preserved, seen in left side view, showing orbit, temporal fenestra, and posterior teeth (Dilkes and Reisz 1996).

and a presumably correlated tendency towards a forward tilt of the very front of the pointed snout. However, they are highly modified for herbivory (Reisz and Sues 2000). The skull is short and low, the temporal fenestra large, and the jaw articulation depressed below the level of the tooth row. These are features relating to enlargement of the adductor musculature to increase the force of the bite. The dentition consists of a homodont marginal dentition in which each individual tooth is stout, blunt, and somewhat bulbous. The palate bears several rows or areas of small palatal teeth which, since there are no comparable teeth on the inner surface of the lower jaw, presumably worked in conjunction with a tough, muscular tongue, an interpretation supported by the presence of a particularly well-developed hyoid apparatus in the floor of the mouth. The skeleton of caseids is also characteristic of a herbivore (Fig. 3.4c). Proportionately the skull is small and the neck vertebrae reduced. The body is barrel-shaped, as indicated by the ribcage, implying that there was a very large alimentary canal with a fermentation chamber somewhere along its length. There is a wide size range in caseids (Olson 1968). *Casea broilii*, for example, had a body length of around a metre, while the largest species known is *Cotylorhynchus hancocki* with a length of at least 4 m and a body weight estimated to have been in excess of 500 kg.

Langston (1965) suggested that *Oedaleops* was a good model for the ancestry of the caseids. The Caseidae appear relatively late in the fossil record, not until into the Early Permian, but on the other hand were one of the last surviving pelycosaur groups, being known from the Late Permian rocks of the San Angelo and Flowerpot Formations of Texas (Olson 1968). They are also amongst the few pelycosaur taxa with representatives from other parts of the world. There is a beautifully preserved Early Permian specimen named *Casea rutena* (Fig. 3.4b) from France (Sigogneau-Russell and Russell 1974), and *Ennatosaurus* is from the Kazanian Zone II of the Late Permian of Russia (Ivakhnenko 1991).

Eupelycosauria: Varanopseidae

All other pelycosaurs are included in the Eupelycosauria, characterised by elongation and narrowing of the snout, the frontal bone forming at least one-third of the dorsal orbital margin, the supratemporal bone narrow and set in a groove, and the pineal foramen set relatively further back in the parietal bone (Reisz 1986; Reisz *et al.* 1992). The members of the family Varanopseidae are the least derived of the eupelycosaurs. Relative to other pelycosaurs, they were small, long-limbed, agile animals with a long row of sharp teeth and little development of caniniforms, and presumably they led an active, predatory mode of life. The best known are *Varanosaurus* itself and *Aerosaurus* (Langston and Reisz 1981), both from the Early Permian (Fig. 3.4(d)). However, the family is actually the longest lived and most widely distributed of all pelycosaur families. Some specimens of *Aerosaurus* also date from the Upper Carboniferous, while other taxa share with the caseids the distinction of surviving into the Late Permian. *Varanodon* occurs in the Chickasha Formation of Oklahoma, and there is also a Russian Late Permian form, *Mesenosaurus* (Reisz and Berman 2001). Even more remarkably, a genus of varanopseid occurs in the southern hemisphere. *Elliotsmithia* (Fig. 3.4e) has long been known from a poorly preserved skull from the Late Permian *Tapinocephalus* Assemblage Zone of the South African Karoo, and Dilkes and Reisz (1996; Reisz *et al.* 1998) confirmed the impression of several earlier authors including Romer and Price (1940) that *Elliotsmithia* is indeed properly referred to the varanopseids. Modesto *et al.* (2001) have described a second specimen further confirming this conclusion. So far this is the only well-established record of pelycosaurs from Gondwana, although another form, *Anningia*, has been claimed to be a South African pelycosaur. Unfortunately, the sole specimen is a skull which is too incomplete and poorly preserved for confirmation to be possible (Reisz and Dilkes 1992).

Eupelycosauria: Ophiacodontidae

The traditional taxon Ophiacodontia was regarded as the most primitive group of pelycosaurs by Romer and Price (1940), but they included the eothyridids as well as the eupelycosaurian family Ophiacodontidae, and it was therefore paraphyletic. Taken alone, the latter family is a monophyletic group of eupelycosaurs containing a number of markedly long snouted forms (Berman *et al.* 1995). The earliest adequately known pelycosaur, *Archaeothyris*

(Fig. 3.2(f)), is actually an ophiacodontid. It is a small pelycosaur with a skull about 9 cm long and a body length of around 50 cm including the tail, that was discovered in fossilised lycopod tree stumps of Florence, Nova Scotia, in Pennsylvanian deposits dated at about 315 Ma. Its long, slender snout bears a row of little, sharply pointed teeth, one of which is distinctly caniniform. The postcranial skeleton of

Archaeothyris is not very well known. The generally similar, but relatively much larger and longer snouted *Ophiacodon* (Fig. 3.5(a)) is one of the best known of all pelycosaurs. It appears in the latest Upper Carboniferous Stephanian of North America and remains common throughout the Early Permian, when some species achieved a very large size, up to about 300 cm in length. There are several indications

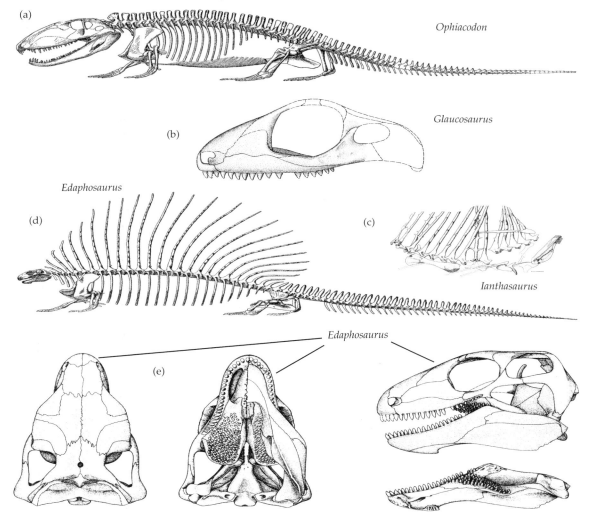

Figure 3.5 Ophiacodontid and edaphosaurid pelycosaurs. (a) reconstructed skeleton of *Ophiacodon retroversus* (Romer and Price 1940). (b) Incomplete skull of *Glaucosaurus megalops* (Modesto 1994). (c) Specimen of *Ianthasaurus hardestii* showing cross-pieces on neural spines, and simple dentition (Reisz and Berman 1986). (d) Reconstructed skeleton of *Edaphosaurus boanerges* (Romer and Price 1940). (e) Skull of *Edaphosaurus boanerges* in dorsal, ventral and lateral views, and internal view of lower jaw (Modesto 1995).

that *Ophiacodon* was an aquatic, fish-eating animal (Romer and Price 1940), with its long, narrow snout, numerous sharp teeth, and absence of a distinct caniniform. The orbits are high up in the side of the skull. The digits are flattened suggesting that the animal indulged in some degree of aquatic paddling.

Varanosaurus is closely related to *Ophiacodon* (Berman *et al.* 1995), but differs in the flattened top and sides of the snout, creating a box-like effect. Its axial skeleton is most unusual, for the neural arches of the vertebrae are swollen, the neural spines alternate in height, and the zygapopophyseal surfaces are close to horizontal, much more so than in other pelycosaurs (Sumida 1989). Thus the vertebrae resemble those of several primitive tetrapod groups such as seymouriamorphs, diadectids, and captorhinids, undoubtedly as a result of convergence. Sumida (1989) proposed that the arrangement of the vertebrae in *Varanosaurus* permitted greater dorso-ventral flexion of the column as well as lateral flexion, giving the animal an enhanced degree of agility of aquatic locomotion.

Eupelycosauria: Edaphosauridae
Edaphosaurids are mainly specialised herbivores which superficially resemble caseids, and indeed were for long classified with them as a taxon Edaphosauria. However, there are significant differences in skull structure between members of the two, and also primitive edaphosaurids lacking the full set of adaptations for herbivory are known. Both these observations indicate that caseids and edaphosaurids independently evolved convergent herbivorous adaptations.

Edaphosaurus itself (Fig. 3.5(d) and (e)) is highly distinctive. It is known from the Late Pennsylvanian well into the Early Permian, and occurs in Europe as well as North America. Species vary in body length from as little as 100 cm up to around 325 cm. The skull is very small compared to the body size, and is short, strongly constructed and has a deep lower jaw indicating powerful adductor musculature. Each of the marginal teeth possesses a cutting edge and is set obliquely across the jaw. The tooth row as a whole would have functioned in cropping vegetation, as indicated by the development of wear facets (Modesto 1995; Reisz and Sues 2000). A second dental mechanism is created by large tooth plates bear-

ing numerous teeth, occurring on both the palate and the opposing inner sides of each lower jaw, which would have had an effective crushing action. The postcranial skeleton is most distinguished by the sail along the back, supported by hugely elongated neural spines from neck to lumbar region. Unlike the comparable sail in *Dimetrodon* (page 26), here there are lateral tubercles or short crosspieces. While usually assumed that the sail of pelycosaurs, where present, had a thermoregulatory role, there has been some doubt about this interpretation in *Edaphosaurus*, because the area of the sail does not scale with body size as predicted (Modesto and Reisz 1990). However, Bennett (1996) conducted wind-tunnel experiments on a model sail and showed that the effect of the transverse projections is to increase the turbulence of airflow and therefore the rate of exchange of heat across the surface of the sail membrane. A relatively smaller sail could therefore remain effective as a heat exchanger in a larger animal.

The Upper Pennsylvanian *Ianthasaurus* (Fig. 3.5(c)) is a member of the family Edaphosauridae by virtue of possessing elongated neural spines with transverse tubercles, and also a lateral shelf above the orbit formed from an enlarged postfrontal bone (Reisz and Berman 1986; Modesto and Reisz 1990). However, it lacks the specialisations for herbivory that make *Edaphosaurus* so distinctive, retaining sharp, slightly flattened and recurved teeth and little enlargement of the temporal fenestra. It is also one of the smallest pelycosaurs known, with a skull length of about 8 cm and a presacral body length of about 35 cm. Such an animal was presumably an insectivore. Modesto (1994) also re-described the Early Permian form *Glaucosaurus*, known only from a single, incomplete skull (Fig. 3.5(b)). It is structurally intermediate between *Ianthasaurus* and *Edaphosaurus*, suggesting perhaps that it was omnivorous.

Eupelycosauria: Sphenacodontia
The sphenacodontians include the large carnivorous pelycosaur *Dimetrodon*, which is much the best known and most studied of all the pelycosaur genera. There is a distinctive dentition in which the upper canine is enlarged and supported by a buttress on the inner surface of the maxilla. Less than five premaxillary teeth are present, of which

the first one is enlarged, often to the same size as the canine. In the lower jaw, it is the second tooth that is enlarged. Another characteristic concerns the keel on the angular bone of the lower jaw, which extends well below the jaw articulation and so appears strongly convex in outline. A number of postcranial features also characterise the group, such as weak development of the ridges for adductor musculature on the ventral side of the femur (Reisz *et al.* 1992).

Haptodus (Fig. 3.6(a)) is the earliest sphenacodontian, occurring in rocks of Upper Pennsylvanian age in North America and Early Permian in Europe (Currie 1979; Laurin 1993). It is also the most primitive, having the lachrymal bone extending all the way from the orbit to the margin of the external

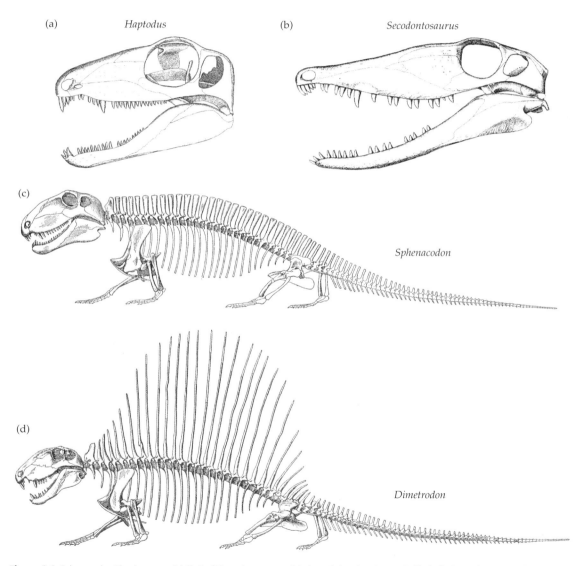

Figure 3.6 Sphenacodontid pelycosaurs. (a) Skull of *Haptodus garnettensis* in lateral view (Laurin 1993). (b) Skull of *Secodontosaurus obtusidens* in lateral view (Reisz *et al.* 1992) (c) Reconstructed skeleton of *Sphenacodon ferox* (Romer and Price 1940) (d) Reconstructed skeleton of *Dimetrodon limbatus.* (Romer and Price 1940).

nostril. The keel of the angular does have the deep form characteristic of the group, but it is not reflected laterally as occurs in other sphenacodontians. The skull is relatively short snouted, and the dentary and incisor teeth only moderately enlarged. *Haptodus* is also a small form, with a skull length around 10 cm in most species described, although the English species *Haptodus grandis* may have reached about 28 cm (Paton 1974).

All other sphenacodontians can be included in the family Sphenacodontidae (Reisz *et al.* 1992; Laurin 1993), in which there is a more powerful dentition. The snout is longer, the lachrymal bone shorter, and the maxilla has a strongly convex margin. The canines are very large and all the teeth develop anterior and posterior cutting edges. *Dimetrodon* (Fig. 3.6(d)) is notorious for its elongated neural spines, presumed to support a sail involved in thermo-regulation in what must be assumed to be an ectothermic animal. Several authors have attempted to model the system and calculate the effect such an increase in surface area would have had, and generally agree that it would have made a very significant difference to the rate of heat exchange with the environment. Possible functional advantages are to achieve more quickly a preferred body temperature in the morning, to lose heat during the day, or to maintain a more or less stable temperature throughout (Bennett 1996). *Dimetrodon* is abundantly preserved in the Early Permian of North America, and also extends into the beginning of the Late Permian in Texas. Some members of the genus achieved very large body size, with *Dimetrodon grandis* about 320 cm in body length and a weight estimated by Romer and Price (1940) to be around 250 kg. *Sphenacodon* (Fig. 3.6(c)) is similar to *Dimetrodon* but has less strongly developed canine teeth. It also lacks the dorsal sail, although the neural spines of the vertebrae are still quite long.

Secodontosaurus (Fig. 3.6(b)) is a specialised sphenacodontid with a low skull and a long, slender snout. The anterior lower teeth are procumbent, and none of the lower teeth are enlarged. Even the upper canine is modest in size. It was long considered to be a secondarily fish-eating animal because of its generally crocodile-like skull. However, Reisz *et al.* (1992) showed that there was a fully developed dorsal sail which they regard as improbable in

a semi-aquatic animal. They proposed instead that it was adapted for catching small tetrapods hiding in crevices and burrows.

The origin and early radiation of the Therapsida

The San Angelo and overlying Flower Pot Formations of Texas, along with the more or less contemporary Chickasha Formation of Oklahoma represent the final stage of the fossil-bearing Permian sequence in North America. They are dated as Guadalupian, which is equivalent to Middle Permian, and pelycosaurs still occur in the form of herbivorous caseids, *Dimetrodon*, and also, in the Chickasha, the varanopseid *Varanodon*. Olson (1962, 1986) claimed that these beds also contain the first fossil record of the more advanced synapsid group Therapsida. The specimens are in fact extremely scarce and consist only of fragments of braincases, vertebrae, and limb bones, and there is considerable doubt about whether any of them at all are actually the remains of therapsids rather than of pelycosaurs, probably caseids (Reisz 1986; King 1988; Sigogneau-Russell 1989). However, approximately contemporaneous continental sediments occur in the *cis*-Uralian region of Russia (Fig 2.2(a)), referred to as the Kazanian and early part of the Tatarian stages (Sigogneau-Russell 1989; Golubev 2000), and contain undisputed primitive therapsids (Ivakhnenko 2003). They belong to several taxa, existing alongside a few lingering caseid and varanopseid pelycosaurs (Modesto and Rybczynski 2000). The fossil record of therapsids in the South African Beaufort Series also extends downwards to beds equivalent to the early Kazanian, and also yields primitive members of therapsid groups (Rubidge 1995; Rubidge *et al.* 1995). This fauna is referred to as the *Eodicynodon* Assemblage Zone. A third area for very primitive therapsids comparable to the previous two has been described in northern China, the Xidagou Formation in Gansu. Although the exact age of these deposits is uncertain, they are presumably approximately the same as the Russian and South African records (Li and Cheng 1995; Li *et al.* 1996). Thus by about 270 Ma, a therapsid fauna had evolved and expanded its range to encompass Gondwanan as well as Laurasian parts of Pangaea.

Even at their initial appearance in the fossil record, therapsids had already diversified into several distinct groups, although the considerable number of characters they share indicates that the Therapsida is a monophyletic group descended from a single hypothetical pelycosaur-grade ancestor. In one of the most recent analyses, Sidor and Hopson (1998) counted as many as 48 possible therapsid synapomorphies. The most prominent single one of these is the reflected lamina of the angular bone of the lower jaw. This is a thin, extensive sheet of bone lying lateral to and parallel with the main body of the angular. It is connected anteriorly but has free dorsal, posterior, and ventral margins. Therapsids also possess a temporal fenestra that is much larger than in the pelycosaurs, indicating an increased mass of jaw closing musculature. Associated with the latter is the presence of a single, enlarged canine tooth in both upper and lower jaws that is sharply distinct from the incisors and the postcanine teeth. The jaw hinge is more anteriorly placed, and therefore the occipital plate is closer to vertical. The whole of the posterior part of the skull is more robustly built, with massive supraoccipital and paroccipital processes extending laterally to brace the squamosal region of the cheek. Other features probably related to strengthening the skull include the loss of the supratemporal bone and the immobility of the basipterygoid articulation between palate and braincase, which in pelycosaurs is constructed as a ball and socket joint between the basisphenoid and the pterygoid bones.

The therapsid postcranial skeleton also has many new features, related to improved locomotory ability. The blade of the scapula is narrow and the shoulder joint is no longer the complex screw-shaped structure that limited the range of humeral movements of pelycosaurs. Instead, the glenoid joint is a short, simple notch and the articulating head of the humerus ball-shaped. An ossified sternum has evolved behind the interclavicle. In the hindlimb, the ilium has expanded and the femur has a slight sigmoid curvature. A trochanter major has developed behind the femoral head, and the internal trochanter of the pelycosaur femur has shifted to the middle of the ventral surface of the bone. The feet, both front and back, have reduced certain of the phalanges to discs so that the digits

are more nearly the same length as each other. In the vertebral column, the intercentra have disappeared from the trunk region, although they are still present in the neck and tail.

That the therapsids are closely related to the pelycosaurs was established by Broom's (1910) classic paper comparing the two, and the affinity has never been seriously doubted since, although the relationship is nowadays acknowledged as a monophyletic Therapsida nesting within a paraphyletic 'Pelycosauria'. From among the known pelycosaur groups, the closest relative of Therapsida is almost universally agreed to be the family Sphenacodontidae, on the basis of several shared characters, including the following (Hopson 1991; Laurin 1993).

- Enlarged caniniform tooth, and differentiation of incisiform teeth in front and postcanine teeth behind.
- Maxilla enlarged at the expense of the lachrymal, so that it contacts the nasal bone, an arrangement that permits accommodation of the upper canine tooth.
- Lower jaw with a high coronoid eminence from which the posterior part of the jaw curves steeply down to a jaw hinge well below the level of the tooth row.
- Sphenacodontids alone among pelycosaurs with a notch between the angular keel and a down-turned articular region, that is an incipient version of the reflected lamina of the angular.
- Occiput with well-developed supraoccipital and paroccipital processes.
- A few characters of the postcranial skeleton, including a degree of reduction of the trunk intercentra, and narrowing of the scapular blade.

One final fossil to consider in the context of the relationships of therapsids is the most mysterious of all (Fig. 3.7(h)). In 1908, W. D. Matthew described the crushed, partial skull of a very peculiar pelycosaur which he named *Tetraceratops* (Matthew 1908). It is small, only about 10 cm in skull length, and has a vaguely sphenacodontid dentition with an enlarged upper canine. However, unlike other sphenacodontids, it has an equally enlarged first upper incisor and is also unique in possessing a row of extraordinarily large palatal teeth on the lateral flanges of the pterygoids, and in the presence of large bosses on the skull-roofing bones. Romer and Price (1940) proposed that it was allied to the basal pelycosaur group Eothyrididae, a view subsequently discarded by Reisz (1986) on the grounds that the two taxa shared no derived characters that he could discover. Recently, Laurin and Reisz (1996) re-studied the specimen and far from regarding it as merely a peculiar pelycosaur, they came to the conclusion that it is, in fact, the most basal therapsid known. Their interpretation is all the more remarkable because the specimen comes from the Clear Fork deposits of Texas, which are Early Permian, Leonardian, in age. Thus *Tetraceratops* predates the first of the Russian and South African therapsids by as much as 10 million years. They listed several characters shared with therapsids, including an enlarged temporal fenestra with signs of a broad, fleshy muscle attachment to its upper edge. In the palate, the interpterygoid vacuity is reduced in size and bounded posteriorly by the meeting of the pterygoid bones. The quadrate is reduced in size, as is the base of the epipterygoid. The braincase is attached to the back of the skull in a therapsid manner. In other respects, *Tetraceratops* has the primitive characters of sphenacodontids and other pelycosaurs, such as a large lachrymal bone, unfused basipterygoid articulation, and differential sizes of the premaxillary teeth.

As interpreted by Laurin and Reisz (1996), *Tetraceratops* is an extremely illuminating fossil that illustrates an intermediate stage in the evolution of the definitive set of therapsid characters. It is all the more unfortunate, therefore, that it is known from such limited and poorly preserved material and so it is hard to avoid the suspicion that *Tetraceratops* is actually a highly specialised member of one of the pelycosaur-grade taxa that has a number of superficial similarities to therapsids.

The diversity of early Therapsida

Leaving aside the very dubious possibility that *Tetraceratops* is a therapsid, a most remarkable feature of the origin of therapsids is the high diversity of taxa present at their earliest appearance in the fossil record. At present, the stratigraphic resolution (Fig. 2.2) is inadequate to distinguish the early Late Permian dates of the Russian Tatarian-Kazanian, the South African *Eodicynodon* Assemblage Zone, or the Chinese Xidagou Formation, all of which have produced very early therapsids. For the time being the respective faunas must be considered at least approximately contemporary, a position supported by their similarity. At least nine lineages are apparent, five of which are identified as basal members of groups that achieved prominence later, as described later in the chapter. The others are representatives of short-lived taxa that did not survive beyond this short period of time. This initial therapsid radiation included a variety of carnivores and herbivores, and at least one possible insectivore.

Biarmosuchia (page 31)

The biarmosuchians are the most sphenacodontid-like therapsids in appearance due to the strongly convex dorsal margin of the skull and the short, broad intertemporal region. The single canine is very much larger than the other teeth and the post-canines are reduced in relative size. Several specimens including postcranial skeletons have been found in the Eshovo (Ocher) and Mezen' faunas of Russia, although not yet in South Africa until younger levels, or in the Chinese sediments.

Brithopian Dinocephalia (page 37)

The earliest dinocephalians were relatively large carnivores, which had retained very prominent canines, and dorso-ventrally elongated temporal fenestrae. Brithopians occur in all the three areas yielding early therapsids, Russian, South African, and Chinese.

Anomodontia (page 39)

Several basal genera of small to medium-sized primitive members of what was to become by far the most diverse of all therapsid herbivores have been found. Unlike the later forms, incisor teeth are still present, but the characteristic shortening of the skull and inferred rearrangement of the adductor jaw musculature was under way. Primitive anomodonts occur in the Russian and the South African early faunas.

Gorgonopsia (page 52)

Some poorly preserved remains of gorgonopsians, the dominant carnivorous group of the later Permian, have been recorded in the South African *Eodicynodon* Assemblage Zone (Rubidge 1993;

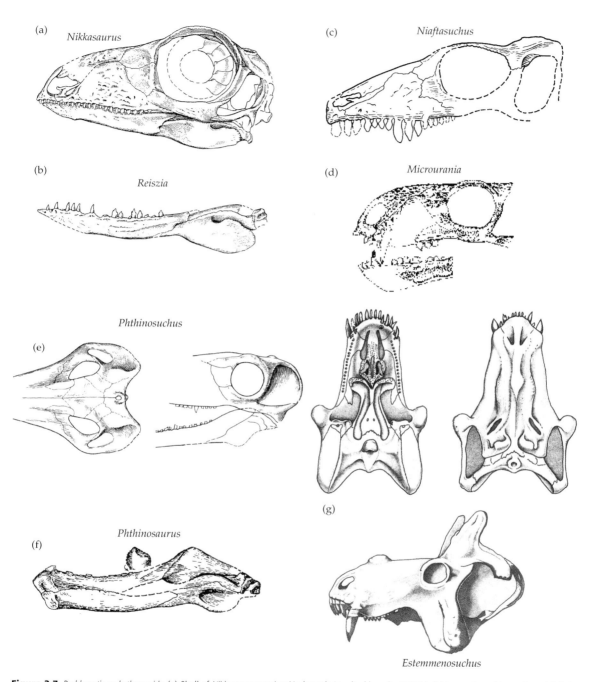

Figure 3.7 *Problematic early therapsids.* (a) Skull of *Nikkasaurus tatarinovi* in lateral view (Ivakhnenko 2000b). (b) Inner view of lower jaw of *Reiszia gubini* (Ivakhnenko 2000b). (c) Lateral view of skull of *Niaftasuchus zekkeli* (Ivakhnenko 1990). (d) *Microurania minima* (Ivakhnenko 1996). (e) *Phthinosuchus discors* in dorsal and lateral views (Sigogneau-Russell 1989). (f) *Phthinosaurus borissiaki* lower jaw in medial view, showing preserved postcanine tooth in lateral and ventral view (Ivakhnenko 1996). (g) Skull of *Estemmenosuchus uralensis* in ventral, dorsal and lateral views. (King 1988). (h) *Tetraceratops* (Laurin and Reisz 1996). *Continued overleaf*

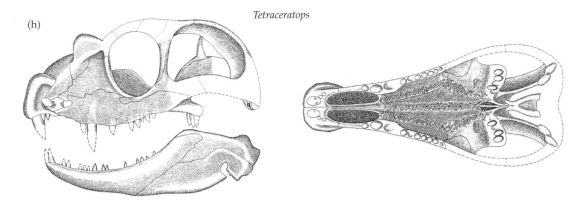

(h)   *Tetraceratops*

Figure 3.7 (continued).

Rubidge *et al.* 1995) but not so far from either the Russian or Chinese contemporary sediments. Among the largely discredited fragmentary specimens from the Guadalupian of North America mentioned earlier as possible therapsids, there is one, *Watongia*, that may be a basal gorgonopsian (Olson 1974). It is represented by two cranial fragments, and several postcranial bones. According to Olson, there is a preparietal bone on the fragment of skull roof, and also somewhat gorgonopsian-like vertebrae, slender limb bones, and some reduction in size of certain phalanges of the hind foot.

Therocephalia (page 56)
An incomplete skull of a member of this group of carnivorous therapsids, that also became prominent later, was described from the South African *Eodicynodon* Assemblage Zone by Rubidge *et al.* (1983).

Estemmenosuchus (page 38)
Estemmenosuchus (Fig. 3.7(g)) was a large herbivore with simple, interdigitating incisors, large canines, and a long row of small, serrated, bulbous postcanines. It occurs exclusively in the Ezhovo (Ocher) fauna, and is usually placed in the Dinocephalia. However, the grounds for this are not convincing, and its relationships may be quite elsewhere as mentioned in the next paragraph.

Rhopalodonta
The genus *Phthinosuchus* (Fig. 3.7(e)) is based on the poorly preserved posterior two-thirds of a skull and

associated lower jaws from the Belebey fauna of the Copper Sandstones of the Urals. Apart from primitive therapsid characters, very little can be made out from the specimen that links it to any other taxon. A second specimen of phthinosuchid consists of a lower jaw, again very poorly preserved, that has usually been referred to a separate genus *Phthinosaurus* (Fig. 3.7(f)). Again, it is largely lacking in diagnostic characters. The two tend to be interpreted rather vaguely as a basal therapsid family Phthinosuchidae, for example by Sigogneau-Russell (1989). Hopson and Barghusen (1986; Hopson 1991) suggested they may be related to biarmosuchians, but otherwise ignored them completely. More recently, Ivakhnenko (1996, 2000*a*) has pointed out that the jaw of *Phthinosaurus* carries a postcanine tooth that is leaf-like, with course serrations on the front and hind edges, and a vertical crest in the middle of the inner side. In these respects it resembles the postcanine teeth of *Estemmenosuchus*. Other forms that have similar postcanine teeth include a small, recently described partial skull, *Microurania* (Fig. 3.7(d)), and several other very fragmentary remains referred to as *Parabradyosaurus* (Ivakhnenko 1995). This led him to suggest that there is a taxon consisting of very early, primitive herbivores characterised by the serrated, leaf-like tooth and also, where it is known, by a canine tooth that is round in section, and by the particular form of the temporal fenestra. The proposed group is named Rhopalodonta and while its reality must be regarded as very tentative at present, it would certainly resolve the question of the relationships of several of these early, incompletely known forms.

Nikkasauridae

The family Nikkasauridae was established by Ivakhnenko (2000*b*) for a number of small, primitive therapsids discovered in the Mezen' River localities (Fig. 3.7(a) and (b)). The skull of *Nikkasaurus* itself is only about 4.5 cm in length. As befits such a small animal, the orbits are relatively huge. Its primitive nature is shown by the small, antero-posteriorly narrow temporal fenestra, the sclerotic ring of bones around the eye, and the possibly still mobile basipterygoid articulation. The dentition, however, is specialised with a long series of small, more or less equally sized teeth and no distinct canine. The cheek teeth are characterised by the development of an anterior and a posterior accessory cusp on the laterally compressed crown. They resemble those of certain later therocephalians, as indeed does the lower jaw. Its slender nature, relatively huge reflected lamina of the angular, and low coronoid process of the dentary all contribute to this impression. At any event, the Nikkasauridae would appear to be adapted to insectivory, adding yet another ecotype to the early therapsid radiation.

Niaftasuchus

Yet another mysterious taxon has been described by Ivakhnenko (1990). *Niaftasuchus* is based on a small, incomplete skull lacking the lower jaw (Fig. 3.7(c)), that was also discovered in the Mezen' River locality. Like the nikkasaurids, its primitive nature is indicated by the small temporal fenestra, although there is little other resemblance between them. The upper dentition consists of three well-developed incisors, four smaller precanines, a larger canine, and then a series of laterally compressed postcanine teeth decreasing in size backwards along the tooth row. Ivakhnenko himself interpreted *Niaftasuchus* as a herbivorous dinocephalian, but Battail and Surkov (2000) suggested that it is a relative of the biarmosuchians. The material is not really adequate to support either contention; it is just as likely to be yet another early diverging specialist herbivore within the initial therapsid radiation.

Biarmosuchia

The most pelycosaur-like and basal therapsids date from the Late Kazanian and Early Tatarian of the lower part of the Late Permian, coming mostly from the classic localities of Mezen', Ezhovo (Ocher) and Isheevo (Modesto and Rybczynski 2000). The exact temporal and palaeoenvironmental relationships of these various localities are not entirely certain yet, but the whole sequence is referred to by Golubev (2000) as the Dinocephalian Superassemblage. *Biarmosuchus tener* (Fig. 3.8(a)), is well known from a number of skulls and partial postcranial skeletons described by Sigogneau and Chudinov (1972) and more recently by Ivakhnenko (1999). The skull bears a striking resemblance to a sphenacodontid pelycosaur due to the strongly convex dorsal margin, and wide but short intertemporal region. However, the temporal region differs from the pelycosaur in the relatively larger temporal fenestra that has expanded both dorsally and ventrally compared to the pelycosaur, and in the more or less vertical orientation of the occipital plate. The single canine of *Biarmosuchus* is very much larger than the other teeth and the simple-pointed postcanines are reduced in relative size. The incisor teeth could probably interdigitate, lowers between uppers. There is a sclerotic ring of bones in the orbit, a feature unknown in pelycosaurs or the vast majority of therapsids. Only isolated bones of the postcranial skeleton have yet been described and no full reconstruction attempted. What is known is that the shoulder girdle is very narrow with a simple glenoid, and in the pelvic girdle the ilium extends forwards. The limb bones are long, gracile, and lack the broadly expanded ends characteristic of pelycosaurs.

A second genus of primitive carnivorous therapsid to consider is *Eotitanosuchus*. It was described from a single crushed skull, incomplete posteriorly, that was found in the Ezhovo (Ocher) locality along with another supposed specimen. It occurs with most of the known *Biarmosuchus* material, to which it bears a strong resemblance, although the several authors who have studied the specimen, from Chudinov's (1960) original account, through Sigogneau-Russell and Chudinov (1972), to Hopson and Barghusen (1986) all found differences. The main one concerned the form of the temporal fenestra, described in *Eotitanosuchus* as larger and showing indications of the invasion of the postorbital region by jaw musculature. One or two minor dental differences have also been pointed out. Despite the very limited material of *Eotitanosuchus*, the genus

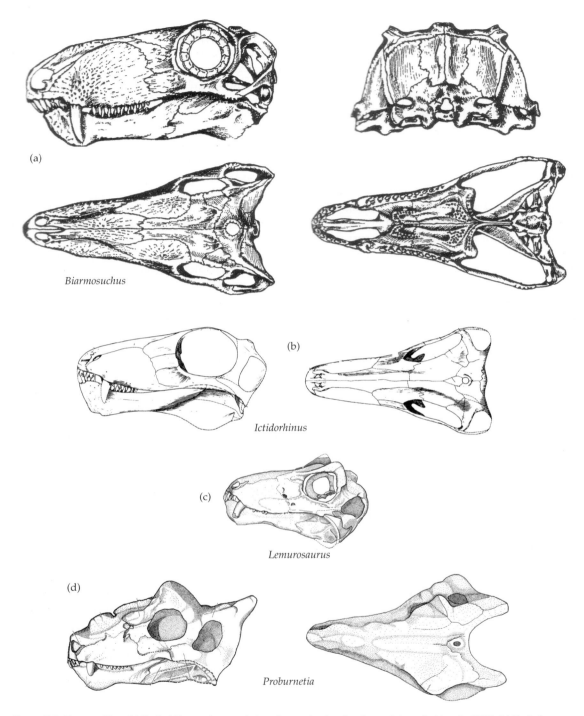

(a)

Biarmosuchus

(b)

Ictidorhinus

(c)

Lemurosaurus

(d)

Proburnetia

Figure 3.8 Biarmosuchians. (a) Skull of *Biarmosuchus tener* in lateral, posterior, dorsal and ventral views. (Ivakhnenko 1999). (b) Skull of *Ictidorhinus angusticep* in lateral and dorsal views (Kemp 1982). (c) Skull of *Lemurosaurus pricei* (Sidor and Welman 2003). (d) Skull of *Proburnetia viatkenensis* in lateral and dorsal views (Rubidge and Sidor 2002).

has figured large in discussions of early therapsid evolution, being regarded as more advanced than *Biarmosuchus*. Kemp (1982) and Hopson and Barghusen (1986) both expressed as cladograms the view that *Biarmosuchus* was more basal, and that *Eotitanosuchus* was more closely related to advanced carnivorous therapsids. In her review of primitive carnivorous therapsids, Sigogneau-Russell (1989) went so far as to place the two genera in different Infraorders, which is to say the least astonishing. At the other extreme, Ivakhnenko *et al.* (1997; Ivakhnenko 1999) reviewed all the material of *Biarmosuchus* and *Eotitanosuchus* and came to the conclusion that the two are the same species, differing only insofar as they are different growth stages. The skull length of the type specimen of *Biarmosuchus tener* is about 16.5 cm while that of the type specimen of *Eotitanosuchus olsoni* is over twice the length, at about 34.5 cm. Other specimens from the same locality lie between these extremes and illustrate a morphological transformation. Whether they really are conspecific may be debated, but there can be little doubt that they are extremely similar and certainly should be treated as congeneric, with *Eotitanosuchus* a junior synonym of *Biarmosuchus*.

Li and Cheng (1995) have described the Chinese genus *Biseridens* as a member of the group, but no biarmosuchians have yet been found in the South African *Eodicynodon* Assemblage Zone. However, there are specimens from later dates in South Africa that are included in the group (Sigogneau-Russell 1989; Hopson 1991). *Hipposaurus* is from the *Tapinocephalus* Assemblage Zone, which lies immediately above the *Eodicynodon* Assemblage Zone, and for which the skull and most of the skeleton has been described (Boonstra 1965; Sigogneau 1970, 1989). It has retained the primitively convex dorsal outline of the skull, relatively small temporal fenestra, broad intertemporal region and dentition of simple sharp teeth dominated by a large canine found in *Biarmosuchus*. A small number of minor derived characters shared between the two have been noted by Hopson (1991), including details of the zygomatic arch and fusion of the fourth and fifth tarsals of the hind foot. There are several other South African biarmosuchians, including *Ictidorhinus* (Fig. 3.8(b)), which occurs in the *Dicynodon*

Assemblage Zone at the very end of the Permian. A well-preserved skull of *Lemurosaurus* (Fig. 3.8(c)) has been described recently by Sidor and Welman (2003) and confirms the presence of interdigitating incisor teeth and a sclerotic ring in the group.

The family Burnetiidae was erected originally for two genera, one South African and the other Russian. *Burnetia* from the *Dicynodon* Assemblage Zone of South Africa, and *Proburnetia* (Fig. 3.8(d)) from the Late Tatarian of Russia are extremely similar primitive carnivores. They are relatively small therapsids, with a skull length around 20 cm and are characterised by heavily pachyostosed skulls bearing numerous bony protuberances and bosses. Taxonomically burnetiids were at one time linked to dinocephalians on the basis of the pachyostosis, and at another to gorgonopsians on the basis of their carnivorous dentition and generally primitive skull. However, appreciation of the absence of any discernable derived characters between burnetiids and gorgonopsians led Sigogneau-Russell (1989) to consider them as biarmosuchians. With the assistance of new South African specimens, Rubidge and Sidor (2002) endorsed the latter view, and a cladogram of all the biarmosuchians and burnetiids by Sidor and Welman (2003) points to *Lemurosaurus* as the sister group of Burnetiidae. This particular genus has incipient bosses on various parts of the skull suggesting an early stage in the full development of the burnetiid pachyostosis.

Dinocephalia

The dinocephalians (Fig. 3.9) are a group of large, heavily built animals that are among the first therapsids to appear in the fossil record in Russia, South Africa, and China, at the beginning of the Late Permian. The very doubtful possibility that fragmentary remains of Guadalupian age in North America may be dinocephalians has been mentioned earlier and isolated teeth have been found in Brazil (Langer 2000). After a brief period as the commonest therapsids, dinocephalians disappeared well before the close of the Permian, not having been discovered later than the *Tapinocephalus* Assemblage Zone of South Africa or its equivalent in Russia. Historically, the carnivorous and herbivorous dinocephalians were separated into two groups, related respectively to more advanced carnivore and herbivore therapsid taxa (Watson and Romer 1956). However, later work made it clear that the similarities between the carnivores are plesiomorphic, and the Dinocephalia have been recognised as a well-defined monophyletic taxon. The characters that define the group include the following.

1. Incisor teeth bearing a slight heel on the lingual face at the base of the crown. Uppers and lowers interdigitate, but this also occurs in biarmosuchians and gorgonopsians, so is probably a primitive therapsids character.

2. Enlargement of the temporal fenestra in such a way that temporal musculature attachment expands upwards and forwards onto the dorsal surface of the parietal and postorbital bones.

3. Forwards shift in the jaw articulation, shortening the length of the jaw.

4. A strong tendency towards thickening of the skull bones, to the point of heavy pachyostosis in many forms.

The relationships of the Dinocephalia to other therapsid groups has also been a matter of some debate, particularly as many of their characteristics are

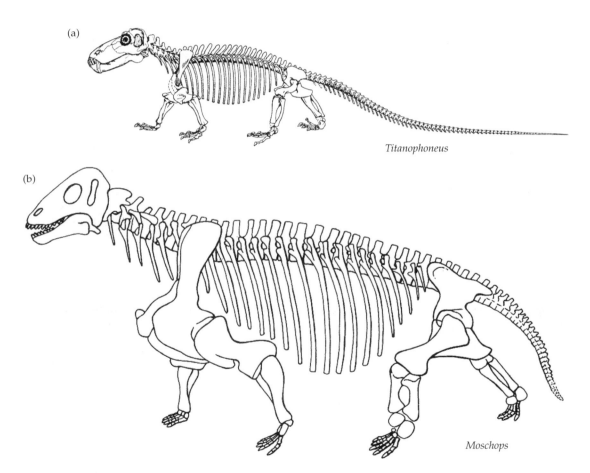

Titanophoneus

Moschops

Figure 3.9 (left) Dinocephalians. Postcranial skeleton of (a) *Titanophoneus potens.* (Orlov). (b) *Moschops,* total length about 2.5 m (Sues 1986*b*, modified from Gregory). (right) *Dinocephalians.* Skulls of (c) *Syodon efremovi* (Orlov, modified by Battail and Surkov 2000) (d) *Jonkeria vanderbyli* with lower jaw of *J. truculenta* (King 1988, modified from Broom). (e) *Anteosaurus magnificus* (King 1988, modified from Boonstra). (f) *Ulemosaurus svigagensis*, with enlarged view of incisor teeth and their interlocking action (King 1988, and Kemp 1982, from Efremov).

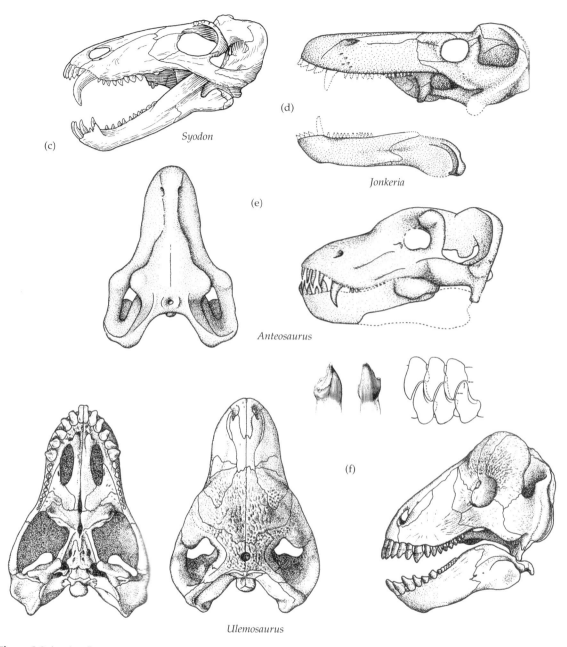

(c) *Syodon*

(d)

(e) *Jonkeria*

Anteosaurus

(f)

Ulemosaurus

Figure 3.9 (*continued*).

primitive for therapsids. In the pattern of skull bones, relatively small temporal fenestra and structure of the lower jaw they are not too dissimilar from pelycosaurs. King (1988) noted characters shared by dinocephalians and the herbivorous therapsid group Dicynodontia. Accordingly, she placed the two together as the Anomodontia, a hypothesis doubted by Kemp (1988) and rejected outright by Hopson and Barghusen (1986) and Grine (1997). The present consensus is that Dinocephalia is basal to both the anomodonts and the theriodonts, the latter two constituting an advanced therapsid taxon (Fig. 3.26).

Not very much has been added to the available knowledge of the phylogeny and biology of this

rather unfashionable group since Kemp's (1982) review. There is a clear division of the dinocephalians into two subgroups (Fig. 3.10). The Brithopia are primitive carnivores and the Titanosuchia are

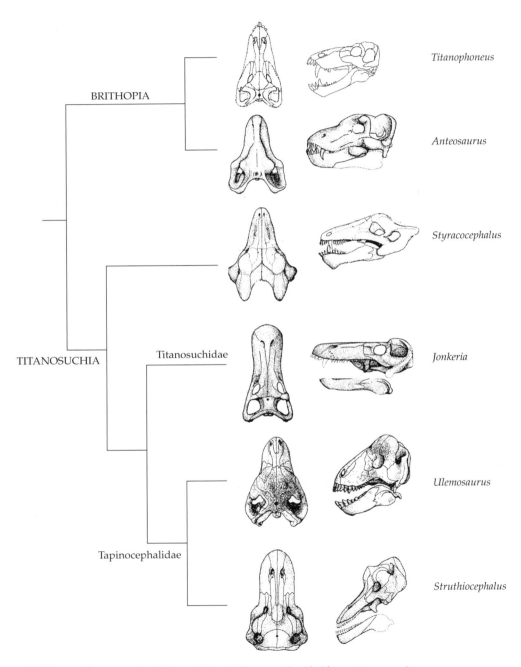

Figure 3.10 Cladogram of representatives of the main dinocephalian groups (modified from various sources).

a mixture of progressively more specialised herbivores. The enigmatic Russian *Estemmenosuchus*, usually included may well not be a dinocephalian at all.

Brithopia
Brithopians (Fig. 3.9(a) and (c)) are the least modified and therefore most pelycosaur-like dinocephalians. They constitute the most abundant group of the earliest Late Permian therapsid fauna of Russia, and there are some superbly preserved, complete specimens. The Isheevo form *Titanophoneus* is a large animal, up to 2 m in length. The canine teeth, both upper and lower are much larger than the rest of the dentition, although the intermeshing incisor teeth are also well developed. The postcanine dentition on the other hand is reduced in both number and individual size of the teeth. The temporal fenestra illustrates the primitive dinocephalian condition. It is enlarged, mainly in a dorso-ventral direction so that the lower temporal bar is depressed, while the area available for muscle attachment extends onto the dorsal surface of the parietal bone and also the posterior-facing surface of the postorbital bone immediately behind the orbit. Seen from above, the intertemporal bar has become quite narrow. Evidently the adaptation for carnivory in *Titanophoneus* consisted of a dorso-ventrally elongated adductor musculature capable of producing a rapid bite, the energy of which would have been dissipated by the anterior dentition. The heels on the incisor teeth are small and would have assisted in the use of the incisors crudely to tear off pieces of the prey's flesh. The postcranial skeleton of *Titanophoneus* is also a primitive version of the therapsid condition. The limbs are relatively long and slender, the shoulder joint simplified, the ilium enlarged, and the femur S-shaped as in other therapsids. However, such pelycosaurian features as undifferentiated dorsal vertebrae and a very long tail are retained. Several other Russian brithopian genera have been known for several decades, but only recently have members of the group been discovered elsewhere. A Chinese specimen was described by Jinling *et al.* (1996), and *Australosyodon* (Rubidge 1994) is known from a single, badly crushed skull from the lowest fossiliferous part of the South African Beaufort sequence, the *Eodicynodon* Assemblage Zone. The overlying *Tapinocephalus* Assemblage Zone of South Africa

contains the more specialised brithopian *Anteosaurus* (Fig. 3.9(e)). This is a massive animal, with a skull length up to 80 cm, composed of heavy bones and a marked boss on the postorbital. The whole postorbital region of the skull is deepened, creating space for an enlarged adductor musculature. The dentition is similar to that of other brithopians except for a further reduction in the postcanine dentition.

Titanosuchia
The titanosuchians are large, heavily built forms that constituted the dinocephalian radiation and dominance of the terrestrial fauna during the *Tapinocephalus* Assemblage Zone of South Africa. The earliest form, *Tapinocaninus*, actually occurs in the *Eodicynodon* Assemblage Zone (Rubidge 1991). A small number of closely related Russian genera occur in contemporaneous deposits. They evolved varying degrees of adaptation towards herbivory, involving reduction of the canines, increase in the size of the heel at the base of the incisor teeth, and an increase in the number of postcanine teeth to as many as twenty. The jaw articulation shifted forwards to reduce the length of the jaws to an even more marked extent than occurs in brithopians. Like *Anteosaurus* among the latter, the thickness of the cranial bones increased to a dramatic extent. As befits gigantic animals, the postcranial skeleton is modified for weight bearing in several respects (Fig. 3.9(b)). The vertebrae are short and wide. The shoulder girdle is massively constructed, the humerus has large, flattened ends, and the radius and ulna are similarly broad, flat bones. The pelvic girdle is relatively short but tall, and again the hindlimb bones very broad. Unexpectedly, the dinocephalians have independently acquired the mammalian digital formula of 2.3.3.3.3. (Hopson 1995). No detailed analysis of the mechanics of locomotion of dinocephalians has been made. Kemp (1982) claimed that the forelimb was probably still used in a sprawling fashion, but that the hindlimb was at least capable of a more erect gait as interpreted in other therapsids. In fact, given the size of these animals it is probable that a more or less fully erect gait was obligatory.

The members of the family Titanosuchidae, namely *Titanosuchus* and *Jonkeria* (Fig. 3.9(b)), are the least modified of the titanosuchians. The canines, though reduced, are still distinct, and the temporal fenestra has not been secondarily reduced in size by expansion of the surrounding bones, as occurs in the more advanced, tapinocephalid dinocephalians. The incisor teeth bear very well-developed heels, and when the lower incisors interdigitate between the uppers, opposing edges of the heels contact. According to Kemp (1982), the effect of this would be to create a continuous slit in the food, rather than simply punch a line of holes, so that tearing up the food would be more effective. The titanosuchid postcanine dentition is very distinctive. There are numerous, small, leaf-shaped teeth that bear serrated edges and although the uppers do not make direct contact with the lowers when the jaws closed, they would presumably have a good grasping function, perhaps of plant material. The diet is not clear but it may be assumed to have been quite catholic.

The Tapinocephalidae are the second family of titanosuchians and are the specialist herbivores (Fig. 3.9(f)). The canine is no larger than the incisors, and the interdigitation of upper and lower teeth extends to the whole tooth row. The way the teeth worked was quite elaborate, as inferred by Efremov (1940) in *Ulemosaurus*. The size of the crown was reduced but the heel is elongated in a lingual direction to form a short, wide shelf. The side edges of the heels continued to have a cutting action as in titanosuchids and the lingual elongation of the heel increased the length of cutting edge available, permitting the animal to cut up vegetation into finer pieces. In addition, the tapinocephalid incisors had a degree of crushing or grinding ability both where the blunt points of the crowns met the opposing heels and perhaps also between the outer faces of the lower incisor against the lingual face of the uppers. The incorporation of the postcanines into the cutting tooth row would have considerably increased the rate of cropping of the vegetation consumed, although it must have been quite soft, leafy material.

The extremes of pachyostosis are found in the tapinocephalids, where the thickness of the skull bones increases to a remarkable extent. Barghusen (1975) found that the frontal and parietal bones in a 32 cm long skull of *Moschops* are no less than 11.5 cm thick. The bones surrounding the temporal fenestra also increased in massiveness, secondarily reducing the size of the fenestra and increasing the intertemporal width. Barghusen (1975) showed how the pachyostosis was designed for the habit of head butting. The occipital condyle lies relatively far forwards on the ventral side, so the skull would habitually have been carried with the nose pointed downwards and the thick intertemporal region facing forwards. Thus head-to-head ritual, competitive contact between individuals would occur over this highly strengthened area of the skull.

There are several genera of tapinocephalids, including *Tapinocaninus*, the earliest Titanosuchian. They differ in the shape of the skull, the extent of the widening of the intertemporal region and the degree of pachyostosis. *Avenantia* is relatively primitive for it has retained a narrow intertemporal roof with an area for external muscle attachment still exposed, and the degree of pachyostosis is slight. *Moschops*, and the very similar Russian *Ulemosaurus* have a greatly reduced temporal fenestra and correspondingly wide intertemporal width, coupled with very heavy pachyostosis. *Riebeekosaurus* combines a long, low snout with a narrow intertemporal region but only moderate pachyostosis. The most highly evolved genus of all is *Tapinocephalus* itself, with its low, broad snout, enormously wide intertemporal width, and massive pachyostosis.

Styracocephalus (Fig. 3.10) has been a problematic genus for it combines the tapinocephalid features of pachyostosis and reduced temporal fenestra with such primitive features as a small but distinct canine and a cluster of palatal teeth. It is also unique in possessing a pair of bony bosses above the orbits and a pair of possibly horn-bearing protuberances at the back of the skull. Earlier workers regarded *Styracocephalus* as a relative of the brithopian *Burnetia*, while King (1988) concluded tentatively that it is was some kind of dinocephalian. With the help of some new material, Rubidge and Van den Heever (1997) confirmed the latter interpretation and their cladistic analysis placed it as the sister-group of the rest of the Titanosuchia.

Estemmenosuchia

The dinocephalians considered thus far form a coherent group with no serious controversy about the interrelationships between the various members. The one major problem concerns the nature of *Estemmenosuchus* (Fig. 3.7(g)). This exclusively Russian form appears in the Ocher/Isheevo fauna as part of the early radiation of therapsids. It is a large animal with interdigitating, but not heeled incisors, large canines that are circular in cross-section, and a long postcanine dentition of around 20 small teeth each with a bulbous crown and serrations on the front and hind edges. There are also many palatal teeth. As in dinocephalians, the occiput slopes forwards so that the jaw articulation is shifted anteriorly. The skull of *Estemmenosuchus* bears two pairs of bony protuberances, a long pair that look like horns on the roof, and smaller laterally extending bosses below the orbits. Until recently, *Estemmenosuchus* was interpreted as the most basal dinocephalian (Hopson and Barghusen 1986; King 1988), but the situation is still unresolved. On the one hand Rubidge and Van den Heever (1997) have published a cladogram of the main dinocephalian groups based on 27 characters. They find that *Estemmenosuchus* is not only a dinocephalian, but that it is more derived than Brithopia and therefore related to the Titanosuchia. At the other extreme, Ivakhnenko (2000*a*) pointed to differences in the design of the temporal fenestra between *Estemmenosuchus* in which the enlargement is postero-dorsal, and dinocephalians in which it is antero-dorsal. As discussed earlier (page 30), he also noted similarities in the teeth between *Estemmenosuchus* and certain other poorly known, primitive Russian therapsids. On these rather nebulous grounds, he proposed that *Estemmenosuchus* is a member of a group of primitive herbivorous therapsids, Rhopalodonta, that evolved independently of the dinocephalians.

Anomodontia

By several criteria, the most successful group of therapsids were the anomodonts, highly specialised herbivorous forms. They consist of the largest number of taxa, with well over 40 genera, and this number is still increasing regularly, particularly from descriptions of new Russian and Chinese material. Once during the late Permian they were the most numerous individual therapsid specimens in all fossil-bearing localities in which they occur, and must have occupied a comparable ecological role to the ungulate mammals of present times as abundant, often herd-dwelling primary consumers within the terrestrial ecosystem. They had a worldwide distribution being the only therapsids yet found in all continents, including Antarctica and Australia. Finally, they were

the longest lived of the major therapsid groups, since they made their appearance at the start of the Late Permian and survived for certain until the Upper Triassic. They may actually have survived for a vastly longer time, for Thulborn and Turner (2003) have described cranial fragments of what appears to be an Early Cretaceous dicynodont from Australia. If correctly interpreted, it extends the temporal range of the group by a huge margin of 110 Ma.

The actual term Anomodontia has varied in meaning. The great majority of forms belong to the advanced group Dicynodontia, but for a time it was believed that the dicynodonts and the dinocephalians were sister groups, and the taxon Anomodontia included both, for example Romer (1966) and King (1988). However, virtually all subsequent phylogenetic analyses have refuted this relationship and the term Anomodontia is now restricted to the dicynodonts, plus a number of more primitive taxa which nevertheless possess some of the dicynodont characters (e.g. Kemp 1988; Hopson 1991).

The main cranial characters defining the Anomodontia are the following (King 1988; Modesto *et al.* 1999):

- shortening of the preorbital region of the skull
- maxillary teeth decrease in size from front to back
- absence of palatal teeth
- zygomatic arch bowed dorsally
- lateral pterygoid processes reduced
- eminence on the dorsal surface of lower jaw formed from dentary and surangular bones
- mandibular fenestra present.

The ultimate adaptations of dicynodonts for herbivory are replacement of most or all of the dentition by a horny beak rather resembling that of turtles, and extreme modification of the adductor musculature of the jaws to produce a powerful backwardly directed slicing bite, an action permitted by a suitably specialised jaw hinge. A number of early anomodonts from South Africa and Russia illustrate stages in the development of the full-blown dicynodont condition.

Primitive Anomodontia

Until about 1990, the only primitive, non-dicynodont anomodonts that had been described were *Venyukovia* and *Otsheria* from Russia and three poorly preserved specimens classified as Dromasauria from the South African Karoo (e.g. Kemp 1982; King 1988). The picture has since improved significantly on the basis of the discovery of South African specimens along with new and newly described Russian material.

Anomocephalus

Anomocephalus (Fig. 3.11(a)) is the most basal anomodont so far known, although not the oldest, having been discovered recently in the *Tapinocephalus* Assemblage Zone of the South African Karoo (Modesto *et al.* 1999). Unfortunately, as it is based so far only on a single, incomplete skull, the full potential of this genus for helping understand anomodont evolution has yet to be realised.

It is quite large for an early anomodont, with a skull length of around 20 cm. Numerous primitive characters have been retained, such as a preorbital length which, while short compared to other therapsids, is still relatively long at about 45% of the total skull length. Of particular interest, the part of the squamosal bone forming the zygomatic arch is still short and rod-like rather than horizontally expanded for the origin of extra adductor jaw musculature, which indicates that there had been rather little development towards the highly characteristic dicynodont arrangement of the jaw musculature. Nevertheless, the bowed zygomatic arch and the coronoid eminence of the lower jaw do indicate a modest degree of enlargement of the lateral part of the adductor mandibulae musculature. The dentition of *Anomocephalus* consists of a row of about eight teeth on either side, upper and lower. The individual teeth are large, peg-like and decrease in size regularly from front to back of the row. Each individual tooth increases in diameter towards its tip, and has a flat, sloping wear facet on its crown indicating a masticating function. The dentition as a whole appears to be adapted for a relatively soft herbivorous diet.

Patronomodon

In 1990, Rubidge and Hopson reported a small anomodont, *Patronomodon* (Fig. 3.11(b)), from the South African *Eodicynodon* Assemblage Zone, which makes it slightly older than *Anomocephalus*, notwithstanding its more derived structure. It is represented by a well-preserved, 6 cm long skull,

with lower jaw and partial postcranial skeleton (Rubidge and Hopson 1990, 1996). Primitive features include the failure of the premaxilla to meet the palatine in the palate, and the absence of a fossa indicating extension of the adductor mandibuli musculature onto the lateral surface of the squamosal. Perhaps most significantly, the jaw articulation did not permit the movement of the lower jaw backwards and forwards, the property of propaliny that is so distinctive a part of the jaw apparatus of other anomodonts. The dentition of *Patronomodon* consists of much smaller teeth than in *Anomocephalus*. In the upper jaw there are probably at least three premaxillary and up to seven maxillary teeth, all about the same size and forming a continuous row. The dentary has about six similar teeth. In both upper and lower jaws, the teeth lie medial to the jaw margin, with a greater medial than lateral exposure and yet there is no indication on the bony surfaces of the development of horny tooth plates alongside the tooth rows. King (1994) on the basis of the dentition, the absence of propaliny, and the primitive, relatively undivided nature of the external adductor mandibulae musculature concluded that *Patronomodon* was not herbivorous, but subsisted on a generalised carnivorous diet. Given the skull size of only about 6 cm, this would imply a diet of insects and other small invertebrates.

Venyukovioidea

There are several genera of primitive anomodonts from late Kazanian or early Tatarian deposits of Russia, the taxonomy of which has until recently been confusing (Ivakhnenko 1996). *Venyukovia prima* was described by Amalitzky in 1922 on the basis of a left dentary and an isolated jaw symphysis, found in an unknown locality within the Copper Sandstones. Subsequently, Efremov (1938) attributed an incomplete skull and jaw fragments from the Early Tatarian Isheevo locality to the same genus, as *Venyukovia invisa*, and Tchudinov (1983), referring to new material from Isheevo, synonomised the two species. Ivakhnenko (1996) reviewed the material and concluded that all the Isheevo specimens are sufficiently distinct from Amalitzky's original specimens for them to require a new generic attribution, *Ulemica invisa* (Fig. 3.11(e)), leaving Amalitsky's

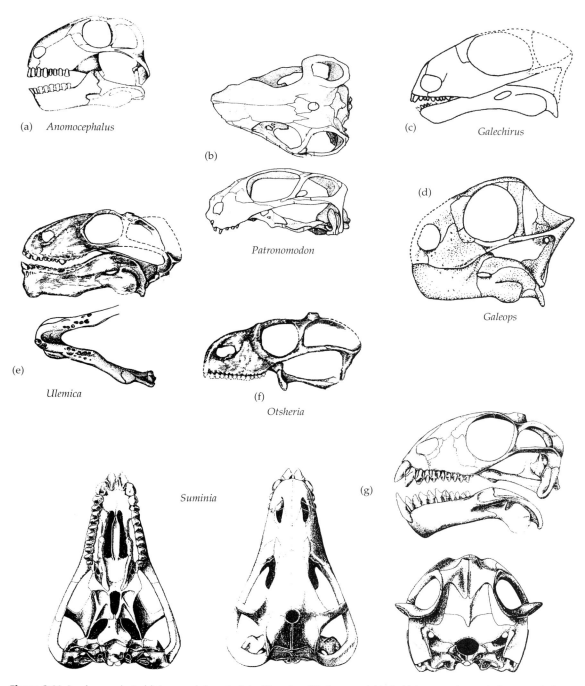

(a)   *Anomocephalus*

(b)

(c)   *Galechirus*

(d)

Patronomodon

Galeops

(e)

Ulemica

(f)

Otsheria

(g)

Suminia

Figure 3.11 Basal anomodonts. (a) *Anomocephalus*: actual size 20 cm long (Modesto *et al.* 1999). (b) *Patronomodon nyaphulii*: approx. skull length 6 cm (Rubidge and Hopson 1996). (c) *Galechirus scholzi*: approx. skull length 6.8 cm. (d) *Galeops whattsii*: approx. skull length 4.5 cm (King 1988 after Brinkman 1981). (e) *Ulemica invisa*: skull length approx. 15 cm (Ivakhnenko 1996 and King 1994). (f) *Otsheria netzvetajevi*: skull length 10.5 cm (Ivakhnenko 1996). (g) *Suminia getmanovi*: skull length 5.8 cm (Rybszynski 2000).

Copper sandstone lower jaw as the type and only safely attributed specimen of *V. prima*. Thus, although the family Venyukoviidae still stands, *Venyukovia* itself is scarcely known.

Meanwhile, Tchudinov (1960, 1983) had described another primitive anomodont as *Otsheria netzvetajevi*, based on a skull without the lower jaw (Fig. 3.11(f)). This specimen is older than the Isheevo material, for it was found at the Ezhovo (Ocher) locality which is dated as late Kazanian. Most recently a fourth, quite different form, *Suminia* (Fig. 3.11(g)), has been described from the Late Tatarian Kotel'nich fauna, which is therefore the youngest of the Russian primitive anomodonts (Ivakhnenko 1994; Rybczynski 2000).

A cladistic analysis by Modesto and Rybczynski (2000; Rybczynski 2000) concludes that the Russian genera are a monophyletic group characterised principally by an elongated postero-dorsal process of the premaxilla, a very broad parietal bone with the pineal foramen situated on a large boss at the front of it, and no preparietal bone.

Otsheria (Fig. 3.11(f)) is thus the earliest of the at least reasonably well-known genera of venyukovioids (Tchudinov 1960, 1983). It has a small, 10-cm long skull that differs mainly from other venyukovioids in its dentition. Although not very well preserved, this can be seen to consist of probably four incisors in the premaxilla and nine maxillary teeth, all of which are short and stout, with laterally compressed points. In the absence of any information at all about the lower dentition, it is not clear whether there was tooth-to-tooth contact. So far as it may be inferred, the form of the individual upper teeth suggests a simple, foliage-cutting action, although a generalised omnivorous diet might well have been the case.

Ulemica (Fig. 3.11(e)) is based on two fairly complete skulls, several partial skulls and lower jaws, but so far no postcranial material, and all from the Early Tatarian Isheevo locality (Ivakhnenko 1996). The skull is 15–20 cm in length and heavily built. The dentition of *Ulemica* is very odd and difficult to understand functionally (King 1994; Ivakhnenko 1996). In the upper jaw, there is a large chisel-shaped first tooth, followed by three of similar size but with heel-like internal extensions. These are followed by four small conical teeth and a much larger, swollen, caniniform tooth. Finally the tooth row is completed by a series of four or five very small pointed teeth. The corresponding lower dentition consists of a chisel-shaped first tooth, followed by three smaller teeth that are laterally flattened in younger specimens, but replaced by blunter, rounded ones in older ones, and apparently lost in fully grown specimens. The tooth row is completed by a series of small, medially displaced teeth. The biting action consists of direct tooth-to-tooth contact between the anterior teeth. The upper caniniform tooth does not meet a lower tooth, but bites into a pit on the bony surface of the jaw, to the side of the lower teeth. Behind this point, the teeth do not appear to meet either. The upper postcaniniform teeth bite towards the lateral side of the lower teeth. The latter, in turn, bite towards the bone of the palate, medial to the upper teeth. King (1994) suggested that horny biting surfaces were present in life on these bony surfaces. Judging both from the form of the dentition and the structure of the articular and quadrate, no propalinal movement of the lower jaw was possible. Instead, the anteriormost teeth would have provided a nipping, food-gathering action, and the posterior teeth acting against their respective opposing tooth plates would have provided a simple crushing function. The diet for which such a system was adapted is not clear, but presumably included various kinds of vegetation.

Suminia (Fig. 3.11(g)), as well as being the youngest of the primitive Russian anomodonts is also the best known, on the basis of several skeletons and skulls (Ivakhnenko 1994; Rybczynski 2000). The skull is small, no more than about 6 cm in length, and unlike *Ulemica* it is lightly built. There are complete upper and lower marginal dentitions of 10 or 11 teeth, which decrease in size gradually from front to back. Each of the teeth has a broad base and a serrated, leaf-like crown, uppers with a concave front edge, lowers with a concave hind edge. Rybczynski and Reisz (2001) have analysed the jaw action, showing that the power stroke of the lower jaw was posteriorly directed and created a cutting action between lower and upper teeth that included direct tooth-to-tooth contact. Their interpretation is supported by the structure of the jaw hinge, which permitted propalinal movement of the jaw, and by the lateral flaring of the zygomatic arch, the squamosal component of which bears a lateral fossa for the origin of a lateral slip of the adductor mandibulae musculature. This, as in the dicynodonts, would have provided the necessary posteriorly directed force on the mandible during occlusion (Rybczynski 2001). The form and evident action of the teeth indicate a committed herbivorous animal.

Dromasauroidea

There are three genera of small, primitive anomodonts from the Late Permian of South Africa generally included in a taxon Dromasauria. They are represented by a total of four specimens, each one of which is preserved as a natural sandstone cast of the skull and postcranial elements (Brinkman 1981).

Galeops (Fig. 3.11(d)) shows features otherwise restricted to the dicynodonts, leading to the view that among the dromasaurs it is the dicynodont sister-group (Modesto and Rybczynski 2000; Rybczynski 2000). The characters in question include further shortening of the preorbital region of the skull, absence of anterior teeth on either premaxilla or dentary, and the remaining posterior dentition reduced to a short row of small, peg-like teeth. The arrangement implies that dicynodont-like horny tooth plates had evolved at the front of the jaws. Other features of dicynodonts that are found are fusion of the paired dentaries at the symphysis, and a quadrate condyle divided by a longitudinal groove into lateral and medial condyles, corresponding to elongated articular condyles. According to King (1994), the structure of the jaw articulation would have permitted propalinal movement of the lower jaw.

The preserved specimens of neither of the other two genera of dromasaurs, *Galepus* and *Galechirus* (Fig. 3.11(c)) are as revealing. They are much the same size as *Galeops*, and look superficially very similar. However, there are differences in the dentition which justifies at least the generic distinction between the three (Brinkman 1981). *Galechirus* possesses premaxillary teeth, which are decidedly procumbent, but the situation of the lower teeth is unclear. *Galepus* has at least six small lower teeth, but in this case the nature of the upper dentition is unclear.

Dicynodontia

The great majority of anomodonts belong to the very well-defined group Dicynodontia, which makes its appearance in the *Eodicynodon* Assemblage Zone contemporary with the earliest of the primitive African anomodonts described above. Indeed, it is the eponymous *Eodicynodon* itself that first represents the group (Rubidge 1984, 1990a).

Eodicynodon

The trends in cranial evolution seen in the primitive anomodonts up to the stage represented by *Galeops* reach their culmination in the fully dicynodont form of *Eodicynodon* (Fig. 3.12(a)). All the incisor teeth are absent, there is an enlarged caniniform upper tusk, and its postcanine dentition is reduced to a small number of relatively minute teeth. (Rubidge, 1990b, described a specimen of *Eodicynodon*, which lacks the caniniform tooth; he attributed it to a different species). The triturating surface of the jaws consisted virtually exclusively of a horny beak. Cox (1998) concluded from the structure of the bony surfaces that there were three parts to the beak. A medial anterior dentary beak bit within the front margins of the palate, presumably with a cropping function. Further back, a pad on the dentary table acted against a palatal pad medial to the caniniform process. Third, a posterior horny blade on the dentary bit against a palatine pad behind the caniniform process. In some specimens the lower blade also carries a row of small teeth. In order to operate these biting structures, the external adductor musculature had a much enlarged lateral component, originating from the horizontally flattened and dorsally bowed zygomatic arch and inserting at least partially on a lateral shelf on the dentary. The temporal fenestra as a whole is elongated, while the lateral pterygoid processes are reduced. These point to the increased importance of a posteriorly directed power stroke during the closing phase of the jaw action, accompanied by a reduction in the importance of the anterior pterygoideus musculature with its anteriorly directed component (Fig. 3.12(c)). The jaw articulation (Fig. 3.12(d)) has the uniquely dicynodont structure of more or less equal lateral and medial convex condyles on the quadrate and corresponding but elongated articular condyles on the lower jaw. King et al. (1989) believed that the feeding mechanism of *Eodicynodon* had achieved the essential dicynodont arrangement but that the degree of propaliny was still less than occurred in later dicynodonts, a view reiterated by King (1994) and Cox (1998).

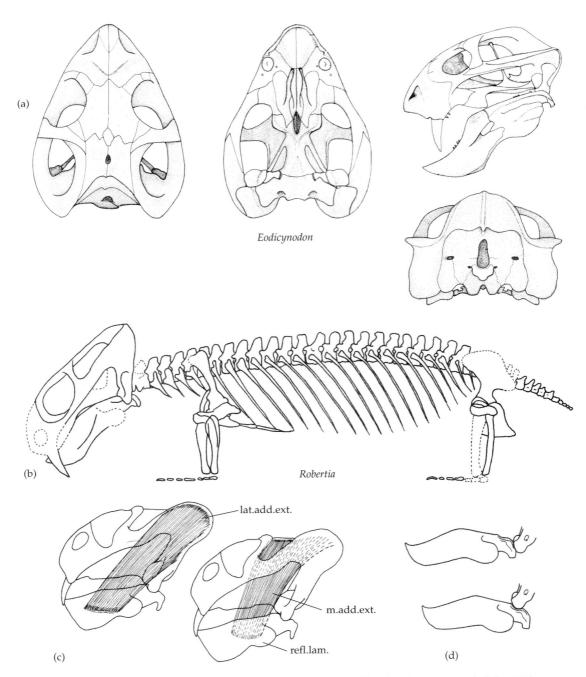

Figure 3.12 Primitive dicynodonts. (a) Skull of *Eodicynodon oosthuizeni* in dorsal, ventral, lateral, and posterior views (Rubidge 1990). (b) Reconstruction of the postcranial skeleton of *Robertia broomiana* (King 1981(*b*)). (c) Reconstruction of the adductor jaw musculature of *Dicynodon*: left, superficial lateral view; right, deep lateral view. lat.add.ext, lateral adductor externus muscle. m.add.ext, medial adductor externus muscle. refl.lam, reflected lamina of the angular. (d) Jaw articulation of a dicynodont showing lower jaw in protracted (above) and retracted (below) postions. (King 1981*a*).

To date the postcranial skeleton of *Eodicynodon* is incompletely known (Rubidge *et al.* 1994). However, it has been described in some detail for the small, relatively unspecialised dicynodont *Robertia* (King 1981*b*), which occurs in the *Tapinocephalus* Assemblage Zone, immediately overlying the *Eodicynodon* Assemblage Zone (Fig. 3.12(b)). There is no reason to doubt that *Robertia* had retained the ancestral form of dicynodont postcranial skeleton, which remained relatively conservative within the group (King 1988). The first strikingly unique feature is the occipital condyle of the skull, which is markedly trefoil-shaped, unlike the kidney-shape of other therapsids. Kemp (1969*a*) showed that in dicynodonts it was part of a complex joint involving the atlas and the axis vertebrae that permitted rotation of the head about a longitudinal and a transverse axis, but by a mechanism sufficiently different in detail from that found in other therapsids, to indicate that it had an independent origin from the pelycosaur state. *Robertia* has a full-length ribcage, with no tendency to develop reduced, immobile lumbar ribs. The tail is very short. The limbs are short compared to therapsids generally, with humerus and femur longer than the respective lower limb bones. On the other hand, the limb girdles are relatively advanced, in the sense of developing mammal-like characteristics (see Chapter 4). The front edge of the slender scapula blade is everted and there is a very well-developed acromion process for attachment of the clavicle. The coracoid plate below the acromion process is reduced so that there is ample space for expansion of the supracoracoideus muscle from its primary site at the front of the coracoid to the anterior part of the internal surface of the scapula. Thus there appears to have been a precocious, independent development of a mammal-like 'supraspinatus' muscle. The pelvic girdle has an enlarged and anteriorly extended ilium coupled with reduction of the pubis, indicating that the ilio-femoralis muscle had expanded and was taking over the role of the presumably greatly reduced caudi-femoralis muscle in femoral retraction. Confirmation is offered by the very distinctive form of the femur, which bears a large trochanter major occupying about a third of the shaft behind the head, but lacks any sign of a fourth or internal trochanter. Finally, the feet are

also precociously evolved. They are very short and the phalangeal formula is reduced to the mammalian condition of 2 : 3 : 3 : 3 : 3.

Despite these superficially 'advanced' features, the mode of locomotion appears to have been relatively primitive. King (1981*a*, 1988) analysed in detail the locomotory mechanics of the later form *Dicynodon trigonocephalus*, and concluded that the forelimb could not have operated in anything other than a sprawling mode. The hind limb had a more complex action, with elements of both a sprawling and a more erect gait, but was not capable of the simple parasagittal gait of other progressive therapsids. The few adequately preserved fossil trackways of dicynodonts generally confirm King's interpretation (Smith 1993; De Klerk 2003). They show a plantigrade foot, with a wide forelimb trackway, but a slightly narrower hindlimb trackway. However, the dicynodont postcranial skeleton can perhaps best be understood as adapted for digging, basically to collect food lying just below the surface of the ground, but in some specialised cases for active burrowing.

The dicynodont radiation

The anatomy and functioning of the skull as seen in *Eodicynodon* remained relatively conservative throughout the subsequent evolutionary radiation of the dicynodonts (Fig 3.13) although, as befits a highly successful, diverse, and widespread group, there is much detailed variation, associated with different diets and habitats, in the size and shape of the skull, the extent to which tusks and postcanine teeth were retained or even elaborated, and the form of the horny beak as inferred from the nature of the bony surfaces of the jaws supporting them. Increasing information about the diversity of the postcranial skeleton is accumulating and can be correlated with habitat, particularly in terms of different degrees of digging and burrowing ability (King 1988; Ray and Chinsamy 2003).

Biogeographically, the dicynodonts are far and away best known as a consequence of the hundreds of specimens recovered from the Late Permian and Lower Triassic of the Karoo of South Africa and other parts of southern Africa. However, by the Triassic the group had a worldwide distribution, occurring in all continents, including *Myosaurus*

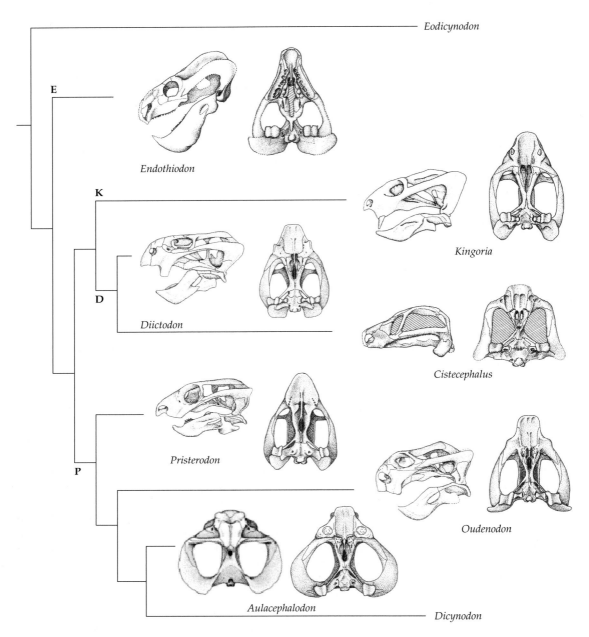

Figure 3.13 Phylogeny of the main groups of Permian dicynodonts, based on King (1988).
D-Diictodontoidea; **E**-Endothiodontoidea; **K**-Kingorioidea; **P**-Pristerodontoidea.

and *Lystrosaurus* from Antarctica (Hammer and Cosgriff 1981; Cosgriff *et al.* 1982) and a couple of fragments of what is probably *Lystrosaurus* from Australia (King 1983; Thulborn 1983).

The complex phylogenetic interrelationships of the dicynodonts are not well understood. The pioneering work of Gillian King and colleagues (Cluver and King 1983; King 1988, 1990; King and Rubidge 1993) recognised four subgroups, based mainly on differences in jaw structure, although there are very few characters defining these taxa so that they are not at all well supported. A more recent and more detailed cladistic analysis is that of Angielczyk (2001), which includes 20 dicynodont taxa and 40 characters. It disagrees with many of King's (1988) relationships, but again most of the clades are very weakly supported. Many of the nodes of his cladogram collapse with only a single additional step, and with as little as three additional steps virtually all the internal structure of the cladogram becomes unresolved. Thus convergent evolution and character reversal was extensive and it should be stressed again that the 40–50 genera accepted nowadays, are all relatively conservative. Were they a modern herbivorous mammal group, dicynodonts might well have been incorporated into a single family with a disparity perhaps equivalent to that of the Bovidae. Given this lack of taxonomic resolution of Dicynodontia, the scheme of King (1988) based on evident differences in the feeding structures is followed, although more from convenience than conviction.

Endothiodontoidea. Endothiodon has a highly atypical dentition. It is a large form found in the *Tapinocephalus* Assemblage Zone of South Africa and also in beds of equivalent age in Zambia, South America, and India (Ray 2000). The skull reaches over 50 cm in length with a distinctively high, very narrow sagittal crest. The lower jaw is distinguished by the very deep anterior part attached to the relatively slender postdentary part. The upturned front tip of the fused dentaries worked against the broad, deeply concave plate formed from the fused premaxillae. Canine tusks are absent, but the postcanine dentition is remarkably and presumably secondarily well developed. There are about 10 long, slender upper teeth, fairly widely spaced and forming a single row that does not oppose the lower teeth, but must have bitten against a presumed horny pad lying in a longitudinal groove lateral to the lower teeth. The lower dentition consists of smaller, tightly packed teeth forming two or three irregular rows. Only the more lateral ones were functional, biting against a horny pad presumably present in life on the heavily rugose palatine bone, internal to the upper tooth row. The more medial of the lower teeth

are young teeth, just erupting and preparing to move laterally to replace existing functional teeth as they are discarded. An area of vascularised bone lying lateral to the upper tooth row has been interpreted by Cluver (1975) as the line of attachment of a muscular, food-retaining cheek. Several authors have speculated on the diet of *Endothiodon*. King (1990) suggested that the teeth had a slicing action, perhaps by direct unilateral occlusion of uppers and lowers. Latimer *et al.* (1995) proposed that they browsed on riverine vegetation using cheek pouches to collect the food. Cox (1998) rather ingeniously speculated that conifer cones were stripped by the action of the tip of the mandibles acting against the premaxillae, followed by maceration of the seeds using the postcanine dentition.

Kingorioidea. Kingoria (Cox 1959) lacks all signs of postcanine teeth which, if its cladistic relationships are correctly inferred, were lost independently of the loss in other lineages. Some specimens of *Kingoria* retain but others have lost the pair of tusks. The horny, biting surfaces inferred from the structure of the bones of the upper and lower jaws differed considerably from those of other dicynodonts. The secondary palate formed from the premaxillae is smooth and flat, and the horn appears to have been restricted to the lateral margins of the adjacent maxillae. Instead of being sharp, these lateral edges are rounded, as are the corresponding anterior parts of the dentaries, implying a crushing rather than a shearing action. However, there are some sharp margins, a short one immediately in front of the caniniform process of the upper jaw, and another at the upturned anterior tip of the joined dentaries. Hotton (1986) attempted to explain this curious combination of features with the suggestion that *Kingoria* fed on a food source that was of small size, soft, and abundant. King (1988) thought that perhaps its food consisted of invertebrates, grubbed up from mud and swallowed more or less without any chewing.

Kingoria also has some distinct specialisations in its postcranial skeleton, reminiscent of the incipiently mammalian condition seen in cynodonts (Cox 1959; King 1985). The shoulder girdle has developed a particularly strongly everted spine extending from the acromion process up the front edge of the scapula. In the hindlimb, the ilium extends forwards and the pubis is reduced and rotated backwards more than in any other dicynodont. King (1985) interpreted these modifications as indicating a forelimb that relied primarily on long-axis rotation of a laterally directed humerus to produce the forelimb stride. The hind limb anatomy indicates that the knee was turned forwards and the foot lay under the body, in mammalian fashion. Quite what use *Kingoria* made of its modified anatomy is not at all clear; she tentatively suggested that the forelimb had a digging function, and the hindlimb was adapted to generate a greater thrust.

Whatever its mode of life, *Kingoria* is uncommon in the fossil record, possibly because it lived in an environment less conducive to fossilisation. The only other known kingorioid is the Lower Triassic *Kombuisia* (Hotton 1974), a small dicynodont representing one of the only three lineages that survived the end-Permian and through to the *Cynognathus* Assemblage Zone.

Diictodontoidea. The diictodontoids share with *Kingoria* a reduction of the palatine bone, and it has been proposed on the basis of this extremely limited evidence that they may be related (King 1988). Diictodontoids are, however, distinguished in their own right by the presence of an embayment, or notch in the margin of the maxilla, immediately in front of the caniniform process, whether a tusk is present or not. The shape of the notch indicates a sharp, horny blade against which a dentary blade would have acted, capable of cutting slender but tough stems, roots, and rhizomes. These would then have been triturated between opposing plates borne by the palate and the front part of the dentaries. Primitive members of the group include *Robertia* which retained a few small postcanine teeth, and *Emydops* which lost them, both from the *Tapinocephalus* Assemblage Zone of South Africa.

Diictodon itself is more advanced, having completely lost the teeth and developed a narrow intertemporal region of the skull to permit further elaboration of the adductor musculature. King (1990) described evidence for a complex form of horny beak. Longitudinal ridges on the bony ventral surface of the premaxillae correspond to the medial edges and to troughs on the dorsal surface of the symphyseal part of the dentaries, indicating a mechanism for shredding up the food that had been collected by cutting between the very prominent caniniform notch and edge of the dentary. Hotton (1986) thought the most likely diet was subterranean roots and tubers, dug up by the strong, clawed hand, and indirect evidence supporting this view is the remarkable discovery by Smith (1987) of *Diictodon* specimens curled up at the bottom of 50 cm deep helical-shaped burrows. The skull shows none of the adaptations typical of powerful, obligatory burrowing animals, but there are signs in the postcranial skeleton of a fossorial ability enhanced beyond that of typical dicynodonts. Ray and Chinsamy (2003) interpreted the cylindrical-shaped body, very short tail, relatively small, broad limbs, and wide manus as those of an adept digging animal using its forelimbs to dig and its hindlimb to push away the soil. Possibly *Diictodon* only made shallow burrows, or adopted existing ones made by other taxa. Whatever its mode of life, *Diictodon* was one of the most successful and widespread of dicynodont genera. It has been found in the northern hemisphere and China, as well as in southern Africa, and in the latter it lasted from the *Tapinocephalus* Assemblage Zone right through to the *Dicynodon* Assemblage Zone at the very end of the Permian.

The idea of burrowing was taken much further in *Cistecephalus* (Cluver 1978) and the closely related *Cistecephaloides* and *Kawingasaurus* (Cox 1972). These are small animals, with a very short, broad skull. The snout is pointed, the intertemporal roof extremely broad and strongly built, and the occiput wide to accommodate powerful neck muscles. The forelimb is mole-like, with a short, powerfully built humerus, large olecranon process on the ulna, and a broad forefoot with enlarged digits. Surprisingly perhaps, the hindlimb is much less modified, indicating that it played a lesser role in digging and, presumably, burrowing (King 1990).

Pristerodontoidea. This is both the most diverse and the longest lived of the four major dicynodont taxa, for in addition to some of the commonest Permian forms it also includes almost all of the Triassic dicynodonts. Pristerodontoids are distinguished by a large, leaf-shaped palatine bone on the palate and the presence of a deep, longitudinal groove along the dorsal edge of the dentary, called the dentary sulcus. The posterior part of the feeding apparatus appears to have been more important than in other dicynodonts, with a slicing, knife-like blade on the lower jaw shearing against a horny palatal pad. In most members of the group, the premaxillary margin is also sharp and fits quite closely against the outer edge of the dentary, indicating a cutting edge perhaps suitable for cropping palate-sized bites of leafy vegetation, with the different species specialised for handling the foliage of different forms of glossopterid trees, shrubs, etc. (Cox 1998).

Pristerodon is a primitive member of the group. It is small in size with a broad intertemporal roof, and possesses small postcanine teeth in both upper and lower jaws (King and Rubidge 1993). Common and widespread in Africa, it existed from the *Tapinocephalus* to the *Dicynodon* Assemblage Zones, a longevity matched otherwise only by *Diictodon* (Angielczyk 2001), and it also occurs in India. The adaptive radiation which occurred from a hypothetical

Pristerodon-like ancestor includes such familiar types as *Oudenodon* which lost its tusks but retained the wide intertemporal region and narrow snout. According to Hotton (1986), *Oudenodon* held its snout forwards when feeding, indicating that its source of food was vegetation well above ground level. This group also includes the largest of the Permian dicynodonts, with skull lengths up to 50 cm. *Aulacocephalodon* is a fairly large, wide-snouted pristerodontoid in which the front edge of the premaxilla formed a transverse cutting blade. In contrast, the even larger *Dinanomodon* possessed a narrow, pointed snout which might perhaps have adapted it to browsing on higher-level foliage. These two inevitably have been compared to white and black rhinos, respectively.

The familiar and once much overused genus *Dicynodon* retained a pair of tusks, evolved a narrow intertemporal region, and tended to shorten the length of the palate. Haughton and Brink (1954) famously listed 111 species of *Dicynodon*, a number whittled down by King (1988)

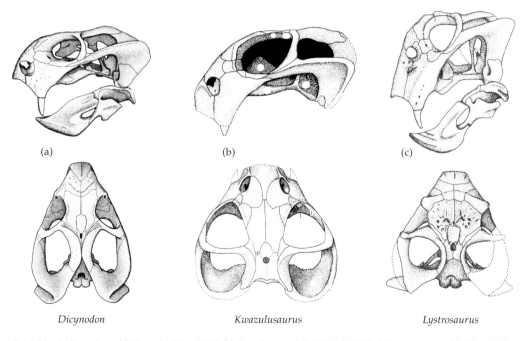

Dicynodon　　　　　*Kwazulusaurus*　　　　　*Lystrosaurus*

Figure 3.14 Triassic dicynodonts. (a) *Dicynodon* (King 1990). (b) *Kwazulusaurus shakai* (Maisch 2002). (c) *Lystrosaurus declivis* (King 1988). (d) Skeleton of *Kannemeyeria* (after Pearson). (e) The shansiodontine *Tetragonius njalilus* (King 1988, from Cruickshank). (f) *Stahleckeria* (King 1990 after Camp). (g) *Ischigualasto jenseni* (King 1988, after Cox). *Continued overleaf*

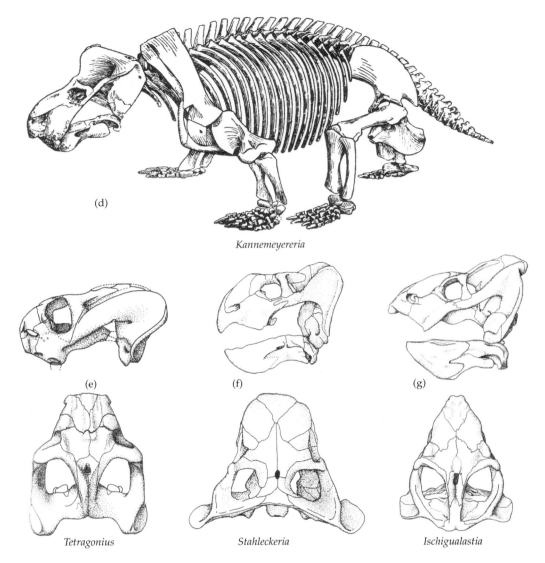

(d)

Kannemeyereria

(e)   (f)   (g)

Tetragonius   *Stahleckeria*   *Ischigualastia*

Figure 3.14 (*continued*).

to 59, with a strong suspicion than several of these are still synonymous.

Lystrosaurus (Fig. 3.14(c)) is the most remarkable pristerodontoid, indeed the most remarkable dicynodont in several ways. It is a tusked form clearly related to *Dicynodon* (Fig. 3.14(a)), but has exaggerated the shortening of both the palate and the temporal region of the skull, resulting in a strongly downturned snout. The Late Permian South African *Kwazulusaurus* (Fig. 3.14(b)) has a skull structure

intermediate in form between *Dicynodon* and *Lystrosaurus* and Maisch (2002) consequently interprets it as a basal lystrosaurid. *Lystrosaurus* itself is the only actual genus of dicynodont that is known to have survived the end-Permian, having been identified in the latest Permian as well as the earliest Triassic deposits (Smith 1995; King and Jenkins 1997), when it became supremely abundant in terms of both numbers and geographical distribution. In South Africa, well over 90% of specimens collected from the *Lystrosaurus* Assemblage Zone are of *Lystrosaurus*, and it has been identified in contemporaneous early Triassic beds worldwide including, Russia, India, China, and Antarctica. Thulborn (1983) identified a few fragments from Australia as the quadrate and tusk of a dicynodont, almost certainly *Lystrosaurus* (King 1983).

For many years, *Lystrosaurus* was believed to have been an amphibious animal, living and feeding in freshwater swamps and lakes. However, King (1991) and Cox (1991) showed that this was probably a misinterpretation of the significance of such anatomical features as the dorsally placed nostrils and eyes, and subsequently King and Cluver (1991) offered an alternative interpretation of its mode of life. They argued that the change from a *Dicynodon*-like ancestral state that had occurred consisted of simultaneously shortening the skull, deepening the temporal region, and elongating the downturned snout. These modifications in geometrical shape resulted in an increased bite force being applied by the adductor muscles to the jaws. Tougher terrestrial vegetation could thus be dealt with, by a combination of the slicing action of the maxillary rim acting against the lateral edge of a dentary beak, and the crushing and shredding that occurred between dentary and palate. Features of the postcranial skeleton such as the short, broad hand, and a rather wide knee joint indicate a degree of digging ability rather than being specialisations for aquatic life as was once supposed.

The Triassic radiation of kannemeyeriids. *Lystrosaurus* was restricted to the narrow time zone either side of the Permo–Triassic boundary, but the clade to which it belongs continued to diversify and produce a range of large forms throughout the Triassic Period, as reviewed by Keyser and Cruickshank (1979), Cox and Li (1983), and King (1988, 1990). These constitute the family Kannemeyeriidae, which never became as diverse or abundant as the Permian dicynodonts, no doubt partly because they shared the large terrestrial herbivore habitat with a number of other taxa, gomphodont cynodont therapsids, rhynchosaurs, and early dinosaurs. Nevertheless, kannemeyeriids had a worldwide distribution, occurring in several parts of Africa, North and South America, Europe, India, and in China where

they are especially abundant (Lucas 2001). They all retained the short snout and basicranial axis characteristic of *Lystrosaurus*, but had developed a high, narrow intertemporal crest. Body size was increased, with skull lengths anything from 25–60 cm. The postcranial skeleton was similar to that of the Permian dicynodonts but, as befits large-bodied herbivores, kannemeyeriids tended to have a relatively short, barrel-shaped trunk, and very stout limbs. The pelvic girdle is remarkable for a huge anterior expansion of the ilium and reduced pubis, which, according to Cruickshank (1978) was an adaptation for rearing up on the hind legs to feed from higher branches of the shrubs and trees.

Kannemeyeria (Fig. 3.14(d)) is characterised by a narrow and pointed snout, very high intertemporal crest, and well-developed tusks. Specimens of this, or very closely related genera rivalled *Lystrosaurus* in their abundance and cosmopolitan distribution. They are common in the Lower-Middle Triassic *Cynognathus* Assemblage Zone of South Africa, and in the equivalent Lower Ermaying Formation of China (Lucas 2001). Very similar forms have been found in, Indian and Russian beds of presumably the same age (Bandyopadhyay 1988; King 1990) and possibly, though less certainly, South American (Renoux and Hancox 2001).

A second kannemeyeriid group is represented by the Chinese Lower Triassic *Shansiodon*, and the closely similar east African *Tetragonius* (Fig. 3.14(e)) and South American *Vinceria*. These are smaller forms with a broad skull, and short, blunt, ventrally directed snout. Throughout the Middle Triassic, kannemeyeriines and shansiodontines, continued to radiate worldwide. New groups also evolved, including stahleckeriines such as the Brazilian *Stahleckeria* (Fig. 3.14(f)) and the Zambian *Zambiasaurus*, which are characterised by the absence of tusks and a very deep posterior part of the skull. One tantalising cranial fragment of kannemeyeriid from Russia that may be a stahleckeriine was named *Elephantosaurus* by V'iuschkov (1969), who estimated that it came from a skull that could have been as much as a metre in length. If so, it was by a considerable margin the largest dicynodont ever, and equal to the largest of the dinocephalians. The last of the kannemeyeriids occur in the Upper Triassic, specifically the early part of the Norian

Stage, and all were by this time restricted to South and North America (King 1988, 1990). There are surviving stahleckeriines, plus one final new group (Fig. 3.14(g)), represented by the North American *Placerias* and the South American *Ischigualastia* (Cox 1991). These possessed large skulls around 50 cm in length, and lacked tusks. The snout is relatively long and the occiput very deep.

The recent description by Thulborn and Turner (2003) of six associated cranial fragments collected in 1914 from Queensland and about which Heber Longman stated 'some slight resemblance might be traced to the Dicynodonts of South Africa' is intriguing in the extreme. One of the fragments consists of a piece of premaxilla, with the stump of a large tusk, that is indistinguishable from the corresponding region of a kannemeyeriid dicynodont. The other pieces are less diagnostic, but all are consistent with that identification. The date appears to be well established as Early Cretaceous, and all the documentary evidence points to a genuine discovery and reliable curation of the specimen. The possibility must therefore be entertained that kannemeyeriids survived in Australia for no less than 110 Ma after their last appearance elsewhere in the world.

Gorgonopsia

The remains of gorgonopsians recorded in the lowermost of the South African Karoo fossil levels, the *Eodicynodon* Assemblage Zone, are very poorly preserved (Rubidge 1993; Rubidge *et al.* 1995) and it is not until the overlying *Tapinocephalus* Assemblage Zone that complete specimens have been found. Here they occur as relatively rare, smallish carnivores, adapted for feeding on prey by greatly enlarged upper and lower canines that are oval in section and carry a serrated hind edge. During the rest of the Late Permian however, gorgonopsians were the dominant top terrestrial carnivores, and they are also found in contemporaneous Russian deposits. So far they are completely unknown outside southern Africa and Russia and, unlike the other latest Permian therapsid groups, not a single gorgonopsian is known to have survived into even the base of the Triassic.

Gorgonopsians (Figs 3.15 and 3.16) varied in body size from that of a small dog to somewhat larger than any living mammalian predator. As well as the enlarged canines, there are five upper and four lower well-developed incisor teeth, while the postcanines are reduced to at most four or five very small, simple teeth. Several cranial characters also diagnose the group, including the following:

- enlarged, flat-topped preorbital region of the skull
- enlargement of the temporal fenestra by lateral and posterior extension while the intertemporal roof remained broad and uninvaded by adductor musculature
- preparietal bone
- vomers fused
- paired palatine bones meeting in the mid-ventral line of the palate, and together with the pterygoids forming a deeply vaulted palate.

The gorgonopsians also retain many primitive therapsid features such as the carnivorous dentition, broad intertemporal region of the skull roof, and very conservative postcranial skeleton. As a consequence there has been a tendency in the past to classify them with other primitive carnivorous taxa, such as the hipposaurids and burnetiids, which are now included in Biarmosuchia. Recent authors now follow the review of Sigogneau-Russell (1989) and limit the Gorgonopsia to the well-defined forms that have the diagnostic characters of the skull temporal fenestra, and palate noted. Taxonomically constrained in this way, there is remarkable conservatism within the group. Genera differ among themselves by little more than size, relative proportions of the skull width, interorbital breadth, etc, and the trivial character of number of postcanine teeth, and all can be accommodated in the single family Gorgonopsidae. Sigogneau-Russell (1989) recognises three subfamilies. The Gorgonopsinae (Fig. 3.15) consists of the majority of genera. The Rubidgeinae (Fig. 3.16(c)) have a very broad, heavily built skull, with wide postorbital bar, and massive zygomatic arches. The Inostranceviinae consists primarily of the Russian genus *Inostrancevia* itself (Fig. 3.16(b)), which is the largest of all gorgonopsians, having a skull length of over half a metre. Its main characteristic apart from this size is the relatively very long preorbital length.

Biologically, the gorgonopsians were superbly adapted for a highly predaceous mode of life.

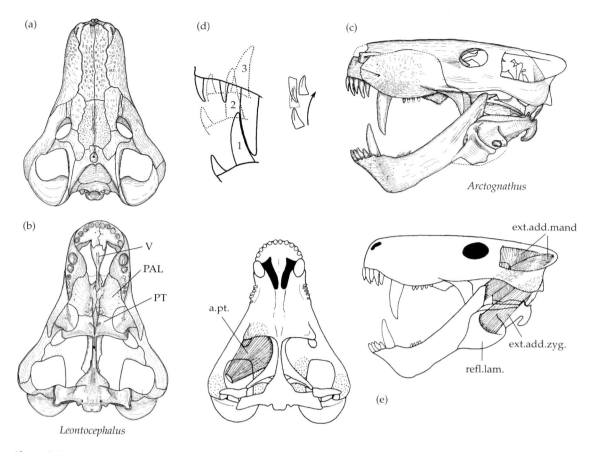

Figure 3.15 Gorgonopsian skull and jaw mechanism. (a), (b) Skull of *Leontocephalus intactus* in dorsal and ventral view. (c) Lateral view of skull and lower jaw of *Arctognathus*. (d) Interdigitation of incisor and canine teeth from positions 1–3, and the action of opposing single upper and lower incisors. (e) Ventral and lateral views of the disposition of the main adductor mandibuli musculature a.pt, anterior pterygoideus muscle. ext.add.mand, external adductor mandibularis muscle. V, vomer. PAL, palatine. refl.lam, reflected lamina of the angular. PT, ptepygoid. (Kemp 1969).

Kemp (1969*b*) analysed the functioning of the jaws in detail and showed from wear facets preserved on the opposing incisor and canine teeth of a spectacularly well-preserved specimen of *Leontocephalus* that the lower jaw was capable of two modes of action. The simplest one consisted of opening the jaws extremely widely, by as much as 90°, to be sufficient for the huge upper and lower canines to clear one another enough. The lower jaw was then powerfully adducted so that the sharp, serrated-edged canines would sink into the prey to disable it. The second mode of bite was more precise and consisted of a propalinal shift forwards of the lower jaw, so that now when it closed the four serrated lower incisors on one side passed between the five similarly serrated upper incisors (Fig. 3.15(d)). The effect of the interdigitated incisor bite would have been to cut a jagged edge in the flesh and therefore enable chunks to be more readily torn off and swallowed. To achieve the necessary antero-posterior movement of the jaw, a mobile articulation between the quadrate and the squamosal had evolved, whereby the ball-shaped body of the quadrate bone rotated in a socket-shaped squamosal recess about a transverse axis. The effect was to permit the lower part of the quadrate to shift forwards or backwards, which in turn caused the whole lower jaw to move forwards or backwards. By this means, the jaw could shift forwards for the incisors to intermesh, or backwards so that the incisors did not get in the way of

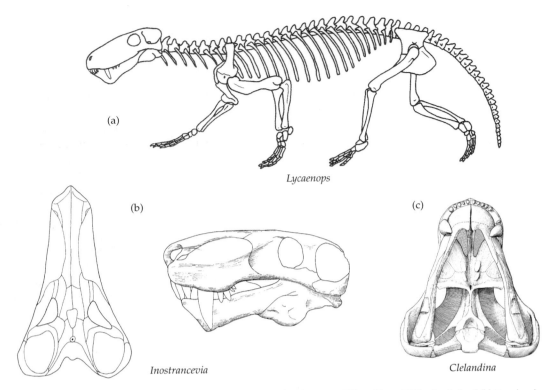

(a)

Lycaenops

(b)

Inostrancevia

(c)

Clelandina

Figure 3.16 Gorgonopsians. (a) Skeleton of *Lycaenops ornatus*. Presacral length approx. 125 cm (Kemp 1982, after Colbert) (b) Dorsal and lateral views of *Inostrancevia alexandri*. Skull length approx. 50 cm (Sigogneau-Russell 1989 after Tatarinov). (c) Ventral view of skull of *Clelandina scheepersi*. Length approx. 18 cm (Sigogneau-Russell 1989 after Sigogneau).

the powerful canine action. Laurin (1998) doubted whether the quadrate of gorgonopsians had this propalinal ability, but he does not comment on how otherwise the wear facets on the sides of the incisors could have formed, since the lower jaw is not long enough for the lower incisors to reach the uppers without an anterior shift. At any event, the gorgonopsian dentition was certainly adapted for dealing with relatively large, active prey that presumably consisted in the main of the abundant smaller dicynodonts.

The enlargement of the temporal fenestra and by inference of the associated adductor musculature was achieved in a uniquely gorgonopsian manner by a posterior extension of the fenestra well beyond the hind limit of the skull roof and occiput, simultaneously with a lateral extension creating a zygomatic arch (Fig. 3.15(e)). Corresponding enlargement of the areas of attachment of the musculature to the

lower jaw was achieved in part by evolution of a well-developed coronoid process extending above the dorsal margin of the jaw. More radically, adductor musculature had also invaded the external surface of the jaw, as indicated by the powerful dorso-ventral ridge and concave area behind it on the reflected lamina of the angular. Musculature from the outwardly bowed zygomatic arch must have attached to this region of the lower jaw, thus creating an analogue of the mammalian masseter muscle, an elaboration of the jaw-closing musculature not otherwise seen until the cynodonts independently evolved a comparable arrangement.

The postcranial skeleton of gorgonopsians (Fig. 3.16(a)) (Colbert 1948; Kemp 1982; Sigogneau-Russell 1989) is that of a relatively long, slender limbed, and agile version of what otherwise illustrates the basic condition of therapsids, discussed in a later chapter.

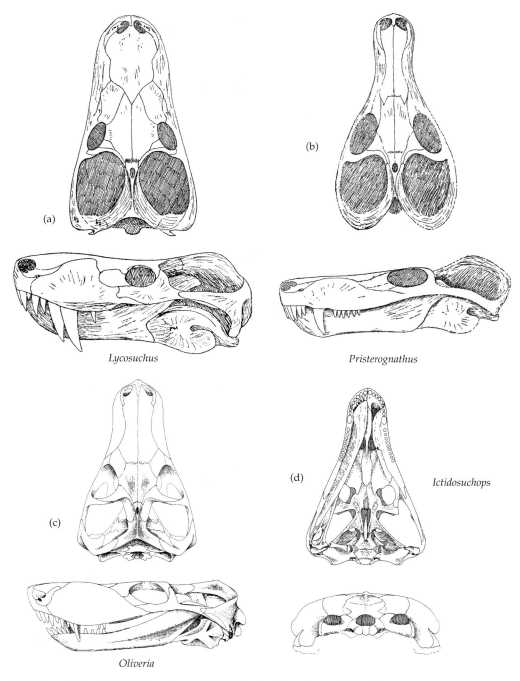

Figure 3.17 Therocephalians. (a) Skull of *Lycosuchus vanderrieti* in dorsal and lateral views. Skull length approx. 23 cm (Broom 1932). (b) Skull of *Pristerognathus minor*. Skull length approx. 25 cm (Broom 1932). (c) Skull of the baurioid *Oliveria parringtoni* in dorsal, lateral, and posterior views, skull length approx. 20 cm. (d) palate of *Ictidosuchops intermedius*, skull length approx. 10 cm (Brink 1965; 1960). (e) Reconstructed skeleton of regisaurid indet. Presacral length approx. 16 cm (Kemp 1986). *Continued overleaf*

(e)

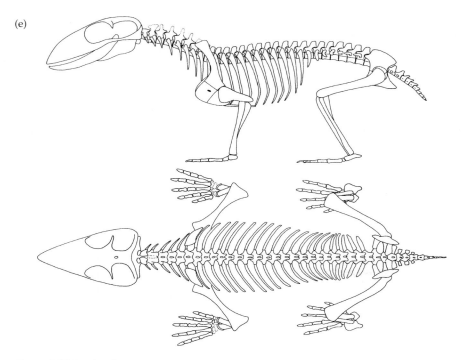

Figure 3.17 (continued).

Therocephalia

The therocephalians were another group of carnivorous therapsids, and many of the earlier ones bear a superficial resemblance to the gorgonopsians. Others, however, were significantly smaller, sufficiently so to be regarded as small-prey carnivores or even committed insectivores, while one specialised late group evolved adaptations for herbivory. Like most other major therapsid groups, the earliest record of the Therocephalia is from the lowest of the South African fossil-bearing horizons, the *Eodicynodon* Assemblage Zone, from which the incomplete skull of a small, primitive form has been described (Rubidge *et al.* 1983). For the rest, the group occurs reasonably commonly throughout the remainder of the Late Permian, and more rarely in the Lower Triassic. Geographically, the majority of specimens are South African, but there are several Russian genera, and also records of the taxon from other parts of southern Africa, China (Sun 1991), and even the Lower Triassic of Antarctica (Van den Heever 1994).

The characters of the Therocephalia include enlargement of the temporal fenestra in a medial direction, greatly reducing the intertemporal skull roof to a narrow bar although, as discussed later, this character is shared with cynodonts and is part of the case for a relationship between these two groups of therapsids. Unequivocally therocephalian characters (Hopson and Barghusen 1986; Van den Heever 1994) include the following:

- paired suborbital vacuities in the palate
- broad vomer
- stapes lacking a foramen and dorsal process
- in the lower jaw, the reflected lamina of the angular combining a free dorsal margin with a series of strong ridges radiating from the central region
- reduced, horizontal lumbar ribs
- anterior process on the ilium and an additional trochanter on the femur
- Digital formula precociously and independently reduced to the mammalian formula of 2.3.3.3.3. (Kemp 1986; Fourie 2001).

The earlier, *Tapinocephalus* Assemblage Zone therocephalians were formerly classified as the Pristerognathidae (Hopson and Barghusen 1986), but in his review, Van den Heever (1994) regards the group as paraphyletic and has abandoned it in favour of two separate families. Lycosuchidae (Fig. 3.17(a)) is a basal group of short, broad-snouted forms, and the longer snouted Scylacosauridae (Fig. 3.17(b)) are related to the rest of the therocephalians, which he refers to as Eutherocephalia. At any event, this paraphyletic primitive group of 'pristerognathids' consists of generally fairly large animals, with skull length of the order of 20–30 cm. As in the gorgonopsians, though to a lesser extent, the dentition is dominated by large upper and lower canine teeth. The number of upper incisor teeth varies from five to seven in different taxa, but there are always only three lower incisors. The postcanine dentition is better developed than in the gorgonopsians, although it does tend to be reduced to as few as five modest teeth in several genera such as *Lycosuchus*. These primitive therocephalian groups disappeared after the *Tapinocephalus* Assemblage Zone, coincidentally with the rise of larger sized, superficially very similar gorgonopsians.

They were replaced in the later part of the Permian by a mixture of both small, generalised therocephalians, and some highly specialised new kinds. Among the specialists, the whaitsiids, such as the South African *Theriognathus* (Fig. 3.18(a)) and the Russian *Moschowhaitsia*, were relatively large animals. They retained well-developed incisors and canines, but the postcanine dentition was greatly reduced or completely absent. The snout is constricted behind the canines, where the palatal exposure of the maxilla forms a broad, heavily ridged and pitted surface that opposes a similarly broad, concave dorsal surface of the dentary bone; the postcanine dentition had evidently been replaced by directly opposing keratinised tooth plates (Watson and Romer 1956; Kemp 1972b). Other unique features of whaitsiids include a partial secondary palate formed by maxillary processes meeting the vomer, partial or complete loss of the paired suborbital fenestrae, closure of the interpterygoid vacuity, and a great broadening of the primitively rod-like epipterygoid connecting the palato-quadrate to the skull roof. All these new characters can be interpreted as devices for strengthening the skull against the action of its own jaw musculature (Kemp 1972b). Whaitsiids were probably highly specialised as hyaena-like scavengers, capable of crushing the bones of abandoned carcasses between nutcracker like jaws.

Moscorhinus (Fig. 3.18(c)) was another fairly large form, about 25 cm in skull length. Here, the skull is very broad and low, with a short, massive snout and extraordinarily expanded vomer in the palate. Like whaitsiids, it too has tended to reduce the number and size of the postcanine teeth (Durand 1991). *Euchambersia* (Fig. 3.18(e)) may be a close relative, having a similarly broad, short-snouted skull. Uniquely in this genus there is an extraordinary, deep recess on either side of the snout, between the canine and the front of the orbit, the walls of which are covered in fine foramina. The canine tooth has a groove down its outer face indicating that *Euchambersia* possessed a pair of maxillary poison glands coupled to fangs for delivering venom. The postorbital bars and zygomatic arches are very weakly developed implying that the jaw musculature was poorly developed and the jaws therefore unable to deal with struggling live prey. Altogether, *Euchambersia* appears to have been a therapsid analogy of snakes.

For the rest, therocephalians are small-sized, generalised carnivorous animals. The smallest ones were at one time referred to as 'scaloposaurs' (Fig. 3.17(d) and 3.18(b)), but Hopson and Barghusen (1986) abandoned this taxon because many and perhaps all of them were believed to be the juvenile individuals of a variety of genera. They erected in its place the taxon Baurioidea, of which the early Triassic, *Lystrosaurus* Assemblage Zone *Oliveria* (Fig.3.17(c)) is a typical, well-described example (Mendrez 1972; Kemp 1986). The skull is about 10 cm in length. There are six upper incisors, a modest canine, and a row of ten small postcanine teeth. A partial secondary palate has formed by contact between the maxillae and the vomer towards the front of the palate, which is continued posteriorly by a pair of ridges that presumably indicates the attachment of a soft secondary palate. The intertemporal roof is narrow and incipiently crested, while the zygomatic arches are slender and not flared laterally as occurs in gorgonopsians and cynodonts. The postcranial skeleton of an immature regisaurid (Fig. 3.17(d)) described by Kemp (1986) is that of a very slender-limbed, agile creature with skeletal proportions comparable to a modern mammal. Several other very similar latest Permian and earliest Triassic baurioids are known that differ from *Oliveria* only in such details as the degree of development of the secondary palate (Hopson and Barghusen 1986),

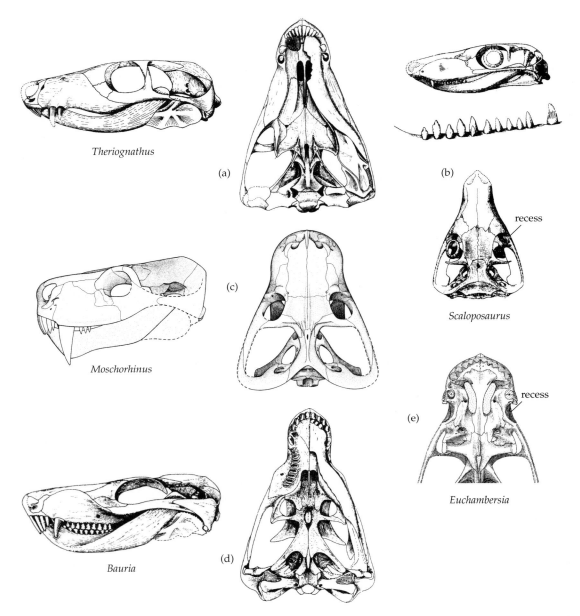

Figure 3.18 Therocephalian diversity. (a) Skull of the whaitsiid *Theriognathus*. Skull length approx. 7.5 cm (Brink 1956*a*). (b) Skull in lateral and dorsal views, and enlarged view of lower post-canines of *Scaloposaurus constrictus*. Skull length approx. 10.5 cm (Mendrez-Carroll 1979). (c) Skull of *Moschorhinus kitchingi* in lateral and dorsal views. Skull length approx. 23 cm (Durand 1991). (d) Skull of *Bauria cynops*. Skull length approx. 12 cm (Brink 1963*a*). (e) Anterior palate of *Euchambersia mirabilis* (Mendrez 1975).

and whether the postcanine teeth bear small additional cuspules or not.

The small animal *Bauria* (Fig. 3.18(d)) is remarkable because of the convergence in tooth structure with the later herbivorous cynodont therapsids. The postcanine teeth are transversely widened and the uppers and lowers meet in true, direct occlusion. The crowns of both the uppers the lowers consist of a large labial cusp and a row of smaller lingual cusps. They intermesh in such a way that the anterior edge of an upper tooth shears against the posterior edge of the corresponding lower tooth to generate a cutting action (Gow 1978). The diet of *Bauria* is assumed to have included tough, fibrous material, although, for all the sophistication of its dental battery, it shows no sign of a particular increase in the size of its adductor musculature. The postorbital bar is in fact absent, and the zygomatic arch not significantly expanded beyond that of earlier, primitive baurioids. *Bauria* is the last occurring South African therocephalian, being found in the Lower-Middle Triassic *Cynognathus* Assemblage Zone. The Chinese *Traversodontoides* is closely related (Sun 1991), as possibly are certain Russian specimens (Battail and Surkov 2000).

Kemp (1972*a*, 1982) analysed the therocephalian jaw mechanism. The more primitive forms retained well-developed incisors and canines but reduced postcanines, and probably had similar jaw mechanics to the gorgonopsians, with the adductor jaw musculature imparting kinetic energy to the jaws that was dissipated by the canines entering the prey. There is no evidence for interdigitation of the incisors as found in gorgonopsians, and therefore dismembering of prey may have been a cruder affair consisting of tearing rather than cutting the flesh. The jaw musculature (Fig. 4.3(b)) was largely restricted to the medial and posterior surfaces of the temporal fenestra, and a connective tissue sheet that may have covered it. Invasion by adductor musculature of the lateral surface of the lower jaw seems to have been at most very limited. On the other hand, the structure of the reflected lamina of the angular of therocephalians is more complex than that of any other therapsid. It consists of a relatively huge, thin sheet of bone, free along the dorsal posterior and ventral margins, and strengthened by three or four broad ridges or corrugations radiating from the front. These ridges on the lamina are difficult to interpret since no comparable structure exists in modern tetrapods. However, they do suggest that there was a complex arrangement of posteriorly and ventrally directed musculature extending from the reflected lamina behind and below the lower jaw. It would presumably have included musculature associated with jaw opening, and with the operation of hyoid apparatus in the floor of the mouth, and therefore complex tongue movements may have occurred. It may be that the considerable range of adaptive types found among the therocephalians compared to the otherwise rather similar gorgonopsians was made possible by greater adaptability of the oral manipulating mechanism in the former. Large and small carnivores, insectivores, scavengers, venom-producers, and herbivores all manifested versions of the therocephalian jaw structure.

The locomotory function of therocephalians has been considered by Kemp (1986). The ratio of limb length to body size is comparable to modern non-cursorial mammals, and like the latter there is a distinct lumbar region with short, stout, immoveable, horizontal ribs behind the functional ribcage (Fig. 3.17(d)). The tail is very reduced and the ilium expanded forwards, presumably to accommodate a gluteal-like muscle for femoral retraction. The mode of action of the limbs is discussed later, since the therocephalians best illustrate an important hypothetical stage in the origin of mammalian locomotion (page 105ff).

Of all the therapsid groups other than the cynodonts with their particular relationship to the mammals, it is the progressive members of the Baurioidea that most give the impression of an advanced level of temperature physiology. As in the cynodonts, they evolved a secondary palate, with the implication that feeding and food manipulation in the mouth needed to occur simultaneously, and which is correlated with an elevated metabolic rate. The triturating teeth of *Bauria* may have increased the rate of food assimilation, and further evidence is the disassociation of the lumbar region from the thoracic, implying the possibility of a diaphragm and thus elevated levels of gas exchange. The fact that several lineages of baurioids of very small body size survived the Permo–Triassic boundary, which is to say survived the greatest mass extinction event in the Earth's history could also be correlated with the relatively high level of independence of environmental stresses such as cold temperature that is associated with endothermy. Nevertheless, despite such intriguing possibilities the therocephalians actually had an extremely modest presence in the Triassic, and it is to their relatives the cynodonts that this history must now turn to see the next great advances in the synapsid story.

Cynodontia

The cynodonts were the last major group of the Late Permian radiation of therapsids to make an appearance, being first recorded just before the close of the period as rare elements of the *Dicynodon* Assemblage Zone fauna of South Africa and equivalent aged beds in Zambia, Russia, and also Western Europe (Sues and Munk 1996). Despite this modest beginning, the cynodont taxon survived the end-Permian and underwent a broad radiation in the Triassic, one branch of which culminated in the mammals. Thus they are fundamental to understanding the transition from basal amniote to mammalian morphology.

Cynodonts share a number of characters with the Therocephalia and the two are usually included in a single group Eutheriodontia. The main similarities are:

• narrowing of the intertemporal skull roof as the temporal fenestrae expanded inwards, creating an elongated, narrow sagittal crest
• reduction of the postorbital and postfrontal bones
• broadening of the epipterygoid
• discrete process of the prootic bone that contacts the quadrate ramus of the pterygoid
• differentiation of the vertebral column into distinct thoracic and lumbar regions, with the lumbar ribs short, horizontal, and immovably attached
• tendency to elongate the ilium and reduce the pubis in the pelvic girdle.

At one time it was thought that among the therocephalians, the Whaitsiidae were closest to the cynodonts, on the basis of a number of shared similarities such as the extremely broad epipterygoid, reduction of the suborbital fenestra, and closure of the interpterygoid vacuity (Kemp 1972*b*). Cladistic analysis, however, indicates that these features are better interpreted as convergences in the two taxa (Hopson and Barghusen 1986), perhaps associated with a similar mode of strengthening of the skull against powerful jaw action.

Cynodonts are a very well categorised group; Hopson and Kitching (2001) noted no less than 27 synapomorphies, mostly cranial, of which the more prominent ones are the following.

• Deepening and lateral flaring of the zygomatic arch, and development of an adductor fossa on the lateral face of the dentary bone, both characters associated with the invasion of the lateral surface of the jaw by adductor musculature.
• Sagittal crest between the paired temporal fenestrae with deep, laterally facing surfaces for origin of the temporalis musculature.
• Reflected lamina of the angular reduced.
• Quadrate and articular bones that constitute the jaw articulation reduced in size.
• Dentition differentiated into unserrated incisors and canines, followed by a series of simple anterior postcanines, and more complex posterior postcanines bearing accessory cuspules.
• Secondary palate formed from crests along the lateral sides of the choanal vault.
• Epipterygoid greatly expanded as a major structural component of the sidewall of the braincase region.
• Occipital condyle double.
• Marked differentiation of thoracic from lumbar regions of the vertebral column.
• Scapula deeply concave and coracoid reduced.
• Ilium expanded well forwards and pubis reduced.
• Femur with inturned head, and strongly developed major and minor trochanters.

Procynosuchia—primitive cynodonts

Procynosuchus

The Late Permian primitive cynodont *Procynosuchus* is very well-known from several South African skulls, and a virtually complete, acid-prepared skeleton from the contemporaneous Madumabisa Mudstone of Zambia described in detail by Kemp (1979, 1980*c*). It has been recorded in Germany (Sues and Boy 1988). By therapsid standards it was a relatively small animal with a skull length of up to 14 cm and a presacral length of about 40 cm. As a primitive-grade cynodont, *Procynosuchus* (Fig. 3.19(a)) illustrates an incipient stage in the evolution of several advanced cynodont and mammalian features. The first important one concerns the dentition. The postcanine dentition has become differentiated into five anterior premolariform teeth, each of which consists of a slightly recurved cusp with a small basal swelling at the back of the tooth, followed by eight molariform teeth which

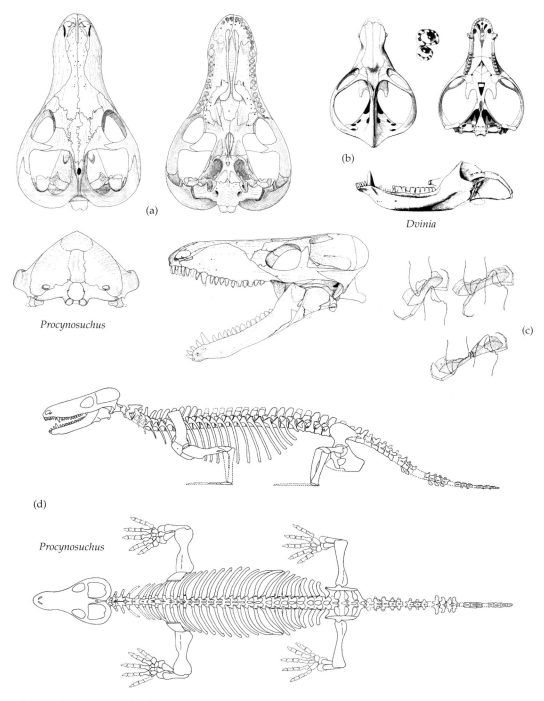

(a)

Dvinia

Procynosuchus

(b)

(c)

(d)

Procynosuchus

Figure 3.19 (a) Skull of *Procynosuchus delaharpaea* in four views. Skull length approx. 11 cm (Kemp 1979). (b). Skull of *Dvinia* (Tatarivov 1968). (c) Action of postcanine teeth of *Procynosuchus* dismembering an insect. (d) Postcranial skeleton of *Procynosuchus delaharpaea*. Presacral length approx. 33 cm (Kemp 1980c).

have a more complex form. There is a single large, main cusp, at the inner base of which is a row of small, but very distinct cuspules. These teeth represent the first step towards the complex multicusped, occluding molar teeth that were eventually to play such a fundamental role in the radiation of mammals. At this stage, however, the upper and lower teeth did not even occlude directly, but appear merely to have increased the general tearing effect as they dealt with food such as insect cuticle (Fig. 3.19(c)). Several features of the skull and lower jaw indicate that there had been further elaboration of the jaw musculature beyond the stage found in the therocephalians, although still to a degree well short of advanced cynodonts and mammals. The temporal fenestra had expanded, both medially to the extent that there is now a deeper, vertical-faced sagittal crest, and laterally to create an outwardly bowed zygomatic arch. The latter is nevertheless still a relatively delicate structure. The lower jaw is comparably modified. The coronoid process has become broadened, and bears a depression on its lateral surface that indicates the first stage of the invasion of the lateral surface of the jaw by adductor musculature. The dentary being relatively enlarged in this way, the postdentary bone complex is relatively reduced in size, a trend to be continued right up to the mammals. Other changes in the skull of *Procynosuchus* include the development of a secondary palate in the roof of the mouth, although at this stage the paired extensions of the maxillae and palatals do not meet in the midline so the osseous secondary palate is incomplete; presumably it was completed by soft tissue.

Despite these considerable modifications to the skull, the postcranial skeleton of *Procynosuchus* has progressed little beyond a basic therapsid condition (Fig. 3.19(d)). There is certainly very clear differentiation of the vertebral column into thoracic and lumbar regions, but the limb girdles illustrate the cynodont characteristics to only a small degree. In fact, the Zambian specimen (Kemp 1980c) shows unexpected specialisations indicating a semiaquatic, perhaps otter-like mode of life. The zygapophyses of the lumbar region of the vertebral column are extraordinarily broad and horizontally oriented. Uniquely among therapsids, they must have permitted very extensive lateral undulation of the hind

region of the vertebral column. The tail is relatively long, with uniquely expanded haemal arches that suggest a laterally operated swimming organ. The limbs themselves are relatively short and the individual bones stout. The hind foot is not preserved, but the forefoot is broad and the individual bones flattened and poorly ossified. The only other postcranial skeleton attributed to *Procynosuchus* was described rather superficially by Broom (1948, as the genus *Leavachia*). This specimen lacks these highly distinctive specialisations of *Procynosuchus* and may actually belong to a different genus, one that was adapted for a fully terrestrial existence.

Dvinia

The only other cynodont of the same primitive grade as *Procynosuchus* is *Dvinia* (Fig. 3.19(b)) from the Russian Late Permian (Tatarinov 1968), for it also has a relatively small coronoid process and adductor fossa of the dentary, and relatively large postdentary bones. Indeed, it is even more primitive than *Procynosuchus* in certain details of its braincase such as a very limited contact between the prootic and the epipterygoid bones and a large post-temporal fossa (Hopson and Barghusen 1986). At the same time, however, *Dvinia* has much more elaborate molariform teeth. In both upper and lower jaws, there is a series of small, simple premolariform teeth, followed by the molariform teeth, each of which is expanded across the jaw so that the crown surface is roughly circular, and bears a complete ring of cuspules. *Dvinia* was similar in size to *Procynosuchus* and its diet can only be speculated upon. Perhaps it was an early cynodont herbivore, although certainly unrelated to the advanced herbivorous members of the group.

Basal epicynodontia—middle-grade cynodonts

All other cynodonts have progressed beyond the procynosuchian grade by enlargement of the coronoid process, and development of a more robust, and dorsally bowed zygomatic arch. There is a further reduction in the size of the quadrate, and the number of incisor teeth is reduced to four uppers and three lowers. The most distinctive feature of the postcranial skeleton is the presence of overlapping costal plates on the dorsal ribs.

They constitute the monophyletic group Epicynodontia, which includes three basal families plus the advanced Eucynodontia.

Galesauridae

The earliest of these more progressive cynodonts is the third kind of cynodont known to have existed in the Late Permian. *Cynosaurus* is from the *Dicynodon* Assemblage Zone of South Africa, and is a member of the family Galesauridae, which is much better known from *Galesaurus* itself (Fig. 3.20(c)), which comes from the overlying, earliest Triassic *Lystrosaurus* Assemblage Zone fauna. Galesaurids are judged to be the most basal of the epicynodonts

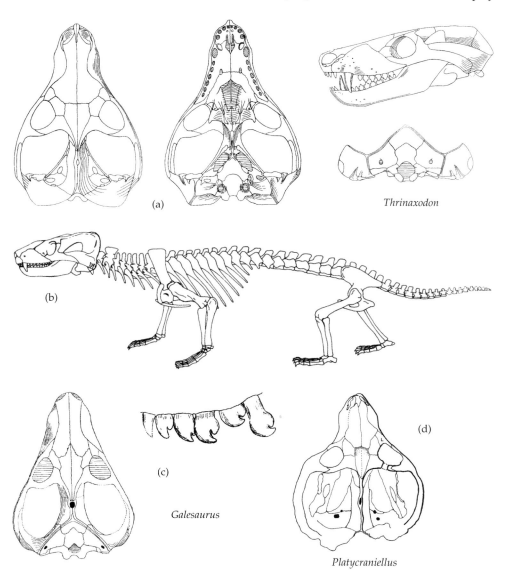

Figure 3.20 Basal epicynodonts. (a) Skull of *Thrinaxodon liorhinus* in four views. Skull length approx. 7 cm (Parrington 1946). (b) Postcranial skeleton of *Thrinaxodon liorhinus*. Presacral length approx. 35 cm (Carroll 1988, from Jenkins 1984). (c) Skull of *Galesaurus planiceps* in dorsal view, with enlarged upper postcanine teeth. Skull length approx. 8.5 cm (Parrington 1934 and Broom 1932). (d) Dorsal view of skull of *Platycraniellus elegans*. Skull length approx. 10 cm (Abdala Unpublished Manuscript).

on the grounds of an incomplete secondary palate. Their most distinctive feature is the structure of the postcanine teeth. Compared to those of *Procynosuchus*, these have lost the cingular cusps. There is also no anterior accessory cusp as occurs in several other Triassic cynodont families, but a well-developed posterior accessory cusp lies immediately behind the somewhat recurved main cusp. Perhaps galesaurids were adapted for dealing with a more robust form of insect prey.

Thrinaxodontidae
Thrinaxodon (Fig. 3.20(a) and (b)) is far the commonest and most studied cynodont of the *Lystrosaurus* Assemblage Zone fauna of the Karoo. It also occurs in the Fremouw Formation of Antarctica (Colbert 1982). Structurally *Thrinaxodon* is the same grade as *Galesaurus*, which is to say intermediate in a number of its features between *Procynosuchus* on the one hand and the eucynodonts to follow. It was somewhat smaller than *Procynosuchus*, and the postcanine teeth were reduced in number. The more posterior, molariform postcanines have a sharp main cusp plus an anterior and a posterior accessory cusp in line with it, which developed from the enlarged end members of the row of cingulum cuspules along the internal base of the tooth. The dentary bone is relatively larger, and its coronoid process rises right up into the temporal fenestra; the adductor fossa has expanded to occupy most of the external surface of the dentary behind the tooth row. The temporal fenestra has enlarged considerably compared to *Procynosuchus*, and the zygomatic arch was much deeper and more robustly built, and bowed dorsally, indicating the presence of a substantial masseter muscle. There are several significant differences in the postcranial skeleton compared to *Procynosuchus*, most conspicuously the appearance of large costal expansions on the hind, proximal part of the ribs, each of which overlaps the one behind and bears a strong ridge for muscle attachment. There are also accessory articulations between adjacent vertebrae, with a peg below each postzygapophysis fitting into a groove below the prezygapophysis of the next vertebra behind. The functional significance of this arrangement of the axial skeleton is obscure. Jenkins (1971*b*) believed that it was a method for

more effectively applying muscle forces causing lateral bending of the vertebral column; Kemp (1980*a*), in contrast, proposed that the effect would be quite the opposite by making the column more rigid. At any event, a full set of costal plates is evidently the primitive condition for Triassic cynodonts generally, and they are variously reduced or completely lost by the Middle and Upper Triassic groups. The limbs of *Thrinaxodon* were probably approaching mammalian in pose, particularly the hindlimb where the knee was evidently turned well forwards. The tail is reduced, indicating the increased reliance on muscles from the ilium and body fascia for the power stroke of the stride.

Platycraniellidae
There is a third well-established genus of *Lystrosaurus* Assemblage Zone cynodont, which is *Platycraniellus* (Abdala unpublished manuscript). It is remarkable for having an extremely short snout and broad temporal fenestrae (Fig. 3.20(d)). In fact, comparatively it has the widest skull of any cynodont. The secondary palate is complete and extends posteriorly as far as the end of the tooth rows, a feature characteristic of later, more advanced forms. Unfortunately, few details of the dentition or the structure of the lower jaw are clear in the one reliably identified specimen. It may be that *Platycraniellus* is an aberrant basal member of the more advanced cynodont taxon Eucynodontia.

Eucynodontia

All the remaining cynodonts form a monophyletic group Eucynodontia, in which there has been further evolution of the temporal fenestra, lower jaw, and by inference the jaw musculature. The dentary has increased in relative size to such an extent that the postdentary bones are reduced to a small, vertically oriented compound sheet or rod of bones set into a recess occupying the medial face of the dentary. The coronoid process rises right up to the level of the top of the sagittal crest, there is a large angular process ventrally, and an articular process posteriorly that reaches towards the jaw articulation. The hinge bones, articular bone at the hind end of the postdentary complex, and quadrate held in a recess in the squamosal are relatively minute.

Although there is no mammal-like contact between the dentary and the squamosal bone, eucynodonts do have a secondary contact between one of the postdentary bones, the surangular, and the squamosal. It lies immediately laterally to the articular–quadrate hinge and presumably functioned to stabilise the jaw articulation. The front ends of the paired dentaries have fused to form a strong, immobile symphysis. The eucynodont postcranial skeleton (Jenkins 1971*b*) shows a number of evolutionary modifications towards the mammalian condition, such as the appearance of a definite acromion process on the scapular for attachment of the clavicle. The ilium is more expanded and the pubis more reduced compared to *Thrinaxodon*. The head of the femur is well turned in and the major and minor trochanters are enlarged and very mammalian in form. In the early eucynodonts from the *Cynognathus* Assemblage Zone of South Africa, such as *Diademodon* and *Cynognathus*, the costal plates on the ribs are still well developed, but by the Middle

Triassic they are either greatly reduced, as in certain traversodontids, or completely absent, assumed to have been secondarily lost, as in probainognathians.

While there are several non-mammalian eucynodont taxa that are well categorised and more or less universally accepted, the interrelationships among them remains a matter of controversy, and at least three recent cladistic analyses have produced results at variance with one another. Profoundly embedded in this taxonomic disagreement is disagreement about exactly which of the non-mammalian cynodonts is the most closely related to Mammalia, an issue taken up later in the section. The more or less unchallenged monophyletic subgroups of eucynodonts are these.

• Cynognathidae: a family of relatively primitive carnivores
• Diademodontoidea: two families of herbivorous eucynodonts, Diademodontidae and Traversodontidae

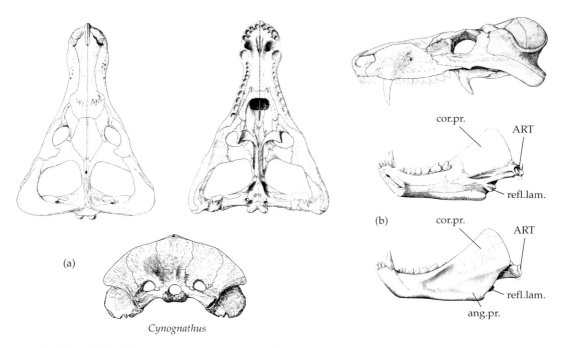

Cynognathus

Figure 3.21 (left) (a) Skull of *Cynognathus* in four views. Approx. skull length 29 cm (Broili and Schröder 1934). (b) Medial and lateral views of lower jaw (Kermack *et al*. 1973). (right) Probainognathian cynodonts. (b) Skull of *Lumkuia fuzzi* in four views. Skull length approx. 6 cm (Hopson and Kitching 2001) (c) Skull of *Probainognathus jenseni* in four views (Carroll 1988, from Romer 1970). (d) *Chiniquodon (Probelesodon)*. Skull length approx. 11 cm (Romer 1969). (e) Lateral view of skull of *Ecteninion lunensis*. Skull length approx. 12 cm (Martinez *et al*. 1996). ART, articular. ang.pr, angular process. cor.pr, coronoid process. refl.lam, reflected lamina of the angular. *Continued overleaf*

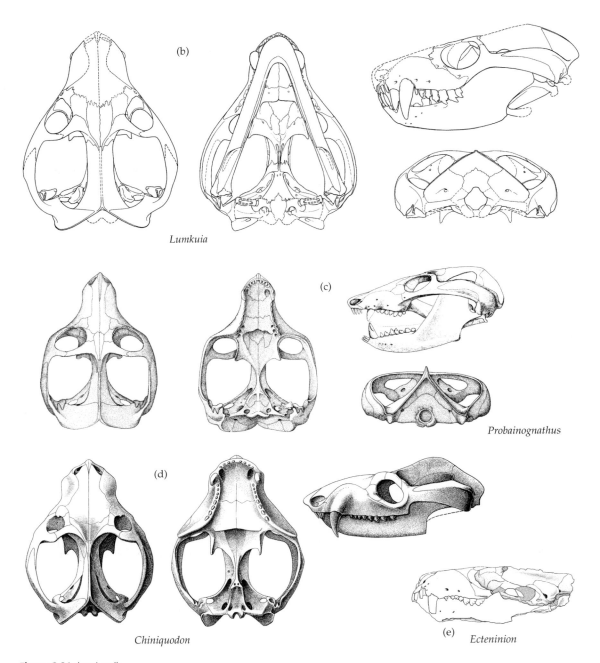

Lumkuia

(b)

(c)

Probainognathus

(d)

Chiniquodon

(e) *Ecteninion*

Figure 3.21 (*continued*).

• Probainognathia: more progressive carnivores, consisting of two families, Chiniquodontidae and Probainognathidae
• Tritylodontidae: a family of highly specialised and progressive herbivores

• Tritheledonta: a possibly paraphyletic series of very small, highly advanced carnivorous forms including Tritheledontidae and Therioherpetontidae
• Mammalia.

Cynognathidae
Cynognathus (Fig. 3.21(a)) occurs in the Lower to Middle Triassic of South Africa and is sufficiently common to have given its name to the *Cynognathus* Assemblage Zone. Specimens of the genus have also been found in beds of the same age in Argentina and Antarctica (Hammer 1995). It is a large form with a robustly built 30 cm long skull and strongly carnivorous dentition. The canines are large and the postcanines blade-like with a recurved main cusp. The latter lack cingulum cuspules around the base but there are sharp accessory cusps in line with the main cusp, increasing in prominence on teeth from the front to the back of the postcanine tooth row. The way in which the postcanine teeth wear indicates that there was no direct, precise occlusion between uppers and lowers, but a more general opposition between the two rows of teeth. While the lower jaw has the fully expressed eucynodont structure, the temporal fenestra is quite short. The hind wall of the fenestra, forming the occiput, does not have the deep embayment between itself and the root of the zygomatic arch that is found in other eucynodonts, possibly an adaptation for extreme carnivory where the major jaw-closing muscle would be a postero-dorsally oriented temporalis.

Diademodontoidea
The diademodontoid eucynodonts include a range of genera characterised by expanded postcanine teeth that engaged in precise tooth-to-tooth occlusion, a design taken to indicate a herbivorous diet even though well-developed canine teeth are also retained. *Diademodon* (Fig. 3.22(a)) was a contemporary of *Cynognathus* in the Lower to Middle Triassic. It has simple incisors, sizeable canines, but a remarkably specialised postcanine dentition. There are three morphologically distinct kinds of postcanine teeth (Fig. 4.12(a)). The anterior three or four consist of simple, conical crowns. Following these there are up to nine, depending on age, transversely widened, multicusped teeth described as gomphodont (Fig. 3.22(b)). The crowns of the uppers are broadly expanded across the line of the jaw. The single main cusp occupies the centre of the outer edge of the tooth and sharp crenulated ridges run backwards and forwards from its apex. The other three margins of the occlusal surface are marked by

a series of small cuspules. The lower gomphodont teeth are less expanded, but morphologically similar to the uppers. These detailed features are only seen in freshly erupted, unworn teeth. Tooth wear was rapid, removing the thin layer of enamel and reducing the teeth to more or less featureless pegs; each smaller lower tooth bit within the basin formed between two adjacent larger upper teeth, in a mortar-and-pestle fashion. Behind the gomphodont teeth there are between two and five sectorial teeth not unlike those of *Cynognathus*. The recurved main cusp is transversely flattened and bears an anterior and a smaller posterior accessory cusp. Comparing a range of *Diademodon* skulls of different sizes and therefore presumably different ages reveals a remarkable pattern of tooth replacement. Teeth are successively lost from the front of the tooth row, and new ones added at the back. In order to retain the same numbers of the three kinds of postcanine teeth, the middle, gomphodont type teeth are shed from the front backwards and replaced by anterior-types. Meanwhile, the posterior, sectorial types are replaced by middle types. The teeth added at the hind end are posterior types.

A number of other members of the family Diademodontidae occur, notably *Trirachodon* which differs in lacking the simple type of anterior postcanines, and in the cusp pattern of the gomphodont teeth. Both the uppers and the lowers have a transverse row of three cusps across the middle of the occlusal surface, of which the central one is the largest. A row of cuspules occupies both the front and the hind edges of the crown. Middle Triassic trirachontines occurred widely, having been discovered in Africa, China, Russia, India, and possibly North America.

The second diademodontoid family Traversodontidae (Fig. 3.22(c)–(f)) are related to diademodontids but have a simpler dentition insofar as both the simple anterior type teeth and the sectorial posterior type teeth are lacking. However, the remaining middle type of gomphodont postcanines are more elaborate, bearing an enlarged transverse crest across the occlusal surface. In the case of an upper tooth it lies well towards the posterior end of the tooth, and in a lower, close to the anterior edge. Crompton (1972*a*) analysed the mode of action of traversodontid teeth in detail, showing how a cutting action occurs between the transverse ridges of the upper and lower postcanines (Fig. 3.22(d)). This is followed by a slight posterior shift of the lower jaw, bringing the basins of the lower teeth opposite the basins of the upper teeth, and so allowing a crushing action between them. Each lower postcanine occludes with one specific upper postcanine in a fashion analogous to that of typical herbivorous mammals. Notwithstanding this precision, the teeth eventually wear down to simple pegs surrounded by enamel as in *Diademodon*. The arrangement of the adductor jaw musculature is such that it

creates the powerful posterior component to the jaw force necessary to activate the movement of the lower teeth against the uppers (Kemp 1980a). The traversodontid snout tends to become broader, and the upper tooth row to be overhung laterally by the maxilla, perhaps indicating the presence of extensive fleshy cheeks for retaining the food being chewed. In the postcranial skeleton of traversodontids (Jenkins 1971b; Kemp 1980a), there is a strong tendency to reduce the costal plates that are characteristic of the more primitive cynodonts. In *Luangwa*, expanded plates are restricted to the more posterior thoracic

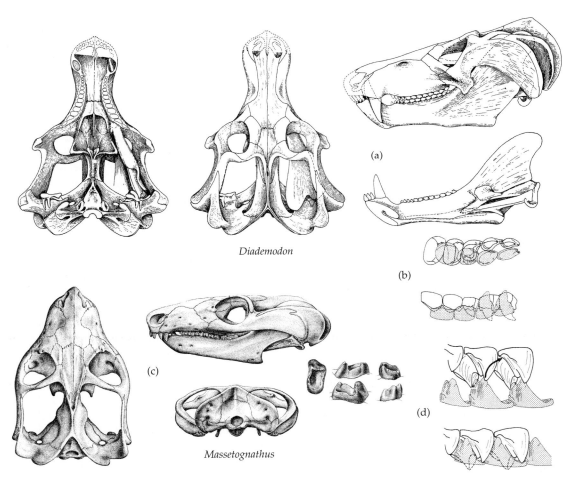

Diademodon

(a)

(b)

(c)

Massetognathus

(d)

Figure 3.22 Diademodontoid cynodonts. (a) *Diademodon rhodesiensis* in ventral, dorsal, lateral, and lower jaw medial views. Skull length approx. 23 cm (Brink 1963(b)). (b) Postcanine tooth occlusion in occlusal and medial views. (Crompton 1972a). (c) The traversodontid *Massetognathus pascuali* in four views and an upper postcanine tooth. Skull length approx. 10 cm (Romer 1967). (d) Postcanine occlusion in *Scalenodon* in medial view, at the start (top) and end (bottom) of the occlusal movement (Crompton 1972). Postcranial skeletons of traversodontids. (e) *Massetognathus pascuali*. Presacral length approx. 36 cm (Jenkins 1970). (f) *Exaeretodon*. Presacral length approx. 125 cm. (Kemp 1982 from Bonaparte 1963).

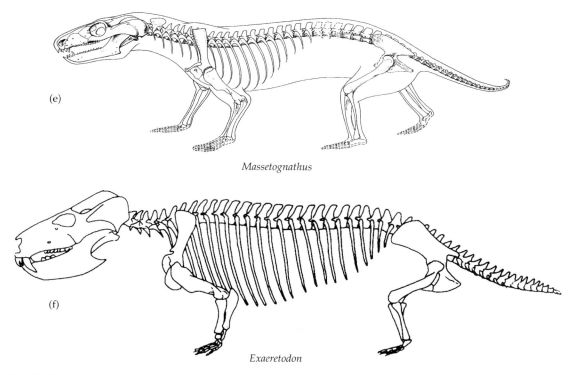

Massetognathus

Exaeretodon

Figure 3.22 (*continued*).

and the lumbar ribs, while in *Massetognathus* (Fig. 3.22(e)) they are very reduced, fine projections, and altogether absent in *Exaeretodon*. The appearance of the vertebral column and ribcage is virtually mammalian. The acromion process of the shoulder girdle is well developed indicating increased invasion of the inner surface of the scapular blade by forelimb musculature. The hindlimb in a form such as *Luangwa* has achieved a structure functionally comparable to mammals. The ilium extends far forwards, although it lacks the mammalian external ridge dividing the blade into discrete upper and lower parts. The pubis is turned back, the ischium is almost horizontal, and the obturator foramen large.

The earliest traversodontids occur in the Lower Triassic of Argentina (Abdala and Ribeiro 2003). *Pascualgnathus,* and *Rusconodon* have teeth in which the transverse ridge is only a little more prominent than in *Diademodon* (Bonaparte 1970). Middle Triassic genera are known from several regions, being represented by the well-known *Massetognathus* from Argentina, and *Luangwa* from Zambia. Other forms have been found elsewhere in southern Africa, North America (Sues *et al.* 1999), and Russia (Battail and Surkov 2000). All the Middle Triassic forms have the fully developed traversodontid version of the gomphodont tooth, with the transverse crest dominating the occlusal surface of the crown (Abdala and Ribeiro 2003). The last and most specialised traversodontids are from the early part of the Upper Triassic. The Argentine *Exaeretodon* (Fig. 3.22(f)) was a large cynodont, with a presacral body length up to 1.4 m. The dentition is remarkable for the extent to which the adjacent teeth along the tooth row interlock. The posterior edge of each one is concave and fits tightly against the convex anterior edge of the next tooth back. The result is a very strongly built triturating surface. A diastema between the canine and the postcanine tooth row, and the parallel arrangement of the left and right tooth rows indicates that the dentition had achieved a highly effective grinding ability.

Probainognathia

The Probainognathia includes a series of Lower to Upper Triassic carnivorous eucynodonts. There are few unique characters uniting the group, the main one being merely an elongated secondary palate. The frontal and palatine bones meet in the front wall of the orbit, there is no pineal foramen, and none have any sign of costal plates on the ribs. However, the situation is confused by the presence of these latter three characters in the more advanced traversodontids. Most authors recognise two probainognathian families, the Chiniquodontidae and the Probainognathidae, but there are two forms that are apparently basal members of the group. One of these is *Lumkuia*, from the Lower Triassic of South Africa and the other *Ecteninion*, from the Upper Triassic of South America.

The skull, jaws, and partial skeleton of *Lumkuia* (Fig. 3.21(b)) were discovered in the *Cynognathus* Assemblage Zone of South Africa (Hopson and Kitching 2001). It is modest-sized, the skull being about 6 cm in length. *Lumkuia* is more derived than the contemporary carnivore *Cynognathus* for it lacks a pineal foramen and has no costal plates on the ribs. Its postcanine teeth resemble quite closely those of chiniquodontids, with high, slightly recurved crowns that are laterally compressed, and bear small anterior and larger posterior accessory cusps. However, it lacks the chiniquodontid features of a greatly elongated secondary palate, and the angulation of the ventral cranial margin.

Ecteninion (Fig. 3.21(e)), from the Upper Triassic Chañares Formation of Argentina (Martinez *et al.* 1996), also lacks these characteristics of the chiniquodontids. Its most distinctive feature is the postcanine dentition, consisting of seven teeth, of which the last three are the best developed. Each of the latter is transversely flattened and overlaps the next tooth behind. There are three sharp cusps in line, the first the largest and the third barely extending above the jaw margin. The more anterior postcanines are similar in form but much smaller. The lower postcanine teeth are not visible in the one known specimen of *Ecteninion*, but it would be a reasonable guess that they are similar to the uppers, and that the whole postcanine dentition was specialised for masticating relatively soft animal matter.

Chiniquodontids (Fig. 3.21(d)) are a carnivorous group, recognisable by the degree of angulation in the ventral margin of the skull where the maxilla and zygomatic arch meet. In a recent review, Abdala and Giannini (2002) concluded that the chiniquodontid

specimens attributed to the familiar genera *Probelesodon* and *Belesodon* are actually different growth stages of the single genus *Chiniquodon*. An adult specimen has a skull length up to about 25 cm, and bears sharp, well-developed incisors and canines. The 7–10 postcanine teeth are bluntly sectorial with a strongly recurved principal cusp and a posterior accessory cusp behind.

Probainognathus (Fig. 3.21(c)) has sometimes been included in the chiniquodontids, but more usually it is placed in its own family. It is a small carnivore with a skull only about 7 cm in length. The postcanine teeth have a linear series of three cusps, the central one being the largest, and there are internal cingular cusps around the base. They are therefore very similar to those of the Lower Triassic *Thrinaxodon* and also the early mammal *Morganucodon*. In the first publication on *Probainognathus*, Romer (1970) described a direct contact between the hind end of the dentary and the squamosal bone, which is technically the mammalian jaw hinge. Taken with the dental structure, this led to the once widely held view that *Probainognathus* was *the* mammal ancestor. Subsequently, however, Crompton (1972*b*; Luo and Crompton 1994) showed that the contact is actually between the surangular bone and the squamosal, a contact that occurs in other eucynodonts. Nevertheless, the dentary of *Probainognathus* extends closer to this pre-existing contact than in any other non-mammalian cynodonts except for tritheledontans described below.

Tritylodontidae

The Tritylodontidae (Fig. 3.23) are a highly specialised and derived family of herbivores that occurred worldwide during the Upper Triassic and Lower Jurassic (Anderson and Cruickshank 1978). Specimens are known from South Africa, Europe, North and more dubiously South America, China, and Antarctica (Lewis 1986). Isolated tritylodontid teeth have also been described from the Early Cretaceous of Russia, which extends the temporal range of the group by a very considerable degree (Tatarinov and Matchenko 1999). They varied in size from the less than 5 cm skull (Fig. 3.23(e)) of *Bocatherium* (Clark and Hopson 1985) to the South African *Tritylodon* and North American *Kayentatherium* with skulls up to 25 cm.

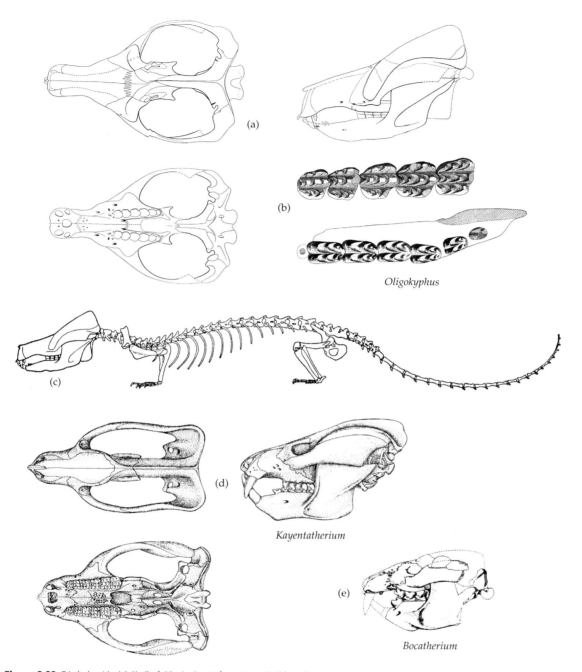

Figure 3.23 Tritylodontids. (a) Skull of *Oligokyphus* in four views. Skull length approx. 9 cm (Kühne 1956). (b) Postcanine dentition of *Oligokyphus major*, uppers (top) and lowers (bottom) anterior to the right (Kühne 1956). (c) Postcranial skeleton of *Oligokyphus*. Postsacral length approx. 28 cm (Kühne 1956). (d) Skull of *Kayentatherium*. Skull length approx. 22 cm (Carroll 1988, from Sues 1983). ((e) Lateral view of skull of *Bocatherium mexicanum*. Skull length approx. 4.2 cm (Clark and Hopson 1985).

The tritylodontid skull has a strong look of a mammal about it because of the loss of the postorbital bar so that the orbit and temporal fenestra are confluent, and the extremely large temporal fenestrae separated from one another by a very deep sagittal crest. The dentary bone occupies almost the whole of the lower jaw and has enormous coronoid and angular processes, while the postdentary bones are reduced to a delicate compound rod set into a groove on the medial side. However, there is still no contact between the dentary and the squamosal, the jaw hinge consisting of the tiny articular and the weakly supported quadrate bone. The superficially rodent-like dentition is highly specialised for a herbivorous diet. Canines are absent but have been functionally replaced by an enlarged pair of second incisors. In primitive members such as *Oligokyphus* (Fig. 3.23(a)) this is the largest of the three pairs present but in other forms such as *Kayentatherium* (Fig. 3.23(d)) it is the sole remaining pair. After a long diastema in both the upper and the lower jaws, there is a row of large, complex postcanine teeth (Fig. 3.23(b)). The crowns of the upper postcanines have three longitudinal rows of sharp, crescentic cusps, while the lowers have two such rows. The tooth rows are parallel and when the jaws closed, occlusion occurred on both sides simultaneously. The two rows of cusps of the lower teeth fitted into the troughs formed by the three rows of cusps of the uppers, and the action consisted of powerfully dragging the lower jaws backwards while the teeth were in contact. The individual crescentic cusps of the upper teeth had the concave edge facing forwards, while those of the lower teeth faced backwards, and so the net effect was a filing action between the two opposing occlusal surfaces, allowing very fine shredding of even quite tough vegetation.

The complete postcranial skeleton of *Oligokyphus* is known (Kühne 1956; Kemp 1983) and is virtually mammalian in form (Fig. 3.23(c)). The shoulder girdle is modified by reduction of the coracoid bone and a large, laterally reflected acromion. The humerus is extremely slender. In the pelvis (Fig. 4.8(b)), the ilium has almost lost the posterior process, and the extensive anterior process has a lateral ridge of mammalian form separating dorsal from ventral regions. The femur has an inturned spherical head

and discrete trochanter major and trochanter minor, and is thus fully mammalian in structure.

Tritheledonta
Tritheledontans (Fig. 3.24) constitute another highly derived group of non-mammalian cynodonts. They are very small, presumably insectivorous forms occurring from the early Upper Triassic Carnian stage through to the Lower Jurassic (Lucas and Hunt 1994). The skull and, to the limited degree it has been described, the postcranial skeleton are remarkably mammalian in general appearance. As in the tritylodontids, the prefrontal and postorbital bones have been lost, and therefore there is no postorbital bar separating the orbit from the temporal fenestra. Unlike the tritylodontid condition, the temporal fenestra is long, the sagittal crest low and broad, and the zygomatic arch slender, characters probably correlated with the small size of the animals. Other mammalian features include the virtual disappearance of the lateral pterygoid processes of the palate. One of the phylogenetically most significant derived features of all is a contact between the articular process of the dentary and the squamosal, and therefore the presence of the new, secondary jaw articulation immediately lateral to the quadrate-articular hinge. This critical feature (Fig. 3.24(d)) was first claimed for *Diarthrognathus* (Crompton 1963) but not universally accepted. However, it has been amply confirmed subsequently in *Pachygenelus* (Allin and Hopson 1992; Luo and Crompton 1994). The tritheledontid dentition also possesses certain otherwise uniquely mammalian characters. One is that the histological structure of the enamel is prismatic. A second is that wear facets are present on the inner faces of the upper and the outer faces of the lower postcanines, indicating not only that they had a direct, shearing occlusion, but also that, as in primitive mammals, there was a unilateral action of the jaws, only one side of the dentition being active during any one instant. However, unlike mammals generally, the upper and lower wear facets do not match exactly, so the occlusion cannot have involved the precise action between specific parts of opposing teeth that distinguishes the more advanced group. Little can yet be said of the postcranial skeleton for want of a description of material whose existence has been

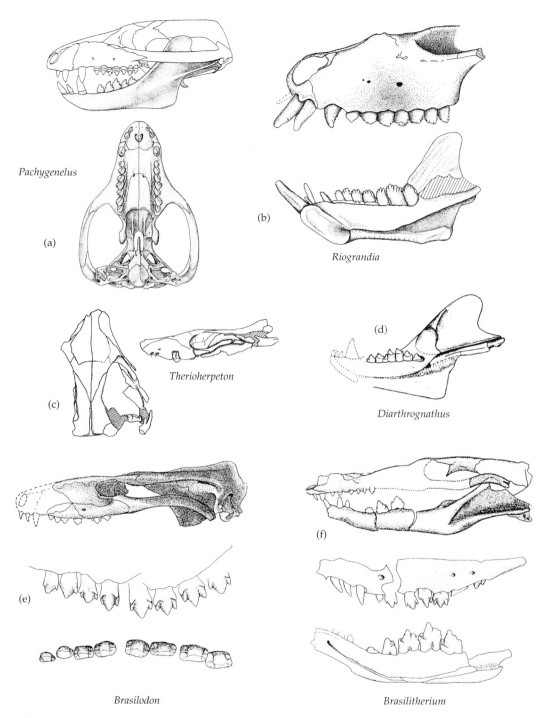

Figure 3.24 Tritheledontans. (a) *Pachygenelus* skull (Bonaparte et al. 2003, from Hopson, in prep). (b) *Riograndia guaibensis* skull in lateral view and lower jaw in medial view. Length of lower jaw approx. 3 cm (Bonaparte *et al.* 2001). (c) Skull as preserved of *Therioherpeton cargnini* in dorsal and lateral views. Skull length approx. 3 cm (Bonaparte and Barbarena 2001). (d) Lower jaw of *Diarthrognathus* in medial view. Jaw length approx. 5 cm (Crompton 1963). (e) Skull in lateral view, and upper dentition in lateral and occlusal views of *Brasilodon quadangularis*. Skull length approx. 2.2 cm (Bonaparte *et al.* 2003). (f) *Brasilitherium riograndensis* skull and lower jaw in lateral view, upper dentition in lateral view, and lower dentition in internal view. Estimated skull length approx. 2.4 cm (Bonaparte *et al.* 2003).

reported. Gow (2001) has described some isolated bones of *Pachygenelus*, showing that the anterior edge of the scapula blade has everted and shifted posteriorly to a sufficient degree as to regard it as a true, mammalian scapula spine. The humerus and femur are very similar indeed to those of the basal mammal *Morganucodon*.

The main character diagnosing the family Tritheledontidae is the dentition, which is quite specialised in some respects. It is best known in *Pachygenelus* (Gow 1980; Shubin et al 1991), where only two incisors are present in the upper and lower jaws (Fig. 3.24(a)). The canines are unique amongst non-mammalian cynodonts in that the lower one works against the antero-lateral rather than the inner side of the upper, a feature that is typical of mammals. There are seven postcanine teeth in both upper and lower jaws, with incipiently divided roots. The uppers are circular in crown view and have a single, dominant main cusp. An anterior and a posterior accessory cusp are present, offset to the inner side of the main cusp and connected by an internal cingulum. The lower teeth are slightly larger, laterally compressed, and the anteriorly positioned main cusp is followed by up to three accessory cusps along the line of the jaw. These also have an internal cingulum. One unexpected feature concerns the relative thickness of the enamel. In the occluding regions of the teeth it is thin and soon wears away, leaving the thicker enamel edges that surround the crown exposed as sharp cutting edges. The postcanine teeth of *Diarthrognathus* are essentially similar to those of *Pachygenelus* though differing in detail (Gow 1980).

Chaliminia (Bonaparte 1980) is a poorly known form from the Upper Triassic of the Rio Grande do Sul in Brazil, but has upper postcanine teeth sufficiently similar to those of *Pachygenelus* for it to be regarded as a member of the Tritheledontidae. *Riograndia* (Bonaparte *et al.* 2001) is represented by several specimens of a very small form of skull from the same locality (Fig. 3.24(b)). It has a very peculiar dentition, but which is comparable to *Pachygenelus* in several respects. There are only three, procumbent upper incisors, of which the second is the largest, and also three lowers, but here the first one is much larger than the other two. The canines are small. Both the upper and lower postcanines are blade-like and bear up to nine cuspules arranged in line. Each tooth is then set at an angle to the line of the tooth row, a condition seen to a small extent in *Pachygenelus* but taken to extremes here. Presumably *Riograndia* is a specialised tritheledontid, perhaps adapted for a diet of softer invertebrates than insects.

In addition to members of Tritheledontidae as diagnosed by the reduced incisors and the post-canine tooth structure, there are several small, superficially tritheledontid-like specimens from the Rio Grande do Sul, Brazil that have been described by Jose Bonaparte and his colleagues. *Prozostrodon* (Bonaparte and Barberena 2001) is a slightly larger form that is represented at present by the anterior half of a 7 cm long skull, with the lower jaw and a few fragments of the postcranial skeleton. It is primitive compared to all other tritheledontans in some respects, notably retention of small prefrontal and postorbital bones, although the postorbital bar is absent and the frontal bone does form the orbital margin as in tritheledontids. The lower jaw has a well-developed articular process of the dentary, but it is unknown whether contact between this bone and the squamosal occurred. The dentition of *Prozostrodon* lacks the derived characters of the members of the family Tritheledontidae. There are five upper and four lower simple incisors, followed by well-developed canines. The seven upper post-canines have up to four linearly arranged cusps and lack a cingulum. There are 10 lower postcanines, of which the more complex posterior ones have a small anterior accessory cusp, the main cusp, and two posterior accessory cusps all in line, and also a well-developed internal cingulum. They have a similar appearance to the postcanine teeth of *Thrinaxodon* and to the basal mammal *Morganucodon*.

Bonaparte and Barberena (1975, 2001) described a very small, poorly preserved Rio Grande do Sul skull as *Therioherpeton* (Fig. 3.24(c)), which they regard as a member of a separate family Therioherpetontidae. Again the postorbital bar is lacking, but as in tritheledontids the postfrontal and postorbital bones have been lost completely and the zygomatic arch has the same very slender construction. The differences from tritheledontids are that the anterior orbital and interorbital region of the skull is more completely ossified by meeting of the frontal, palatal, and orbitosphenoid bones, and the more derived palatal structure involving reduction of the pterygoid bones and closure of the interpterygoid vacuity. The upper postcanine teeth of *Therioherpeton* are laterally compressed and have incipiently divided roots. Unfortunately, no lower jaw or dentition is yet known for the genus, although enough of the postcranial skeleton of *Therioherpeton* is preserved to indicate features that are closely comparable to the tritylodontid *Oligokyphus* and the early mammal *Morganucodon* (Kemp 1983). The vertebrae have a relatively large neural canal, and the cervical vertebrae are broad. The ilium has a long anterior process but no posterior process, and the femur is very similar in its slender build, bulbous inturned head, and the appearance of the trochanters.

The most recent additions to the Rio Grande do Sul tritheledontan fauna are *Brasilodon* and *Brasilitherium* (Bonaparte *et al.* 2003). These are of particular interest because their dentitions are close in structure to the presumed ancestral mammalian condition. They also differ slightly from one another. In the case of *Brasilodon* (Fig. 3.24(e)), the crowns of the postcanines are symmetrical about the main cusp, with equally well-developed anterior and posterior accessory cusps in line, and an anterior and a posterior cingulum cusp on the inner face. The upper postcanines of *Brasilitherium* (Fig. 3.24(f)) are like those of *Brasilodon*, but the lowers differ. They are asymmetrical, with the main cusp set further forwards, one small anterior accessory cusp, and up to three posterior accessory cusps. There is also an anterior cingulum cusp. The authors consider these two genera to be more closely related to Mammalia than are tritheledontids such as *Pachygenelus*, a question taken up below.

In addition to the reasonably well-defined small, insectivorous-type advanced families Tritheledontidae and Therioherpetontidae of the Upper Triassic, there are several records of jaws and isolated postcanine teeth of superficially comparable forms. Several of these hail from North America and historically have been referred to as dromatherians; various European forms have been considered members of this same group (Battail 1991; Hahn *et al.* 1994). Sues (2001) reviewed the material of *Microconodon*, which is the most complete dromatherian, but which even so is represented only

by three incomplete dentaries and a few isolated teeth. Beyond the conclusion that it is, indeed, an Upper Triassic eucynodont with basically triconodont-like postcanine teeth, he was unable to interpret its relationships. None of the other taxa referred to this group fare any better. It may be that they are all part of the Upper Triassic radiation of Tritheledontans, or conceivably that there were other, as yet ill-understood lineages of advanced, near-mammalian eucynodonts at that time.

Interrelationships of Cynodontia and the phylogenetic position of Mammalia

For all the importance of establishing exactly where within the Cynodontia the Mammalia cladistically nest, with the implications it holds about its mode of origin, there remains considerable disagreement about both this issue, and about other fundamental aspects of the interrelationships of the cynodont groups described. A brief history of the situation may help clarify the debates. For many years it had been accepted that there were two lineages of cynodonts (e.g. Hopson and Kitching 1972; Crompton and Jenkins 1973; Battail 1982). One consisted of a carnivorous group including *Procynosuchus*, *Thrinaxodon*, cynognathids, chiniquodontids, and tritheledontids and from which the mammals had supposedly evolved. The other consisted of the herbivorous, or 'gomphodont' cynodonts, namely the diademodontids and traversodontids, and also the tritylodontids which were therefore inferred to have evolved their otherwise mammalian characters convergently. Consequent upon a detailed description of *Procynosuchus*, Kemp (1979; 1982) questioned what amounted to a diphyletic origin of advanced cynodonts. He proposed that cynognathids, diademodontids, and all the other cynodonts with greatly enlarged dentaries, carnivores and herbivores alike, constituted a monophyletic group and that *Procynosuchus* and *Thrinaxodon* were its successive plesiomorphic sister groups. The hypothesis of a monophyletic Eucynodontia (Kemp 1982) for these advanced cynodonts has been corroborated by all subsequent authors apart from Battail (1991).

The next area of dispute concerned the position of Cynognathidae within Eucynodontia, and this has not yet been resolved. Kemp (1982) argued that cynognathids were the plesiomorphic sister group of all the rest of the eucynodonts on the basis of certain primitive characters, such as the lack of a deep embayment between occiput and zygomatic arch. The analyses of a number of authors have agreed with this, notably Rowe (1986) and Martinez *et al.* (1996). However, others, primarily Hopson and Barghusen (1986; Hopson 1991; Hopson and Kitching 2001) argued for a monophyletic clade Cynognathia, consisting of cynognathids plus diademodontoids but excluding the more advanced carnivorous eucynodonts. The latter, formally including Mammalia, constitute their Probainognathia which is thus the sister group of Cynognathia.

A third dispute, which also has yet to be resolved, concerns the position of the tritylodontids. This issue is related to the broader issue of which eucynodontian taxon is the closest relative of Mammalia, but for the moment will it be treated separately. When Kühne (1956) gave the first detailed, comprehensive description of a tritylodontid, based on the fragmentary but virtually complete remains of *Oligokyphus* from England, he had no doubts about the prevailing view that they were closely related to the mammals, on the basis of such characters as the loss of the prefrontal and postorbital bones and postorbital bar, the multirooted, complex postcanine teeth, and the huge dentary bone almost, but not quite contacting the squamosal. At about the same time, however, Crompton and Ellenberger (1957) proposed that tritylodontids were highly derived 'gomphodont' cynodonts, and opinion shifted to this view when Crompton (1972a) published his detailed study of the form and function of the teeth of cynodonts, showing that both traversodontids and tritylodontids had an occlusal mechanism based on transverse crests on the postcanine teeth and a posteriorly directed bilateral power stroke of the lower jaw. From this standpoint it must be inferred that those mammalian characters of tritylodontids not found in traversodontids were acquired independently of the mammals. In 1983, Kemp listed the numerous derived characters shared by tritylodontids and the early mammal *Morganucodon*, including not only several cranial features but also a virtually identical postcranial skeleton, and accordingly proposed a return to the older view that the two groups are related, and

therefore that traversodontids and tritylodontids had evolved a specialised herbivorous dentition convergently. This view gained qualified supported shortly thereafter from a detailed cladistic analysis produced by Rowe (1986, 1988). Since then, the majority of analyses have agreed that tritylodontids are related to mammals rather than to diademodontoids, including Wible's (1991) re-evaluation of Rowe's data, Luo's (1994) cladistic analysis of dental and cranial characters, Luo and Crompton's (1994) extremely detailed study of quadrate characters, and Luo's (2001) analysis of the structure of the inner ear. The recent analyses of eucynodontian interrelationships by Martinez, May, and Forster (1996), and of gomphodont cynodonts by Abdala and Ribeiro (2003) also conclude that the tritylodontids are not 'gomphodonts'. On the other hand Sues (1985) and Hopson and his colleagues (Hopson and Barghusen 1986; Hopson 1991; Hopson and Kitching 2001) and Battail (1991) continued to support the diademodontoid relationship of tritylodontids. Surprisingly and despite considerable efforts, no proposed resolution of this question has been universally accepted, as discussed shortly.

The final question concerning cynodont interrelationships is which group is most closely related to the monophyletic taxon Mammalia, as usefully represented by *Morganucodon*, by far the best known of the early mammalian taxa. For those authors who support the diademodontoid–tritylodontid relationship, such as Hopson and Kitching (2001) and Rubidge and Sidor (2001), the only contender is a tritheledontan, taking this group to include now the various Rio Grande do Sul genera such as *Therioherpeton*, *Brasilodon*, and *Brasilitherium* as well as Tritheledontidae. However, among those who accept that tritylodontids are related to mammals, there are differing views about whether the tritylodontids or the tritheledontids are actually the closer. The problem is that these two advanced non-mammalian cynodont taxa differ in the combination of mammalian and non-mammalian characters that they exhibit, so whichever view is correct, convergence of certain mammal characters must have occurred.

Characters that support a sister group relationship between tritylodontids and the mammals include several braincase characters (Wible 1991;

Wible and Hopson 1993). The interorbital region is largely filled in by expanded orbitosphenoid, frontal, and palatal bones. The paroccipital process is bifurcated and the quadrate is attached directly to the distal end of its anterior process without the squamosal intervening. Also, the postcanine teeth are multi-rooted. Impressive similarities are also found in the postcranial skeleton (Kemp 1983), which is almost indistinguishable from that of *Morganucodon*. For example, in the shoulder girdle the virtual loss of the coracoid plate and the extensive reflection of the acromion process of the scapula; and in the pelvis, loss of the posterior process of the ilium and very long anterior process with the external ridge separating upper from lower regions so characteristic of mammals, and also an epipubic bone. However, as more information emerges about the postcranial skeleton of tritheledontids (Gow 2001), the more it looks as if these characters are true of that group as well, and not exclusively of mammals and tritylodontids. Hopson and Kitching (2001) actually include seven postcranial characters in their list defining a tritheledontid—mammal clade, of which six are also found in tritylodontids.

The characters of tritheledontids that support their sister group relationship with mammals include the secondary jaw articulation between the dentary and squamosal, although the fully formed dentary condyle of the mammal had not evolved, and the detailed structure of the quadrate (Luo and Crompton 1994). The mode of action of the post-dentary teeth of tritheledontids (Shubin *et al.* 1991), involving unilateral contact between teeth of one side of the jaw at a time approaches the mammalian condition, even though there is not the precise occlusal relation of specific lower and upper teeth. The teeth themselves have prismatic enamel, and the upper postcanines have a buccal cingulum. The zygomatic arch is much more slender than in tritylodontids even in specimens of the same skull size, and there is a longer secondary palate. The basicranial region of the skull is shortened, and the fenestra rotunda and jugular foramen are completely separated.

Agreement on the interrelationships of the major eucynodont groups might be expected to come from cladistic analyses based on large morphological data sets. In recent year there have indeed been several of these, and yet there is still no consensus on any of the contentious issues. Luo (1994) used 82 dental and cranial characters and found tritheledontids to be the sister group of mammals but only by a very narrow margin over tritylodontids (Fig. 3.25(a)). Martinez *et al.* (1996), on the basis of 68 characters found tritylodontids to be the immediate sister group of Mammalia, and tritheledontids the sister group of these two (Fig. 3.25(b)). Like Luo, they also found cynognathids to be the most basal eucynodont group. Hopson and Kitching (2001) coded for 101 characters including dental, cranial, and postcranial. Their cladogram continued to support Hopson and Barghusen (1986; Hopson 1991) by relating cynognathids to the diademodontoids, placing tritylodontids with the latter, and recognising tritheledontids as the sister group of mammals (Fig. 3.25(c)). In part these inconsistent results may be due to the different taxa used. But the main cause lies firmly in the choice of unit characters made. In their total of 101 characters, Hopson and Kitching (2001) included 28 dental characters of which no less than 17 were concerned with the morphology of the individual postcanine teeth. Twenty of their characters were postcranial. In contrast, out of 68 characters, Martinez *et al.* (1996) included only 13 dental characters of which a mere five concerned individual

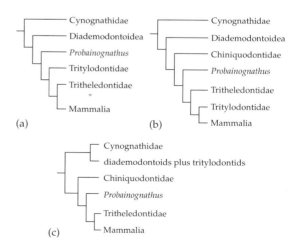

Figure 3.25 Three cladograms of the main cynodont groups. (a) Luo 1994). (b) Martinez *et al.* 1996. (c) Hopson and Kitching 2001.

postcanine structure, and no postcranial characters at all. Luo (1994) from his perspective makes use of the inordinate number of 33 characters of the articular and quadrate, and 20 of the petrosal bone, out of his total of 82. It is certainly hard to believe that these characters are all phylogenetically independent of one another. Using a large number of postcanine characters seems bound to bias the analysis towards a relationship of diademodontoids and tritylodontids, while postcranial characters generally favour a relationship of tritylodontids to mammals rather than to diademodontoids but may be poor resolvers of the tritylodontid, tritheledontid, and mammal trichotomy.

A possible source of resolution of the particular question of the precise origin of Mammalia from within the non-mammalian cynodonts lies in the series of small forms, described above as the Rio Grande do Sol tritheledontans from the early Upper Triassic of South America. These differ from one another in small details of the cranial structure and dentition, and tentatively reveal detailed patterns of the emergence of mammalian characters (Bonaparte *et al.* 2003). With more knowledge of their structure, morphological sequences showing in finer detail the acquisition of fully mammalian characters may be determined.

To complicate, or perhaps to elucidate matters even further, Bonaparte and Crompton (1994) described a juvenile specimen of possibly *Probainognathus*, which has a number of more derived, mammal-like character states, compared to mature specimens. The prefrontal and postorbital bones are small and the frontal bone borders the orbit. The zygomatic arch is relatively more slender and the braincase relatively larger. In the lower jaw, the dentary extends backwards very close indeed to the squamosal, and at the front end the symphysis is horizontal and unfused. Even the postcanine teeth resemble those of *Morganucodon* much more than do adult *Probainognathus* teeth.

More will be said later about the possible role of miniaturisation of the body in the process of the origin of mammals. For the moment, this mosaic of ancestral and derived mammalian characters seen among the most progressive eucynodont groups, tritylodontids, tritheledontids, therioherpetontids, juvenile probainognathids, and presumably also dromatherians and their like, reveals the possibility that several lineages of Middle Triassic eucynodonts were independently evolving reduced body size with a concomitant convergence of several basic characters associated with small body size, but superimposed upon divergent characters associated with differing diets etc. It is perhaps significant that, as described in a later chapter, members of the taxon Mammalia almost at their first appearance consisted of at least four distinct groups. At a far lower taxonomic level than was believed 50 years ago, maybe the 'reptilian–mammalian' boundary *was* crossed several times, if such a cladistically incorrect statement may be forgiven.

Overview of the interrelationships and evolution of the Therapsida

Rubidge and Sidor (2001) reviewed the interrelationships of the principal therapsid groups and

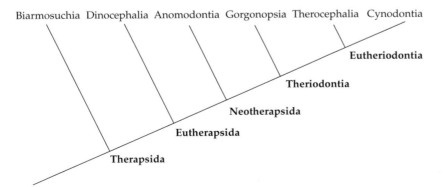

Figure 3.26 Rubidge and Sidor's (2001) cladogram of the principal therapsid groups.

published a cladogram based on recent literature that reflects the current majority view (Fig. 3.26). It may reflect the most parsimonious arrangement of the characters considered, have relatively little homoplasy, and would perhaps measure acceptably high on tests such as consistency index and boot-strapping, but what it does not in itself reveal is how few, minor, and sometimes ill-defined are the characters supporting some of the nodes:

Therapsida: very strongly supported by around 40 characters of the dentition, skull, and postcranial skeleton. Not since Olson (1962) proposed an independent origin of anomodonts from caseid pelycosaurs has anyone seriously challenged the monophyly of the therapsids.

Eutherapsida: in contrast, very few diagnostic characters supporting it. Rubidge and Sidor (2001) quote just three: lateral bowing of the zygomatic arch indicating an increase in the size of the temporal fenestra; loss of the olecranon process of the ulna; only three phalanges in the fifth pedal digit.

Neotherapsida: also very little support, with four diagnostic characters mentioned, all uninspiringly trivial-looking: the vaguely described ventral expansion of the squamosal obscuring most of the quadrate from hind view; epipterygoid broadly contacting the underside of the parietal; epiphyses on the atlas vertebra; enlarged obturator foramen in the pelvic girdle.

Theriodontia: given its long acceptance, at least since Watson and Romer (1956), surprisingly little evidence for the relationship of these three carnivorous groups, with just four characters: free-standing coronoid process of the dentary; quadrate and quadratojugal bones reduced in height and lying in a recess in the squamosal; dorsal process of premaxilla reduced; skull reduced in height.

Eutheriodontia (Therosauria of Kemp 1982): slightly better diagnosed group than the previous three. Hopson (1991) noted five unique characters: very narrow intertemporal region; reduction of the postorbital bone; posteriorly elongated zygomatic arch; dentary thickened postero-ventrally below angular bone; broadened epipterygoid. Several other characters found in this group are shared with at least some anomodonts, such as a reduced postfrontal bone and loss of palatal teeth, though the cladistic analysis implies that they are convergent.

The great majority of the characters found in therapsids are uninformative, being either plesiomorphic for Therapsida, or autapomorphic for the individual taxa. Although this may be the best cladogram, and therefore the basis of the best classification available, it does not inspire great confidence in its truth. The situation is reminiscent of the relationships of the placental mammal orders, where supraordinal groupings based on similarly weak morphological evidence have been shown to be highly inconsistent with the ever-better supported groupings based on molecular data. Unfortunately, no such alternative source of evidence is accessible for non-mammalian therapsids.

There is, however, another possible approach to the problem, which is to consider a functional scenario. A cladogram of relationships is evaluated by noting the implied evolutionary changes between nodes, and then assessing the likelihood of those transformations in terms of their structural and functional plausibility. As a methodology this is fraught with difficulties, relating less to the principle than to the practical extent to which degrees of plausibility can be objectively evaluated. However, the case of the interrelationships of therapsids is possibly one where this approach is powerful. Much of the case for the systematic arrangement of therapsids subgroups concerns the organisation of the temporal fenestra. Increase in its size is the most prominent character of the Eutherapsida, but it is highly improbable that this really is a homologous state in all the eutherapsid groups (Kemp 1988). The structure of the temporal fenestra, associated lower jaw, and inferred adductor musculature of *Biarmosuchus* is sufficiently similar to that of a sphenacodontid pelycosaur to render it safe to assume that it represents the ancestral therapsid condition. Compared to it, the modification of the temporal fenestra found in a primitive dinocephalian such as *Titanophoneus* consists of a dorsal expansion in such a way that adductor musculature could invade a broad dorso-lateral exposure of the postorbital and squamosal bones. Turning to the Anomodontia (Fig. 3.11), the genera *Anomocephalus*, *Patronomodon*, *Ulemica*, and *Eodicynodon* constitute a sequence illustrating the morphological transition in structure of the temporal fenestra from the broad temporal roof and short, rod-like zygomatic arch of the first stage, to the fully dicynodont condition of the last. Now the initial stage in the sequence, represented by *Anomocephalus*, is quite similar to *Biarmosuchus* except for an antero-posterior expansion and there is no extension of the attachment area of the temporalis muscle dorsally onto the external surface of the intertemporal region. Furthermore, at no subsequent stage does the temporal fenestra resemble the basic dinocephalian condition, or that of any of the other supposed eutherapsids. The morphology corroborates the hypothesis that the Anomodontia were not related to any other derived subgroup of Therapsida, but modified their adductor jaw musculature independently via a different route from the ancestral condition.

A unique arrangement of the temporal fenestra also evolved in the gorgonopsians, this time by extensive posterior and lateral expansions of the temporal fenestra, but virtually no medial expansion, leaving a very broad, flat intertemporal roof. It was associated with modifications of the lower jaw, jaw articulation, and adductor musculature. These included an invasion by musculature of the lateral surface of the reflected lamina of the angular of the lower jaw, in a manner not matched at all in the Therocephalia. Functionally, it is difficult to see how either the gorgonopsian or the therocephalian arrangement could have been ancestral to the other, and therefore enlargement of the temporal fenestra and associated remodelling of the adductor musculature must have occurred independently in the two, by different routes from an ancestral, *Biarmosuchus*-like condition. It further follows that since the discrete coronoid process of the dentary is a functionally integral part of the modified arrangement of the adductor musculature, it too must have been evolved separately in these two respective taxa. The same statement is likely to be true of the reduced quadrate bone, and in any case the detailed structure, mode of attachment, and function of the quadrate is quite different in gorgonopsians compared to therocephalians. The unavoidable conclusion from this approach is that the association of gorgonopsians and therocephalians in Theriodontia is based on convergent characters. As yet there is no morphological sequence of increasingly derived gorgonopsian skulls, or therocephalian skulls comparable to the anomodont sequence, but surely

when they are discovered, they will demonstrate an independent transformation of the two lineages all the way from a *Biarmosuchus*-like ancestor.

Only the Eutheriodontia, consisting of the Therocephalia plus Cynodontia stands up to functional scrutiny of the temporal fenestra and associated structures. The mode of enlargement of the fenestra is similar in the two, consisting mainly of a medial expansion, though followed later by a lateral expansion in the cynodonts alone. However, discovery of a morphological sequence of primitive to derived therocephalians might well produce a surprise on this front and demonstrate convergence of the feeding structures in the two.

In the meantime the picture that emerges is one of the rapid diversification of several therapsid lineages from a hypothetical *Biarmosuchian*-grade ancestor, each developing a more sophisticated feeding mechanism, but doing in a variety of different ways.

Although Rubidge and Sidor's (2001) cladogram may still be the best, given the taxonomic characters to hand, it should not be forgotten how very weakly supported are several of its groupings. In this light, the reference classification of Table 2.1 is fairly non-committal.

The palaeoecology and evolution of synapsida

The rise of the pelycosaurs as the first group of tetrapods to dominate the terrestrial environment; their mid-Permian complete replacement by therapsids; the subsequent dominance by the latter of the terrestrial tetrapod biota during the Late Permian; the huge, sudden drop in diversity at the end of the Permian; and the eventual replacement of therapsids by other taxa in the later part of the Triassic. These high points of the synapsid story all illustrate one way or another the profound, intimate relationship between the evolution of the group and its environment.

Palaeogeography and palaeoecology of the pelycosauria

The pelycosaurs appear in the fossil record in Pennsylvanian (Upper Carboniferous) deposits of North America during Westphalian times

310–320 Ma. By this time, the great southern super-continent of Gondwana had drifted northwards and all but completed its coalescence with Laurasia, forming the huge land mass of Pangaea, most of which lay in the southern hemisphere and extended over the South Pole. Only the small landmasses of Siberia, Kazakhstan, North China, and South China were separate, forming an arc of islands on the eastern side (Scotese and McKerrow 1990). The contact of Laurasia and Gondwana had created a major range of mountains right across the middle of Pangaea, almost exactly following the equator, and of the order of size of the modern Himalayas (Zeigler *et al.* 1997). The climate of the Westphalian world is indicated by several kinds of characteristic sediments preserved in different regions (Parrish *et al.* 1986). Coal measures are found in the equatorial regions either side of the mountain range, indicating hot, humid, non-seasonal conditions, covered in freshwater swamps with their associated rich flora. This plant life was still dominated by the spore-bearing pteridophytes and lycopsids, although seed-bearing plants, conifers, and cordaites in particular, were already begin to increase in prominence (Behrensmeyer *et al.* 1992). To the north and the south of this equatorial band there are evaporite deposits extending to a latitude of around 30°, which are indicative of warm conditions with at least seasonal and possibly permanent aridity. Finally, much of the southern regions from 30° south to the pole have tillites, which are associated with glaciation. A very large ice-sheet must have covered a significant proportion of the southern hemisphere.

All the Pennsylvanian and Early Permian pelycosaurs so far found have come from the equatorial region, either side of the mountain range (Berman *et al.* 1997), and were therefore part of the equatorial coal measures habitat, which indicates that they were adapted for hot, permanently wet conditions (Fig. 3.27(a)). In fact, this particular ecosystem was quite different from any that has existed since, because there were very few herbivorous terrestrial tetrapods. The energy flow was still evidently strongly dependent on freshwater productivity, with fish a major element in the ecosystem (Milner 1987, 1993). On land, the main primary consumers were the terrestrial invertebrates. Under these conditions, the tetrapods had three broad ecological roles. One was as insectivores, carried out by various small amphibian groups and the protorothyridid amniotes. The second was as piscivores, a guild that included larger amphibians and also the ophiacodontid pelycosaurs. The third was as secondary carnivores, able to consume the smaller amphibians and the protorothyridids, as manifested mainly by sphenacodontid pelycosaurs. Terrestrial tetrapod primary consumers did appear in the Pennsylvanian in the form of the pelycosaur *Edaphosaurus*, but only as a relatively insignificant component of the community: the evolution of what was to become the fundamentally important position of tetrapods as fully terrestrial herbivores had to await a change in environmental conditions, though when it did happen in the Early Permian, pelycosaurs were profoundly implicated.

By the lowest part of the Early Permian, conditions in this equatorial region of the world had begun to change (Olson and Vaughn 1970). The coal measure deposits gradually disappeared to be replaced increasingly by the Red Beds that form the huge fossiliferous area of Texas and adjacent states, and in which the vast majority of pelycosaur specimens have been found. Red Beds indicate a drier, and seasonally arid climate, and the increase to dominance of seed-bearing plants over the spore-bearing groups reflects the environmental shift. Initially there was little change in the fauna, but gradually during the course of the Early Permian, some of the elements disappeared and others appeared (Milner 1993). Fossilised lungfish burrows similar to those of modern forms strongly support the idea of seasonal aridity, and different groups of amphibious tetrapods, perhaps better adapted to periodic dry circumstances, appeared. A fauna distinctly less dependent on freestanding bodies of water all the year round seems to have evolved. The pelycosaurs evidently reflect these changing circumstances, and in the process became the dominant tetrapod component of the fauna. Edaphosaurids, primarily the genus *Edaphosaurus*, were the first terrestrial tetrapod herbivores to became abundant, and therefore the first to adopt a role continuously filled thereafter by a succession of taxa, culminating in the deer and antelopes of the present day. The sphenacodontids radiated to be the correspondingly dominant large carnivores as this source of large prey increased.

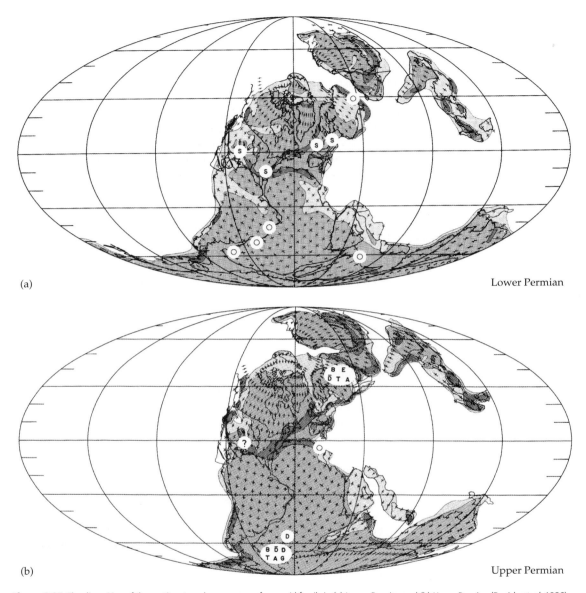

(a)   Lower Permian

(b)   Upper Permian

Figure 3.27 The disposition of the continents and occurrences of synapsid fossils in (a) Lower Permian and (b) Upper Permian (Parrish *et al.* 1986). S-occurrence of pelycosaurs, O-occurrence of tetrapads but synapsids absent. Other letters-therapsid taxa present.

Yet further changes occurred later in the Early Permian, to judge from the North American sequence (Fig. 2.2a), as seasonal aridity continued to increase. Ophiacodontids last occur in the Clyde Formation, and the edaphosaurids only shortly afterwards in the Arroyo Formation. It is at this time that the caseids appeared, and they persisted right through into the highly arid San Angelo and Flower Pot Formations that mark the start of the Late Permian. Olson (1975) proposed that during the Early Permian there existed what he termed a caseid chronofauna, a fauna adapted to drier, seasonal conditions. The only evidence for it was the occasional, erratic occurrence in the fossil record of rare forms such as *Casea* itself, varanopseids and certain amphibians. Only in one locality may the caseid chronofauna have been sampled directly, Fort Sill in Oklahoma. This is an upland fissure-fill deposit in the Arbuckle Limestone, probably of Arroyo Formation age, and containing fragmentary remains of several small pelycosaurs, some apparently caseid-like. Olson's view is that as increasingly seasonally arid conditions spread during the Early Permian, so this hypothetical caseid chronofauna expanded and by the end of the Early Permian it had replaced the pre-existing Permo-Carboniferous fauna that had been adapted to wetter conditions. From this argument, it would follow that the caseids and varanopseids were physiologically adapted for drier conditions, and it would also follow that the sphenacodontids underwent a parallel physiological change, since they spanned the transition from coal measure to full-blown red bed environments.

Thus pelycosaurs were the major faunal element in the development for the first time of a fully terrestrial vertebrate ecosystem that was no longer embedded directly in freshwater productivity. The palaeoecological evidence leads to the conclusion that the important physiological features ultimately characterising mammals made their initial, incipient appearance within the pelycosaur grade. The relevance of this possibility to the question of the origin of the therapsids will be discussed next.

Biogeography and palaeoecology of the origin of the Therapsida

Irrespective of whether or not the origin of the taxon Therapsida lay in the earliest Permian as very doubtfully suggested by *Tetraceratops* (page 27), the fossil evidence is of a rapid radiation of therapsids about 270 Ma, soon after the start of the Late Permian, or Guadalupian. Within a very narrow time band, a considerable variety of ecotypes appeared, carnivores and herbivores varying considerably in body size. This radiation coincided with the decline almost to extinction of pelycosaurs. It was also associated with a major increase in biogeographic range. The restriction of pelycosaurs to the tropical region, within 30° latitude of the palaeoequator is markedly contrasted by the initial

therapsid radiation, which included the temperate zones between 30° and 60°, in both the northern and the southern hemispheres (Fig. 3.27(b)). Admittedly, few tetrapod bearing beds have been discovered yet within the tropics from this time period apart from the extremely poor North American Guadalupian, so it is not known whether the early therapsid distribution really was disjunct, or whether it spanned the intervening equatorial zone. The taxonomic similarity between the respective northern fauna of Russia and the southern fauna of South Africa strongly suggests that the two could readily interchange, and therefore that there was a more cosmopolitan distribution that included the intervening tropics (Parrish *et al.* 1986; Milner 1993).

The question of what could have triggered this initial radiation of the therapsids, apparently at the expense of the pelycosaurs, hinges on whether there were any environmental changes at the time that opened up new opportunities for the nascent group, or whether therapsids possessed some evolutionary innovation in their biology that both created a competitive advantage over pelycosaurs and allowed their radiation into areas hitherto unavailable to synapsids.

There are no indications in the geological record of any large, or abrupt palaeo-geographical or climatic events at this time. The start of the Late Permian coincided approximately with the attachment of the eastern landmasses of Kazakhstan, Siberia, North China, and South China to Pangaea, increasing to a small extent the area potentially available for a cosmopolitan fauna. More importantly, it also coincides with the withdrawal of the glaciation that had covered much of Gondwana from the Carboniferous through the Early Permian, although this was merely the continuation of a process that had been going on through much of the Early Permian. Zeigler *et al.* (1997) reconstructed the climate and climatic zones of the Permian, and according to their analysis, the transition from the Early Permian to the Late Permian was not remarkable apart from this glacial retreat to high altitudes and latitudes. The palaeoclimate of the tropical regions increased in aridity as illustrated in most detail by the southwestern USA sequence (Behrensmeyer *et al.* 1992). The flora indicates a gradual but eventually considerable increase in the proportion of seed plants such as gymnosperms and a corresponding reduction in spore-bearing, fern-like lepidodendrids and sphenopsids. There was also an increase in plants with xeromorphic adaptations, such as leathery leaves, hairy surfaces, and spiny leaves. Meanwhile, increasingly rich, diverse floras were developing in the temperate areas (Cuneo 1996). The actual environments of the two areas in which the majority of early therapsids have been found, Russian in the north and South African in the south, were both moist, temperate zones.

The consequences of these gradual changes in the environment on the tetrapod fauna are not surprising. There was a reduction and eventually a virtual disappearance from the North American deposits, apart from the extremely few pelycosaurs and possible, but very doubtful, early therapsids found. On the other hand, a wide range of tetrapods spread into the southern and northern temperate regions, including various kinds of amphibians and diapsid amniotes (Milner 1993). In fact, there is considerable taxonomic continuity between the Early and the Late Permian non-therapsid, terrestrial faunas (Anderson and Cruickshank 1978; Milner 1993). The pelycosaurs also underwent this biogeographic shift, for a few relict members of the group occurred briefly for the first time outside the tropics: caseids in the north and varanopseids in the north and the south.

Looking at the whole picture, the transition from Early Permian to Late Permian did not appear to involve any abrupt changes. The climate, flora and fauna all shifted relatively gradually over the course of the Permian, with one exception: the sudden appearance of the diverse therapsid fauna 270 Ma. The absence of any sign of a large change in climate or biota in the middle of the Permian leads to speculation that it was the evolution of some biological innovation that caused the rapid radiation of the therapsids and their replacement of the pelycosaurs. The idea of a particular key evolutionary innovation leading to instant competitive superiority is notoriously difficult to test (Kemp 1999), as in reality any change of sufficient magnitude to have the effect is likely to have involved the subtle interplay between many attributes of both organism and environment. Nevertheless a number of interesting

possible pointers can be considered. It has been suggested earlier that the pelycosaurs played a fundamental role in the establishment of a modern style of fully terrestrial ecosystem involving tetrapod herbivores, and therefore the end of direct dependence on aquatic productivity. Compared to the pelycosaurs, the therapsids were undoubtedly more highly evolved along the same road. The structure of their skeletons indicates that they were more active organisms. The carnivorous forms had much larger canine teeth and indications of more elaborate musculature to operate the jaws. The herbivores had a more specialised and presumably therefore a more efficient mode of dealing with plant food. The therapsid postcranial skeleton, with its relatively longer, more slender limbs and less constraining shoulder and hip joints, indicates greater mobility and agility. As discussed at length later, these modifications must have been accompanied by an incremental shift in metabolic and neurological physiology, which in turn imply a more mammal-like organisation, more capable of surviving the fluctuations of the terrestrial environment. Two associated consequences might have followed the origin of therapsid biology. One is that, unlike most cases of co-existing taxa, the pelycosaurs and therapsids were direct competitors for essentially the same niches, those requiring physiological tolerance of warm and dry conditions, and that the therapsids were the better adapted. The other is that the higher level of homeostatic organisation of therapsids opened up a range of niches not hitherto available to pelycosaurs, or indeed to any other contemporary tetrapods, and hence their rapid radiation into new kinds of terrestrial organisms. This may be one of the few cases where a convincing hypothesis of competitive replacement of one taxon by another may be entertained.

Therapsida and the end-Permian mass extinction

During the Late Permian, the therapsid radiation is far and away best illustrated by the great Beaufort sequence of the South African Karoo and equivalent-aged strata in other southern African areas, such as in Zambia and Tanzania (King 1990). The environment consisted of broad, slow-flowing rivers with lakes and swamps in the low-lying levels and forested higher levels. The climate was cool and seasonal, with dry and wet seasons. The flora is described as the *Glossopteris* flora after the hardy seed-fern of that name which was the most abundant form. It was a genus of woody plants, the largest species being 4 m in height and others lower and shrubbier. Other seed ferns, a variety of horsetails, and primitive conifers were also present. It was in this habitat that the dicynodonts flourished, radiating into dozens of species, adapted in different ways to exploit different aspects of this rich flora, during the times of the successive Assemblage Zones of the Late Permian (Fig. 2.2(c)). Gorgonopsians and the larger therocephalians preyed on them, and smaller carnivores, the baurioid therocephalians and the rare primitive cynodonts occupied insectivore or small-prey carnivore roles. The other elements of the fauna were far less diverse. There were pareiasaurs, which were large-bodied herbivores superficially like the by then extinct tapinocephalid dinocephalians, and the small-bodied herbivorous, procolophonids, both probably related to the turtles. A few superficially lizard-like diapsid reptiles, and a handful of temnospondyl amphibians more or less complete the tetrapod faunal list. During the course of the Late Permian, there certainly were faunal changes. The dinocephalians as a whole did not survive beyond the *Tapinocephalus* Assemblage Zone, and neither did the carnivorous lycosuchid and scylacosaurid therocephalians. At the other extreme, whaitsiid therocephalians and cynodonts did not appear until the *Dicynodon* Assemblage Zone, just before the end of the Permian. However, this degree of faunal turnover is within the normal, background rates for terrestrial tetrapods generally.

The same cannot be said of the Permo–Triassic boundary. The end-Permian marked the largest mass extinction in the Earth's history with an estimated loss of 90–95% of all species; the terrestrial biota was as much affected as the marine. The discovery of a large, sharp shift in the ratio of the stable carbon isotopes ^{13}C to ^{12}C (expressed as a negative shift of the $\delta^{13}C$ value) in several parts of the world, coincident with the biotic changes, indicates that the event in question was synchronous worldwide, and on the geological time scale was at least relatively brief if not actually catastrophic (Erwin *et al.*

2002). The cause of the end-Permian mass extinction continues to be extensively discussed (Erwin *et al.* 2002; Benton and Twitchett 2003). One of the most significant environmental features is the negative shift in the $\delta^{13}C$ value itself, indicating a large rise in organically derived carbon in the atmosphere. This is presumably due at least in part to decreased levels of plant productivity, but estimates suggest that the level of CO_2 was too high for that alone to explain it. One possibility is that there was a massive release of methane from methane hydrates trapped within the polar ice sheets because of a severe greenhouse warming of the environment due to the initial increase of CO_2. A positive feedback process would have followed, leading to a rise of the Earth's average surface temperature by as much as 6°C. Indeed, there is direct evidence for such a global warming episode, including changes in the oxygen isotope ratios, and the nature of the floral changes. Other geochemical signals of the time indicate that anoxic conditions in both deep and shallow water settings occurred. A change in the strontium isotope ratios indicates a possible increased rate of weathering of continental rocks. Concerning possible triggers for these changes, there is some evidence quoted for an extraterrestrial impact, but this is limited and ambiguous, and does not approach the convincing nature of the evidence that exists for an end-Cretaceous bolide impact. In contrast, there is increasingly convincing evidence for a volcanic trigger. The Siberian Traps are a huge deposit of basalt produced by vulcanism. They have an estimated volume of 3×10^6 m^3 and a thickness of up to 3,000 m. The date these rocks began to form coincides with the end-Permian, 251 Ma. Hallam and Wignall (1997; Wignall 2001) have developed a plausible sequence of events beginning with the eruption of the Siberian Traps that accounts for all the geochemical and biotic signals described. The volcanic outgassing increased CO_2 levels, causing global warming, and this in turn released methane trapped in gas hydrates in the polar ice sheets. At the same time, the release of SO_2 and chlorine caused acid rain. The effect on the biota was devastating, both directly and by the effects of increasingly anoxic conditions as photosynthesis levels fell. Survivors were fungi and algae, as indicated by a brief increase in the presence of preserved spores in the sediments, known as a fungal spike (Visscher *et al.* 1996). A handful of presumably particularly hardy land plants also survived. Among animals, it must be assumed that despite the breakdown of the ecosystem worldwide, some species had sufficiently cosmopolitan diets, and were capable of occupying sufficiently protected microhabitats to withstand extreme seasonal conditions. Perhaps some were already adapted by having evolved the ability to aestivate.

The effect on the terrestrial environment is most clearly and dramatically seen in the transition from the latest Permian *Dicynodon* Assemblage Zone to the overlying basal Triassic *Lystrosaurus* Assemblage Zone of South Africa, where the $\delta^{13}C$ shift has been recorded (MacLeod *et al.* 2000). There is a marked change in the palaeoenvironment at this time. The latest Permian geography consisted of slow flowing, meandering rivers with extensive seasonal floodplains, laying down greenish-grey mudstones. In the immediately overlying basal Triassic, the river systems were much less sinuous and faster flowing, and the floodplains had largely dried up (Smith 1995). The deposits are now dominated by reddish sandstones indicative of extreme drought conditions and faster flow of the rivers. The change can be interpreted as the consequence of a severe loss of vegetation, reducing the forests and therefore removing the stabilising effect of the plant life. Erosion would increase and the flow of the river systems be less checked. The shift in sediment types coincides with a dramatic change in the vegetation, with the disappearance of the *Glossopteris* dominated flora, and its replacement by a much lower diversity flora consisting of the seed fern *Dicroidium* and a few genera of conifers and lycopods.

The accompanying change in the fauna was also stark. The *Dicynodon* Assemblage Zone contains approximately 44 reptilian genera of which only three occur in the *Lystrosaurus* Assemblage Zone (Rubidge 1995). Of these, 36 are therapsids, and two occur in both of the adjacent Assemblage Zones, namely the dicynodont *Lystrosaurus* and the therocephalian *Moschorhinus*. A few other lineages must also have survived the end-Permian because their descendants occur in the Triassic, although no Permian specimens of these genera have yet been found. At present, the full list of known or inferred surviving lineages includes four dicynodont ones, *Lystrosaurus* itself, the kannemeyeriids which, although related to *Lystrosaurus*, must have differentiated from that genus before the end of the Permian, the diictodontoid *Myosaurus* which occurs in the *Lystrosaurus* Assemblage Zone, and the kingorioid *Kombuisia* that occurs in the succeeding *Cynognathus* Assemblage Zone. Of the therocephalian lineages, there was *Moschorhinus*, several baurioids, for example *Regisaurus*, from the *Lystrosaurus* Assemblage Zone, and the possibly separate lineage that includes the *Cynognathus* Assemblage Zone *Bauria*. Finally, two cynodont lineages made the transition. These are the galesaurids represented by *Cynosaurus* in the latest Permian and *Galesaurus* in the *Lystrosaurus* Assemblage Zone, and *Thrinaxodon* from the latter zone. Set against this dozen or so survivors was the disappearance of all the gorgonopsians, the great majority of the hugely diverse dicynodonts, the last of the large-bodied therocephalians, and the procynosuchian cynodonts.

Smith and Ward (2001) have studied high-resolution stratigraphic sections in the South African Karoo which include the actual Permo–Triassic boundary. As well as confirming the lithological change from mostly greenish-grey mudstone to mostly reddish sandstone dominance, the section contains a non-fossiliferous layer within which is found the $\delta^{13}C$ shift that marks the actual boundary. Below this layer, the fossil record indicates a geologically abrupt extinction event, and the only taxon they found that occurred both below and immediately above the Permo-Triassic boundary is *Lystrosaurus*, already abundant in the latter. Other *Lystrosaurus* Assemblage Zone forms do not appear until higher levels in the zone. It is difficult precisely to estimate the time represented by the unfossiliferous transitional bed, but at most it was 50,000 years and could have been considerably less, especially if associated with an increased sedimentation rate consequent upon the loss of vegetation; by geological standards the change in biota was undoubtedly a catastrophic and not a gradual process.

The surviving taxa found in the *Lystrosaurus* Assemblage Zone must have been able to tolerate very arid conditions. *Lystrosaurus* was probably an adept digger, capable of protecting itself by burrowing and subsisting on a diet of subterranean roots and tubers. The other surviving therapsid taxa were small in size, and may have existed initially in less harsh habitats such as uplands and fed primarily on insects. Of particular interest, the one known and the one inferred lineages of cynodonts that lived through the end-Permian event may perhaps have owed their survival to a higher level of homeostasis combined with small body size, and possibly burrowing (Damiani *et al.* 2003).

The Triassic decline of the Therapsida

The vegetation soon recovered from the effects of the extinction and during the Lower Triassic an ecosystem basically structured like that of the Permian had been re-established. There was an increase in diversity of various primitive conifer groups, but seed ferns remained abundant. On the whole, the flora reflected a warm, seasonal climate (Behrensmeyer *et al.* 1992). Of the therapsid survivors of the end-Permian mass extinction, only two lineages were of much significance. One is the kannemeyeriid dicynodonts and the other the eucynodontian cynodonts. Both groups radiated through the Lower and Middle Triassic worldwide, but therapsids never again dominated as they had in the Late Permian. They shared the terrestrial amniote world from the Lower Triassic onwards with increasing numbers of diapsid reptiles, notably the herbivorous rhynchosaurs and a variety of thecodontan archosaurs. Then during the Upper Triassic, they disappeared. Correlation of the ages of terrestrial Triassic strata is notoriously difficult, but the main extinction appears to have occurred during or at the end of the Carnian stage, by about 220 Ma. This included the disappearance of the kannemeyeriids and the chiniquodontid

cynodonts, while the herbivorous traversodontid cynodonts were greatly reduced in diversity, although a few occur in the following Norian Stage (Benton 1994).

There has been a long dispute about whether the disappearance of the therapsids and the radiation of the dinosaurs in the Upper Triassic resulted from competitive interactions in which the dinosaurs were somehow superior, or from a change in the environment that caused a decline in therapsids and allowed an opportunistic radiation of dinosaurs into the large terrestrial tetrapod niches (Benton 1986). The traditional view was the competitive one, and a number of authors speculated on what might have been the basis for the dinosaurs's better general adaptation. Charig (1984), for example, believed that dinosaurs simply evolved a more effective locomotory ability, while Bakker (1968) argued that dinosaurs evolved endothermy before the mammal-like reptiles. Benton (1986, 1994) argued forcefully that there was actually a significant mass extinction in the late Carnian, correlated with an environmental change. There is evidence for an increase in the extinction rate of plants at least approximately at the same time (Boulter *et al.* 1988; Simms *et al.* 1994). From this perspective, the decline of the therapsids resulted from the environmental change associated with this extinction event, and the increase in dinosaur diversity from the Norian onwards occurred as the latter opportunistically invaded the habitats vacated by the former.

Whatever the truth of the matter, the last part of the Triassic record of therapsids consisted only of the highly specialised tritylodontids, the tritheledontans and the small, relatively rare, and insignificant mammals (Lucas and Hunt 1994). The phase of synapsid evolution represented by the Mesozoic mammals was about to commence.

CHAPTER 4

Evolution of mammalian biology

There are large biological differences between the mammals and the primitive living amniotes as represented by turtles, lizards, and crocodiles

- Differentiated dentition with occluding post-canine teeth, and radical reorganisation of jaw musculature to operate them
- Differentiation of vertebral column and limb musculature, and repositioning of limbs to bring feet under the body, increasing agility of locomotion
- Relatively huge brain and highly sensitive sense organs
- Endothermic temperature physiology, with very high metabolic rates, insulation, and high respiratory rates
- Precise osmoregulatory and chemoregulatory abilities using loops of Henle in the kidney and an array of endocrine mechanisms

Incomplete as it is, the fossil record of the mammal-like reptiles, or 'non-mammalian synapsids' permits the reconstruction of a series of hypothetical intermediate stages that offers considerable insight into how, when, and where this remarkable transition occurred.

Deriving these stages starts with a cladogram of the relevant fossils (Fig. 4.1) that is then read as an evolutionary tree, with hypothetical ancestors represented by the nodes. The characters that define a node, plus the characters of the previous nodes, constitute the reconstruction. The differences in characters between adjacent nodes represent the evolutionary transitions that by inference occurred, and the whole set of successive nodes generates all that can be inferred about the sequence of acquisition of characters. If a hypothetical ancestral synapsid is placed at the base of the cladogram, and a hypothetical ancestral mammal as the final node, then the set of nodes in between represents everything

the fossil record is capable of revealing about the pattern by which mammalian characters evolved: the sequence of their acquisition, the correlations between characters, and possibly the rates of their evolution. Of course, the inferred pattern of evolution of characters is only as reliable as the cladogram which generated it, and that in turn is only as realistic as the model of evolution used in its construction from the character data. And of course, there must have been many intermediate stages in the transition than cannot be reconstructed for want of appropriate fossil representation of those particular grades. Nevertheless, limited as it may be, this is what can be known from the fossil record.

The respective stages that can be reconstructed are named after the fossils or fossil groups that reveal them, though it must not be forgotten that the actual fossils are members of taxa with their own autapomorphic characters that do not constitute part of the reconstruction of the hypothetical ancestors. The known fossils certainly do not themselves constitute an ancestral-descendant series.

- Ancestral amniote grade
- Basal pelycosaurian grade
- Eupelycosaurian grade
- Biarmosuchian grade
- Therocephalian grade
- Procynosuchian grade
- Basal epicynodont grade
- Eucynodont grade
- Tritylodontid grade
- Tritheledontid grade
- Mammalian grade

In the following sections, it is not always necessary to refer to all these stages, because in some features adjacent ones differ little from one another. It is also useful on occasion to refer to other taxa of fossils

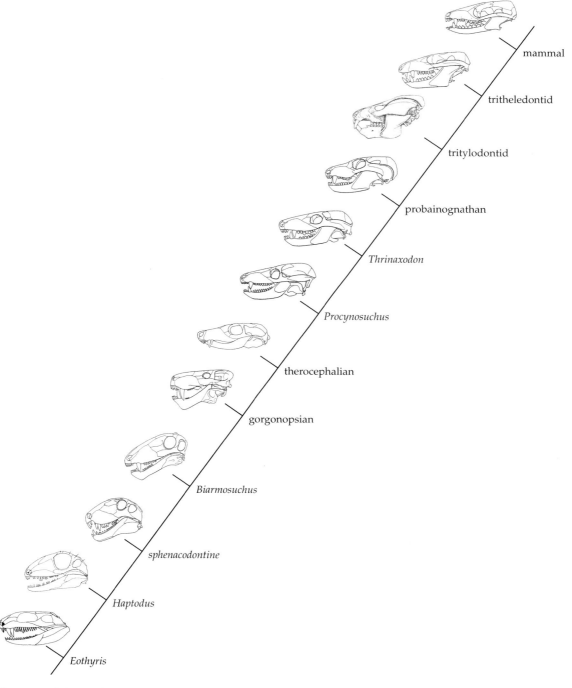

Figure 4.1 Cladogram of synapsids (Hospson 2001; Carroll 1988; Kemp 1988c).

whose exact cladistic position is uncertain, such as gorgonopsians.

Once an acceptable sequence of hypothetical ancestral stages has been established, a decision needs to be made about whether to adopt an atomistic or an integrated approach to the biology of these reconstructed hypothetical ancestral organisms. Discussing the transition to mammals functional system by functional system, which is the atomistic approach, eases description and clarifies explanation of particular morphological changes from an existing state to a new one. The alternative is the integrated approach of considering each stage in its entirety, before moving on to the next. This focuses on the interdependence of all the structures and processes in the individual organisms, and therefore clarifies the role of each functional system in the overall biology of a particular organism. Without this appreciation, it would be impossible to explain the evolution of 'mammalness' as a whole, even though this is the property that actually evolved. An atomistic approach has been adopted here, but the chapter concludes with a section in which the evolution of the mammal as an integrated, whole-organism is reviewed. This should be seen as the culmination of the chapter and not merely an appendage.

Feeding

Ancestral amniote grade

The ancestral amniote skull (Fig.4.2(a)) is reconstructed on the basis of *Westlothiana* (Smithson *et al.* 1994). There was a homodont dentition of numerous single-cusped teeth suitable for small prey such as insects. No temporal fenestra had evolved, and a relatively undifferentiated capiti mandibularis muscle ran from its origin on the inner surfaces of the temporal region of the skull behind the orbit, to an insertion on the medial and dorsal surfaces of the posterior part of the lower jaw. Pterygoideus musculature from the dorsal surface and hind edge of the palate was inserted on the medial face of the lower jaw. The jaw articulation was a simple, transversely oriented hinge joint, between the roller-like quadrate condyle above and the trough of the articular bone below. Although, with the abandonment of suction feeding in water, kinetic movements of

the cheeks of the skull no longer occurred, the pattern of skull bones still indicates the position of the old hinge lines. The feeding mechanism by this stage was a relatively weak and simple orthal closure of the jaws, which would have disabled the prey. Lauder and Gillis (1997) suggested that manipulation of the food may have involved elaborate tongue and hyoid musculature even by this early stage, though there is little evidence about any details.

Basal pelycosaur grade

The major feature of the feeding system that had evolved in the ancestral pelycosaur was the temporal fenestra, but how exactly it developed remains uncertain. The standard explanation for the function of temporal fenestrae in amniotes is that of Frazzetta (1969), who argued that the edges of the bone surrounding a fenestra offer a stronger connective tissue attachment for muscles than does the flat, internal surface of the temporal bones of the skull. The function of the fenestra is for the adductor musculature to be larger and more powerful, without tearing itself away from its anchorage on the skull bones by its own force. Kemp (1980*b*, 1982) proposed that the synapsid fenestra evolved by expansion of the connective tissue that connected the cheeks to the skull table, across the old kinetic lines (Fig. 4.2(b)). As the connective tissue expanded, it became the aponeurotic sheet covering the fenestra, and more and more adductor mandibuli muscle fibres would have attached to its under surface. Thus the relationship between the temporal fenestra and adductor musculature was established from the start. The increased force produced by the jaw-closing musculature required strengthening of the skull to resist it, and powerful braces between the braincase, quadrate region, and skull roof evolved. The supraoccipital expanded dorsally and the paroccipital processes laterally, creating the characteristic structure of the synapsid plate-like occiput.

This stage of cranial evolution had been reached by *Eothyris* (Fig. 3.2(c) and 4.2(b)), the pelycosaur with the most primitive known arrangement of the temporal region. By the more progressive, basic eupelycosaur grade represented by varanopseids (Fig. 3.4(d)), there had been additional remodelling of the bones of the temporal region of the skull. The supratemporal and postorbital bones were reduced

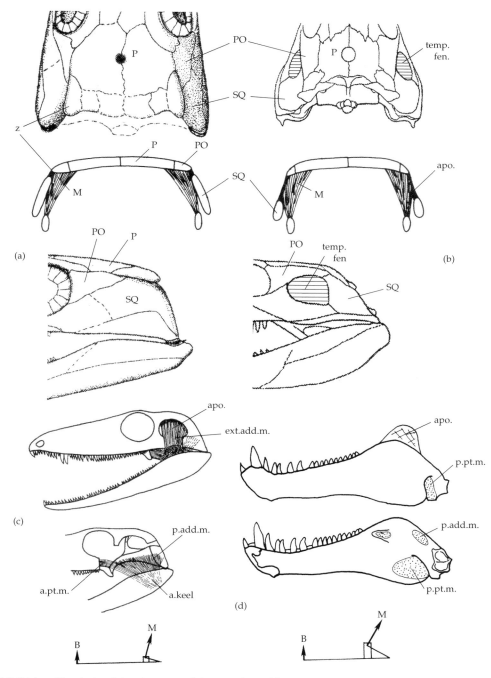

Figure 4.2 Origin and functioning of the pelycosaur-grade jaw musculature. (a) Dorsal view, coronal section and lateral view of the posterior region of an ancestral amniote skull based on *Westlothiana*. (b) The same for the basal pelycosaur *Eothyris*. (c) Lateral view of a eupelycosaur skull with the lower temporal bar cut away (top). The same to show the deeper musculature (middle). Orientation of force of adductor mandibuli muscle (M). (d) Lateral and medial views of lower jaw of the sphenacodontine *Dimetrodon*. Orientation of force of adductor mandibuli muscle (bottom) a.keel, angular keel; apo, aponeurotic sheet; a.pt.m, anterior pterygoideus muscle; ext.add.m, external adductor muscle; M, muscle; P.parietal; PO, postorbital; p.add.m, posterior pterygoideus muscle; SQ, squamosal; temp.fen, temporal fenestra; z, zone of weakness.

and the skull table as a whole narrowed. Reduction in cranial width reduced the medially directed component of the adductor mandibuli, and so increased the net bite force available.

Sphenacodontine grade
Several important mammal-like innovations are evident in *Dimetrodon* and the other sphenacodontines. To start with, the dentition had become differentiated to some extent. The anterior, incisor teeth are large and round in cross-section for grasping and possibly ripping up prey. They are followed by an enlarged canine in both upper and lower jaws for an initial disabling bite. Behind these are the smaller postcanines, laterally compressed, sharp and slightly recurved, which would have aided in retaining the still struggling prey of these top-carnivore animals.

The morphology and functioning of the jaw musculature of eupelycosaurs (Fig. 4.2(c)) has been analysed in most detail by Barghusen (1973). By analogy with modern primitive amniotes, he assumed that the main jaw-closing muscle, the adductor mandibuli, was divided into separate slips. An external adductor mandibuli originated from the edges of the temporal fenestra and from an aponeurotic sheet of connective tissue covering the fenestra. It inserted into both faces of a vertical aponeurotic sheet attached to the dorsal edge of the jaw. A posterior adductor mandibuli originated from the posterior part of the temporal fenestra and inserted into the hind region of the vertical aponeurosis and adjacent areas of the jaw. Third, an anterior pterygoideus muscle originated from the dorsal surface and posterior edge of the palate, especially from the large lateral pterygoid processes and massive quadrate rami of the pterygoid further back. Insertion of the anterior pterygoideus muscle was low down on the internal surface of the lower jaw. While this basic arrangement no doubt also occurred in more primitive pelycosaurs, important innovations in the form of processes on the lower jaw had evolved in the sphenacodontines, which increased the effectiveness of the muscle action (Fig. 4.3(d)). The dentary and surangular bones had extended dorsally to form a prominent coronoid eminence, the forerunner of the mammalian coronoid process; the angular had expanded ventrally as a large keel-like structure below the level of the jaw articulation. These processes have a dual effect. One is simply to increase the area available for the insertion of muscle. The other is to create lever arms so that the torque generated by the muscles on the jaw increased. In the case of the coronoid eminence, the adductor mandibuli muscle ran from it in a postero-dorsal direction and so had a longer lever arm with respect to the jaw articulation than if it had been attached to the dorsal edge of a jaw without a coronoid eminence. The effect is further increased by the ventral reflection of the articular bone, depressing the position of the jaw hinge. In the case of the enlarged angular keel, here a ventral extension of the lower jaw creates a larger lever arm for a muscle that runs antero-dorsally, which is the orientation of the anterior pterygoideus muscle. A curious morphological effect follows from the combination of the development of the large angular keel with the ventral deflection of the articular region of the jaw. A space exists between the keel laterally and the articular region medially. It is unencumbered below, and part of the pterygoideus muscle may have wrapped around the ventral edge of the main part of the jaw to insert within it. Although not very prominent at this stage, the space is the forerunner of the therapsid condition discussed shortly, where a much larger recess is bounded by the reflected lamina of the angular.

Given this arrangement of the major jaw-closing muscles, a further consequence arises, although not of great importance at this stage of evolution. The adductor mandibuli has a posteriorly directed component to its line of action, while the anterior pterygoideus has an anteriorly directed component. These will tend to cancel out, and therefore the tendency for the lower jaw to be pulled either backwards or forwards during the bite is reduced. Not only would this have relieved some of the stress acting at the jaw hinge, but it also endows the potential to control precisely any antero-posterior shifts of the jaw. As it happens, the jaw articulation of *Dimetrodon* indicates that there were still large stresses, and no possibility of such propalinal movement of the jaw, but this potential property was to be realised in the subsequent evolution of the mammalian jaw mechanism.

The sphenacodontian jaw articulation is complex. There are two quadrate condyles in the form of elongated cylinders. The outer one is at a higher

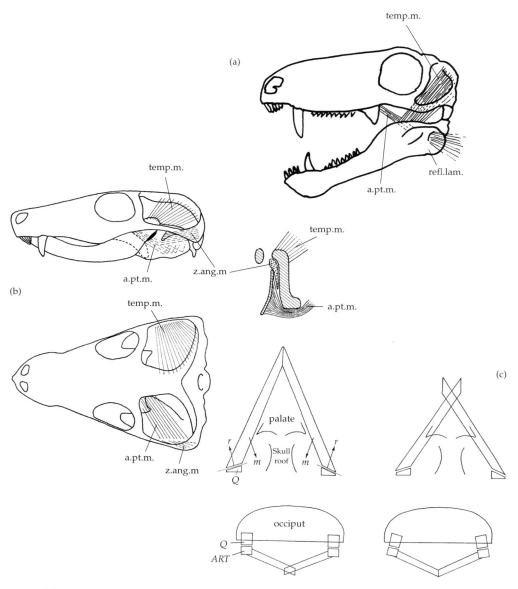

Figure 4.3 Basal therapsid jaw musculature and function. (a) Lateral view of the skull and main jaw muscles of a basal therapsid. (b) The jaw musculature of a therocephalian, based on *Therioognathus* in lateral view, dorsal view, and a coronal section throughy the zygomatic arch and reflected lamina of the angular (c). The streptostylic action of the therocephalian quadrate: top left, dorsal view of the paired lower jaws showing direction of net muscle force (m) and jaw hinge reaction (r); top right and bottom left, the effect of attempting to open the jaws without allowing quadrate movement; bottom right, the rotation of the quadrate as seen from behind that enables the jaws to open. (Kemp 1972*a*). a.pt.m, anterior pterygoideus muscle; ART, articular; Q, quadrate; refl.lam, reflected lamina of the angular; temp.m, temporalis muscle; z.ang.m, zygomatico-angularis muscle.

level in the skull than the lower. They correspond to two equally elongated glenoid fossae of the articular bone. Thus the jaw hinge was strongly built, and restricted movements to a simple orthal opening and closing of the jaw. It was also designed to resist the tendency for the lower jaw to be pulled inwards towards the midline of the skull by the medially directed net component of the muscle force.

Biarmosuchian grade

The next stage is the primitive therapsid, or biarmo-suchian grade (Fig. 3.8(a) and 4.3(a)). Relatively little innovation had occurred, but rather an exaggeration of the novelties that had evolved by the sphenacodontine-grade condition. The principle of well-developed incisors, large canines, but less-prominent postcanines was retained, but with even greater distinction between them. The canines were relatively enormous and must have been used in a kinetic fashion in which the lower jaw was accelerated from a widely open position and the kinetic energy so generated was dissipated by the teeth sinking into the prey. The reason for thinking this, apart from the size of the canines, is that the jaw adductor musculature still acted at the posterior end of the jaw, so no great static force could have been generated between teeth situated towards the front of the jaw. The incisors, in addition to being well developed, were probably capable of interdigitating, lowers between uppers, which increased the effectiveness of their biting action.

The overall mass of the adductor musculature had certainly increased considerably, as indicated by the increased size of the temporal fenestra. It extends further dorsally, ventrally, and posteriorly than in the sphenacodontine stage, thereby providing a larger area of connective sheet over it, and bony edging around it for the attachment of muscle. There is still no development of a discrete coronoid process of the dentary rising above the dorsal margin of the jaw, although the coronoid eminence is prominent. The most striking change in the lower jaw was the evolution of a full-sized reflected lamina of the angular. Instead of being merely the keel of that bone separated by a notch from the depressed articular region of the jaw as in the sphenacodontines, it has become a very large, thin sheet enclosing laterally a deep, narrow space between itself and the main body of the jaw. The exact function of this prominent structure in basal therapsids has never been fully agreed upon. Despite a number of earlier suggestions that it housed a gland, or an air-filled diverticulum associated with hearing, there is little doubt now that its primary function was for muscle insertion. In *Biarmosuchus* (Fig. 3.8(a)), Ivakhnenko's (1999) reconstruction indicates a pattern of ridges on the external surface of the lamina indicative of

strengthening against muscle-generated forces. The most plausible general explanation for the form of the reflected lamina is that it served as the insertion site for a complex set of muscles running in a variety of different directions. At the biarmosuchian grade, this may have included tongue and hyoid musculature inserted on the lower part of the lamina, and jaw-opening musculature running from the posterior region backwards towards the shoulder girdle region. In the more advanced carnivorous groups, gorgonopsians, and therocephalians, the strengthening ridges are more strongly developed, although in quite different patterns respectively. Certainly in gorgonopsians and possibly in therocephalians, there is evidence for an invasion of part of the reflected lamina by adductor mandibuli musculature. However, this could not have been true of biarmosuchians because their zygomatic arch was not expanded laterally to create room for the attachment of such a muscle.

The jaw articulation of *Biarmosuchus* is a simple roller and groove system, restricting the jaw to strictly orthal movements. Thus, the biting mechanism at this stage was little more sophisticated, although more powerful than in pelycosaurs, with a disabling bite delivered by the canines, a tearing action using the incisors, and a finer tearing, or food retention by the modest-sized postcanines.

Therocephalian grade

The next definable stage is the therocephalian grade (Fig. 4.3(b)), in which the principal innovation is enlargement of the temporal fenestra by extreme medial extension. This has occurred to such an extent that the intertemporal region of the skull roof is reduced to a narrow girder, the sagittal crest. The midline of the crest tends to be sharp and the sides face dorso-laterally, providing an area for the origin of the innermost part of the adductor mandibuli musculature. This area of muscle attachment continues smoothly round on to the front face of the occiput, while the almost horizontal temporal fenestra, covered by its connective tissue sheet, gives further origin for the muscle. By this stage, a large part of the adductor musculature is distinguishable as a medial component, namely a true temporalis muscle. A coronoid process of the dentary had evolved, as a postero-dorsal extension of the dentary bone above the level of the rest of the jaw, for increasing the attachment area of this newly enlarged temporalis muscle. The masseter muscle is the lateral component of the adductor mandibuli found in later evolutionary stages, where it takes its origin from an expanded zygomatic arch bounding the temporal fenestra laterally. It is possible that there was an incipient masseter muscle in therocephalians, but it would have been restricted to the posterior root of the zygomatic arch, and have inserted on part of the reflected lamina of the angular (Kemp 1972*a*). Even if it did exist, it could not have had an important functional role at this stage.

A consequence of the very considerable increase in the size of the medial and posterior parts of the jaw-closing musculature is that there was a corresponding increase in the postero-medially directed component of its net force, that required a remarkable specialisation of the jaw articulation with far-reaching consequences (Kemp 1972*a*, 1982). The articular bone of the lower jaw faces postero-medially and articulates with an antero-laterally facing condyle of the quadrate (Fig. 4.3(c)). This geometry resists to the tendency of the back of the jaw to be disarticulated by the postero-medial component of the jaw-closing force pulling it inwards. But to achieve this effect, the axis of the hinge joint has to be aligned obliquely to the transverse axis of the skull. As the right and left lower jaws are firmly attached at the symphysis, it would be impossible for the jaws to open without disarticulation between articulars and quadrates unless the latter were capable of adjusting their orientation. The nature of the attachment between the quadrate and the squamosal bone against which it lies allows exactly this to happen. The two are not sutured and the contacting surfaces are smooth and covered with cartilage in life, forming a mobile joint. The quadrate could rotate about a longitudinal axis which allowed the necessary adjustments during jaw opening and closing for the articulating surfaces of the articular and quadrate to remain fully in contact. Improbable as such a mechanism might seem at first sight, not only is it an intelligible mechanical adaptation for a therapsid whose jaw adductor musculature is concentrated internal and posterior to its insertion on the jaw, but it also helps to explain the extraordinary history of the jaw articulation

that was to follow from this stage. It was, indeed, this evolution of quadrate mobility that permitted the eventual evolution of the mammalian jaw and hearing mechanisms.

Procynosuchian grade

Within the cynodont grades that follow, all the mammalian features of the feeding mechanism reach their full expression step by step, some as further development of the incipient characters seen so far, others as innovations. The procynosuchian grade was a significant advance in a number of features. The dentition achieved the mammalian pattern of heterodonty (Fig. 3.19(a)). Simple incisors are followed by canines as in earlier stages. However the postcanine dentition has become the main region for food manipulation, with around a dozen teeth, divided into a series of simpler premolariform teeth with little more than a swelling around the base of the main cusp, and a series of more complex molariform teeth behind. In the latter there is a large main cusp and a lingual cingulum of small, but distinct accessory cusps around the base; for the first time, incipiently multicusped teeth had evolved, teeth that ultimately were to evolve into the array of highly elaborate, precisely occluding teeth that characterise the mammalian radiation. At this stage, there was no direct, precise occlusion between upper and lower postcanines, although the nature of the wear facets indicates that there was general, abrading contact. The function of the incisors and canines was to seize and ingest insects and other small prey. The postcanines, particularly the molariforms, masticated the food prior to swallowing. The effect of the accessory cuspules of the molariform teeth would have been particularly important when dealing with tough insect cuticle (Fig. 3.19(c)). As opposing teeth entered the material, they would become anchored by the cingulum, after which further relative movement between opposing teeth would tend to cause tearing of the stiff cuticle (Kemp 1979).

Use of a dentition in this way requires a larger bite force than does crude disablement and tearing of prey with the anterior teeth, and it was provided by further increase in the size and effectiveness of the adductor musculature (Fig. 4.4(a)). The medial expansion of the temporal fenestra characteristic of

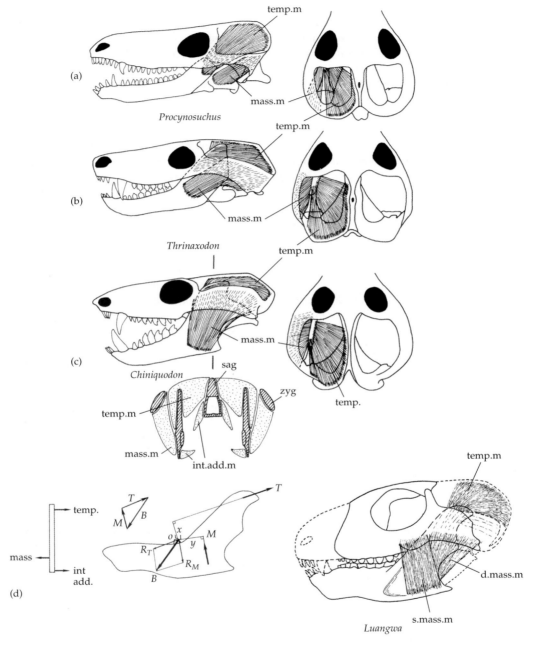

Figure 4.4 The jaw musculature of cynodonts. (a) Lateral and dorsal views of the skull and jaw musculature of *Procynosuchus*. (b) the same for *Thrinaxodon*. (c) the same for *Chiniquodon* (*Probelesodon*), with a coronal section through the temporal region. (d) Lateral view of the jaw musculature of *Luangwa*, with outline of the lower jaw showing the vectors of the masseter muscle (M), temporalis muscle (T), and reaction force generated at the teeth (B). The triangular of forces of these three is shown, and at the left, the balance in a transverse direction achieved by the temporalis muscle (temp) plus internal adductor muscle (int.add) acting medially and the masseter (mass) muscle acting laterally. (Kemp 1980a; 1982). d.mass.m, deep masseter muscle; int.add.m, internal adductor muscle; mass.m, masseter muscle; sag, sagittal crest; s.mass. m, superficial masseter muscle; zyg, zygomatic arch.

the therocephalian grade was not only retained but the face of the sagittal crest became deeper, increasing the area for the origin of the temporalis muscle. The posterior part of the zygomatic arch is also deeper and its internal face indicates that it gave origin to a small, but definite masseter muscle. The principal innovation of the musculature at this stage was the invasion of the lateral surface of the lower jaw by adductor musculature. At the therocephalian grade, at best only a very small fraction of the total muscle mass had gained an external insertion on the jaw. In *Procynosuchus*, the temporalis had extended its insertion to a shallow fossa, the adductor fossa, which occupies part of the external surface of the broadened coronoid process. There is a similarly broad, shallow concavity occupying most of the external face of the angular bone onto which the enlarged masseter inserted. The increased overall size of the adductor musculature increased the magnitude of the potential bite force available. Moreover, the particular way in which it was arranged introduced yet another highly significant property. The temporalis had a medially directed component as in therocephalians, but now there was also a masseter muscle, running between the external face of the jaw and the more laterally placed zygomatic arch, that had a laterally directed component of its force. To some degree at least, these respective medial and lateral components acting on the jaw would have cancelled each other out, reducing the net tendency of the back of the jaw to be forced medially, and therefore reducing the net stress between the articular and quadrate bones of the jaw hinge. Furthermore, this arrangement increased the potential ability to control movements of the lower jaw in the transverse direction. Insofar as even primitively multicusped teeth like those of *Procynosuchus* required accurate placing of the lower teeth relative to the upper teeth during closure, an increase in the control of the precise positioning of the lower jaw during mastication was necessary. By evolving the masseter muscle laterally as well as increasing the temporalis muscle medially, the procynosuchian grade saw the initiation of an essential requirement of musculature for operating large, complex, occluding teeth: a combination of a large bite force with fine precision of movement of the lower jaw in the horizontal plane.

Two other points about the jaw musculature should be noted. The first is that the reflected lamina of the angular is greatly reduced in size, and could no longer have supported significant jaw musculature. Any tongue, hyoid, or jaw-opening muscles that may have inserted on the reflected lamina in the therocephalian grade had presumably shifted to the main body of the angular and adjacent bones. There is a longitudinal groove low down on the inner face of the angular of *Procynosuchus*, which probably marks part of the attachment area for such musculature. The second point is that the pterygoid processes of the palate were further reduced and by now the pterygoideus musculature would have been involved in fine control of jaw movements, rather than contributing significantly to the production of bite forces.

The structure of the jaw articulation of the procynosuchian grade corresponds to the new arrangement of the adductor muscles. As in the therocephalian grade, the axis of the joint is still oblique rather than transverse, indicating that the net adductor muscle force still had a medially directed component. However, the angle of the jaw hinge to the transverse is less, indicating that the net medially directed component of the jaw-closing force had been reduced. While the quadrate retained its ability to rotate in order to maintain the patency of the jaw hinge, the movement was less extensive (Kemp 1979; Luo and Crompton 1994).

Basal epicynodont grade
The next cynodont grade is represented by *Thrinaxodon* (Fig. 4.4(b)), in which there are differences in degree rather than innovations compared to the procynosuchian grade. The temporal fenestra is larger, having expanded posteriorly and laterally, and the sagittal crest deeper. Therefore, there is attachment for a much larger temporalis muscle from the medial and posterior walls of the fenestra as well as from the connective tissue aponeurosis covering it. The corresponding increase in area for the muscle's insertion on the lower jaw comes from a large expansion of the coronoid process, which now extends up into the temporal fenestra, almost to the height of the skull itself; the medial and lateral surfaces of the process were both available for the insertion. The increase in size of the masseter

muscle was even more impressive. The zygomatic arch has expanded to form a broad, dorsally arched sheet of bone providing origin for masseter muscle fibres all along its length. The more posterior part of the masseter muscle still inserted onto the external surface of the angular bone. By this stage, however, the dentary had expanded ventrally to form a large angular region. The greater part of the masseter muscle inserted on this new region of the dentary, as indicated by the smooth, shallow masseteric fossa occupying most of its surface. The functional effect of the continued evolution of the adductor musculature was threefold. First, the magnitude of the laterally directed component of the force generated by the masseter muscle more closely matched the medially directed component of the temporalis, and therefore a closer approach to an actual balance between these two was achieved. Second, the increasingly horizontal line of action of the temporalis muscle from its area of origin high up in the coronoid process, and the increasingly vertical orientation of the masseter muscle as it expanded forwards along the zygomatic arch, together created a tendency to concentrate the net adductor muscle force on the teeth, and away from the jaw articulation. This is better explained in the context of the next evolutionary stage, the eucynodont grade, where the effect had become fully expressed. The third consequence arose from the capture by the enlarged dentary of an increased proportion of the total jaw muscle fibres, leaving very few still inserting on the reduced postdentary bones. Stresses across the sutures between the dentary on the one hand and the attached postdentary bones on the other were reduced, permitting the latter to be less firmly attached.

Apart from these important quantitative modifications to the jaw musculature, little else of great significance occurred at the *Thrinaxodon* stage. The one that should be mentioned is a minor technicality at this stage. The surangular bone, which lies immediately lateral to and above the articular bone of the jaw hinge, has developed a small boss at the back which faces towards, but does not quite contact a process on the squamosal bone adjacent to the quadratojugal bone. At this stage, a ligamentous connection occurred, helping to stabilise the jaw hinge and relieve some of the stress on the quadrate

bone (Crompton 1972b; Crompton and Hylander 1986). The significance of this new feature is that it is actually the incipient forerunner of what was eventually to become the new, mammalian jaw hinge. Apart from this detail, the jaw articulation of *Thrinaxodon* differs little from *Procynosuchus* except for a relative reduction in size of the participating bones (Luo and Crompton 1994).

Compared to *Procynosuchus*, there is a reduced number of teeth in *Thrinaxodon*, which has only four upper and three lower incisors, and around eight postcanines. The anteriormost and posteriormost cuspules of the cingulum of the molariform teeth are enlarged to the extent that they can be described as accessory cusps. This elaboration is no doubt correlated with the greater force and accuracy of the available bite. Although there is a range of molar tooth-form in basal epicynodonts, that of *Thrinaxodon* in particular resembles several eucynodont and early mammal groups, so it is probably a good model for this hypothetical ancestral stage.

Eucynodont grade
The eucynodont grade is characterised by completion of the trend towards balancing the adductor muscle forces so that the net force is concentrated at the point of bite and the stress at the jaw hinge reduced, theoretically to zero. A form such as *Chiniquodon* (=*Probelesodon*) illustrates this (Fig. 4.4(c)). The dentary has enlarged to the extent that it now completely dominates the lower jaw. The coronoid process reaches the level of the top of the temporal fenestra, and the angular region below has also expanded, enlarging yet further the available area for insertion of the masseter muscle. The dentary has also for the first time expanded posteriorly to form an articular process that overlies and supports the relatively minute postdentary bones. While impossible to be sure, it is likely that by this stage the magnitude of the laterally directed component of the masseter muscle force generated during the bite equalled the magnitude of the medially directed component of the temporalis so that the two virtually cancel out. Therefore, any tendency for the lower jaw to be forced inwards or outwards was abolished, and there was zero transversely directed stress generated between the articular and the quadrate at the jaw articulation.

A balance was also achieved in the sagittal plane (Fig. 4.4(d)). The lines of action in the plane of the lower jaw of the two main muscles were arranged geometrically so that the total force generated by them was concentrated on the posterior teeth, with zero reaction force at the hinge. Bramble (1978) explained this by means of the principle of the triangle of forces, which states that if the vectors of three forces acting on an object form a closed triangle, that object is in equilibrium. Relative to the skull roof, the vector of the temporalis muscle of a eucynodont was oriented horizontally backwards, and that of the masseter was oriented vertically upwards. Imagine the jaws were clamped on a piece of food held between posterior teeth. If the force applied by the teeth to the food was oriented antero-ventrally, which it perfectly well could have been, then this reaction force completes a triangle of three vectors, and the lower jaw would be in equilibrium. No other forces acted on the jaw, not even a reaction force at its hind end, visualised perhaps by noting that the back of the jaw tended to move neither upwards nor downwards. All the force generated by the two muscles was applied as a reaction force between upper and lower teeth, namely the bite force, and any item of food held between the teeth at this point would have suffered a very strong bite. Support for this theory about the muscle reorganisation comes from the specialised herbivorous postcanine teeth of traversodontid eucynodonts. The basined occlusal surfaces of the upper teeth face antero-ventrally towards the postero-dorsally facing basins of the lowers (Fig. 3.22(d)). The most effective direction for the bite force between them was therefore antero-ventral to dorso-lateral, as predicted from the reconstructed muscle arrangement (Kemp 1980a). Furthermore, wear facets on the opposing crests of the teeth indicate that the power stroke of the lower jaw included a slight backward movement, which is another expectation from the theory. Crompton and Hylander (1986) were able to use the model to explain a number of variations in different eucynodonts. In *Probainognathus* (Fig. 3.21(c)), the postcanine dentition is not well developed, and most of the active biting must have involved food capture by the incisors and canines. By shifting the insertion of part of the masseter and pterygoideus muscles forwards, and part of the temporalis muscle

ventrally, the hinge reaction can again be more or less abolished. At the other extreme, the exceptionally high coronoid process of tritylodontids (Fig. 3.23) allowed the muscles to be arranged so that the hinge reaction was abolished under the conditions of a very powerful bite acting towards the hind end of the tooth row.

Thus the lower jaw of the eucynodont grade simultaneously achieved two ends. There was a balance between the components of the muscle forces acting in the horizontal plane, and so no tendency for the jaw to be displaced antero-posteriorly or medio-laterally. Therefore, only very minute modifications to the pattern of muscle activity were needed in order to make small, finely controlled adjustments to the horizontal position of the jaw and lower teeth relative to the uppers. Second, the muscle forces were concentrated at the point of occlusion between the teeth, so bite forces were very large. These are the twin requirements of a muscle system designed for precise occlusion between large, multicusped teeth.

The further reduction of the relative size of the postdentary bones and accompanying quadrate, reflects this removal of significant stresses between the dentary and postdentary bone complex, and between the articular and the quadrate at the jaw hinge. By the eucynodont stage, the maximum reaction forces which this part of the skull could withstand, must have been exceedingly small. Even the development of the secondary bony contact between surangular and squamosal alongside the articular-quadrate hinge added little strength. However, there is an apparent anomaly to be explained. Reorientation of the adductor musculature to concentrate the force at the teeth required an overall change in the silhouette of the lower jaw, but this could have been achieved equally well without altering the relative sizes of the dentary and postdentary bones. Simply strengthening the sutural attachments between the individual jaw bones could have occurred, as suffices in other tetrapods with a powerful bite, such as turtles, and even other therapsids, namely dicynodonts. A different reason for the spectacular reduction of the postdentary bones must be sought, and is found when the evolution of the mammalian hearing mechanism is considered: the reptilian jaw hinge bones became the mammalian sound transmitting ear ossicles, as explained later in the chapter.

Given the general similarity between the teeth of *Thrinaxodon*, the eucynodont *Probainognathus*, the tritheledontan *Brasilitherium*, and the basal mammal *Morganucon*, it is safe to assume that the postcanine teeth of the eucynodont grade hypothetical ancestor would have been triconodont, with a main cusp flanked by smaller anterior and posterior accessory cusps, and an internal cingulum of cuspules.

Tritylodontid grade

The taxonomic uncertainty about the interrelationships of tritylodontids, tritheledontans, and mammals has been discussed, and this spills over into ambiguity about the details of the next stages towards the fully mammalian condition that can be inferred from the fossils. The only useful contribution to the sequence of evolutionary events illustrated by the tritylodontids is the inference that loss of the postorbital bar preceded the development of the new jaw articulation between the dentary and the squamosal. Otherwise, the extreme specialisation of the jaw and dentition of tritylodontids obscures any other possible innovations that appeared at this evolutionary stage.

Tritheledontan grade

The tritheledontids (Fig. 3.24) represent the stage at which the new jaw articulation occurred. It had evolved as a consequence of an increase in length of the articular process of the dentary to the point where it invaded and replaced the pre-existing contact between the surangular bone and the squamosal, and lies immediately alongside the primary hinge between articular and quadrate. At this stage the contact between dentary and squamosal is simple, there being no development of a prominent dentary condyle or squamosal glenoid fossa. The function was mainly to free yet further the primary hinge bones for their evolving hearing function. But the new arrangement may also have increased the degree of fine control of horizontal position of the jaw. Wear facets on the postcanine teeth of *Pachygenelus* indicate that a significant change in the occlusal mechanism had occurred; the outer sides of the lower teeth wear against the inner sides of the uppers. The only way this could happen is by a slight initial lateral shift of

the lower tooth row, placing it below the upper tooth row. As the jaw closed and the teeth met, the lower teeth moved upwards but also shifted slightly medially during the tooth-to-tooth contact. Given this geometry, occlusal biting could only occur on one side at a time, right or left, but not both simultaneously. At this stage the postcanine teeth lack precise patterns of tooth wear, indicating that the occlusal was a general contact between opposing tooth rows, rather than the precise action of a specific lower tooth with a specific upper tooth that characterises mammals.

Mammalian grade

Morganucodon represents the ancestral mammalian stage best (Fig. 5.3(b)). The principal innovation is the evolution of a definite ball-like condyle on the dentary that articulates with a well-defined glenoid cavity in the squamosal, alongside but functionally replacing the articular-quadrate hinge. This is at first sight paradoxical. The eucynodont jaw articulation was exceedingly weak, but did not have to bear large reaction forces. Why now replace it with a new articulation that promptly evolved into a robust structure capable once again of withstanding large forces? Crompton and Hylander (1986) answered the question by pointing out that in primitive mammals only one side of the dentition is used at a time during mastication, as in tritheledontans. The difference in the mammals is that the adductor jaw musculature on both sides of the head contributes simultaneously to the bite force between the teeth on one side. However, while the ipsilateral musculature can in principle behave as described in eucynodont-grade skulls and avoid generating a jaw hinge reaction, the contralateral muscles cannot do so. They can only contribute their muscle force to the bite force of the teeth of the opposite side by transmitting it via a firm symphysis between the front ends of the respective jaws, plus a jaw articulation strong enough to withstand a significant component of the force. The benefit of the arrangement is a yet greater bite force, and therefore this final evolutionary innovation of a strengthened jaw articulation was the last step in achieving not only a combination of extremely large, but also extremely accurately applied, bite forces.

The final fate of the primary jaw hinge bones and other postdentary bones is discussed in the context of the evolution of mammalian hearing.

Locomotion

Ancestral amniote grade

Amniote locomotion is compounded from several separate movements of the limbs and body. Lateral undulation of the vertebral column contributes a significant fraction of the overall stride length: as waves of contraction pass down alternate sides of the body, the limbs are passively protracted and retracted. Added to this, the limbs actively protract and retract relative to the vertebral column, which increases the length of the stride. The transversely oriented humerus and femur also undergo rotation about their long axes, which has the effect of shifting the foot forwards and backwards relative to the body. Finally, the limbs as a whole are extensible struts so the stride can be increased yet further by extension at the joints once the foot is behind the level of the limb girdle. The change from this design to that of mammals involved altering the relative contributions of these four elements of the stride. Lateral undulation and long-axis rotation of the propodials were lost, while active retraction–protraction, and extension of the limbs were retained and developed. Additionally, two new elements were added: movement of the shoulder girdle on the ribcage, and dorso-ventral bending of the lumbar region of the vertebral column. The reconstructions of the hypothetical ancestral stages of mammal-like reptiles illustrate much of how and by inference why this radical remodelling of the locomotor system occurred.

No understanding of the functioning of locomotor systems, or the transition from primitive amniote to mammalian is possible without appreciating that in all non-specialised tetrapods the function of the forelimb differs in important respects from the hindlimb, which accounts for several differences in their design. The forelimb is primarily to maintain the front of the animal off the ground during locomotion, and produces virtually no net locomotor force, analogous to the wheel of a wheelbarrow. The humerus has a simple, predetermined stride pattern and there is about the same volume of protractor as retractor musculature. It is the hindlimb that generates the necessary thrust. Therefore, the hindlimb is the larger, movement of the femur is much less constrained, and the retractor musculature is far larger than the protractor musculature.

Sphenacodontine grade

As reviewed by Kemp (1982), Romer's (1922) classic reconstruction of the musculature and locomotor mechanism of the pelycosaur *Dimetrodon* is still the basis for the sphenacodontine-grade ancestor, although subsequent studies, especially of the joints, have added further detail.

Vertebral column. A significant modification of the vertebral column towards the eventual mammalian condition occurred, even at this early stage in which otherwise the locomotor apparatus is little modified from the ancestral amniote. The articulating surfaces of the zygapophyses of adjacent vertebrae are no longer horizontal, but oblique in orientation. Even in more primitive pelycosaurs, such as ophiacodontids, the prezygapophyses face inwards at an angle of about 30° from the horizontal, while in *Dimetrodon* the angle is increased to about 45°. This indicates that the lateral undulation component of locomotion was reduced in pelycosaurs, and probably virtually abolished in the latter genus. The small size of the intercentra, small bones between the ventral margins of adjacent vertebrae, also relates to the reduction of mobility between adjacent vertebrae, and presumably the loss of lateral undulation permitted certain pelycosaurs to evolve hugely long neural spines. Otherwise, the axial skeleton (Fig. 4.5(a)) has changed little from a primitive amniote condition. Moveably attached and ventrally directed ribs occur all the way back to the pelvic region, with no abrupt distinction between dorsal and lumbar regions, indicating that prevention of the body from sagging while walking still depended on the intercostal muscles and tendons between the ribs, rather than on the specialised lumbar musculature that evolved later.

Forelimb. The action of the forelimb of pelycosaurs was heavily constrained by the design of the shoulder and elbow joints. The pectoral girdle (Fig. 4.5(b)) is a massive structure. The clavicles and interclavicle together form a U-shaped arch around the thorax that would have prevented any extensive movements of the shoulder girdle relative to the ribcage. The scapula is very broad, and the coracoid plus procoracoid bones together form a broad ventral plate. The glenoid fossa is elongated from front to back and its articulatory surface is described as

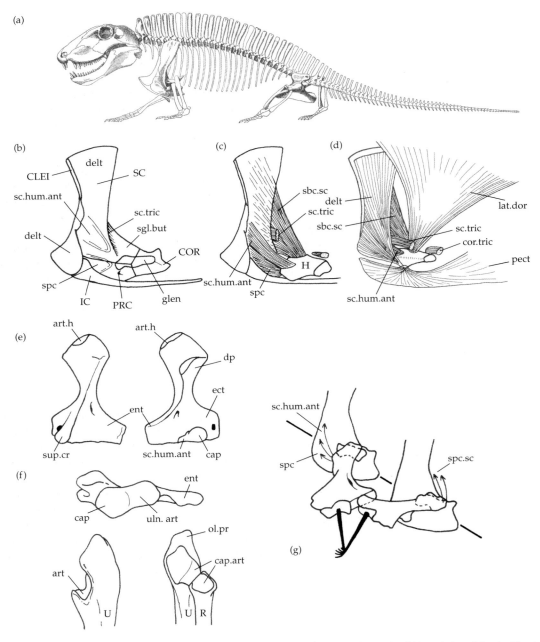

Figure 4.5 Pelycosaur-grade forelimb. (a) Lateral view of the sphenacodontine pelycosaur *Sphenacodon*. (b) Lateral view of the shoulder girdle of *Dimetrodon*. (c) The same with deep shoulder musculature. (d) The same with superficial shoulder musculature. (e) Dorsal, ventral and distal views of the left humerus, and (below) anterior and medial views of the upper part of the left ulna and radius. (g) The position of the humerus at the start and completion of a stride, in postero-lateral view. Kemp 1982, after Romer and Price 1940 and Jenkins 1971*b*. art, articulating surface; art.h, articular head; cap.art, articulating surface for capitulum of humerus; CLEI, cleithrum; COR, coracoid; cor.tri, coracoid slip of triceps muscle; delt, deltoideus muscle; dp, deltopectoral crest; ect, ectepicondyle; ent, entepicondyle; glen, glenoid fossa; IC, interclavicle; lat.dor, latissimus dorsi muscle; ol.pr, olecranon process; pect, pectoralis muscle; PRC, procoracoid; sbc.sc, subcoraco-scapularis muscle; SC, scapular; sc.hum.ant, scapulo-humeralis anterior muscle; sc.tric, scapular slip of the triceps muscle; sgl.but, supraglenoid buttress; spc, supracoracoideus muscle; sup.cr, supinator crest; uln.art, articulation for the ulna.

screw-shaped: the front part faces backwards and downwards, the middle part faces directly outwards, and the back part faces forwards and upwards. This shape matches that of the articulating area on the head of the humerus (Fig. 4.5(e)), which occupies the end of the expanded proximal region of the bone. It too is elongated from front to back and is in the form of a spiral ribbon. At the front it faces antero-dorsally, in the centre medially, and at the hind end postero-ventrally. The movement of the humerus in the glenoid fossa (Fig. 4.5(g)) was wholly constrained to a single track (Jenkins 1971a). At the start of an active stride, the spiral of the humerus head fitted completely into the screw-shaped glenoid. The humerus extended laterally and its distal end was elevated and faced antero-ventrally. This caused the lower leg to extend forwards and forefoot to be placed on the ground a little in front of the rest of the limb. As the humerus was retracted from this position, the anterior part of the head passed forwards beyond the confines of the glenoid. As retraction continued, the shape of the middle and posterior parts of the articulating surfaces forced the distal end of the humerus to lower and the humerus as a whole to rotate about its long axis so that the lower limb comes to project postero-ventrally.

The movements and stresses that had to be accommodated by the rest of the limb during the locomotory cycle were complex, and explain several features of the design of the sphenacodontine grade forelimb that had implications for its subsequent evolution. The elbow joint (Fig. 4.5(f)) must act as a hinge joint so that the lower leg can extend and flex on the end of the humerus. Also, if the foot is to remain stationery on the ground while the humerus is retracted in a horizontal plane, then a relative rotation between the foot and the distal end of the humerus about a vertical axis must be accommodated via the lower leg. Third, with rotation of the humerus about its long axis playing a role in the stride, torsional stress has to be resisted by the elbow joint. It is a principle of vertebrate skeletal design that no single joint can have too many degrees of freedom, and so for complex, multiple movements, several associated joints, each designed to control one or at most two specific movements, tend to evolve. Nowhere is this principle better

illustrated than in the relationship between the distal end of the humerus and the foot in sprawling-gaited tetrapods. The two epipodial bones, radius and ulna, participate in four joints altogether, two at the elbow and two at the wrist, and these joints have differing functions. At the elbow, the radius has a concave surface that articulates with the ventrally facing, hemispherical capitulum of the humerus. The function of this joint is to control the rotation of the humerus relative to the forefoot. It is capable of passively accommodating an applied hinging movement but is not designed to control it, and it provided little resistance to the torsion between humerus and lower leg. The ulna, on the other hand, is designed to control the extension–flexion movements at the elbow, and also to transmit the torsion stress from the humerus to the lower leg, when the humerus was rotating about its long axis. To achieve these two functions, the articulating surface of the ulna is in the form of a deep sigmoid notch into which fits the articulating facet of the humerus. This joint is, however, incapable of accommodating the rotation of the ulna about its long axis on the humerus. Instead, the rotation of the ulna occurs at the joint it makes with the forefoot. Thus, as far as the rotation is concerned, the radius and ulna behave independently of one another. The radius rotates at the top on the humerus; the ulna rotates at the bottom on the forefoot. This is a strong arrangement, resistant to disarticulation during powerful locomotory activity. The forefoot has a large number of separate ossifications indicating a general flexibility rather than precise functions at specific joints, and the digital formula is still the primitive amniote 2-3-4-5-3.

The musculature of the forelimb (Fig. 4.5(c) and (d)) has been reconstructed on the basis of comparison with living primitive amniotes along with the morphology of the bones (Romer 1922). The main depressor, or adductor muscle complex was the pectoralis, originating on the massive ventral parts of the shoulder girdle, interclavicle, sternum, and clavicle, and inserting on the huge delto-pectoral crest of the humerus. Retraction of the limb was brought about by the subcoraco-scapularis, a muscle complex originating on the inner face of the scapulo-coracoid and emerging behind to insert on the hind part of the humerus head. Its action pulled the back part of the humerus head inwards and forwards, which imparted the anterior shift of the head of the humerus in the glenoid and therefore the posterior retraction movement of the shaft of the bone. These two muscles were primarily responsible for the power stroke of the forelimb. The accompanying rotation of the humerus about its long axis was caused by the exact points of insertion of the muscles. The pectoralis inserts in front of the line of the axis of the humerus; the subcoraco-scapularis inserts behind that line. Both therefore impart an anticlockwise rotation, as viewed from the left. The final component of the power stroke, extension of the limb at the elbow, was the result of contraction of the triceps muscle that ran from the girdle and dorsal surface of the humerus to an insertion on the olecranon process of the ulna.

The recovery phase of the limb cycle required flexion of the elbow by means of the biceps muscle from the coracoid to the flexor surface of the radius and ulna. Elevation of the humerus resulted from the contraction of several muscles running dorsally from the humerus. The deltoideus connected the clavicle and dorsal edge of the scapula to the inner end of the delto-pectoral crest of the humerus, and the latissimus dorsi was a broad sheet of muscle from the tissue facia of the sides and back of the animal to a transverse ridge distal to the head of the humerus. The main protractor muscle was the supracoracoideus, originating from the lateral face of the coracoid and lower part of the scapula. Its insertion was on the anterior part of the head of the humerus and by imparting an inwardly directed force to the head, it caused the head to move posteriorly and the shaft to protract anteriorly, exactly the reverse of the effect of subcoraco-scapularis during retraction. Again, the rotation of the humerus about its long axis, this time clockwise as viewed from the left, resulted from the points of attachment of the recovery-phase muscles relative to the line of axis of the bone.

Hindlimb. The ilium, pubis, and ischium are comparable to one another in size and together constitute a broad, plate-like pelvis with the acetabulum occupying the middle (Fig. 4.6(a)). This is in the form of a simple, quite shallow concavity into which the convex articulating surface on the proximal end of the femur fits. The femur (Fig. 4.6(c)) was constrained to a mainly horizontal plane, but within that, its

movement was quite free. It could undergo different combinations of protraction and retraction, elevation and depression, and rotation about its long axis. As with the equivalent forelimb bones, the tibia and fibula of the hindlimb participated in four joints to control and accommodate the various relative movements between the distal end of the femur and the foot, although the details are quite different. The knee joint (Fig. 4.6(c)–(e)) was a very strongly built hinge joint, by virtue of the wide articulating head of the tibia connecting to both of the two articulating condyles on the ventral side of the end of the femur. In contrast, the fibula is a far more slender bone whose head articulates only with the side of one of the femoral condyles. Most of the hinging action at the ankle was controlled by the articulation of the lower end of the fibula with both the proximal ankle bones, calcaneum, and astragalus. The tibia, so broadly expanded proximally, has a relatively slender distal end, which only articulates with the side of the astragalus. The relative rotational movement between the end of the femur and the foot occurred by rotation between both the tibia and the fibula independently of each other on their articulations with the underside of the femur (Fig. 4.6(d)). The importance of this arrangement may have been to restrict the ankle joint exclusively to a hinging action rather than allowing rotation as well, which was better suited to transmitting large forces to the ground during locomotion. The pes is interpreted as digitigrade, with the large astragalus and calcaneum held vertically and the several tarsal bones and digits spreading the weight of the animal over the ground (Fig. 4.6(e)). As with the forefoot, there was considerable flexibility between the bones, and the primitive hind foot digital formula of 2-3-4-5-4 was retained.

The musculature of the hindlimb (Fig. 4.6(b)) was quite simple in principle, with four major muscles for each of the four major movements of the femur at the hip joint. The pubo-ischio-femoralis externus ran from the outer surface of the pubo-ischiadic plate to the ventral side of the femur and caused adduction. The huge caudi femoralis muscle was inserted on to the fourth trochanter on the underside of the femur, and extended the entire length of the tail. This, the largest muscle in the primitive tetrapod's body, caused the powerful retraction of the femur that provided the main locomotory force.

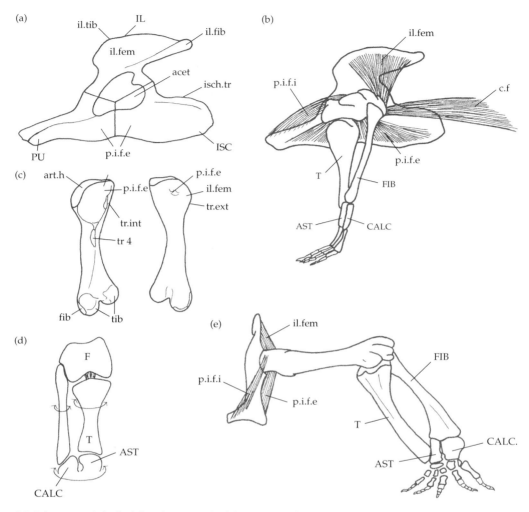

Figure 4.6 Pelycosaur-grade hindlimb, based on *Dimetrodon*. (a) Lateral view of the left pelvic girdle. (b) lateral view of pelvis and hindlimb, with the four major muscles reconstructed. (c) Ventral and dorsal views of left femur. (d) distal view of the femur, crus, and proximal tarsus to show rotation action of tibia and fibula. (e) Anterior view of left hindlimb. acet, acetabulum; art.h, articulating head of femur; AST, astragalus, c.f, caudi femoralis muscle; CALC, calcaneum; F, femur; FIB, fibula; fib, articulating facet for fibula; IL, ilium; il.fem, ilio.femoralis muscle; il.fib, origin for ilio-fibularis muscle; il.tib, origin for ilio-tibialis muscle; ISC, ischium; isch.tr, origin for ischio-trochantericus muscle;p.i.f.e, pubo-ischio femoralis externus muscle; p.i.f.i, pubo-ischio femoralis internus muscle; PU, pubis; T, tibia; tib, articulating surface for tibia. tr.ext, trochanter externus; tr.int, trochanter internus; tr.4, fourth trochanter (Kemp 1982).

For the recovery phase, the relatively modest ilio-femoralis muscle, running between the posteriorly directed ilium and the dorsal surface of the femur, elevated the limb, while the pubo-ischio-femoralis internus was the protractor. It ran forwards from the inner face of the pubo-ischiadic plate, wrapped around the front edge of the pubis, and turned backwards to insert on the front of the femur. As in the case of the forelimb, rotation of the femur about its long axis was due to the exact points of attach-

ment of these four muscles on the bone relative to the long axis. Also as in the forelimb, there were strong extensor muscles, particularly the triceps, attaching to the outer parts of the tibia and fibula. A complex of antagonistic flexor muscles attached to the flexor surfaces of these bones.

Basal therapsid grade
Sphenacodontine locomotion is characterised by massive limb girdles and short, heavy, sprawling

limbs, and was doubtless slow and clumsy by modern standards. Nevertheless it was the starting point for the evolution of what became the agile, long-limbed parasagittal gait seen in typical mammals. Indeed, one hint of the mammalian design has already been seen in pelycosaurs, namely the reduction of the lateral undulation of the vertebral column. However, even in the most basal of therapsids, numerous significant new developments in the direction of the mammals are to be found. Ideally at this point the postcranial skeleton of the biarmosuchian grade should be described and interpreted, but unfortunately it is very poorly known. Sigogneau and Chudinov (1972) and Sigogneau-Russell (1989*b*) have described several isolated postcranial bones of *Biarmosuchus*, as has Boonstra (1965) for the South African *Hipposaurus*. As far as it is known, it shares a general similarity to the far better understood gorgonopsian skeleton. Several complete and many partial specimens of the latter taxon have been described (Sigogneau-Russell 1989) and aspects of the functional anatomy considered (Colbert 1948; Kemp 1982; Sues 1986*a*). Furthermore, they are not too dissimilar from a basal, brithopian dinocephalian such as *Titanophoneus*. Taken together, these groups can be combined to create a general picture of the structure and functioning of the primitive therapsid-grade postcranial skeleton.

Axial skeleton. Considering first the axial skeleton, the very attachment of the vertebral column to the skull was modified to increase the mobility of the head relative to the body (Kemp 1969*a*). The occipital condyle broadened to a kidney shape, which permitted the head to rotate about the longitudinal axis ('shaking') through many more degrees without the first vertebra, the atlas, damaging the spinal cord. It also increased the extent of dorso-ventral rotation of the skull about a transverse axis ('nodding') without undue stretching of the spinal cord because each of the paired neural arches of the atlas articulated to the side rather than directly below the foramen magnum. Behind the atlas and axis, there are five cervical vertebrae. These have broad, horizontal zygapophyses, and intercentra are retained between adjacent vertebrae, both features indicating that large lateral movements of the head were accommodated by the intervertebral joints of the neck. The dorsal region of the vertebral column between the pectoral and pelvic girdles is relatively undifferentiated at this stage (Fig. 3.16(a)). There is no significant anatomical distinction between thoracic and lumbar vertebrae, and moveably attached, ventrally directed ribs extend for the full length. The zygapophyses are close to vertically oriented, and intercentra are absent, both indicating that lateral undulation was virtually eliminated, but as yet dorso-ventral bending of the vertebral column had not evolved. Three sacral vertebrae are present, behind which the tail was probably very much reduced in length compared to pelycosaurs, although no completely preserved specimen is yet known.

Forelimb. The basal therapsid shoulder girdle and forelimb (Fig. 4.7(a)) have undergone a profound change. The dermal shoulder girdle arch, formed from the interclavicle and clavicles, is far less massive and the contact between the clavicle and the scapulo-coracoid looser. The scapula blade is much narrower and the coracoid plate shorter. Significant movements of the scapulo-coracoid relative to the ribcage had evolved by this stage, adding to the total stride length of the forelimb. Although to a limited degree at this stage, the change anticipates the fuller expression of scapulo-coracoid mobility in later cynodonts and most extremely in mammals.

The most remarkable evolutionary change in the forelimb concerns the shoulder joint (Fig. 4.7(a)). The glenoid no longer has the elongated screw shape found in the pelycosaurs, but is a short notch formed equally between the scapula above and the coracoid below. It faces postero-laterally, and the surface is concave from top to bottom, but convex from front to back. The corresponding articulating surface of the humerus is curiously very different, for it is in the form of a hemicylinder on the proximal surface of the head that is much longer than the glenoid. The extreme incongruity between the respective articulating surfaces of the shoulder joint can only be explained by a very subtle mechanism (Kemp 1980*c*, 1982). The shoulder joint must have functioned by a roller action between the two opposing articulating surfaces, analogous to a wheel passing over the ground, rather than the sliding action between the opposing surfaces found in most joints (Fig. 4.7(b)). Beginning at the start of the

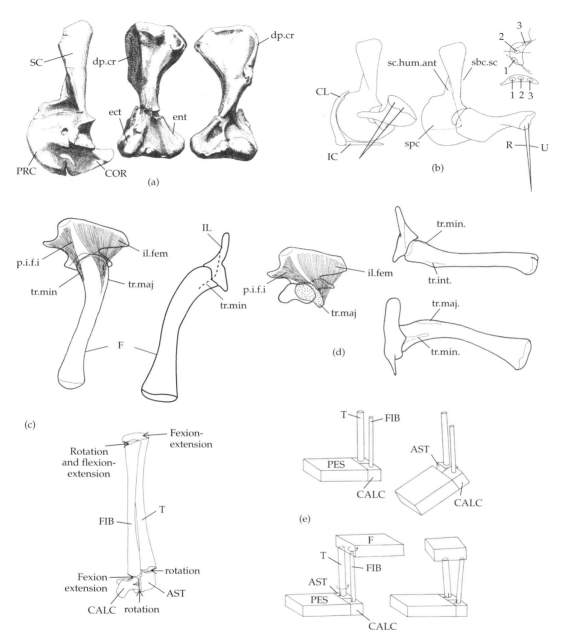

Figure 4.7 Basal therapsid-grade limb function. (a) Lateral view of left scapulocoracoid and dorsal and ventral views of left humerus of gorgonopsian (Kemp 1982, from a drawing by David Nicholls). (b) Diagram of the orientation of the humerus at the start and the end of a stride, as seen in lateral view, with illustration of the successive points of contact, 1 to 3, between the glenoid cavity and the head of the humerus. (c) Therocephalian ilium and femur in the parasagittal mode, in lateral, and posterior views. (d) The same in sprawling mode, in distal, anterior and dorsal views. (e) Hind view of the lower hind leg, indicating the distribution of possible movements between the five joints involved. (e) Diagrammatic illustration of the action of the ankle joint in the parasagittal mode (above) and sprawling mode (below). AST, astragalus; CALC, calcaneum; CL, clavicle; COR, coracoid; dp.cr, delto-pectoral crest; ect, ectepicondyle; ent, entepicondyle; F, femur; FIB, fibula; IC, interclavicle; il.fem, ilio-femoralis muscle; p.i.f.i, pubo-ischio femoralis internus muscle; PRC, procoracoid; sbc.sc, subcoraco-scapularis muscle; SC, scapula; sc.hum.ant, scapulo-humeralis anterior muscle; spc, supracoracoideus muscle; T, tibia; tr.int, trochanter internus; tr.maj, trochanter major; tr.min, trochanter minor; U, ulnax (Kemp 1982).

stride, the front part of the articulating surface of the humerus contacted the front of the glenoid, making contact at two points, one on the upper and one on the lower part of the glenoid. In this position, the shaft of the humerus extended laterally and horizontally and its leading edge was elevated so that the radius and ulna ran antero-ventrally to the forefoot, which was placed on the ground further forwards. The humerus was then retracted while remaining in the horizontal plane, and simultaneously rotated about its long axis. The movement was controlled by the rolling of the cylindrical humerus surface across the convex glenoid surface, there always being two points of contact between the two. By the final position, the humerus was still horizontal but retracted by about 45°, and the rotation about its long axis had placed the forefoot further back relative to the end of the humerus. The structure of the joint prevented adduction of the humerus below the horizontal, but did permit elevation to occur at any stage. This, along with a certain amount of freedom over the extent of the long axis rotation gave the humerus a much wider variety of movement than the highly constrained movement found in the pelycosaurs. The reason for this unusual joint mechanism may be related in a very particular way to the musculature that activated it (Kemp 1982). With the freeing of the shoulder girdle, there was a need to separate the muscles responsible for movement of the humerus relative to the girdle from those responsible for movement of the girdle relative to the ribcage. The roller mechanism causes the front and the hind points of the humerus head to move medially and laterally during the stride cycle. Therefore muscles attached to these points can run medially and so attach exclusively to the girdle, rather than to the body wall. Under these circumstances, the shoulder girdle was left unhampered to undergo its own independent movements relative to the body, controlled by muscles running between the shoulder girdle and the ribcage, body fascia and back of the skull. This unexpected design of the shoulder joint is manifestly weak. First, with so little area of direct contact between the articulating faces of the glenoid and humerus, even after allowing for a generous layer of cartilage the joint could not have withstood particularly large imposed forces. Second, mobility of the pectoral girdle on the body wall using only soft

tissues for attachment would also limit the size of locomotor forces that could be imposed without disarticulation. All this points again to the fundamental principle of tetrapod locomotion mentioned, that the primary function of the forelimb was to support the weight of the front part of the animal, and not to generate locomotor forces.

The actual musculature of the shoulder girdle did not differ greatly from the sphenacodontine grade. The main retractor was the subcoraco-scapularis, originating on the internal face of the scapulo-coracoid, emerging immediately above and behind the glenoid, and inserting on the posterior part of the proximal head of the humerus. The main retractor musculature consisted of muscles from the procoracoid and lower part of the scapula inserting on the anterior part of the humerus head. The probable origin of a supracoracoideus muscle is indicated by a shallow concave area in front of the glenoid, and of a scapulo-humeralis anterior by a large area in the antero-ventral region of the scapular blade. Adduction remained the function of a large pectoralis muscle running ventrally from the still massive delto-pectoral crest of the humerus, and elevation was the consequence of the action of deltoideus and latissimus dorsi muscles.

The functioning of the lower forelimb was not radically different from the sphenacodontine stage. As there, here also the radius and ulna were as stout as one another, and shorter than the humerus. The design of the joints was similar, with the sigmoid notch of the ulna making a strong hinging joint at the elbow and also resisting the powerful torque arising from long-axis rotation of the humerus. The radius, by contrast, accommodated the rotation at the elbow. At the wrist, the roles were reversed, with the ulna forming a rotatory joint and the radius a hinge joint with the manus. The manus itself is short and robust, with approximately equal lengths of digits although this was achieved by reduction in size of some of the phalanges: the phalangeal formula of biarmosuchians and gorgonopsians is still 2-3-4-5-3. Although never studied in detail, the relatively short metacarpals indicate that manus was probably plantigrade.

Hindlimb. The pelvic girdle and hindlimb of the primitive therapsid grade of evolution was also radically modified from the sphenacodontine grade, and Kemp (1978) proposed that the anatomy of the therocephalian pelvis and hindlimb could be accounted for by a dual-gait hypothesis. This is the idea that the hindlimb could operate facultatively in either a sprawling mode (Fig. 4.7(d)) or a parasagittal, mammal-like mode with the knee turned forwards and the foot beneath the body (Fig. 4.7(c)), as is seen today in crocodiles and large varanid lizards. Most, although not all of the therocephalian features of the pelvis and hindlimb occur in the more primitive therapsids, and Kemp (1982) extended the hypothesis to gorgonopsians. The basal therapsid hip joint is completely unlike the shoulder, for it is a conventional ball-and-socket arrangement whereby the circular, regularly concave acetabulum accepts the head of the femur congruently, both having the same radius of curvature. In fact, the femur can fit comfortably into the acetabulum in two different orientations. It can extend laterally and swing backwards in a horizontal plane. Alternatively, because of the inturned head and sigmoid curvature that it possesses, the femur can extend antero-ventrally and only slightly laterally and retract in a more nearly vertical plane (Fig. 4.7(c)).

The structure of the knee and ankle joints of therocephalians, including the evolution of a new intratarsal joint, is one of the strongest pieces of evidence for the dual-gait hypothesis and, although not studied with this functional possibility in mind, the ankle of the gorgonopsian *Lycaenops* described by Colbert (1948) appears to match the therocephalian in the essential features. The two different gaits involve two different respective movements at both the knee and the ankle joints (Fig. 4.7(c) and (e)). During the sprawling mode, there was relative rotation between the end of the femur and the foot, about a vertical axis. This movement was accommodated by rotation of the tibia on the astragalus of the ankle and rotation of the fibula on the underside of the femur. During the parasagittal gait, the movement at the joints were hinging rotations about approximately transverse axes. The wide upper end of the tibia forms a strong hinge joint with the underside of the distal end of the femur, though the lower end does not hinge on the foot. Instead, a new intratarsal joint had evolved between the astragalus and the calcaneum that allowed the calcaneum-plus-pes as a unit to flex and extend relative to the tibia-plus-astragalus as a unit. By this means, the joint between the tibia and the astragalus

does not have to undergo both rotation and hinging, but only the former. The latter is taken care of by the new intratarsal joint. Separating the functions increased the strength of the ankle and its ability to withstand the two different patterns of stress associated with the two different gaits. The nature of the astragalus and calcaneum, along with the shortening of the foot indicate that the therapsid hindfoot as well as the forefoot was plantigrade.

The inferred musculature of the hip and hindlimb also indicates significant changes in the mammalian direction (Fig. 4.7(c) and (d)). In mammals, the ilio-femoralis muscle has completely taken over from the caudi femoralis the role of the principal retractor, becoming the gluteal muscle complex. In primitive therapsids, the combination of enlarged and somewhat anteriorly extended ilium, development of a distinct trochanter major on the hind edge of the femur, and reduction of the tail, point to an early stage in this transition. Nevertheless, the retention of an internal trochanter on the underside of the femur indicates that a caudi femoralis muscle from the tail vertebrae, and the posterior section of the pubo-ischio-femoralis externus from the ischium were still important in retraction. The second major change in the arrangement of the hip musculature in the mammals is an antero-dorsal extension of the pubo-ischio-femoralis internus muscle, which has become the principal retractor of the femur, and is named as the psoas–iliacus muscle complex. This evolutionary transition had also commenced in the primitive therapsids. The lower part of the ilium in front of the acetabulum is occupied by a broad, shallow depression interpreted as the site of origin of what was to become the iliacus muscle. The significance of the tendency to shift the main hip muscles dorsalwards relates to the facultative adoption of the parasagittal gait at this stage, for it keeps them well away from the femur as it passes to and fro much closer to the ventral part of the pelvic girdle.

Blob (2001) has tested the dual-gait hypothesis by estimating the nature and magnitude of the stresses that would have arisen in the therapsid limb bones and comparing these with the situation in living crocodiles and iguanas. He concluded that the hypothesis is indeed a plausible explanation for the anatomy of the limb bones of animals transitional between sprawling pelycosaurs and those with fully mammalian locomotion.

To summarise the primitive therapsid grade of locomotion as inferred mainly from biarmosuchians and gorgonopsians, the head was more mobile on the body, and lateral undulation of the vertebral column was completely lost. The forelimb operated in sprawling mode, but its amplitude of movement was increased by some mobility of the shoulder girdle and by a rolling type of shoulder joint. It was adapted only for support of the front of the animal and not for production of any significant locomotor thrust. The hindlimb was more versatile and capable of at least two different gaits. For slow, low-energy movement it operated in a primitive, sprawling gait. For more active, faster locomotion the knee was turned forwards, bringing the foot below the body, and the limb was operated in a mammal-like parasagittal mode. The ilio-femoralis muscle, primitively an elevator of the femur had started to expand forwards over the ilium to act as a retractor muscle. The feet were plantigrade.

All the progressive features of these primitive therapsids are also found in therocephalian-grade therapsids and the functional anatomy of the limbs was probably fundamentally the same. In fact, the principal difference concerns the vertebral column. In the therocephalians, it has developed a true lumbar region in which the ribs are short, immobile, and horizontally disposed. The complete skeleton of *Regisaurus* (Fig. 3.17(d)) indicates a virtually mammalian degree of vertebral differentiation, and also mammalian proportions of limb length to body size.

Basal cynodont grade

Relatively little change in the postcranial skeleton and locomotion had occurred within the basal cynodont grades. *Procynosuchus* (Fig. 3.19(d)) was adapted for an amphibious mode of life (Kemp 1980c) and therefore not very representative. The relatively flattened limb and forefoot bones, wide-open glenoid and acetabulum, and very long tail bearing elongated haemal are evidently specialisations for swimming. *Thrinaxodon* (Fig. 3.20(b)) is a better model for this stage. The most striking innovation was the development of overlapping costal plates on the inner parts of the ribs. These are accompanied by accessory zygapophyses, the anapophyses, immediately below the normal pre- and postzygapophyses of adjacent vertebrae. The effect of both structures was to strengthen and stiffen the vertebral column,

although it is far from clear what the functional significance of this was. Kemp (1980a) supposed that it conveyed resistance to bending of the column in the face of increased locomotory thrust from the hindlimb. Others (Brink 1956; Jenkins and Bramble 1989) thought that it might relate in some way to the evolution of a diaphragm, though the nature of the functional connection between the two structures is difficult to see. At any event, the presence of the costal plates in the less-derived eucynodonts such as *Diademodon* and *Cynognathus* indicates that they were indeed a stage in the evolution of the mammalian axial skeleton, subsequently reduced and finally lost in more progressive eucynodonts and the mammals. By these later stages, the complex of intervertebral muscles and tendons, had presumably taken over whatever function the osteological system of interlocking ribs originally had (Jenkins 1971a).

The structure of the limbs and girdles of *Thrinaxodon* were sufficiently similar to those of primitive therapsids to assume that locomotory function was not greatly modified. The scapula blade was concave in lateral view, indicating larger muscles originated from the medial surface to pass out laterally and insert on the humerus. The shoulder joint had remained basal-therapsid in structure, and the forelimb continued to operate in a strictly sprawling mode. The hindlimb was still capable of adopting sprawling and parasagittal gaits (Blob 2001).

Eucynodont grade

There was a tendency to reduce the costal plates within the eucynodonts. They were minute processes in the traversodontid *Massetognathus* (Fig. 3.22(e)) and the chiniquodontids (Romer and Lewis 1973) have lost all sign of them, leaving an axial skeleton exactly as in mammals. Kemp (1980a) analysed the shoulder joint of the traversodontid *Luangwa*, showing that the humerus must still have acted in a strictly sprawling mode because any attempt to lower the distal end of the bone below the level of the glenoid leads to disarticulation (Fig. 4.8(c)). Indeed, the mechanical action of this joint was very similar to that of primitive therapsids, although there were significant changes in the associated musculature (Fig. 4.8(a)). The anterior edge of the scapula blade is strongly everted, at the base of which is the acromion process to which the outer end of the clavicle is attached. The gap between the

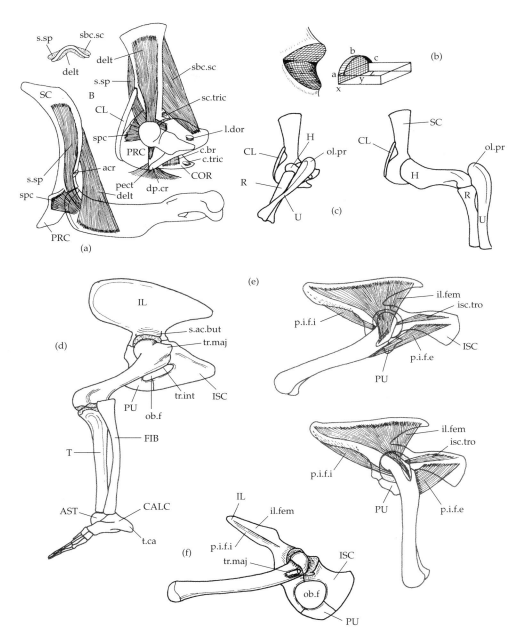

Figure 4.8 Cynodont-grade limb function. (a) Left scapulo-coracoid and humerus of a eucynodont in anterior and lateral views showing orientation and major shoulder musculature, and (inset) section through the scapula blade. (b) Diagram of the shapes of the corresponding glenoid and humerus articulating surfaces. Successive points of the humerus head in contact with the glenoid are x-a, y-b, and z-c as it rolls backwards. (c) Orientation of the forelimb at the start and finish of a stride. (d) Left pelvis and hindlimb of the eucynodont *Cynognathus* in lateral view. (e) Left pelvis and femur of the traversodontid cynodont *Luangwa* in side view at the start (upper) and end (lower) of the stride, showing the mammal-like musculature and gait. (f) Left pelvis and femur of the tritylodontid *Oligokyphus*. acr, acromion process; AST, astragalus, CALC, calcaneum; c.br, coraco-brachialis muscle; CL, clavicle; c.tric. coracoid slip of triceps; delt, deltoideus muscle; dp.cr, deltopectoral crest; FIB, fibula; H, humerus; IL, ilium; il.fem, ilio-femoralis muscle; ISC, ischium; isc.tro, ischio-trochantericus muscle; l.dor, latissimus dorsi muscle; ol.pr, olecranon process; pect, pectoralis muscle; p.i.f.e, pubo-ischio femoralis externus muscle; p.i.f.i, pubo-ischio femoralis internus muscle; PRC, procoracoid; PU, pubis; R, radius; s.ac.but, supra-acetabular buttress; sbc.sc, subcoraco-scapularis muscle; SC, scapula; sc.tric, scapular slip of the triceps muscle; spc, supracoracoideus; s.sp, supraspinatus muscle; T, tibia; tr.int, trochanter internus; t.ca, tuber calcis; tr.maj, trochanter major; U, ulna (Kemp 1982).

acromion process above and the coracoid bone below leads to the forward-facing internal face of the scapula. The area of origin of the supracoracoideus muscle of primitive therapsids had expanded upwards and forwards through the gap, to spread over this part of the internal face of the scapula, which is the equivalent of the supraspinatus fossa of mammals. A similar eversion of the hind edge of the scapula blade has also occurred, and so the posterior part of the inner scapula blade faces largely posteriorly. This is the area of origin of the retractor muscle, the subcoraco-scapularis. Even though the shoulder joint, and also the humerus, are still primitive in structure, the musculature controlling the limb had increased in size. This is explicable by recalling that the function of the forelimb is to support the front of the animal off the ground; enlargement of both the retractor and protractor muscles would have increased the strength of the support, and also the manoeuvrability of the forelimbs in an animal increasingly able to accelerate and rapidly change direction.

In contrast to the primitive nature of the forelimb, the hindlimb of eucynodonts demonstrates the transition to the more or less fully mammalian mode of action, involving loss of the ability to undergo the sprawling gait. The bulbous, inturned head of the femur fits comfortably into the deep, hemispherical acetabulum only when the femur is held in a virtually parasagittal orientation (Fig. 4.8(d)). In this position, the very prominent, mammal-like trochanter major lies on the postero-lateral part of the femoral head, in exactly the right position to accept the insertion of a large ilio-femoralis muscle. To a varying extent in different eucynodont taxa, the ilium has extended further forwards, the pubis been reduced and turned backwards, and the ischium turned more horizontally (Fig. 4.8(e)). These pelvic girdle features are correlated with an increasing development of the gluteal (ilio-femoralis) and psoas–iliacus (pubo-ischio-femoralis internus) musculature as the prime motivators of the locomotory cycle, in the mammalian fashion.

Tritylodontid and tritheledontid grade
The postcranial skeleton of tritheledontids is as yet insufficiently known to base a reconstructed stage on that group. Although, as far as it goes, the partial postcranial skeleton of *Pachygenelus* described by Gow (2001) is similar to those of both tritylodontids and early mammals. The postcranial skeleton of tritylodontids is known in practically every detail from the description of *Oligokyphus* (Fig. 3.23(c)) by Kühne (1956), and the as yet unpublished material of the North American form *Kayentotherium*. It closely resembles the basal mammalian stage represented by the morganucodontans (Jenkins and Parrington 1976). At last the forelimb had evolved a mammalian mode of action. The coracoid plate is very small and the glenoid fossa is widely open. The humerus is relatively slender with a bulbous head that can move quite freely in the glenoid as a ball-and-socket joint instead of the rolling joint characterising the earlier stages. Adduction–abduction as well as protraction–retraction movements were freely possible. The elbow was turned backwards and the forelimb is placed more or less below the body, as in modern small mammals (Jenkins 1971a). The anterior edge of the scapula and acromion process are even more extremely everted, and the supraspinatus muscle was very well developed, as indicated by a large depression on the now fully anterior-facing part of the internal scapula surface.

In the hindlimb, the trend towards losing the posterior process of the ilium and elongating the anterior process was completed, with the latter bearing the characteristically mammalian longitudinal ridge separating gluteal musculature above from iliacus musculature below (Fig. 4.8(f)). The pubis is fully turned back to a level entirely behind the acetabulum, and the obturator fenestra in the pubo-ischiadic plate is full sized. In short, the pelvic girdle is completely mammalian in form, as too is the femur, with the head separated by a short neck from the distinct trochanter major and trochanter minor. Actually, the two trochanters and femoral head are in line with each other, rather than set off at an angle as in typical mammals, a feature that Gow (2001) attributed to fossorial adaptations specific to tritylodontids.

Mammalian grade
At this rather generalised level of discussion, there is little to add to the story of the origin of mammalian locomotor mechanics, as directly inferred from the preserved skeletons. The morganucodontans added

no significant novelties to what had evolved by the tritylodontid–tritheledontid grade (Jenkins and Parrington 1976), but a number of refinements were added subsequently and are found as basic features of modern mammals. These include the fusion of the atlas intercentrum and neural arches to form the ring-shaped atlas vertebra, which rotates on the odontoid process of the axis, increasing the amplitude of head movements (Kemp 1969a). Fusion of the cervical ribs to the vertebrae may also be associated with increased head movements.

In the shoulder girdle, a vertical spine appeared, that is the homologue of the anterior edge of the primitive scapula blade. It separates the supraspinatus muscle in front of it from the infraspinatus muscle behind that is attached to what was the original lateral face of the scapula blade, and also gives origin to the deltoideus muscle. This increase in size and elaboration of the protractor musculature, which is functionally the main weight-supporting musculature of the forelimb, yet again points to the support rather than thrust-generating function, even in mammals. In the hindlimb, one of the most unexpected adaptations was the evolution of superposition of the astragalus on the calcaneum, a process initiated in later cynodonts (Jenkins 1971a) but not coming to full expression until the mammals. The intratarsal joint between the astragalus and the calcaneum that was associated with the therapsid-stage dual gait had allowed rotation between these two bones about a transverse axis. With adoption of an obligatory parasagittal gait, the extra joint was no longer necessary, but instead of losing it, it became modified to allow pronation–supination movements of the foot relative to the lower leg. The astragalus shifted on to the top of the calcaneum, losing its contact with the ground. The axis of rotation therefore shifted to longitudinal, which permitted the foot to raise and lower its inner and outer edges. Thus another element was added to the increased manoeuvrability of mammals compared to primitive, sprawling-limbed tetrapods.

Functional significance of mammalian locomotion
The functional significance of the profound evolutionary changes in the anatomy of the postcranial skeleton and mechanics of locomotion from pelycosaur to mammal is open to debate. Modern non-cursorial mammals have neither greater maximum speed, nor greater mechanical efficiency than sprawling-gaited lizards of the same body weight, as demonstrated long ago by Bakker (1974). The mammal does have much greater locomotor stamina, being able to maintain a given speed for far longer before the oxygen debt accrues, but as discussed later this is a function of the animal's physiology, not its morphology. There are two current hypotheses to explain the anatomical changes. One is that by getting the feet underneath the body and the main musculature high up at the top of the limb, agility of locomotion was enhanced (Kemp 1982). This is a difficult property to measure, but it can be argued that the repositioning of the feet closer together decreased stability and therefore increased manoeuvrability. Mammals certainly seem more capable than reptiles of rapid changes in speed and direction of movement, and of coping more effectively with a highly irregular terrain including tree climbing. The second hypothesis is that by abandoning lateral undulation of the vertebral column, and getting the belly permanently off the ground, the rate of respiration could increase (Carrier 1987). Lateral undulation supposedly prevents breathing and running simultaneously, because as the lung on one side of the body is expanded, that on the other is compressed. Furthermore, a diaphragm at the back of the ribcage could not work so effectively if the belly was on the ground. This issue is discussed in a broader context later, but for the moment it may be noted that these two explanations are not mutually exclusive.

Sense organs and brain

Of the many features that distinguish mammals from reptiles, the extraordinary hearing mechanism utilising the old reptilian jaw hinge bones as ear ossicles, the vast expansion in range and sensitivity of olfactory ability, and the monumental increase in brain volume by some ten times, mean the neuro-sensory system was as dramatically modified as any other. Most of the evolutionary changes affected soft tissues and so little is known of them, but several aspects can be inferred from the fossil record.

Hearing
The enlargement of the dentary bone, and concomitant reduction in size of the postdentary bones and

the quadrate have been described earlier and the functional significance of the implied changes in size and orientation of the adductor jaw musculature discussed. Establishing that two of the chain of three mammalian ear ossicles (Fig. 4.9(a)), namely the malleus and incus, are the homologues of the respective reptilian jaw hinge bones, the articular and the quadrate, and also that the mammalian ectotympanic bone supporting the tympanic membrane is the homologue of the reptilian angular bone was one of the triumphs of pre-Darwinian comparative anatomy (Reichert 1837). A satisfactory functional explanation for the implied transition from the one to the other took a little longer, in fact not until 1975, when Allin's hypothesis incorporated several previously inexplicable details (Allin 1975; Allin and Hopson 1992).

Apart from mammals, all living amniotes that possess a tympanic membrane for air-borne hearing transmit the vibrations of the membrane to the fenestra ovalis in the otic capsule, and thus to the cochlea canal of the inner ear, by means of the stapes, or columella auris bone. This middle ear apparatus provides an impedance matching mechanism whereby the low pressure of the air-borne sound waves is amplified to the higher pressure necessary for water-borne sound waves. The ideal level of amplification of the pressure is about 200 times. In lizards, birds, and crocodiles, it is achieved partly by the high ratio of the area of the tympanic membrane to that of the fenestra ovalis membrane, a 'stiletto heel' effect. There is also a lever effect by which the amplitude of movement of the tympanic membrane at the outer end of the stapes is geared down, and therefore the pressure increased at the inner end of the stapes where it contacts the fenestra ovalis. Mammals have an analogous mechanism. Again the tympanic membrane has a much larger area than that of the fenestra ovalis, around 100 times. A lever effect also exists, but it is completely different and involves the chain of three middle ear ossicles: malleus, incus, and stapes (Fig. 4.9(b)). The big problem to understanding the origin of mammalian hearing was always how the reduced jaw hinge bones became interpolated between a presumed functioning tympanic membrane and stapes, of reptilian design. It is also true that the acuity of sound perception in modern

reptiles and birds can be just as high as in mammals, and therefore there seemed no apparent reason to shift from one design to the other. Allin's (1975) proposal, not entirely new in principle but for the first time convincingly argued, was that tympanic sound reception evolved independently in mammals from other amniotes, and that the mammalian version of a tympanic membrane originated from superficial tissues overlying the postdentary bones in a pre-mammalian stage (Fig. 4.9(c)–(g)). It was held largely by the angular bone and its reflected lamina, which is the homologue of the mammalian ectotympanic bone that supports the modern mammalian tympanic membrane. It also received some support at the back of the jaw from the articular, which in turn articulated with the quadrate of the upper jaw as the functional jaw hinge. The quadrate itself made contact with the external end of the stapes, and at the end of the chain the inner end of the stapes was inserted into the fenestra ovalis. In other words, the anatomical relationships of tympanum, postdentary bones, quadrate, and stapes in pre-mammals were the same as in modern mammals. Allin's proposal, then, is that at some mammal-like reptile stage, a crude form of tympanic-activated hearing already occurred, no doubt limited to detection of low-frequency and high-amplitude sound. Subsequent evolutionary reduction of the postdentary bones and quadrate through the cynodont stages caused a decrease in the inertia and therefore an increase in the sensitivity of the system, a process which finally culminated in the full mammalian system where, once the new dentary-squamosal jaw hinge had evolved, contact between the angular, articular, and quadrate with the lower jaw was lost altogether.

Many features of cynodont jaw evolution support this theory. The reorientation of the jaw muscles to reduce the stress on the jaw articulation not only increased the bite force at the teeth, but also had the effect of allowing reduction in size and increase in mobility of the hinge bones. On top of this, Crompton (1972b) demonstrated that long before the new dentary–squamosal joint evolved in mammals, it was preceded in cynodonts by a secondary contact between the surangular bone of the lower jaw and a lateral flange of the squamosal. The effect of this would have been to reduce the stress

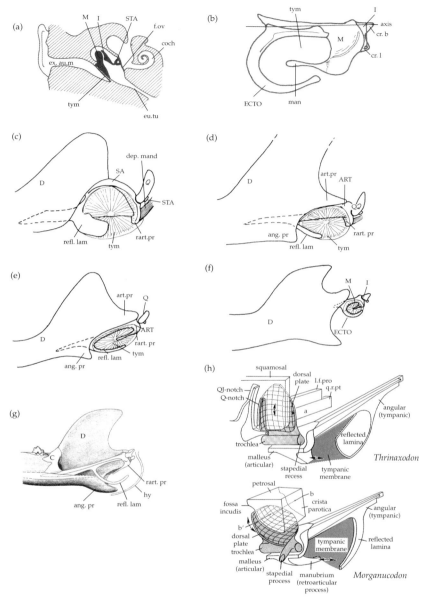

Figure 4.9 Evolution of the mammalian middle ear. (a) Basic structure of the modern mammalian middle ear. (b) Function of the mammalian ear ossicles as a lever system. The lever arm from the axis of rotation to the manubrium of the malleus in the centre of the tympanic membrane is approximately twice the length of the lever arm from the axis to the crus longis of the incus to which the stapes attaches. (c) Reconstruction of the tympanic membrane in a primitive cynodont such as *Thrinaxodon*. (d) The same in a eucynodont. (e) The same in a primitive mammal such as *Morganucodon*. (f) The same, in principle, in a modern mammal. (g) Reconstruction of the internal view of the hind part of the dentary and the postdentary bones in a primitive mammal. (h) Comparison of the sound-conducting system of the cynodont *Thrinaxodon* and the primitive mammal *Morganucodon*, showing the basic similarity but change in orientation of the axis of rotation of the quadrate from transverse to longitudinal. ((a) and (b) after Hopson 1966; (c) to (f) after Allin 1975; (g) Allin and Hopson 1992; (h) Luo and Crompton 1994). ang.pr, angular process of the dentary; ART, articular; art.pr, articular process of the dentary; coch, cochlea; cr.b, crus brevis; cr.l, crus longis; D, dentary; dep.mand, depressor mandibuli; ECTO, ectotympanic; eu.tu, eustachian tube; ex.au.m, external auditory meatus; f.ov, fenestra ovalis; hy, hyoid; I, incus; M, malleus; man, manubrium; Q, quadrate; q.r.pt, quadrate ramus of the pterygoid; rart.pr, retroarticular process; refl.lam, reflected lamina of the angular; SA, surangular; STA, stapes; tym, tympanum.

at the jaw articulation that was borne by the quadrate and articular, so allowing them even greater mobility. The great reduction in size of the reflected lamina of the angular of cynodonts, but not its actual loss was always a mystery, but can now be seen to be a reduction to a suitably delicate bone supporting the tympanum without adding excessively to its mass. In the course of a detailed review of the evolution of the quadrate in cynodonts and early mammals, Luo and Crompton (1994) have provided very detailed evidence for the supposed transition. From a primitive condition represented by *Thrinaxodon*, through a sequence of increasingly advanced forms to the early mammal *Morganucodon*, there is a reduction in size of the quadrate. This was coupled with a shift in the orientation of its contact with the squamosal, creating an increasingly effective axis for transmitting vibrations between the presumed tympanic membrane and the stapes. Indeed, a direct comparison of their reconstructions of the morphology of this region in *Thrinaxodon* and *Morganucodon* shows a remarkable similarity (Fig. 4.9(h)). The final step occurred in later mammals, where the tympanic bone and ear ossicles lost all contact with the dentary bone in the adult, being now supported only by the crista parotica, which is a small process of the periotic bone. They are protected within a partial bony housing formed from the surrounding bones, alisphenoid or periotic.

The implication of Allin's theory for pre-cynodont stages of mammal-like reptiles is that they did not possess a tympanic membrane for sound reception, and therefore had very poor impedance matching for converting air-borne sound waves to water-borne sound waves in the inner ear. In pelycosaurs, the massiveness of the stapes indicates that it still served its primitive function of mechanical support between the braincase and palate. Nevertheless, even at this stage the stapes was associated with an open fenestra ovalis so must have been involved in sound reception. There is no difficulty postulating detection of low-frequency, high-amplitude sound, received via the ground or the side of the skull. It would have been transmitted by intermolecular vibrations within the bony tissue, rather than by vibration of a light, low-inertia stapes. In the gorgonopsians the stapes is lightened by the presence of a very large stapedial foramen, suggesting that

a mechanism involving vibration of the bone as a whole had evolved, a mechanism potentially capable of detecting higher-frequency sound. Even so, there is no indication of the presence of a specialised tympanic membrane, so the lower jaw as a whole was probably still the initial sound receptor. Therocephalians do not possess a stapedial foramen, an unexpected and unexplained fact, but certainly not one that suggests acute high-frequency hearing ability in this group. The reflected lamina of the angular of primitive therapsids is large and primarily a site of muscle attachment so any role it may have had as an incipient eardrum would have been at best highly inefficient. Thus the anatomical evidence points to the cynodonts, with their reduced reflected lamina, as the first to develop a dedicated tympanic membrane on the lower jaw.

The cochlea of mammals is considerably longer than that of cynodonts, indicating enhanced sensitivity to a range of frequencies (Kielan-Jaworowska *et al.* 2004). It has become housed within a swollen promontorium on the ventral side of the skull, formed entirely from the periotic bone, a feature which, Luo *et al.* (1995) suggest, increased the sound insulation of the organ from the surrounding cranial structures.

Olfaction

Most modern mammals have a very acute sense of smell in terms of range of discrimination and acuity of reception. Little can be inferred about the sense of smell of non-mammalian synapsids, but what signs there are strongly suggest early elaboration of the faculty. In modern mammals, the nasal cavity is large and the area of the olfactory epithelium is greatly increased by thin, bony turbinals extending into it from the roof and sides (Fig. 4.10(a)). The naso-turbinals in the roof, and the ethmo-turbinals in the postero-dorsal region of the cavity are the main olfactory areas; the maxillo-turbinals lying laterally along the course of the airflow through the nasal cavity are associated with mucous epithelium related to the respired air and endothermy, as discussed elsewhere. No turbinal bones as such have been discovered in any non-mammalian synapsid, but fine ridges on the internal surfaces of the nasal and frontal bones are universally present from pelycosaurs onwards (Fig. 4.10(b)–(e)). These are

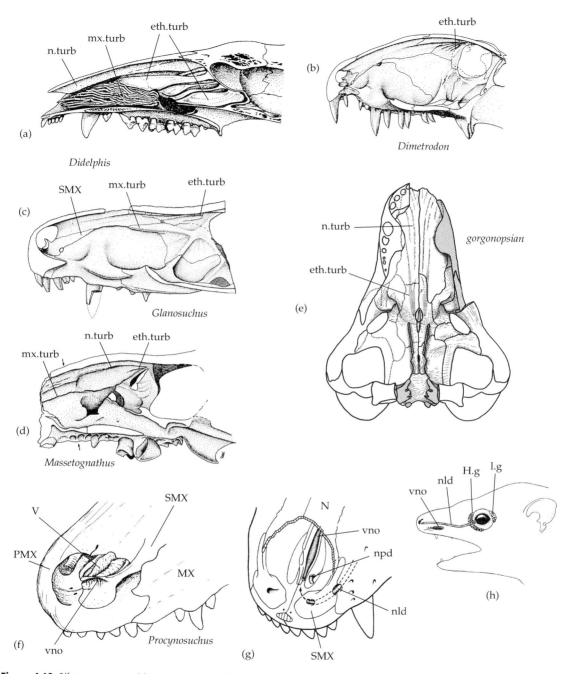

Figure 4.10 Olfactory structures. (a) Internal structure of the snout of a mammal *Didelphis* to show turbinal bones (Hillenius 1994). Internal views of snouts to show possible turbinal attachment sites of (b) the pelycosaur *Dimetrodon* (c) the therocephalian *Glanosuchus* and (e) the eucynodont *Massetognathus* (Hillenius 1994). (e) Internal surface of the skull roof of a gorgonopsian *Leontocephalus* (Kemp 1969b). (f) Duvall's interpretation of the septomaxilla and Jacobson's organ in *Procynosuchus* (Duvall 1986). (g) Hillenius' interpretation of the position of the septomaxilla and Jacobson's Organ in a generalised therapsid (Hillenius 2000). (h) The Harderian gland, Jacobson's organ, and naso-lachrymal duct in a mammal. (Hillenius 2000). eth.turb, ethmo turbinal; H.g, Harderian gland; l.g, lachrymal gland; MX, maxilla; mx.turb, maxillo turbinal; N, nasal; nld, naso-lachrymal duct; npd, nasopalatine duct; n.turb, nasoturbinal; PMX, premaxilla; SMX, septomaxilla; V, vomer; vno, vomero nasal organ.

interpreted as the sites of attachments of sheets of turbinals that were cartilaginous rather than bony, and which presumably enlarged the area of olfactory epithelium available several-fold.

Jacobson's, or the vomero-nasal organ, is a characteristic tetrapod olfactory organ situated in the floor of the nasal cavity, utilised primarily in social signalling by detecting pheromones (Duvall 1986). Undoubtedly the organ was present in all the mammal-like reptiles since it is still present and often well developed in modern mammals, but there has been much debate and confusion about its position and possible relationship with the prominent foramen and adjacent canal in the septomaxillary bone of the snout region of synapsids (Fig. 4.10(f)). In a detailed comparison with living tetrapods, Hillenius (2000) concluded that in non-mammalian synapsids, Jacobson's organ was a paired tubular structure lying in the floor of the front part of the nasal cavity, and that it had an association with the naso-lacrimal, or tear duct. He proposed that this duct ran from the glands in the orbit, particularly a Harderian gland, through the snout and opened at the septomaxillary foramen (Fig. 4.10(g)). He assumed that, as in many living tetrapods, the exudate of the Harderian gland is a serous fluid that moistens the nasal area, so that it collects odour molecules, which then pass backwards to Jacobson's organ for detection. In living mammals, the situation has changed (Fig. 4.10(h)). Jacobson's organ receives molecules directly from the oral cavity, usually via a naso-palatine duct. The naso-lacrimal duct no longer has a direct association with Jacobson's organ, but discharges directly to the external nostril and fleshy rhinarium of the snout. The nature of the fluid derived from the Harderian gland has also altered, and is high in lipids. It serves two functions, one is to deliver pheromones to the body surface, and the other is to provide waterproofing material to be spread over the fur. This changed function in mammals is associated with reduction of the septomaxilla bone and loss of its foramen and canal, and therefore this osteological modification may be an indication of the presence of insulating fur, and of more complex social behaviour. Related to this, Duvall (1986) speculated that social signalling by pheromones was an essential precursor of the evolution of lactation and mammalian levels of maternal

care. Cynodonts still possessed the primitive form of septomaxilla, foramen, and canal, whereas mammals from *Morganucodon* onwards have the reduced version lacking the foramen. Perhaps this is evident that cynodonts were uninsulated and lacked complex behaviour, though it is hard to be convinced by such tenuous reasoning.

Brain

The external features of the brains of mammals and birds are indicated very well by the impressions on the inner surfaces of the braincase because the brain is almost entirely enclosed within bone. The relatively much smaller brains of other amniotes are not so enclosed and therefore it is impossible to reconstruct completely the size, or the relative sizes of the different parts from cranial material alone. Typically the hindbrain, medulla oblongata, and cerebellum are determinable, and also the form of the dorsal surface. But the sides and floor of the midbrain and forebrain, including the cerebral hemispheres are not. Therefore, there is scope for considerable disagreement about the anatomical evolution of brains in synapsids.

An endocast of the brain of a specimen of the pelycosaur *Dimetrodon* was described long ago by Case (1897; Hopson 1979), and indicates a primitive, tubular structure lacking evidence for large expansions of any of its regions.

For the basal therapsids, Kemp (1969b) reconstructed the brain of gorgonopsians on the basis of a number of acetic acid-prepared braincases of large specimens (Fig. 4.11(a)). One of these has sheets of a crystalline material that may have replaced cartilaginous sheets present in life, and if this interpretation is correct, then a fairly complete representation of the external form of its brain is possible. Compared to the pelycosaur endocast, the cerebellum was significantly larger and there is an impression of a relatively large optic lobe. There is no room for the cerebrum to have been greatly enlarged, but it is possible that it was significantly larger than the pelycosaur tubular form.

Several authors have attempted to reconstruct the brain of cynodonts, with inconsistent results. According to Hopson's (1979) review of endocranial casts, all cynodonts possessed a tubular brain at the upper end of the size range of those of

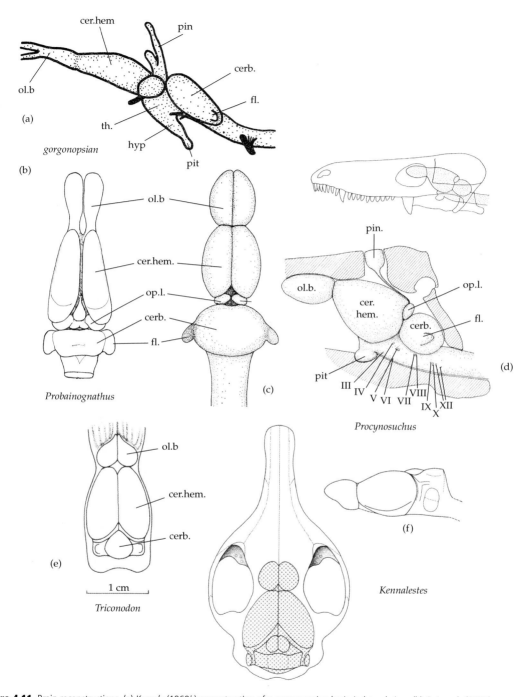

Figure 4.11 Brain reconstructions. (a) Kemp's (1969*b*) reconstruction of a gorgonopsian brain in lateral view. (b) Quiroga's (1980) reconstruction of the brain of the eucynodont *Probainognathus* in dorsal view. (c) Kemp's (1979) reconstruction of the brain of the basal cynodont *Probainognathus* in dorsal view. (d) The same in lateral view, and in situ in the skull. (e) Simpson's (1927) recnstruction of the brain of the Jurassic mammal *Triconodon*. (f) Kielan-Jaworowska's (1986) reconstruction of the brain of the Cretaceous placental mammal *Kennalestes* in dorsal and lateral views. cerb, cerebellum; cer.hem, cerebral hemisphere; fl, flocculus; hyp, hypothalamus; ol.b, olfactory bulb; op.l, optic lobe; pin, pineal; pit, pituitary; th, thalamus.

comparable-sized modern reptiles, and with no indication of the beginning of the enlargement of the forebrain that was later to dominate mammalian brains. Quiroga (1980) described the cranial cast of the Middle Triassic eucynodont *Probainognathus* (Fig. 4.11(b)). According to his interpretation, there actually was some expansion of the cerebrum and the beginning of the development of the neocortex. Even more extreme, Kemp (1979) reconstructed a much larger brain in the basal cynodont *Procynosuchus*, based on an acid-prepared skull (Fig. 4.11(c) and (d)). His interpretation starts with a confident reconstruction of the cerebellum, because the markings on the internal walls of the hind part of the braincase indicate that the cerebellum filled the cavity completely. Given that in all vertebrates the cerebrum is larger than the cerebellum, this gives a minimum size for the cerebrum that is substantially greater than Hopson's and other's tubular structure. Therefore the forebrain, although still relatively narrow as indicated by the impression of its dorsal surface, but must have been considerably deeper than other authors allow. At the maximum estimate possible, the brain volume would have been at the lower end of the size range for mammals.

There is, however, little consensus about the evolution of the brain within the cynodonts, important a topic as it is. Kielan-Jaworowska (1986) inclined to the view that it was still very small compared to mammals, although accepting Quiroga's (1980, 1984) interpretation that some enlargement had occurred within the group. Rowe (1996) also believed that little brain enlargement had occurred in cynodonts, and indeed that the neocortex did not develop significantly until much later, in the common ancestor of modern mammal groups. On the other hand, Allman (1999) accepted Kemp's reconstruction of a considerably larger, and especially deeper brain for the group.

What is beyond dispute, however, is that the earliest mammals themselves did have significantly enlarged brains. As shown by a famous endocast (Fig. 4.11(e)) of the Jurassic *Triconodon* (Simpson 1927), and by the reconstruction of the *Morganucon* brain by Kermack *et al.* (1981), and those of Cretaceous multituberculates (Fig. 5.12) and placentals (Fig. 4.11(f)) by Kielan-Jaworowska (1986), brain size in Mesozoic mammals lay within the lower part of the size range of the brains of living mammals. This represents an overall increase of some four or more times the volume of basal amniote brains, and presumably involved the evolution of the neocortex, the complex, six-layered surface of the cerebral hemispheres that is one of the most striking of all mammalian characters (Kielan-Jaworowska 1986; Allman 1999).

Growth and development

The ancestral pattern of growth of amniotes is described as indeterminate, because it is continuous throughout life and there is no absolute adult size. It is associated with polyphyodont tooth replacement, in which there are several to many successive replacements of each tooth. This process provides the necessary increasing size of teeth and length of tooth row as growth proceeds. In mammals, the growth is determinate, with a rapid phase of juvenile growth ending in adult size, after which no further growth takes place. This is associated with diphyodont tooth replacement, in which there is a single juvenile, deciduous, milk dentition, followed by a permanent adult dentition (e.g. Luckett 1993; Kielan-Jaworowska *et al.* 2004). The mammalian growth pattern is only possible with an extremely high rate of parental provision of nutrition to the young, in their case by lactation, although by comparison with the similar growth pattern found in birds, direct provision of foraged food can achieve the same end.

Many direct studies of size ranges of specimens have revealed an indeterminate growth pattern in all non-mammalian synapsid groups (e.g. Abdala and Giannini 2000, 2002). The pattern of tooth replacement corresponds to this, for in all there is polyphyodonty, although a specialised version is present in the diademodontoid eucynodonts (Fig. 4.12(a)) in which the rate of replacement was reduced, and the anterior postcanines were not replaced as they were shed. The total tooth number was maintained by the addition of new teeth at the posterior end of the tooth row (Hopson 1971). However, this is related to the specialised form of the gomphodont teeth, which are designed for true, accurate occlusion between specific uppers and lowers, rather than reflecting the evolution of lactation or mammalian growth pattern.

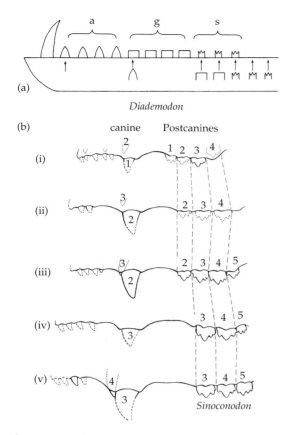

(a)

Diademodon

(b)

canine   Postcanines

(i)

(ii)

(iii)

(iv)

(v)

Sinoconodon

Figure 4.12 Tooth replacement (a) *Diademodon* (Kemp 1982) and (b) *Sinoconodon*. a-anterior teeth; g-gomphodont teeth. s-sectorial teeth (Crompton and Luo 1993).

The extreme version of this condition is seen in the tritylodontids, which do not replace any of their postcanines at all, but only discard them from the front and add to them at the back (Kühne 1956). Other cynodonts, including the basal form *Thrinaxodon* and the highly derived tritheledontid *Pachygenelus* (Crompton and Luo 1993), retained the primitive pattern of polyphyodonty, consisting of waves of replacement from front to back affecting teeth alternately along the jaw.

It is not until the basal mammal *Morganucodon* that the combination of determinate growth and diphyodonty is known to have evolved, as was demonstrated by Parrington (1971), who found specimens amongst the hundreds of fragments that were either juvenile growth stages or, the great majority, identical-sized adults. The incisors, canines,

and anterior postcanines are replaced once, and posterior postcanines are added sequentially at the back, not replaced, and therefore can properly be referred to as molar teeth. Given its correlation with growth pattern, it is assumed by this stage that lactation had evolved. However, the story is complicated by the situation in *Sinoconodon*, which is basal to *Morganucodon*. It still had indeterminate growth, for specimens are found that range in skull length from 2.2 to 6.2 cm, corresponding to an estimated body mass range of 13–517 g (Kielan-Jaworowska 2004). The tooth replacement pattern is also more primitive in *Sinoconodon*, as indeed it had to be in order to allow for the very considerable growth in size of what must have been sub-adults not dependent on lactation for their growth. The incisors and canines still show alternate, multiple replacements. The postcanine teeth are not replaced, and there is loss of anterior and addition of new posterior postcanines maintaining the relatively short postcanine tooth row of only three or four teeth (Fig. 4.12(b)). This condition in *Sinoconodon* is therefore intermediate between the primitive tritheledontid and the fully mammalian conditions. It may be speculated that the state of evolution of lactation was also intermediate, with maternal provision of milk limited to an early neonate stage only, after which the juvenile was weaned and relied on its own foraging, or perhaps on a more limited conventional food supply provided by the mother.

Temperature physiology

Nothing is more fundamental to the life of mammals than their endothermic temperature physiology, if only because it entails a 10-fold increase in daily food requirements. Such a huge cost must be balanced by an equally large benefit for endothermy to have evolved and been maintained. Yet surprisingly there is no consensus about exactly how, why, or when endothermy evolved in the course of the evolution of the mammals. The fact that the birds share a virtually identical mode of endothermic temperature physiology with the mammals adds little elucidation: the same contentious issues apply to them. The problem arises because of the complex nature of endothermy. It has two distinct primary functions in modern mammals, and it also involves a considerable array of structures and processes, including a regulating system for the high metabolic rate, variable conductivity of the skin by use of hair and cutaneous capillaries, neurological mechanisms for bringing about panting and shivering, and so on.

Before addressing the question of the origin of endothermy, it will be helpful to review the mechanisms and functions associated with it in living mammals. There are four physiological features associated with, and therefore effectively defining, endothermic temperature physiology:

- High basal or resting metabolic rate: normally between 6 and 10 times the basal metabolic rate (BMR) of an ectotherm of the same body mass.
- Elevated body temperature: somewhere between 28 and 40 °C.
- Constant body temperature: regulated to within about 2 °C of the thermostat setting.
- High aerobic scope: the maximum aerobically sustainable metabolic rate of tetrapods is generally around 10–15 times the BMR, so it is up to 10 times higher than an ectotherm of the same body mass.

The high BMR has no biological function per se, as is clear from several observations. To begin with, the BMR is related to body mass by the well-known Kleiber or 'mouse to elephant curve', which shows that the total metabolic rate varies as body mass to the power of 0.75. Thus the metabolic rate per unit mass of the elephant is only a few percent of the mouse, yet all the other aspects of the temperature physiology of these two mammals are similar. There are also differences in BMR between closely related species that are attributable to such factors as habitat or food type. For example in temperate, semi-arid, and desert species of hedgehog, the BMR varies in the ratio, respectively, of about 1 : 0.75 : 0.5. The relatively high body temperature of endotherms does not have any direct function either, as indicated for example by the many ectothermic reptiles that are active during the daytime at a body temperature equal to, or often higher than the endotherms. Therefore, the first two features of endothermy listed above must be interpreted as the mechanisms causing the second two features: the constant body temperature and the high level of maximum sustainable aerobic activity are the two respective biologically significant functions.

The constancy of the body temperature is maintained to the very high level of accuracy by controlling the rate of heat loss from the body surface on a moment-by-moment basis. To achieve this, it is first necessary to maintain a temperature gradient between the body and the environment, and the role of the high metabolic rate is to raise the body temperature to a value that will normally be higher than the ambient temperature. It is also essential, or at least vastly less wasteful, to have insulation of the body surface. The third necessity is for the conductance of the body surface to be instantly variable, so that the rate of heat loss can increase or decrease rapidly and at low metabolic cost, and this is achieved by the familiar thermoregulatory devices of vasodilation–vasoconstriction of the skin capillaries, piloerection–pilodepression of the hairs, and postural changes exposing different areas of the skin to the environment. A fourth requirement is to have what might be called emergency devices in place. If the maximum possible conductance of the skin is still too low to allow a high enough rate of heat loss under conditions of high ambient temperature, or high levels of activity, evaporative cooling by panting or sweating can temporarily relieve the problem. Conversely, if under cold conditions the minimum level of conductance is still too high to prevent excessive heat loss, the temporary measure available is an increase of metabolic heat production by shivering, or by switching on non-shivering thermogenesis in certain specialised tissues.

The adaptive significance of a constant body temperature is hard to describe succinctly because it so permeates the total biological organisation of a mammal. In one respect, this mode of thermoregulation expands the niche so that activity can continue under a wider range of ambient temperatures, both cooler and hotter. In a second respect, accurate thermoregulation increases the possible complexity of the organism because the various processes of enzymatic reaction, diffusion, muscle contraction, and so on take place at a constant, and therefore predictable rate. Therefore, a greater number of elemental biochemical, physiological, and physical processes can be reliably linked together into the more complex functional networks that underlie more complex levels of biological activity. It is significant that when a mammal is subjected to excessive heat or cold stress, the part that usually fails first is the central nervous system, and this is the structure most obviously associated with a highly complex network of interacting elements.

The second function of endothermy is the very high rate of sustainable aerobic activity that is possible. For reasons not readily explained, there is a roughly constant ratio between the basal or resting metabolic rate and the maximum sustainable aerobic metabolic rate of all vertebrates. The latter is typically 10–15 times the former, although there are exceptions. If the BMR of a typical ectotherm such as a lizard is taken as one unit of energy per unit time, then its maximum sustainable aerobic metabolic rate is about 13 units. For a typical mammal of the same body weight, the figures are a BMR of about 7 units and an expected maximum sustainable aerobic rate of 91 units, a huge increase in the latter property over the ectotherm. The mechanism behind the increase primarily involves a far larger number of mitochondria in the skeletal muscle, coupled with a greatly enhanced oxygen delivery system to them. The enhanced aerobic capacity does not affect the total maximum power output, or the top running speed attainable, because ectotherms can achieve similar values by anaerobic metabolism. However, ectotherms can maintain this level of exercise for no more than a very few minutes, after which time activity has to cease as the oxygen debt is repaid and lactic acid removed, a process that can take some hours to complete. In contrast, the maximum power output, and therefore maximum speed that can be sustained indefinitely, or at least until the body's food reserves are exhausted, is far greater in mammals than reptiles. The biological functions of this enhanced endurance are fairly obvious: food capture, predator avoidance, size of territory, vagility, and energy available for courtship all spring immediately to mind.

There are several hypotheses about the evolutionary origin of endothermy in mammals (Bennett 1991; Hayes and Garland 1995; Ruben 1995). They are all predicated on the assumption that endothermy in living mammals is too complex a character to have evolved in a single step. Therefore, one of the two major functions must have evolved first, whilst the other evolved secondarily later on. Consequently, there are two categories of hypothesis currently on

offer. One regards thermoregulation as the initial function, the other that high aerobic capacity was the first to evolve. Even within this dichotomy, there is more than one version of each, differing in what they take to be the initial selection pressure.

Miniaturisation: the physiological thermoregulation hypothesis
McNab (1978) pointed out that there was a general evolution in body size through the mammal-like reptiles to the first mammals. From the ancestral synapsids, which are assumed to have been small, there was a tendency to increase in size through the pelycosaurs and primitive therapsids. There then followed a trend of decreasing body size through the cynodonts, which culminated in the extremely small first mammals. His proposal was that temperature physiology tracked these changes in body size. The ancestral starting point was a small tetrapod with lizard-like ectothermic temperature physiology. The large-bodied therapsids had acquired the ability to maintain a relatively constant body temperature, not by increasing the metabolic rate, but by virtue of their large mass. The low surface area to volume ratio of a large organism reduces the relative rate of heat exchange with the environment and therefore of temperature change—a process termed inertial homeothermy. He suggested that some degree of insulation of the skin might also have existed, since this would enhance the effect. In the course of size reduction in the lineage that led ultimately to the ancestral mammal, the process would be expected to have reversed as the surface area to volume ratio increased, and the animals gradually to have returned to ectothermy. However, McNab argued, the benefits of the relatively constant body temperature having once been gained, they could not be secondarily lost. Instead, as body size reduced selection for continuance of the constant body temperature ensured the evolution of increasing relative metabolic rate and effective insulation in the form of fur. Thus metabolically driven thermoregulation gradually took over from heat inertial endothermy as the means of maintaining the constancy of the body temperature.

Attractive as it is, direct evidence in favour of McNab's miniaturisation hypothesis is limited. The hypothesis predicts that the origin of morphological characters associated with increased metabolic rates should be correlated with decreasing body size. The earliest cynodonts such as *Procynosuchus*, *Dvinia*, and *Thrinaxodon* confirm this since, along with their relatively small size, they possess a secondary palate and possibly a diaphragm. However, looked at in more detail the trend in size through the mammal-like reptiles is not at all clear-cut, and there are large, medium, and small members at every stage. Many pelycosaurs are large-bodied, but even amongst later pelycosaurs, varanopseids, and some caseids have body lengths less than 1 m. There are small dicynodonts and therocephalians that could not possibly have been inertial homeotherms, existing alongside large forms. While the earliest cynodonts are indeed relatively small, there are plenty of large-bodied genera throughout the Triassic. Even the highly advanced, extremely mammal-like tritylodontids include species with skull lengths as much as 25 cm, which are far from miniature synapsids. In fact, the only part of the record that clearly supports the hypothesis is that of the earliest mammals, which were very small, shrew-sized animals.

Arguments against the miniaturisation hypothesis are mainly those against any version of the thermoregulation-first hypothesis. Bennett *et al.* (2000) attempted to test it experimentally. They increased the metabolic rate of a resting lizard by introducing a large meal directly into its stomach, and then measuring whether the body temperature increased and the rate of cooling declined, both of which would be predicted by the hypothesis that endothermy is caused by a simple increase in metabolic rate. In fact they found that although the metabolic rate was elevated by three to four times, there was only a very small (0.5 °C), increase in temperature and no significant decrease in rate of cooling. Of course, like all such experiments, the results must be treated with considerable reservation since so many of the possible differences between the experimental circumstances and the real evolutionary event cannot be controlled for.

Nocturnalisation: the ecological
thermoregulation hypothesis
Crompton *et al.* (1978) offered a different version of the thermoregulation-first view, which stressed the significance of maintaining a constant body temperature for nocturnal activity, rather than for general biological organisation. Having observed that certain primitive nocturnal mammals such as monotremes, tenrecs, and hedgehogs have significantly lower metabolic rates and body temperatures than other similar-sized mammals, they proposed that these forms represent the primitive condition for mammals. From this perspective, endothermy arose initially as an adaptation for entering a nocturnal, insectivorous niche, as the first mammal is believed to have done. All that was necessary was to evolve an insulating layer to reduce the rate of heat loss. At the low ambient temperatures of night-time, a relatively low constant body temperature of around 30 °C would have created an adequate temperature gradient for thermoregulation, which in turn would have allowed the animal to remain active. This relatively low level of body temperature could be maintained by means of the still relatively low metabolic rate. Only with a subsequent shift to diurnal activity in various later lineages of mammals was a higher body temperature needed in order to maintain a large enough temperature gradient with the environment. Therefore, only at this later stage did the metabolic rate have to evolve the typical modern mammalian level.

It has since become appreciated that the various living mammals with low BMR are unrelated and therefore that the condition probably evolved convergently as a specialisation, although this does not actually refute the hypothesis. Tenrecs, for example, could nevertheless be regarded as suitable analogues for the ancestral mammalian mode of temperature physiology, although even they are now known to have metabolic rates some four times those of comparable-sized ectotherms and not the very low, reptilian-type energetics originally attributed to them (Ruben 1995). The main argument against the ecological thermoregulation hypothesis is that, contrary to McNab's miniaturisation hypothesis, it requires no increase in metabolic rate prior to the origin of the mammals themselves. This leaves unexplained the evidence pointing towards elevated metabolic rates in the cynodonts, namely the secondary palate, possible diaphragm, and increased masticatory ability. Many years ago Cowles (1958) conducted the experiment of wrapping a lizard in a fur coat and noting whether it maintained a higher body temperature. It did not, which suggested to him that evolving insulation alone does not immediately confer thermoregulatory ability upon an ectotherm, insofar as such a crude experiment can be trusted to show anything significant.

The aerobic capacity hypothesis

A number of arguments have been cited against any thermoregulation-first hypothesis. Bennett and Ruben (1979; Bennett 1991) asserted that the level of thermoregulatory benefit gained from a small initial increase in BMR would be far outweighed by the increased cost of food collection. This would be especially manifest if the organism in question was a competent ectothermic regulator, for modern reptiles are able to maintain their diurnal body temperature to a remarkable degree of accuracy by entirely behavioural means, and therefore incurring a very low metabolic cost. It can also be pointed out that endothermic regulation only works adequately if the accessory structures and processes such as insulation and finely variable conductivity have also evolved, which is unlikely to have been the case in the initial stages of increase in BMR. These criticisms are part of the argument in favour of the hypothesis that the initial selection pressure was for increased sustainable levels of aerobic activity.

Bennett and Ruben (1979) noted the roughly constant relationship between an animal's resting metabolic rate and the maximum level of sustainable aerobic metabolism. Without a clear understanding of why this physiological relationship should hold, they nevertheless proposed that an increased level of sustainable aerobic metabolism was the primary advantage of evolving an increased metabolic rate, and that any thermoregulatory advantage was initially insignificant. This was because even a small increase in BMR would confer an immediate advantage in terms of an incremental increase in the level of sustainable activity, and there would have been no need for the additional adaptations such as insulation that are necessary for thermoregulation to have evolved at the same time. Support for the hypothesis is also claimed from the fossil record (Ruben 1995). Carrier (1987) argued that the change in therapsids to a more erect

gait was to permit an increase in ventilation rates by decoupling the locomotory from the breathing functions of the axial skeleton. Hillenius (1994) claimed that maxillo-turbinals, whose function in living mammals is warming and humidifying inspired air, and reducing evaporative water loss from expired air, are first found in therocephalian therapsids. Both these claims point to the existence of endothermy in relatively large primitive therapsids, supposedly with a sufficient degree of inertial homeothermy for metabolically driven thermoregulation not to have been necessary. However, neither of them stands up well to scrutiny, as discussed shortly in the context of the timing of the appearance of endothermy.

Testing the aerobic capacity hypothesis experimentally consists of investigating the relationship between the resting and the maximum sustainable metabolic rates. The hypothesis implies that natural selection acted upon the maximum aerobic rate but that there is a physiological linkage between that and the resting rate such that the latter also necessarily increased as a correlation. Hayes and Garland (1995) reviewed studies comparing the resting and maximum aerobic rates between species and found some supported while others failed to support the correlation. The ratio between the two individual species, although typically around 10–15, can be as low as six and as high as 35. More unexpectedly, and hard to explain, different tissues are primarily involved in the two respective rates (Ruben 1995). Of the total resting heat production in mammals, around 70% is generated by the visceral organs: liver, kidneys, intestine, brain, etc., which amount to only about 8% of body weight. Compared to an ectotherm, the increase is achieved by a combination of a twofold increase in the volume of tissue, and a twofold increase in the density of mitochondria within it. In contrast, practically all the increased aerobic metabolism during exercise is due to activity in the skeletal and cardiac muscle tissue, which occupies around 45% of the total body mass. Again compared to an ectotherm, there is a relatively higher percentage of tissue and of mitochondrial density, but in this case the individual mitochondria have a larger membrane surface and therefore a higher level of metabolic activity as well. Given this difference in the source of heat related to, respectively, basal and activity levels of metabolism, there is no immediately obvious functional linkage between the basal and the maximum rates. Several somewhat vague suggestions have been made that high levels of routine maintenance activities, such as maintaining plasma membrane gradients, and high rates of protein and phospholipid recycling are required for the high levels of muscle activity (Ruben 1995).

Thus, like the thermoregulatory-first hypothesis, the aerobic capacity hypothesis rests largely on assertions of plausibility, rather than firm physiological or palaeontological evidence.

Parental provision hypothesis
A relationship between endothermy and reproduction in mammals (and birds) has long been suggested. This can be in terms of either the need in an endothermic species for the parent to care and provide for its juvenile offspring because it has too small a body size to behave as an endotherm itself (Hopson 1973; Kemp 1982), or in terms of the enhanced growth rate of the juvenile that is possible via lactation (Pond 1984). Two authors have presented updated scenarios along these two respective lines of thought. Farmer's (2000) version is no more than an application of both thermoregulation and aerobic scope views simultaneously to a particular proposed selective advantage, namely the control of incubation temperature and provision of food, respectively, to the juvenile. It offers no view on which function, if either, came first.

Koteja's (2000) hypothesis is somewhat different. He proposed that the initial selection was for increased parental provision of nourishment. The result was an enlargement of the viscera in order to cope with the necessary increased rate of assimilation of the food collected during an increased length of time spent foraging. Neither thermoregulation nor enhanced maximum activity levels were initially involved. However, he argues, a passive effect of enlargement of the viscera was an increase in BMR due to the increased mitochondrial activity of these organs, which in turn, by a positive feedback process, required a further increase in the rate of food collection. Therefore, selection for an increased level of maximum aerobic metabolism to achieve this followed. The hypothesis does therefore address the issue of the distinct tissues involved, respectively, in basal and maximum aerobic metabolic rates, but it is silent about when thermoregulation evolved.

Timing of the appearance of endothermy in synapsids
Some constraint on hypotheses about the mode of origin of endothermy may be applied by identifying the grade within the synapsids that it can first be identified as present. Several morphological and palaeo-ecological characteristics have been proposed as indicators of temperature physiology, originally in the context of the great debate about hot-blooded dinosaurs, but also applicable to the synapsids. A number of them that are allegedly correlated with homeothermy or endothermy do not stand up to scrutiny because the correlation is at best weak and often there is no satisfactorily established functional reason to explain the supposed connection (Bennett and Ruben 1986). The most unambiguous evidence would be the presence of insulation because there is no other conceivable primary function for a furry covering other than maintenance of an elevated body temperature produced by internally generated heat. Unfortunately, because it is a protein, hair is rarely preserved, and as yet no mammal-like reptile has been shown by direct fossil evidence to have possessed a pelt. Convoluted and not overly convincing arguments for its presence have been offered, such as that the presence of foramina in the snout of cynodonts indicates the presence of vibrissae, which in turn indicates that hair had evolved, and therefore may have covered the body (Watson 1931; Brink 1956b).

Enlargement of the brain to mammalian size has been correlated with the evolution of endothermy (Hopson 1979; Allman 1990), and in view of the extreme sensitivity of recent mammalian brains to temperature fluctuations, it is undoubtedly true that, to function, a large brain depends on endothermic thermoregulation. However the converse need not be true; endothermy is possible without a large brain. At any event, as discussed earlier in the chapter, it is difficult to assess when the brain evolved mammalian size and degree of complexity. Some authors have reconstructed a significantly enlarged brain in cynodonts, but others believe it to be reptilian in both size and structure, and that it was not until the mammals themselves that the enlarged cerebral hemispheres and neocortex of the forebrain evolved significantly.

Another possible source of direct morphological evidence for endothermy are features associated directly with a large increase in metabolic rate, for necessarily this requires greatly enhanced rates of gas exchange. The mammalian diaphragm bounding the posterior wall of the ribcage greatly increases the maximum rate of gas flow to and from the lungs and its presence would be good evidence for an elevated level of metabolic rate and maximum aerobic activity. The form of the ribcage in therocephalians and cynodonts is very suggestive. There is a distinctive thoracic region of long, ventrally curving ribs, followed by a lumbar region where the ribs are short, horizontally oriented, and immovably attached to the vertebrae (Figs 3.17(d) and 3.20(b)). Certainly a diaphragm could have been accommodated at the transition between the two regions, although, as Bennett and Ruben (1986) indicate, the development of the distinctive lumbar region might have been related to modifications of the locomotor rather than the respiratory mechanics. On this latter point, Carrier (1987) considers the evolution of the advanced gait in therapsids as itself a pointer to an elevated metabolic rate. Primitive amniote locomotion involving lateral undulation of the axial skeleton supposedly prevents breathing during running, due to the simultaneous compression of the lung on one side of the body and expansion of the one on the other. Getting rid of the lateral undulation and placing the feet beneath the body frees the two lungs to expand and compress in synchrony. On this argument, the most basal therapsids were already indulging in high rates of gas exchange and were presumably therefore endothermic. However, modern sprawling-gaited lizards are actually capable of increasing their rate of oxygen consumption during running by some three times (Ruben 1995), which rather detracts from the argument.

In modern mammals, the turbinal bones inside the nasal cavity play an important role in endothermy. These are very thin, often scroll-shaped bones covered in epithelium that serve two functions. The more dorsal and posterior ones, the naso-turbinals and ethmo-turbinals, respectively, have olfactory epithelium and increase the surface area available for chemoreception. The more ventral ones are the maxillo-turbinals, which are covered with mucous epithelium and are used to warm and humidify the air prior to entering the lungs so as to reduce evaporative stress to the cells of the alveoli. They also provide a cool surface area for condensation of water vapour from the expired air, which significantly reduces water loss by reclaiming most of the water vapour that had been added to the air on its way to the lungs. Ossified turbinals have never been discovered in mammal-like reptiles, but fine ridges on the internal snout surfaces have often been interpreted as the sites of attachment of either very delicate bony, or more likely unossified cartilaginous turbinals (Fig. 4.13). In pelycosaurs and primitive therapsids such as gorgonopsians and dicynodonts, there are well-developed ridges suggesting that naso-turbinals and ethmo-turbinals were present, but none lying along what would have been the direct route of the inspired air from nostril to oral cavity. Hillenius (1992; 1994) recognised a small extra ridge in the snout of the therocephalian *Glanosuchus* (Fig. 4.10(c)), in about the equivalent position to the line of attachment of maxillo-turbinals in living mammals which he interpreted as the attachment of a maxillo-turbinal in this form, and therefore as evidence for an enhanced metabolic rate. Impressed by this evidence as some authors have been (Ruben 1995), the fact is that this is a single, short ridge that may or may not have been the site of attachment of a turbinal, and if it was, it may have been olfactory rather than respiratory in function. Within several cynodonts there is similarly limited evidence for maxillo-turbinals (Fig. 4.10(d)).

Increased food assimilation must also have been associated with increased metabolic rate, but the evidence for this is bound to be even more ambiguous than for increased oxygen uptake: however efficient the feeding system might have been, there can be no way of knowing for how long the animal used it during the course of its day. Thus the dentitions of the cynodonts which appear to have achieved a degree of mastication comparable to modern mammals, certainly suggest a high rate of food preparation. The secondary palate in mammals is also associated with increased assimilation because one of its functions is the separation of mastication by the teeth in the oral cavity from the passage of air to the epiglottis at the back of the throat. The presence of a secondary palate in dicynodonts, later therocephalians and especially cynodonts might then indicate a greater extent of oral preparation of the food and therefore a greater rate of food assimilation. However, secondary palates have other functions not associated with endothermy, notably as a physical surface for the tongue to act against, as a mechanical device to increase the strength of the skull (Thomason and Russell 1986), to aid suckling in the neonate mammal (Maier *et al.* 1995), and to aid a semi-aquatic existence as in modern crocodiles.

Other supposed indicators of temperature physiology are yet more open to alternative explanations than those associated directly with enhanced respiratory and food assimilation rates. The bone histology, whereby the presence of fibro-lamellar bone and extensive Haversian systems is taken to indicate endothermy, is subject to far too many exceptions amongst both living and fossil vertebrates to be reliable (Bennett and Ruben 1986). At best it may indicate high rates of growth and bone remodelling during ontogeny (Chinsamy 1997). For what it is worth, pelycosaurs had a basically 'ectothermic' lamellar-zonal type of bone, whereas therapsids possessed the mammal-like fibro-lamellar type, suggesting that therapsids had evolved some mammalian characteristics of growth and development, although this does not necessarily prove that endothermy had evolved.

Claims that certain extrinsic, palaeo-ecological factors correlate with endothermy are the most ambiguous of all. Bakker (1975, 1980) argued that the predator–prey ratio of a community indicated the metabolic status of at least the predators, for an endotherm requires about 10 times as much food as a comparable ectotherm. In the case of the pelycosaurs, the ratio of carnivorous forms to herbivorous forms is about 45%, which would be typical of ectotherms, and of therapsids around 10–15%, which is closer to, but still in excess of, the value expected for endotherms. However, there are so many unknowns as to render any simple interpretation in terms of metabolic rates meaningless. To mention but two, the pelycosaur community almost certainly included freshwater fish as significant low-level consumers in the food web, while in

the case of the therapsids, it is very likely that large-bodied herbivores amongst the dinocephalians and later dicynodonts were not available as food for the carnivores. Once into the Triassic, the picture is completely distorted by the presence of increasing numbers of non-therapsid amniotes.

While it is easy and right to be sceptical about all these claims of simple correlations between an observable anatomical character of the fossils and the presence or otherwise of endothermy, taken together there is a reasonably convincing case to be made that it was the cynodonts that first manifested a significant increase in the BMR. The cumulative evidence for a diaphragm, a secondary palate, efficient food mastication, maxillo-turbinals, and a skeleton with the build of a highly active, agile animal supports the conclusion. The possibility that it was in this group that brain size started to increase further supports the probability that they were endothermic.

A synoptic hypothesis of the origin of endothermy
Despite its fundamental importance for understanding the origin of mammals, no consensus exists on either why or when the mammalian version of endothermy evolved. The problem lies in the integrated nature of organisms, and therefore the integrated nature of characters during an evolutionary transformation (Kemp 1982). As described, endothermy in living mammals serves two principal functions simultaneously, thermoregulation and

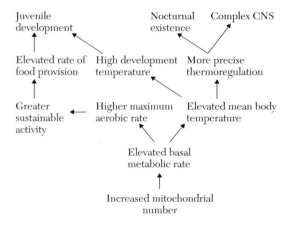

Figure 4.13 Synoptic hypothesis of the origin of mammalian endothermy.

elevation of maximum aerobic activity. Virtually all authors have made the assumption that such a complex arrangement could not have evolved in its entirety in one step, and therefore must have arisen by a simpler route consisting of two successive steps. The argument then shifts to which function was initially selected for, and which only subsequently. The underlying assumption is arguably false, however, and therefore the ensuing question is the wrong question. The correct question is not a matter of the order in which the parts of the complex whole changed, but rather of how the various evolving characteristics are interrelated such that every stage in the transition from fully ectothermic organism to fully endothermic organism remained a viable, integrated entity. This approach to understanding the origin of mammals is taken up *a fortiori* later, but for the moment the particular case of the origin of endothermy can be illustrated by way of a plausible scenario (Fig. 4.13). Suppose a mutation occurred that caused a small increase in number of mitochondria in all the cells. This would increase the maximum level of sustainable aerobic activity a trifle, permitting a few extra minutes in the chase. Simultaneously and unavoidably, the greater heat produced by the extra mitochondria would incrementally increase the animal's average body temperature by a few degrees, perhaps enough to permit an extra half an hour's activity before torpor set in at nightfall. Not much else can happen now until maybe a small incremental increase in oxygen uptake, sufficient to support another increment in BMR, or perhaps a modification to the skin vascularisation reducing its conductivity slightly. A hypothetical model like this indicates that both the main functions of endothermy can in principle evolve simultaneously and incrementally, with neither having primacy over the other. Furthermore, within the constraints imposed by the functional integration of organisms, it is actually a much simpler explanation than that based on the serial accumulation of separate functions.

Meanwhile, in this light the question of when endothermy evolved becomes transmuted into the question of what level of endothermy had been achieved by this or that particular stage. There is little doubt that the pelycosaur-grade synapsids were ectothermic, as witness particularly the dorsal sail of *Dimetrodon* and *Edaphosaurus*. At the other extreme, a case has already been made that cynodonts had achieved a relatively high level of endothermy. The largest problem is assessing the status of pre-cynodont therapsids such as gorgonopsians, therocephalians, and dicynodonts. The impressive modification of the locomotory mechanics at the basal therapsids level supports the view that there were enhanced aerobic activity levels. Their widespread abundance in temperate regions of the world suggests a reasonable degree of thermoregulation.

As to the temperature physiology of the earliest mammals themselves, the story is complicated by the process of miniaturisation that occurred in the lineage which led to them. Very small organisms face an array of biological constraints and potentials differing in several respects from those of otherwise comparable larger ones, as will be discussed presently (page 135). Nevertheless, it is hard to doubt that animals with the complete anatomy of modern mammals did not share their complete general physiology as well.

An integrated view of the origin of mammalian biology

Having reviewed what can be inferred from the fossil record about the evolution of the mammalian condition of various separate functional systems, it is now appropriate to take an integrated view of the evolution of mammalian biology as a whole. Comparison of a typical modern reptile such as a lizard or crocodile with a primitive living mammal reveals immediately just how great was the transition from their last common ancestor to the first mammal, and how it affected virtually every physiological and anatomical feature. The evolutionary paradox implied by this observation has been long and widely appreciated; indeed, ever since evolution by natural selection became accepted as the overarching explanation for the diversity of life. On the one hand, evolutionary change is caused by mutations in genes that affect discrete features, yet on the other hand organisms must remain complex, tightly integrated individuals in which all the structures and processes are designed to interact with one another to generate the overall biological nature of that individual. How can atomistic change in separate biological features occur independently of one another, while the integrated functioning of the organism as a whole is maintained?

Most commentaries on the paradox have been of a philosophical or theoretical, rather than an empirical nature, for want of adequate illustrative examples of what actually happens in the course of the evolution of a radically new kind of organism (e.g. Wake and Roth 1986; Kemp 1999). As it happens, the fossil record of the mammal-like reptiles is still far the best palaeontological documentation of the origin of a major new taxon, notwithstanding recent discoveries of transitional grades of tetrapods (Ahlberg and Milner 1994; Clack 2002) and birds (Chiappe and Witmer 2002). The 100 million years or so of their history, from Upper Carboniferous pelycosaurs to the end of the Triassic when the earliest mammals appear, give a fairly comprehensive picture of the sequence and pattern of acquisition of the main mammalian characters. It is certainly adequate to suggest answers to several questions about the evolutionary processes by which the major new taxon, Mammalia arose.

What was the ancestral mammalian habitat and mode of life?
Mammalian biological design was in due course to prove adaptable for a great array of different habitats: large grazing herbivores, polar carnivores, desert burrowers, powered fliers, permanent sea dwellers, huge-brained, highly social bipeds, and so on. But this potential is irrelevant to the actual origin of the group. To what habitat and in what fashion was the ancestral mammal adapted, and what therefore can be described as the end point of the process of the origin of mammals? All the systematic evidence indicates that the ancestral mammal was very like the known early fossil forms such as *Morganucodon* and *Megazostrodon*. These are characterised by very small body size compared to the majority of non-mammalian synapsids and they fall at the lowest end of the size range of modern mammals. The estimated body mass of the order of 5–20 g equates them with the smallest shrews, mice, and opossums of today. They show all the osteological signs of being endothermic with an elevated metabolic rate, and must surely have been insulated by fur, although this is not actually known

for certain. Sharp-cusped, occluding molar teeth that are analogous to the teeth of small, modern insectivorous mammals dominate the dentition. The postcranial skeleton indicates that they were highly agile animals, with the feet beneath the body, a mobile lumbar region of the vertebral column, and a reduced tail. The brain was considerably enlarged relative to body mass.

From a comparison with modern mammals of similar size and body form, it is more or less universally accepted that the ancestral mammals were nocturnal, insectivorous and at least facultatively arboreal. More specifically, the evidence that olfaction and hearing were more important senses than vision in basal mammals supports the argument for nocturnality. The forebrain of vertebrates was originally associated with the olfactory region, and therefore the enlargement in mammals of the cerebral hemispheres and neocortex to form the main controlling centre of the whole central nervous system indicates the fundamental importance of the sense of smell in the ancestral mammal. In the case of hearing, the evolution of the ear ossicles to increase the acuity of sound reception is evidence for the importance of that function. In contrast, the lack of colour vision in primitive living mammals, and the relatively small size of the optic lobes of the brain suggest that originally mammals depended less on sight. The argument that the ancestral mammal was insectivorous is supported by a functional interpretation of the shearing action of the molar teeth, coupled with body size. The more tenuous assumption of arboreality is based on the evidence of a high level of mobility of the limb joints and generally gracile nature of the postcranial skeleton.

Various authors have made proposals about which of the many mammalian characters is the most important adaptation. A common suggestion is that endothermy is the most significant because, by allowing temperature regulation, it permitted activity at night in the absence of a source of external heat. Others have proposed that the most important mammalian characteristic is lactation and the consequent increase in the level of juvenile care and provision, thus increasing the chances of survival of the offspring. A third idea is that it is the increased brain size and associated increase in the variety and complexity of behaviour that defines the essence of a mammal. The difficulty with all such suggestions is the difficulty that arises in any kind of key-adaptation hypothesis (Kemp 1999). As long as more than one characteristic has appeared in a descendant organism, it is impossible to test by correlation which of them is the 'key character' and which are subsidiary ones. And as long as there are several characters that are functionally interdependent on one another, it is meaningless to attribute paramountcy to one and subserviency to others: all are necessary parts of the single working whole.

Kemp (1982) published a diagrammatic scheme to illustrate the functional interrelationship between all

the fundamental mammalian attributes (Fig. 4.14). He proposed that this was most intelligible when looked at from the focal point of homeostasis. This is the maintenance by processes of active regulation of gradients between the internal environment of the organism and the external environment. To a degree it is a *sine qua non* of life of any kind, of course. There are inevitable limits to the magnitude of the gradients that can be so maintained by any particular organism, and this dictates the physico-chemical boundaries of that organism's tolerable habitat. Mammals may be seen as the organisms that have evolved the highest capacity for regulation of

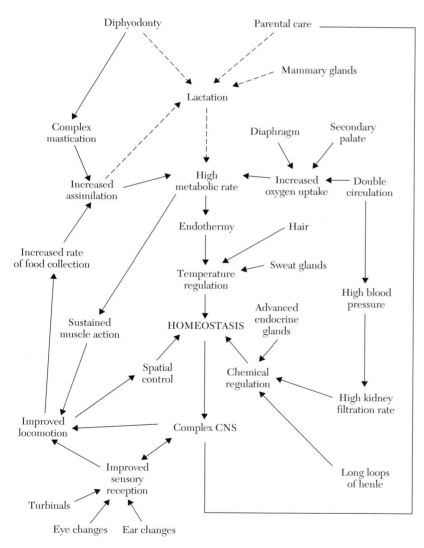

Figure 4.14 A schematic interpretation of the interrelationships between the structures and functions of a mammal (Kemp 1982).

their internal environment, which is to say, that have the highest degree of homeostatic ability. The most profound challenge to the potential of homeostasis was the shift in habitat from water on to land. If the internal environment is to be maintained constant in dry air, then large water and chemical gradients between the animal's tissues and the external environment have to be maintained by suitable regulatory mechanisms of differential intake and excretion. On top of this, the daily temperature fluctuation in air is huge in the absence of the high heat capacity and consequent buffering effect of water, which challenges the temperature regulation mechanism to maintain the internal body temperature within the much narrower limits of viability. The very physical patency of an organism on land, and its ability to control its own movement is challenged by the absence of the upthrust of water. Aside from these three major problems of life on land there are several lesser ones such as the absence of suction as a means of food intake, and of external water as a medium for gamete transfer and embryo development.

Different grades of tetrapods have adapted to terrestrial existence to different extents. Modern amphibians, with their permeable skin and small size, are unable to regulate their internal environments physiologically to any great extent against water, chemical, or temperature gradients, and are therefore restricted to humid, nocturnal habitats. Thus they may be described as avoiders in that they avoid places where such gradients would be high. Living reptiles by and large have solved the water gradient problem by a strategy of reducing loss. They have also solved the temperature problem by a regulatory ability that depends on differential uptake of environmental heat during the daytime. However at night, this process of ectothermic thermoregulation is not possible and therefore inactivity is imposed upon the animal.

The mammals (and equally the birds) have achieved the highest levels of regulation of the internal environment of all, and are therefore the tetrapods most highly adapted to the habitat of dry land. They are able to regulate the chemical composition and the temperature of the body in the face of higher gradients. Chemoregulation is achieved mainly by the ability of the kidney tubule to create hyperosmotic urine by means of the mechanism of the loop of Henle. By concentrating the urine, the mammal can afford to utilise the soluble urea as its prime nitrogen-excreting molecule without incurring an unacceptably high rate of water loss. The liquid ultrafiltrate entering the proximal end of the kidney tubule passes down the full length of the tubule and as it does so the concentration within it of water, the various ions, urea, pH, etc. is adjusted by a balance between reabsorbtion and secretion. Under fine hormonal control, the level of each of these constituents is adjusted to that which is necessary on a moment-by-moment basis to maintain the plasma concentration at the optimal composition.

Mammalian endothermic temperature regulation works by an analogous mechanism. An excess of heat is generated and the rate at which it flows out of the body is finely, and almost instantaneously, adjusted by varying the conductivity of the skin so as to keep the internal temperature constant. For the mechanism to operate, there has to be a gradient from a higher body temperature to a lower ambient temperature, so that the heat flow is continuous and therefore continuously adjustable. Thus, there has to be a high enough permanent metabolic rate to raise the body temperature to a thermostat setting above that of the environment. In case, on occasion, the gradient is temporarily lost or reversed because of hot conditions, the emergency expedient of evaporation by panting or sweating has to be available which is effective for a while but inconvenient and short term due to the need to replace the lost water. Conversely, if, under cold conditions, the temperature gradient becomes too large for the basic level of heat production to maintain, there are several ways in which extra heat can be generated by elevating the metabolic rate, again on a temporary and expensive basis, such as by shivering, exercise, and non-shivering thermogenesis.

But regulation is metabolically expensive. Maintaining chemical gradients requires the energetic process of active transport of molecules at the cellular level. Maintaining the temperature gradient requires a high level of aerobic respiratory activity by the mitochondria. Together these dictate the need for the 6–10 times greater BMR of endotherms over ectotherms. In order to achieve this, the rate of gas exchange and the rate of food assimilation need to increase proportionately. Efficiency of food detection, collection, ingestion, and assimilation must rise, with implications for the design of the sense organs, the locomotor system, and the central nervous control. Increase in gas exchange requires more effective ventilation such as is provided by a diaphragm and freeing of the ribcage from a simultaneous locomotor function. These add to the requirements of the actual regulatory systems, such as elaborate internal nervous and endocrinal monitoring systems and high blood pressure to increase the kidney filtration rate. For thermoregulation, variable insulation, cutaneous blood flow rates, and evaporation mechanisms are just some of the necessary components of the system.

Organisms maintaining high chemical and temperature gradients with the environment cannot be very small because of the surface area to volume consideration. Therefore, a juvenile of an already small mammal cannot exist independently, relying on its own regulatory mechanisms, which in any case take a significant time to develop fully. Therefore, parental maintenance of what amounts to a regulated external environment become necessary. In the first mammals this was presumably in the form of a nest, or conceivably a maternal pouch in which the egg and neonate existed in a controlled temperature and humidity, with the molecular requirements provided by lactation.

Seen in this light, there is no identifiable, single key adaptation or innovation of mammals because each and every one of the processes and structures is an essential part of the whole organism's organisation. To regard for example endothermy, or a large brain, or juvenile care as somehow more fundamental is arbitrarily to focus on one point in an interdependent network of causes and effects. Endothermy is necessary for maintained elevated levels of aerobic activity, but the activity itself is simultaneously essential for collecting enough food to sustain the high metabolic rate. The large brain causes high levels of learning and social behaviour, but the latter are necessary for the parental care that allows the offspring time to develop the large brain in the first place. Lactation is on the one hand necessary for mammalian development, yet on the other can only exist by virtue of the high metabolic rates and efficient food collection. Which has ontological priority?

To return to the question posed at the start of this section, what was the habitat, or better perhaps the niche to which the ancestral mammal was adapted? It must have involved a significantly fluctuating temperature range over which activity was maintained. It would have been dry at least at times. There would have to be abundant highly nutritious food but which required a particularly agile locomotory ability to acquire, and which suited a small animal weighing only 5–10 g. The physical habitat would have been very heterogeneous and complex to negotiate. A small, nocturnal insectivore living on the forest floor and capable of tree climbing sounds about right! Certainly there seems to have been little competition for this habitat at the end of the Triassic. Small lizards and their lepidosaurian relatives were presumably diurnal insectivores as now, and the archosaurs were starting to flourish as the large tetrapods of the day. Birds had barely even started their evolutionary journey down what was to prove, 60 million years later, to be a remarkably convergent route.

How was organism-level integration maintained during the transition?

There is a paradox when matching an evolutionary mechanism based on single, small changes in discrete characters to a long term, large evolutionary change in very many, fully integrated characters. The fossil record of the mammal-like reptiles and the transition to mammals supports the resolution to the paradox termed 'correlated progression'. In this model, all the major processes and structures of an organism are integrated such that each is both necessary for, and permitted by the rest. No one attribute can evolve and yet remain functionally useful, without being accompanied by appropriate changes in the others. However, the functional linkage is presumed not to be completely tight. A small amount of change in one attribute is possible and can be adaptive while still remaining adequately integrated with the rest. For example, the plausibility of a small rise in metabolic rate caused by a few per cent increase in mitochondrial numbers without needing any immediate increase in the ventilation, feeding, or vascular systems has already been proposed. The sensitivity of the middle ear might well increase a little, even within the constraints of the existing level of central nervous organisation. However, only small degrees of change in single characters are possible in this way. There is a limit to how much the metabolic rate, or the hearing acuity can increase in the absence of compensatory changes elsewhere in the organism. Further progress in those particular characteristics must await the time when other attributes have themselves undergone their own small changes. The overall evolutionary progression can be likened to a row of people walking forwards hand in hand. Any one person can get a little ahead of the rest, but further forward movement must wait until the rest of the members of the row have all progressed forwards in their own good time. The line as a whole moves forwards and all the individuals remain part of the interconnected whole, although at any instant some lag slightly behind and some have pulled slightly ahead.

Correlated progression predicts that in a sequence of ancestors and descendants reconstructed from the fossil record, each successive stage will have undergone changes in several biological systems. No one system will have evolved very much in the absence of changes in the others. This is exactly what the fossil record of the mammal-like reptiles shows insofar as the preserved characters are concerned. At every hypothetical stage that can be reconstructed from known fossils, there are modifications in the mammalian direction of dentition, jaw structure, and aspects of the postcranial skeleton. Less detailed evidence at least suggests that incremental changes to the ventilation mechanism, sense organs, and brain size similarly occurred over a succession of stages. Furthermore, it is implicit that if the evolutionary sequence of changes in other, non-preservable characters and processes could be determined, they too would fit into the scheme. The model also predicts that, because of the correlated progression of all characters, over a significant period of time each recognisable biological system of the organism will evolve at least approximately at a similar rate, insofar as such a parameter can be estimated.

What drove the trend towards mammals?
Having postulated what the first mammals were adapted for, and having concluded that all the changes that led up to them evolved in a coordinated fashion, the final question is what drove this trend from primitive, ectothermic, sprawling-gaited, simple-toothed amniotes to mammals? Of course,

the trend to mammals is a single lineage picked out from a highly branched phylogeny for no better reason than that there is a special interest in mammals as the taxon containing humans, although as it happens it is the longest branch on the tree measured both by the number of relevant grades of known fossils and the morphological distance spanned. There is no reason to suppose that trend culminating in, say, the Upper Triassic dicynodonts had a different kind of evolutionary cause.

The characters of mammals emerged gradually over the whole of the 100 Ma history of the mammal-like reptiles, mostly in the context of a habitat and mode of life rather different from those of the reconstructed hypothetical ancestral mammal. Furthermore, at several of the various levels in the evolving sequence, there was a radiation into a range of different kinds of animals, large and small, herbivore, omnivore, and carnivore. However, when the hypothetical sequence of ancestral forms is reconstructed from the distribution of characters on the cladogram, there are two consistent features. All of them have a carnivorous dentition; and all of them have a body size that is towards the small end of the size range of the members of the radiation to which they gave rise. This is reasonably confidently inferred for the respective hypothetical ancestors of the basal pelycosaur grade, sphenacodontid grade, therapsid grade, therocephalian grade, and basal cynodont grade. In short, the trend towards mammals consisted of a sequence of relatively small carnivores with an ever-increasing level of homeostatic regulatory ability.

The normal explanation proposed for a morphological trend that has been revealed in the fossil record is the neodarwinian one, that natural selection drove the change in a direction of ever-increasing fitness. In the present case of the trend towards mammals, it is supposed that each increase in homeostatic, regulatory ability adds an increase in adaptedness to the terrestrial habitat. There is an anomaly though: if it was indeed true that increasing homeostatic regulation was advantageous in the habitat of mammal-like reptiles, why were only relatively small carnivores evidently selected for it? Over the timescale of the evolution of the mammal-like reptiles, increasingly mammalian herbivores are found to replace one another. In the Early Permian there were edaphosaurid and caseid pelycosaurs; in the Late Permian tapinocephalid dinocephalians and dicynodonts; in the Triassic diademodontoid cynodonts. But these replacements did not happen by evolution of increasing mammalian characters in any of those actual lineages, but by extinction of the old groups and their replacement by new forms that had newly evolved from the succession of hypothetical small carnivorous ancestors. The same is true of large carnivores, where sphenacodontid pelycosaurs, brithopian dinocephalians, therocephalians, gorgonopsians, and chiniquodontid cynodonts replaced one another over time, but are members of different lineages. Kemp (1982, 1999) proposed an alternative explanation to neodarwinism, based on the theoretically sound but rarely demonstrable phenomenon of species selection. This concept requires that some species have a higher probability than others of splitting into daughter species, on the basis of characteristics of the species as a whole, not of the individual organisms. Some possibly relevant species-level characters are population size, population structure, and rate of dispersal. The argument for a species-selection explanation for the trend towards mammalian biology assumes that for ecological reasons, a species of synapsid that consisted of relatively small, relatively more homeostatic carnivores had a higher probability of speciating than other kinds. If so, then over a timespan measured in tens of millions of years, a statistically significant trend would occur consisting of just such kinds of species.

A species-selection model such as this could explain the bias towards small carnivores in the hypothetical sequence of ancestors and descendants because of a combination of three characteristics. First, carnivores tend to have smaller populations than herbivores. Second, organisms with higher metabolic rates potentially have smaller populations than those with lower metabolic rates. Third, small organisms tend to have lower dispersal abilities than larger ones. Therefore relatively small carnivores, possessing relatively higher, more mammalian energy budgets have a higher probability of forming geographically isolated sub-populations, which is supposedly the commonest cause of speciation.

This model can also help to explain another overlooked enigma concerning the origin of mammals.

For all that the advantages of homeostasis may be extolled in terms of freedom of activity of the organism, the energetic cost is huge, and it is not easy to see how simple, competitive advantage alone permitted the evolution of animals eventually requiring 10 times as much food each day simply to live. However, if the very effect of organisms with a higher energy budget on its population size and structure were to increase the probability of speciation, a positive feedback between degree of homeostasis and speciation rate would occur. This is of considerable theoretical interest.

The significance of miniaturisation

The final stage in the evolution of mammals involved the process of miniaturisation, because the tritheledontans, which are probably the closest relatives of mammals, as well as the earliest preserved mammals themselves all had a body weight of 5–15 g, a skull length around 2 cm, and a presacral length around 8 cm. There are numerous well-studied allometric relationships in modern mammals, where the dimension of some particular feature scales differently to body mass in different-sized organisms (e.g. Eisenberg 1981, 1990; Calder 1984). Samples of some of the possibly more significant cases are:

• BMR scales as Body Mass to the power of 0.75. This is the famous Kleiber relationship, and states that smaller animals have a higher BMR per unit mass;
• Brain size scales as Body Mass to the power of 0.67, so smaller animals have a relatively larger brain volume;
• Total skeletal mass scales as Body Mass to the power of 1.1, so smaller animals have relatively lighter skeletons;
• Individual limb bone mass scales as Body Mass to the power of about 1.35, so smaller animals have relatively lighter limbs;
• Lifespan scales as Body Mass to the power of 0.2, so smaller animals live for far less time.

Relationships like these are more or less in accord with simple physical relationships between such parameters as surface area, volume, mass, and energy. The main question concerning miniaturisation in the origin of mammals is whether it was solely an adaptive response to a particular habitat, or whether the process of miniaturisation was a structurally necessary way of evolving some of the characteristics of mammals.

Several suggestions have been made to the effect that miniaturisation was structurally necessary for a mammal to evolve. One discussed earlier is McNab's (1978) hypothesis for the origin of endothermy, proposing that is was the reduction of body size that necessitated the evolutionary increase in BMR, and therefore the shift to endothermy. Another possibility is that the relative increase in brain size accompanying miniaturisation was necessary for remodelling the anatomy of the braincase and its relationship to the rest of the skull in such a way as to be able to house an absolutely larger brain in mammals. A further consequence of the enlarged brain is that the distance between the quadrate and the fenestra ovalis that is spanned by the stapes decreased, which may have been a requirement for the final development of the mammalian ear ossicle system of hearing. It is very likely that the relative reduction in the mass of the limb bones was instrumental in the final stages of evolution of the characteristic mammalian locomotor mechanism.

However, none of these speculations are overwhelmingly convincing, because so many of the mammalian characters first evolved in earlier mammal-like reptiles that were not at all miniaturised. Many eucynodonts, and even some of the tritylodontids, with their large number of mammalian features, were medium to large in body size. It is simpler to accept that the miniaturisation was simply an adaptive response to the habitat of nocturnal, insectivorous tetrapod.

CHAPTER 5

The Mesozoic mammals

The expression 'Mesozoic Mammals' refers to more than simply the mammals of that particular period of time; it also stands for an extraordinary and quite mysterious concept. From the first appearance in rocks of Late Triassic times of the small, obviously highly active, large-brained animals thought of as mammals, through the following 145 million years of life on earth culminating in the great end-Cretaceous mass extinction that saw the end of the dinosaurs, these animals remained small. Although probably far from rare at the time, the great majority of species of Mesozoic mammals were of the size of shrews, rats, and mice. A tiny handful managed to evolve to the body size of foxes or beavers, but there were no representatives at all of mammals the size of the prominent mammals of today, the herbivorous horses, antelopes, and elephants, the lions and wolves that feed upon them, or the specialist apes, whales, and anteaters. Two points highlight just how odd this restriction in body size is. The first is that the Mesozoic mammals represent no less than two-thirds of mammalian evolution from their origin to the present, so there was plenty of time for evolution, and an extensive radiation did indeed occur producing a plethora of taxa. The second is that somewhere along the line, the potential for evolving large body size certainly existed because within, metaphorically speaking, moments of the end of the Mesozoic Era, middle-sized and soon thereafter large mammals had arisen and were flourishing.

Since their very earliest recognition by Dean William Buckland (Buckland 1824) from the Middle Jurassic Stonesfield Slate of Oxfordshire, Mesozoic mammals have generated controversy (Desmond 1985). Transformationists like Robert Grant denied that they were mammals, because it disturbed their accepted temporal sequence of Mesozoic reptiles preceding the exclusively Tertiary mammals. On the other hand, establishment figures like Buckland himself and Sir Richard Owen welcomed this apparent refutation of transformationism and had no doubt that they were indeed opossum-like mammals. In the end, the true nature of these fossils was accepted, and by 1871, a good number of undoubtedly Mesozoic localities had yielded undoubtedly mammalian fossils. In that year, Owen (1871) published his monograph covering all the known forms, dating as they did from Late Triassic through to latest Cretaceous, and derived from several localities in Europe and North America. Yet, while clearly more abundant than hitherto imagined, every single specimen consisted of no more than at best an incomplete jaw and teeth, or the odd isolated postcranial bone. In fact, little had changed by the time G. G. Simpson re-monographed all the world's material in two volumes (Simpson 1928, 1929) and the definitions of the groups and the interpretation of their interrelationships were still necessarily based solely on characters of the teeth and dentary bone. The prevailing view was that the mammals had been very rare, tiny animals totally dominated numerically and taxonomically by the dinosaurs. Simpson also epitomised the widely held opinion that the early mammals were highly polyphyletic, with anything up to five lineages having achieved the mammalian grade independently from the mammal-like reptiles (Simpson 1960). This concept was due to the practical necessity of relying on the divergent characters of the teeth, coupled with a faith in the efficacy of natural selection to drive similar kinds of organisms to considerable lengths in the same evolutionary direction.

Historically, rather little happened in the study of Mesozoic mammals until the remarkable discovery by Walter Kühne (1950) of abundant remains of latest Triassic mammals contained in clay-filled fissures of the limestone quarries of South Wales. Although extremely fragmentary, these specimens were the singularly well-preserved bones of little animals, not only mammals but several kinds of reptiles as well, that apparently fell into the potholes and underground streams of the limestone hills of that time. These, plus one or two complete skulls and skeletons of contemporaneous specimens from elsewhere allowed for the very first time the full description of the dentition (Mills 1971; Parrington 1971), skull (Kermack *et al.* 1973, 1981), and post-cranial skeleton (Jenkins and Parrington 1976) of one of the very earliest kinds of Mesozoic mammals, *Morganucodon*. Modern study and understanding of Mesozoic mammals dates from this period.

The next important overview was the edited volume of Lillegraven *et al.* (1979), by which time a reasonably clear picture of the main groups and their supposed interrelationships had emerged. Most authors had come to regard the mammals as monophyletic, having achieved their grade via a single lineage from one or other of the advanced cynodont taxa. However, interpretation of the interrelationships of the various groups was by this time dominated by a belief in a single, early dichotomy of the mammals into two groups. On the one hand were the therian mammals, which include the modern marsupials and placentals along with their more primitive Mesozoic relatives; on the other hand were the nontherian mammals, which included the living monotremes amongst many fossil groups (e.g. Hopson and Crompton 1969; Crompton and Jenkins 1978).

Kemp (1983, 1988*a*) first questioned this simple phylogenetic division at the base of mammal evolution on the grounds of the inadequacy of the characters used to support it, and before long it had been largely abandoned. Since then it has become increasingly clear that the evolution of mammals in the Mesozoic produced a very complex phylogenetic pattern. Several quite new kinds of Mesozoic mammals have been discovered, and the detailed knowledge of several known forms has increased enormously with the description of a number of virtually complete skeletons. This phase of study has culminated in the publication of the three-authored monograph of Kielan-Jaworowska *et al.* (2004).

Although the story is a good deal clearer as a result of this new information and several detailed cladistic analyses, a number of important aspects of the interrelationships are still very much debated. Meanwhile, the underlying conundrum of the group remains unsolved: why were the Mesozoic mammals so diverse in taxa, yet so conservative in body size and form?

The diversity of the Mesozoic mammals

In Chapter 3, it was established that a well-corroborated group Mammalia could be defined by a number of unique characters, which excludes the tritylodontids, tritheledontids, and other 'near-mammalian' forms. Omitting them is necessary for clarity because their precise interrelationships are still unclear. On this basis, the main diagnostic characters of Mammalia are the dentary condyle articulating in a glenoid fossa of the squamosal; the formation of a large percentage of the side wall of the braincase by the anterior lamina of the petrosal, which encloses the foramina for the maxillary and mandibular branches of the trigeminal nerve; and several detailed features of the periotic region of the braincase, and occipital condyles. Using these characters, ten major groups of Mesozoic mammals can be recognised, although, as will become clear, some of them are not strictly monophyletic. The nomenclatural issue of whether the term Mammalia or Mammaliaformes should be applied to the monophyletic taxon including all of these ten groups was also addressed earlier, when it was decided to follow Luo *et al.* (2002; Kielan-Jaworowska *et al.* 2004) by adopting the former.

Adelobasileus

The earliest contender to be a member of the strictly defined Mammalia is an incomplete and poorly preserved specimen from the Upper Triassic of Texas (Lucas and Luo 1993). It is from rocks of Carnian age, which dates it at about 225 Ma, some 10 Ma earlier than any other known mammal. Unfortunately, only the hind part of this little skull is actually present, and even that is badly crushed (Fig. 5.1(a)). There is no indication at all of the jaws or dentition. Nevertheless, the preserved portion is certainly mammalian. An anterior lamina of the periotic enclosing the foramina for the V_2 and V_3 branches of the trigeminal nerve is present, there is at least an incipient promontorium for the cochlea, and the

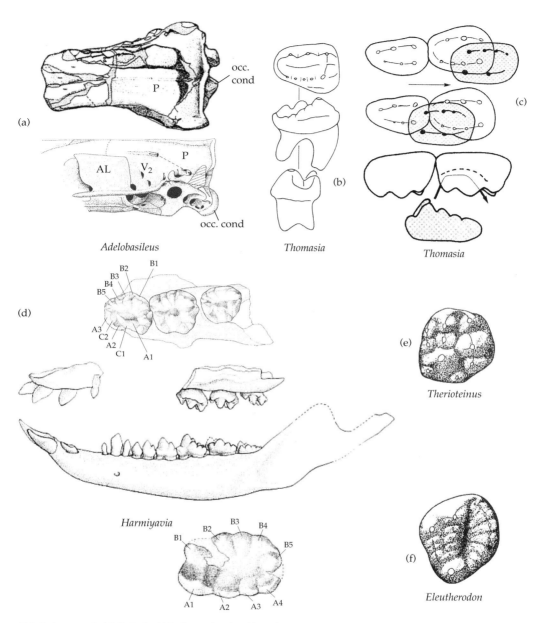

(a)

occ. cond

P

AL V₂ P

occ. cond

Adelobasileus

Thomasia

(b)

(c)

Thomasia

(d)

B1
B2
B3
B4
B5

A3
C2
A2
C1 A1

(e)

Therioteinus

Harmiyavia

B3
B2 B4
B1 B5

A1 A2 A3 A4

(f)

Eleutherodon

Figure 5.1 Early mammals. (a) Skull of *Adelobasileus* in dorsal and lateral views as preserved. Length of fragment approx. 1.5 cm (Lucas and Luo 1993). (b) Single molariform tooth of *Haramiyia moorei* in occlusal, side, and end views. Maximum length approx. 2.2 mm (Butler and MacIntyre 1994). (c) Occlusion of lower tooth (stippled) against two upper teeth of *Thomasia* (Butler 2000). (d) *Haramiyavia clemmenseni* premaxilla, maxilla, and dentary fragments in lateral view, with enlarged crown views of three upper molars and one lower molar. Length of jaw approx. 3.1 cm (Jenkins *et al*. 1997). (e) Isolated tooth of *Theroteinus* (Butler 2000 from Hahn *et al*. 1989). (f) Isolated tooth of *Eleutherodon* (Butler 2000). AL, alisphenoid; occ.cond, occipital condyle P, parietal;

condylar foramina are separate from the jugular foramina. As befits its early age, *Adelobasileus* is also demonstrably more primitive in some respects than any other mammals. The squamosal contributes a relatively large proportion of the braincase wall, and small basisphenoid wings of the parasphenoid bone have been retained on the ventral side of the braincase.

It is tantalising that so little is yet known of this form, since it is likely that it will prove to be of considerable importance in finally understanding the relationships between tritylodontids, tritheledontans, and the earliest groups of mammals. Further specimens are awaited with interest.

Haramiyida

The next earliest known mammals are the haramiyidans, which first appear in the Upper Norian part of the Upper Triassic of Germany. This gives a date for their appearance in the fossil record of about 212 Ma (Lucas and Hunt 1994), and they survived until at least the Upper Jurassic of Tanzania (Heinrich 1999). Like *Adelobasileus*, the haramiyidans are tantalisingly poorly understood, although in their case it is for want of cranial material. Until recently they were known only as isolated teeth, incisors, premolars, and relatively large multicusped, double-rooted molars (Sigogneau-Russell 1989a). Each individual molar tooth consists of a broad crown carrying a row of three large cusps along one edge and a row of five smaller cusps along the opposite one (Fig. 5.1(b) and (c)). The rows are connected at one end by a transverse crest, and between them lies a concave basin. For a long time it was assumed that all the molariform teeth, both uppers and lowers, were identical, and that the uppers must have been simply reversed versions of the lowers. In this arrangement, the three cusps of an upper tooth would have occluded in the basin of a lower. Simultaneously, the three cusps of the lower tooth would have occluded in the basin of the same upper tooth (Parrington 1947).

Most specimens were put into one of two genera, either *Haramiya* or *Thomasia*, on the basis of small differences in the teeth, although Sigogneau-Russell (1989a) suggested that the two genera actually represented upper and lower teeth, respectively. Subsequently, Jenkins *et al.* (1997) described the first known skeletal remains of a haramiyidan. *Haramiyavia* comes from a Late Triassic locality in East Greenland, and includes of a pair of dentaries and a maxilla, all bearing teeth (Fig. 5.1(d)), along with some of the postcranial bones. It shows that Sigogneau-Russell's (1989) interpretation was correct: teeth of the *Haramiya* kind are the lowers, teeth of the *Thomasia* kind the uppers. *Haramiyavia* also demonstrates that the group has a fully mammalian dentition. The dentary is not completely preserved so it is uncertain whether it had a dentary condyle, although the shape of the hind region of the bone is very similar to that of other early mammals so that it seems likely that a fully developed condyle was indeed present in life.

The new material allows the tooth action to be interpreted. There are four incisors, which are rather specialised, for the first three lowers are strongly procumbent, producing a near-horizontal surface against which the four short, vertical upper incisors could work. The incisors are followed by a diastema and then a small canine tooth. The four increasingly complex premolariform teeth are followed by three molariforms, which made up the main chewing surfaces. The wear facets on the opposing teeth indicate that the action was completely orthal, with virtually no horizontal movement of the lower teeth against the uppers, either side-to-side or to-and-fro.

Haramiyavia (Fig. 5.1(d)) was a little larger than most other mammals of the time, with a skull length estimated to be about 4 cm. Its skeleton was lightly built and agile, to judge by the handful of postcranial bones preserved. It seems that the haramiyidans were already specialised as small herbivores, or at least omnivores, with a simple, crushing action of the dentition that might have dealt with seeds, for example.

The relationships of the Haramiyida have always been a mystery, because they are so specialised yet occur so early in mammalian history. There had even been doubts expressed about whether they were true mammals at all, although the discovery of *Haramiyavia* certainly dispels them. Comparing them with the other mammal groups, the commonest concept is that haramiyidans are basal relatives of the multituberculates, as argued by Butler and McIntyre (1994; Butler 2000). The most important evidence for this is the form of the molariform teeth, with their two multiple rows of cusps, and the evidence for an element of antero-posterior occlusal action in some genera, particularly *Thomasia* (*Haramiya*). However, the similarities in molar structure is not close, and the force of the occlusal character is considerably weakened by the absence of horizontal movement of the teeth in *Haramiyavia* (Jenkins *et al.* 1997). There is really no good evidence to relate the haramiyidans to any other mammal group.

Having stated this, there are other groups of peculiar early mammals that may themselves be modified haramiyidans. Sigogneau-Russell (1983; Sigogneau-Russell *et al.* 1986) described some isolated teeth from the Rhaetic of France. They have been named *Theroteinus*, and consist of a broad occlusal surface bearing two or three rows of low, rounded cusps (Fig. 5.1(e)), and no sign of a propalinal component in occlusion. The root was probably divided into two or three, and the histology of the enamel resembles that of other mammals. Sigogneau-Russell's (1989*a*) tentative suggestion was that they are the milk teeth of haramiyidans, since normal teeth of the latter are abundant at the locality. Butler (2000) agreed that *Theroteinus* is a haramiyidan and suggested that it was a form whose molar teeth had adapted for crushing hard, brittle food. Kermack *et al.* (1998) described *Eleutherodon* (Fig. 5.1(f)) on the basis of some isolated Middle Jurassic teeth that resemble *Theroteinus*, but

have an extra row of cusps on the uppers and more cusps on both. They also have wear patterns indicating a propalinal action.

Whatever their relationships may be, haramiyidans were certainly a very early, specialised and short-lived, derivative from the hypothetical mammalian ancestor that nevertheless underwent a modest radiation.

Sinoconodon

Crompton and Sun (1985) described a complete skull of the critically important Chinese mammal *Sinoconodon* (Fig. 5.2), dating from the Lower Jurassic Sinemurian Stage, about 200 Ma. It possesses a number of characters suggesting that it is the most basal form attributable to the Mammalia. Further specimens available to Crompton and Luo (1993) amply confirmed this conclusion, indicating that it is something of a mosaic of primitive and derived features. Confirmation of its mammalian status is

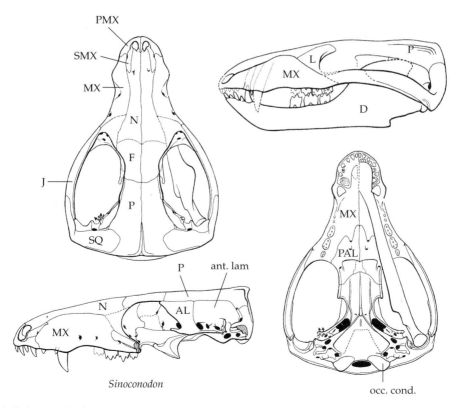

Figure 5.2 Skull of *Sinoconodon changchiawaensis*. Skull length approx. 6 cm (Crompton and Sun 1985 and Crompton and Luo 1993). AL, alisphenoid; ant.lam, anterior lamina; D, dentary; F, frontal; J, jugal; MX, maxilla; N, nasal; P, parietal; PMX, premaxilla; SMX, septomaxilla.

provided by the dentary condyle and glenoid fossa of the squamosal complete with a postglenoid ridge, the complete ossification of the internal wall of the orbit, and the expansion of the brain case. Set against these is a range of characters presumed to be primitive since they do not occur in contemporary mammals, but are found in tritheledontans and/or tritylodontids. The most important is the dentition. There are only five, multicusped postcanine teeth, which are fairly simple and lack anything more than a trace of a cingulum. Furthermore, there is no evidence of true occlusion between the upper and lower teeth. Comparison of a series of specimens of differing sizes shows that the mode of tooth replacement was cynodont-like rather than mammal-like (Zhang *et al.* 1998; Kielan-Jaworowska *et al.* 2004). The incisors and canines were replaced several times in an alternating pattern, while anterior postcanines were shed but not replaced. New postcanines were added on sequentially at the back of the tooth row.

Sinoconodon had a relatively larger body size than contemporary mammals, with a maximum skull length of about 6 cm. Interestingly, a wide range of skull sizes has been found, indicating that the growth from juvenile to adult was gradual, as in the primitive amniote type, rather than the brief, rapid pattern of juvenile growth that is typical of mammals (Luo 1994). No postcranial skeleton has yet been described, which might confirm this interpretation, but in any event, the *Sinoconodon* growth pattern is presumably correlated with its multiple tooth replacement.

Phylogenetically, *Sinoconodon* was undoubtedly close to the base of the mammalian radiation, as indicated by its retained primitive characters. Indeed, leaving aside the poorly known *Adelobasileus* and the poorly understood haramiyids, it is more or less universally accepted as the sister-group of the rest of the Mammalia (Crompton and Luo 1993; Luo 1994; Kielan-Jaworowska *et al.* 2004).

Morganucodonta

Parrington (1941) described two isolated teeth, one premolariform and the other molariform, that had been discovered by Walter Kühne in a Late Triassic, probably Norian fissure-fill deposit from a limestone quarry in Somerset, south-west England. They were undoubtedly mammalian because of their multiple cusps with a cingulum at the base, and their double roots, and he referred them to a new genus *Eozostrodon*. The main characteristic of the molariform tooth was a linear row of three main cusps on the crown, in which respect it resembled both carnivorous cynodonts such as *Thrinaxodon*, and certain Middle and Upper Jurassic mammals known as triconodonts. This, plus the fact that they were by some way the earliest mammalian remains discovered, apart from haramiyidan teeth, pointed to their considerable significance in the early history of mammals. Unfortunately, little more could be said about such extremely limited material. However, Kühne went on to announce the discovery of further, similar material in a quarry in South Wales (Kühne 1949). This proved to be extraordinarily abundant, yielding many thousands of teeth, jaws, and other skeletal fragments, which in due course formed the basis of the first serious appreciation of the nature of a very early mammal. The exact date of the South Wales fissure material is not certain and may be latest Triassic or earliest Jurassic. There was some early dispute about the nomenclature. Mills (1971) described the teeth, and Kermack *et al.* (1973) the lower jaw, and dentition under Kühne's name of *Morganucodon*. Parrington (1971) also described the dentition, but under the name *Eozostrodon*, in the belief that the new material was of the same genus as his original Somerset teeth. Reviews at time were divided. Jenkins and Crompton (1979) and Kemp (1982) referred to the material as *Eozostrodon*, but Kermack and Kielan-Jaworowska (1971) as *Morganucodon*. Eventually, consensus was reached that the two teeth of *Eozostrodon* are an inadequate basis upon which to make a satisfactory diagnosis of a genus, and *Morganucodon* is now universally used for the Welsh form. Fossils referred to various genera and species of morganucodontans have now been found in China, South Africa, North America, and India, as well as other parts of Europe, indicating a virtually worldwide distribution (Lucas and Hunt 1994).

Thanks primarily to the work of Mills (1971) on the dentition, Parrington (1971) on the jaws and dentition, Kermack *et al.* (1973, 1981) on the jaws and skull, Jenkins and Parrington (1976) on the postcranial skeleton, and Graybeal et al (1989) on the inner ear structure, the osteology of *Morganucodon* is extremely well studied (Fig. 5.3(a)–(c)). It was a small animal with a skull 2–3 cm in length and a presacral body length of about 10 cm. In general appearance it would have looked like a shrew or mouse. The skull has all the superficial appearance of a small, modern mammal characterised by the absence of a postorbital bar, slender zygomatic arch, and expanded braincase consisting largely of the parietal bone above. The sidewall of the brain case is formed by a prominent anterior lamina of the periotic behind that is pierced for the V_2 and V_3 branches of the trigeminal nerve, and an epipterygoid, or alisphenoid in front. As in the non-mammalian cynodonts, the latter bone is a broadened, vertical pillar from the base of the skull right up to the roofing bones.

The lower jaw is also characteristically mammalian in form, with fully developed coronoid, angular, and articular processes of the dentary. However, unlike most subsequent fossil and modern mammals, there is a double jaw articulation (Fig. 5.3(b)). The fully formed mammalian jaw joint between a dentary condyle and a squamosal glenoid fossa is visible laterally. Medially to, and supported by the dentary and squamosal bones respectively, lie the reduced but still easily recognisable articular and quadrate hinge bones of the reptiles. Indeed, the postdentary bones are altogether cynodont-like, though reduced in size, and they lie in a medial trough of the dentary. Crompton and Luo (1993) have demonstrated the presence of a slender reflected lamina closely attaching to the hind edge of the angular process in a specimen of the Chinese species *Morganucodon oehleri* (Fig. 4.9(h)). Unfortunately it is not possible to be sure whether or not or a retroarticular process of the articular was also present, as predicted by Allin's theory of the evolution of the middle ear. Certainly there was a large promontorium on the ventral side of the ear region of the skull, housing the enlarged cochlea and implying that Morganucodon had a significantly enhanced hearing system (Luo *et al.* 1995).

The dentition (Fig. 5.3(b)) is what would be expected in an animal living primarily on insects and other small, easily dealt with invertebrates. The cusps of all the teeth tend to be sharp, and the molariform teeth are primarily adapted for a cutting or shearing action rather than crushing and grinding. There are usually four incisors in both the upper and the lower jaw, followed by modest canines. All of them have wear facets indicating that the upper and lower front teeth actively worked against each other in the course of food collection. There are five upper and four lower premolars, each with a large, laterally flattened and sharp main cusp, a smaller posterior accessory cusp behind it, and practically no trace of a cingulum.

The three molariform teeth in both upper and lower jaw together formed a shearing system designed to shred up food, as first analysed by Crompton and Jenkins (1968). Upper and lower molars have the same basic structure of a large, laterally compressed, and sharp main cusp. Two posterior accessory cusps and a single anterior accessory cusp are also present, and lie along a line parallel to the jaw. The upper molars possess a cingulum of fine cuspules that completely surrounds the crown; the lower molars differ in that their cingulum is restricted to the inner, lingual edge of the crown, and the individual cuspules are larger but less numerous. The wear facets on the teeth show that accurate, extensive occlusion occurred between the opposing molar teeth. (Fig. 5.3(c)). The main cusp of a lower molar worked against the gap between the main cusp and the anterior accessory cusp of the corresponding upper molar. At the same time, the main cusp of the upper tooth worked against the equivalent space between the main cusp and the posterior accessory cusp of the lower. As a consequence of the wearing effect of tooth-to-tooth contact, flat facets with sharp free edges developed on the opposing teeth that come to behave like the opposing blades of a pair of scissors. Any resilient food caught between them is liable to be sliced, or sheared into two. As was described earlier, the essence of the mammalian design of postcanine teeth relates to the direction of movement of the lower molars during occlusion. As one of them shears against its opposing upper tooth it is forced to move very slightly inwards as well as upwards, while retaining contact. To achieve this, the lower jaw as a whole must follow a triangular pathway as viewed from behind: upwards and inwards during occlusion, downwards as the jaws opened, and laterally ready for the next bite. Only the molar teeth on one

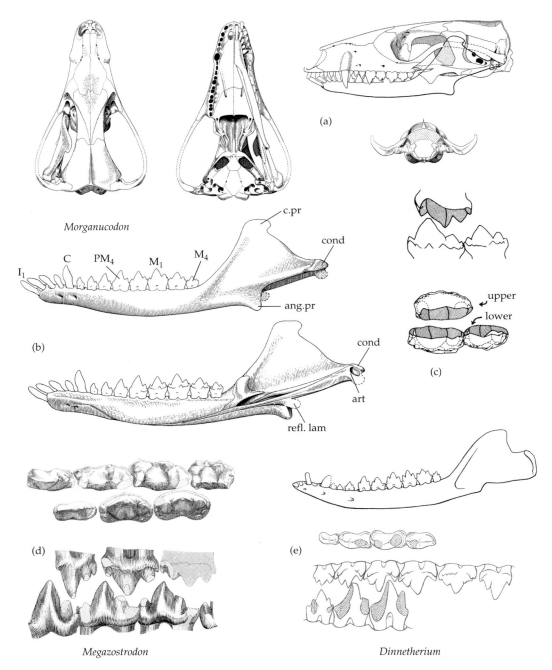

Morganucodon

Megazostrodon

Dinnetherium

Figure 5.3 Morganucodontan skull and dentition. (a) Skull of *Morganucodon watsoni* in dorsal, ventral, lateral, and posterior views. Skull length approx. 2.6 cm (Kermack *et al*. 1982). (b) lower jaw in lateral and medial views (Kermack *et al*. 1973). (c) Occlusal relationships between one upper and two lower molars in side and occlusal views. (d) Molariform dentition of *Megazostrodon rudneri* in occlusal and side views (Crompton 1974). (e) *Dinnetherium* lower jaw in lateral view. Length of jaw approx. 5 cm (Crompton and Luo 1993) and *Dinnetherium* dentition, occlusal view of uppers (top) and external view of uppers and lowers. Shaded areas are wear facets (Crompton and Luo 1993). ang.pr, angular process; art, articular; cond, dentary condyle; c.pr, coronoid process; refl.lam, reflected lamina of the angular.

side of the head can be in action at any one time, but this permitted the forces from the adductor musculature of both sides of the head to be applied simultaneously to the teeth on only one side, thereby increasing the potential bite force available, as is discussed elsewhere (page 100). Neither of the more primitive mammalian dentitions described so far, haramiyidans or *Sinoconodon*, had evolved this mode of occlusion, but all phylogenetically more derived mammals had, or were descended from forms that had such a system. In this respect morganucodontans represent a very important step in the evolution of fully mammalian biology.

The mode of tooth replacement of morganucodontans also exhibits for the first time a state ancestral to all subsequent mammals. Amongst the many hundreds of *Morganucodon* fragments of the fissure-fill material studied by Parrington (1971), he found specimens of jaws containing replacement teeth for incisors, canines, and premolariform teeth (Fig. 4.12(c)). He found none at all for the posterior three molariform teeth. Consequently, he inferred

that *Morganucodon* had achieved full mammalian-type diphyodonty, with only a single replacement of the deciduous anterior teeth, and no replacement of what can properly be termed molars. Crompton and Luo (1993) have expressed some doubt about whether Parrington's sample was really adequate to support this conclusion, although it seems certain that at the very least tooth replacement was greatly reduced in morganucodontans and approached the definitive mammalian condition. The functional relationship of diphyodonty, growth pattern, and lactation in these early mammals is mentioned elsewhere (page 120).

The postcranial skeleton of morganucodontans (Fig. 5.4) was the first of any Mesozoic mammal to have been properly described, on the basis of both the excellently preserved *Morganucodon* fragments from the Welsh fissures, and the complete though less finely preserved skeletons of the South African genera *Megazostrodon* and *Erythrotherium* (Jenkins and Parrington 1976). While characteristically mammalian in most features, morganucodontans still

Figure 5.4 Postcranial skeleton of *Megazostrodon rudneri*. Presacral length approx. 7 cm (Jenkins and Parrington 1976).

retained a number of more primitive features. The first two vertebrae, the atlas and axis respectively, are cynodont-like insofar as the elements of the atlas have not fused to form a ring-shaped bone. Nevertheless, they may well have functioned as in the modern mammals, with extensive rotation of the head plus atlas about the anterior odontoid process of the axis vertebra behind (Kemp 1969a). The ribcage is fully mammalian in appearance, with the individual ribs connecting ventrally to a row of sternebrae rather than with a flat, continuous sternal plate. There is an abrupt transition to the lumbar region of the vertebral column, indicated by a change in orientation of the neural spines from posteriorly directed to vertical, as well as a complete absence of distinguishable lumbar ribs. This arrangement indicates both the presence of a diaphragm at the back of the thorax, and also the development of dorso-ventral flexibility at, and behind the point of transition between the thoracic and the lumbar regions. Only three sacral vertebrae are present, compared to the five of typical later mammals.

The structure of the morganucodontan pectoral girdle and forelimb indicates the evolution of a greater degree of flexibility and, probably, amplitude of movement compared to the primitive cynodont condition. The coracoid plate is reduced and the acromium process at the base of the scapula blade is large, which together indicate the presence in life of a large supraspinatus muscle originating from the inner face of the scapula blade. However, there is no development of a scapula spine or of a supraspinatus fossa in front of the acromium, which are so characteristic of modern mammalian scapulae. An interclavicle is still present. The glenoid fossa is more open than in cynodonts, which permits a greater range of movements of the humerus at the shoulder joint. The humerus itself is a more slender version of the cynodont humerus. The pelvis and hindlimb are very similar indeed to typical small, modern mammals. The ilium is expanded forwards and is divided by a longitudinal ridge into an upper area for the gluteal musculature and a lower area for the iliacus. It no longer extends posteriorly behind the level of the acetabulum. The pubo-ischiadic plate has rotated backwards and gained a large obturator fenestra. The femur has developed the inturned head, large trochanter major, and a trochanter minor replacing the primitive internal trochanter.

The locomotion of morganucodontans must have resembled a modern small, non-cursorial mammal (Jenkins 1971a, 1974; Jenkins and Parrington 1976). The knee and elbow would have been turned inwards at least halfway towards the body and both fore and hind feet placed underneath it. Both humerus and femur were oriented approximately horizontally and in both cases the principal movements would have been in oblique vertical planes, with large extension–flexion movements of the lower limb at the elbow and knee, respectively. At least half of the amplitude of the stride of the smaller forelimb would have resulted from movement of the shoulder girdle on the ribcage, controlled by the clavicle. The larger hindlimb would have produced virtually all of the locomotory force. As discussed elsewhere (page 113), possible benefits of re-modelling of the limb action in mammals, particularly getting the feet close to the midline, include both increased agility, and increased ventilation for sustained aerobic activity.

There are several genera of morganucodontans known, distinguished from one another mainly on differences in molar tooth form. Most are still represented only by isolated teeth from various localities in Europe (Clemens 1986; Kielan-Jaworowska et al. 2004), but two virtually complete skeletons have been found in the Red Beds of the Stormberg Series of South Africa (Crompton 1974; Crompton and Jenkins 1978), which are dated as Late Triassic, possibly Norian but otherwise Rhaetian. In *Megazostrodon* (Fig. 5.3(d)), the three main cusps of the upper molar teeth form an obtuse triangle, and the surrounding cingulum is better developed and bears far larger cingular cusps than in *Morganucodon*. The other South African form is *Erythrotherium*, represented by a juvenile specimen that is very similar to *Morganucodon*.

North American morganucodontans have been discovered in the Kayenta Formation of Arizona, which is probably Lower Jurassic in age (Sues 1986b). There are specimens of *Morganucodon* itself, but the genus *Dinnetherium* is more interesting (Jenkins et al. 1983; Crompton and Luo 1993). It is very similar to other morganucodontids in most respects, but the central cusp on both the upper and the lower molars is relatively higher (Fig. 5.3(e)). The wear facets on the teeth indicate that there was a radical difference in the way in which occlusion occurred. Those on the lower teeth are almost vertical, but those on the upper tooth almost horizontal. This means that as a lower tooth worked against an opposing upper, the lower jaw must have rotated about its long axis to a much greater extent than in *Morganucodon*, causing the outer surface of the lower tooth to face almost upwards as it moved in a medial direction. The morphology of the lower jaw was modified in several ways to accommodate this rotation. First, the symphysis was mobile and almost horizontally orientated to permit a degree of differential movement between the paired jaw rami. Second, the ventral edge of the dentary is complex, with a process referred to by Jenkins *et al.* (1983) as a pseudoangular process lying well in front of the jaw hinge. There is also a flange developed below the jaw hinge, which may be equivalent to the true angular process of later mammals. These modifications indicate an elaboration of the masseter musculature in order to exaggerate the tendency that muscle already has to rotate the lower part of the jaw laterally.

Docodonta

Docodontans are characterised by molar teeth whose crowns have expanded to form complex occluding surfaces (Fig. 5.5(c)). Indeed, they are comparable to the tribosphenic teeth of the modern mammal groups, which achieved similar functional ends by entirely separate means. Docodontans are believed to occur only in Jurassic sediments, the earliest from the Middle Jurassic of the Isle of Skye (Waldman and Savage 1972), and the latest from several Upper Jurassic localities of North America and Europe. Pascual *et al.* (2000) described a jaw fragment bearing three teeth as a latest Cretaceous docodontan from South America. Named *Reigitherium*, it would indicate a remarkably long Gondwanan radiation of the group in isolation of the northern continents. However, the pattern of cusps of its very worn teeth is quite different from that of other docodontans, and therefore its inclusion in the group is regarded as very dubious by Kielan-Jaworowska *et al.* (2004). As with most

Mesozoic mammal groups, docodontans were until relatively recently known only from teeth and jaw fragments. The discovery by Walter Kühne of a largely complete specimen of *Haldanodon*, from the Guimarota lignite mine in Portugal (Henkel and Krusat 1980) was actually the first discovery of any Upper Jurassic mammalian skeleton. Its postcranial bones have not yet been fully described, but Lillegraven and Krusat (1991) have given a very detailed account of the skull (Fig. 5.5(a) and (b)).

The molar teeth of docodontans are expanded across the jaw to increase the area of occlusion between the uppers and the lowers. The result is that there is a crushing basin between the anterior lingual parts of the molars that worked analogously to the protocone-talonid system of tribosphenic teeth described later. A comparison of the dentition of *Morganucodon*, the primitive docodontan *Haldanodon*, and the advanced, Upper Jurassic *Docodon* (Fig. 5.5(d)) illustrates how docodontan teeth may have evolved from a hypothetical primitive condition (Crompton and Jenkins 1968). What in effect occurred was that the upper molars evolved a lingual extension that acted against the inner part of the crown of the lower molars. Instead of occlusion and therefore shearing being restricted to the sides of the main cusps, now the whole of the crown was involved. Exactly what diet this was adapted for can only be guessed at. Kron (1979) speculated that it involved a greater degree of simple crushing and grinding, associated perhaps with an omnivorous or frugivorous diet. This suggests that docodontans were one of the several superficially rodent analogues of the Mesozoic, incorporating a significant percentage of high-energy plant material into their hitherto largely insectivorous food. The limb bones of *Haldanodon* are stout and apparently horizontally oriented, which led Krusat (1991) to suggest that the animal was adapted for burrowing. Kielan-Jaworowska *et al.* (2004) proposed an alternative interpretation, that it was semiaquatic, living in a swampy environment that is preserved as the Guimarota Lignite mine.

Views about the phylogenetic relationships of docodontans have shifted widely over the years. At one time the specialised, complex molar teeth were considered homologous with the triangulated molars of therians, and at another they were compared to

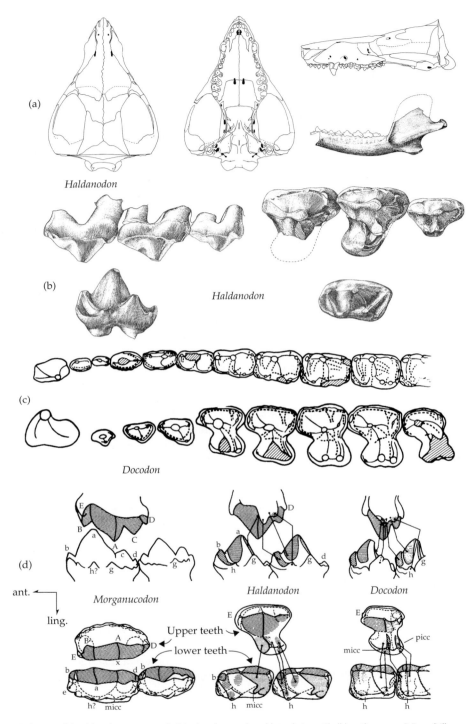

Figure 5.5 Docodontans. (a) *Haldanodon expectatus* skull in dorsal, ventral, and lateral views. Skull length approx. 3.5 cm (Lillegraven and Krusat 1991) and Lower jaw in lateral view (Krusat 1980). (b) Upper molars of *Haldanodon expectatus* in buccal view. Lower molar in lingual view; Upper molars in occlusal view. Lower molar in occlusal view (Krusat 1980). (c) Occlusal view of complete lower and upper postcranial dentition of *Docodon*. Approx. length of tooth rows 1.6 cm (Jenkins 1969). (d) Side and occlusal views of a hypothetical transition series from a *Morganucodon*-like ancestor through a primitive docodontan such as *Haldanodon* to an advanced form such as *Docodon*. Shaded areas are wear facets (Kron 1979, after Hopson and Crompton 1969).

morganucodontan teeth. However, despite the highly modified dentition and its Upper Jurassic age, the skull of *Haldanodon* proved to be rather surprising in the primitive features it possesses. These are relatively large articular and quadrate bones and also stapes, a large orbital fissure, which is the space anterior to the epipterygoid and posterior to the interorbital septum, and a relatively large septomaxilla bone in the snout. This led Lillegraven and Krusat (1991) to argue that docodontans must have diverged at a very early stage of mammalian evolution, prior to the Morganucodonta and possibly even the *Sinoconodon* lineages. However, the more recent cladistic analyses, for example, Rougier, Wible and Hopson (1996), and Luo *et al*. (2002; Kielan-Jaworowska *et al*. 2004), find that Docodonta are more derived than Morganucodonta, although still occupying a basal position relative to all other mammal taxa (Fig. 5.23).

Hadrocodium

Hadrocodium (Fig. 5.6) is represented by the almost complete sub-adult skull of a tiny mammal from the Lower Jurassic of the Lower Lufeng Formation of Yunnan in China, the locality that has also yielded *Sinoconodon* and the Chinese *Morganucodon*. As described by Luo *et al*. (2001b), *Hadrocodium*

combines yet another suite of primitive and more derived characters. The skull is a mere 1.2 cm in length from which the body weight is estimated to have been about 2 g, which compares closely with the smallest of the living mammals. The dentition is generally morganucodontan-like with the main cusps of the molars in a linear row. However, unlike morganucodontans there is no postdentary trough in the dentary, which presumably indicates that the postdentary bones and quadrate had lost their contact with the dentary and become independently attached ear ossicles. The promontorium marking the housing of the cochlea organ is more prominent. In two other important characters, *Hadrocodium* is more advanced than morganucodontans. The jaw articulation is relatively more anteriorly placed, and brain is distinctly enlarged, even after making allowance for the allometric relationship of brain to body size in mammals. These are all features of mammals that otherwise do not appear in the fossils record for another 45 million years. Thus *Hadrocodium* is an important form in reconstructing the pattern of evolution of modern mammalian characters, combining as it does a relatively primitive dentition with a derived structure of the ear, brain, and braincase. It suggests that convergent evolution of advanced auditory characters occurred during the early radiation of the group.

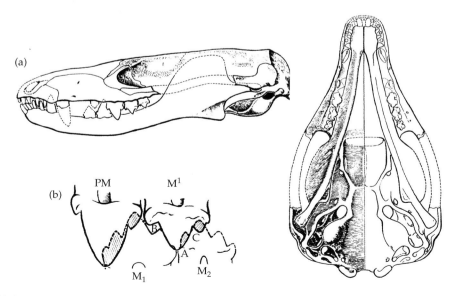

Figure 5.6 (a) Skull of *Hadrocodium* in lateral and dorsal views. Skull length approx. 1.2 cm. (b) Enlarged view of last upper premolar and first molar in occlusion with first two lower molars (Luo *et al*. 2001a).

Eutriconodonta

The taxon Triconodonta was erected in the nineteenth century for those Mesozoic mammals whose molar teeth possessed three main cusps arranged in a longitudinal row along the length of the crown. For a brief period after its discovery in the 1950s, *Morganucodon* was included in the group but it soon became clear that morganucodontans have many characters, including several primitive ones, not found in other tricodonodont-toothed mammals.

There is still considerable doubt about whether the remaining forms included in the group, now renamed Eutriconodonta, are monophyletic (Kielan-Jaworowska *et al.* 2004).

The family Triconodontidae lies at the core. It occurs from the Middle Jurassic through to the Late Cretaceous, and is characterised by the approximately equal size of the three main cusps, the well-developed cingulum, and particularly by a unique tongue and groove mechanism for interlocking

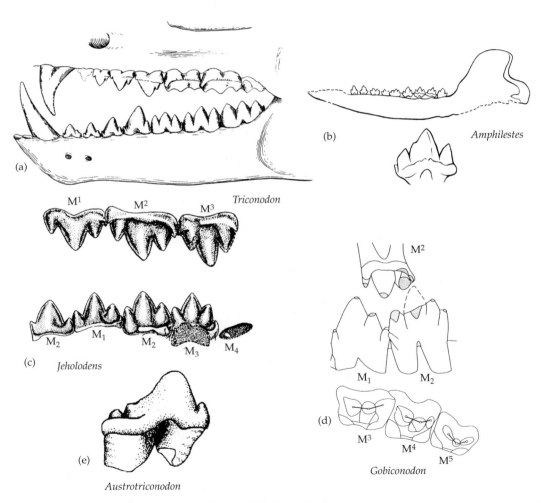

Figure 5.7 Eutriconodontan dentitions. (a) *Triconodon* (Simpson 1928). (b) *Amphilestes* lower jaw and molar tooth (Jenkins and Crompton 1979, after Owen 1971). (c) Upper molars and lower last premolar and molars of *Jeholodens jenkinsi*. Length of upper row approx. 4.2 cm (Ji *et al.* 1999). (d) *Gobiconodon* an upper and two lower molars in lateral view, and last three upper molars in crown view (Kielan-Jaworowska and Dashzeveg 1998). (e) Isolated molar of *Austrotriconodon* (Bonaparte 1992).

adjacent lower molars. A small extension, or cusp at the back of one molar fits into a corresponding concave area at the front of the molar behind. Presumably this was a way of helping to stabilise an individual lower tooth, while it occluded powerfully against an upper. *Triconodon* itself is typical (Fig. 5.7(a)). It occurs in the Upper Jurassic of Great Britain, and was amongst the larger of the Mesozoic mammals. Although not known from more than fragmentary material, there is a complete lower jaw of *Tricondon ferox*, which is about 8 cm in length, suggesting that it was a carnivorous animal about the size of a modern genet.

While the great majority of tricondontids are known only as isolated jaw and dental fragments, a complete skeleton from the Late Jurassic or Early Cretaceous of China, *Jeholodens*, has come to light (Ji *et al.* 1999). It is one of several beautifully preserved kinds of mammal skeletons found in the extraordinary Yixian Formation in Liaoning Province, which is renowned for early bird skeletons (Lucas 2001; Zhou *et al.* 2003). The skeleton of *Jeholodens* (Fig. 5.8(b)) is that of a small animal,

a mere 8 cm in presacral length with a long, slender tail. Its limb and foot structure suggest it was ground dwelling rather than arboreal in life, because there is no sign of marked pedal flexibility or long, curved claws. The structure of the molar teeth (Fig. 5.7(c)) differs from typical triconodontids. Each molar, upper and lower, has the triconodont condition of three major cusps on the crown arranged linearly from front to back, but in this case the middle cusp is distinctly higher than the anterior and posterior accessory cusps. The cingulum is weak and incomplete in the upper molars, and both upper cingulum and the better-developed lower cingulum lack cuspules. These are probably primitive features of the teeth, but *Jeholodens* does have the interlocking arrangement of the lower molars found in tricondontids. The postcranial skeleton of *Jeholodens* reveals an unexpected mosaic of characters, most notably a combination of primitive features of the hindlimb with characters otherwise found in more derived mammals in the forelimb. The pelvic girdle is very similar to that of morganucodontans, there is only a weakly developed

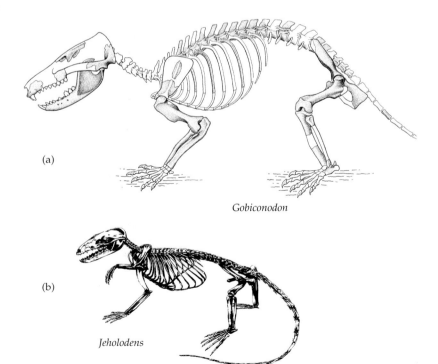

(a)

Gobiconodon

(b)

Jeholodens

Figure 5.8 Eutriconodontan skeletons. (a) *Gobiconodon ostromi*. Presacral length approx. 33 cm (Jenkins and Schaff 1988). (b) *Jeholodens jenkinsi*. Presacral length approx. 4 cm (Qiang *et al.* 1999).

patella groove on the end of the femur, and the two proximal ankle bones, astragalus and calcaneum lie side by side, rather than the astragalus superposed on the calcaneum to create a joint for supination of the foot. However, the shoulder girdle has evolved a true supraspinatus fossa anterior to a scapular spine, a character of mammal groups more progressive than morganucodontans and docodontans.

Amongst the original jaws known from the classic Middle Jurassic Stonesfield Slate of Oxfordshire, first described early in the nineteenth century, there are two kinds that have the three cusps of the molar teeth in a line, but as in *Jeholodens*, the middle one is higher than the anterior or posterior ones. They have symmetrical premolars, and lack the tongue and groove interlocking between the molars that is found in triconodontids. Thus *Amphilestes* (Fig. 5.7(b)) and *Phascolotherium*, are placed in the separate family Amphilestidae.

Forms with similar teeth have been described from North America, including two incomplete skeletons of *Gobiconodon* (Jenkins and Schaff 1988). *Gobiconodon* (Fig. 5.8(a)) was a relatively very large Mesozoic mammal, even more so than *Triconodon*. The skull length was about 10 cm, and the presacral body length 35 cm and quite robustly built. Like the amphilestids, the central cusp of the molar teeth of *Gobiconodon* is taller than the accessory cusps (Fig. 5.7(d)), but differs in being slightly offset medially from the latter, causing a slight degree of obtuse triangulation of the three cusps. The adjacent lower molars have evolved a different interlocking system from that found in triconodontids, for here the back of one molar tooth bears a cingulum cusp that inserts between a pair of cingular cusps on the front of the next tooth. Isolated jaw fragments and teeth attributed to *Gobiconodon* have been found in Mongolia (Kielan-Jaworowska and Dashzeveg 1998), and a somewhat younger, Early Cretaceous Chinese genus, *Repenomamus*, has been described that is very similar to *Gobiconodon* but was of even larger body size, with a 11 cm long skull (Wang *et al.* 2001). As a whole, this group seems to have consisted of the nearest thing to specialised carnivores that evolved amongst the Mesozoic mammals. With a body size that of a smallish cat and sharp almost carnassial-like teeth, they were no doubt capable of capturing and consuming other, smaller mammals.

The third group of triconodont mammals to consider is represented only by *Austrotriconodon*, from Late Cretaceous times in South America (Bonaparte 1994). At present they are known from isolated teeth that occur quite commonly in the Los Alamitos Formation of Patagonia, which is probably Campanian in age. The molariform teeth (Fig. 5.7(e)), while basically triconodont in form, have the middle cusp very much larger, and the other cusps relatively very small. Despite the obvious biogeographical interest, it is not possible to be at all confident about their relationships to other triconodont-toothed mammals.

Multituberculata

Multituberculates are unambiguously recognisable as a specialised, monophyletic group defined by very clear-cut characteristics of the skull, dentition and, as more specimens are described, by the postcranial skeleton as well. They are superficially rodent-like in structure and inferred habits, and range in size from that of a small mouse to that of a marmot. They make their first appearance in the fossil record relatively late for a major Mesozoic mammal group, in the Middle Jurassic of Europe and Upper Jurassic of North America, but having remained abundant throughout the Cretaceous, several lineages survived the end-Cretaceous and occur well into the Tertiary. Indeed the group reached considerable diversity in the Palaeocene, while the youngest record is from the North American Late Eocene (at one time believed to be Oligocene, Krishtalka *et al.* 1982, but see Swisher and Prothero 1990). Geographically, the vast majority of fossils are from Europe, North America, and Asia. Gondwanan representation is limited to one or two possible fragments from the Late Cretaceous of South America, and isolated teeth from the Early Cretaceous of Morocco (Sigogneau-Russell 1991).

Krause *et al.* (1992) described some highly unusual, isolated teeth from the Late Cretaceous (Campanian) Los Alamitos Formation of Argentina, which they named *Ferugliotherium*, and regarded as a specialised group of multituberculates. However, when Pascual *et al.* (1999) described another form, the Palaeocene *Sudamerica* (5.9e), in which they showed that there are four molars teeth, and no

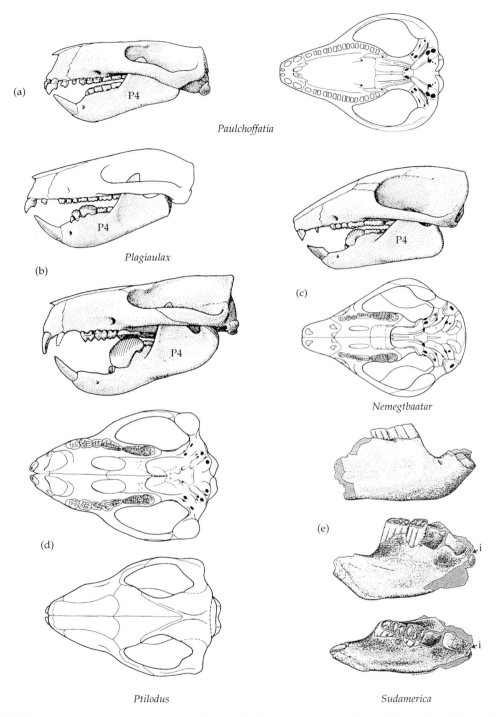

Paulchoffatia

Plagiaulax

Nemegtbaatar

Ptilodus

Sudamerica

Figure 5.9 Multituberculate skulls and Gondwanatheria. (a) *Paulchoffatia*. (Kielan-Jaworowska and Nessov 1992 and Maio 1993). (b) *Plagiaulax*. (Kielan-Jaworowska and Nessov 1992) (c) *Nemegtbaatar* (Kielan-Jaworowska and Nessov 1992 and Maio 1993). (d) *Ptilodus* in lateral, ventral and dorsal views (Kielan-Jaworowska and Nessov 1992 and Maio 1993). (e) *Sudamerica* fragment of jaw and teeth (Pascual *et al.* 1999).

blade-like lower premolar tooth, it became clear that these are not multituberculates at all, but a quite separate group, Gondwanatheria. They have since been described from Madagascar and India as well (Krause *et al.* 1997), and there is a possible, but virtually indeterminate jaw fragment from Tanzania (Krause *et al.* 2003). Thus the group is restricted to the Gondwanan continents. The molar teeth of gondwanatheres are hypsodont, and have transverse enamel ridges that develop on the occlusal surface, once the softer dentine and enamel has worn down. They were presumably yet another group of Mesozoic mammals adopting the rodent-like habitat of small herbivore, feeding on tough, high-energy vegetable matter. In the absence of more material their relationships remain totally obscure.

Diversity and evolution

The skull of multituberculates (Fig. 5.9) is broad with a short, blunt snout, and laterally facing orbits. The jugal, long thought to be absent, is now known to consist of a very thin plate attached to the inner face of the otherwise robust zygomatic arch (Hopson *et al.* 1989). Other cranial characters in which they differ from other mammals are the absence of a component of the palatine bone in the orbital wall, although Hurum (1994) noted some exceptions, and the nature of the specialised jaw joint permitting antero-posterior movements of the lower jaw. Perhaps to free the jaw for this propalinal movement, multituberculates possessed ear ossicles free of any connection to the dentary bone (Fig. 5.12(b)), and which were therefore more or less identical to those of living mammals (Miao and Lillegraven 1986; Hurum *et al.* 1996). The dentition is the most strikingly unique character of the group. The upper incisors are reduced in number to a maximum of three, the second of which is enlarged, and worked against the single, large, procumbent lower incisor. The lower premolars are equally distinctive, with serrated, blade-like crowns that had a shearing or chopping action against the multicusped upper premolars. The molars are reduced to two in number. Both the uppers and the lowers have two rows (three in the upper molars of more advanced forms) of up to eight blunt cusps arranged longitudinally, and with no connecting ridges between them. When in occlusion, the rows of cusps of the upper

molars fit into the valleys between cusp rows of the lowers, and vice versa, and the posterior shift of the lower jaw then caused a very effective grinding action, as discussed shortly.

The traditional classification of the approximately 70 genera of multituberculates was into three superorders, the Plagiaulacoidea for the more primitive ones of the Upper Jurassic and European Early Cretaceous; the Ptilodontoidea for the advanced, predominantly North American forms; and the Taeniolabidoidea for the advanced, predominantly Asian forms (e.g. Clemens and Kielan-Jaworowska 1979; Hahn and Hahn 1983). However, subsequent attempts at formal cladistic analyses have amply confirmed the suspicion that this classification is a very poor reflection of the phylogeny of the group (Simmons 1993; Rougier *et al.* 1997; Kielan-Jaworowska and Hurum 1997). On the other hand, there is far from a consensus amongst these authors on what the true interrelationships are. In the most recent review, Kielan-Jaworowska and Hurum (2001) analysed 62 characters amongst 32 taxa and obtained 17,783 equally parsimonious trees. This extremely low resolution results from a combination of conservatism in many characters, extensive convergence and mosaic evolution particularly in the dentition, and the inevitable problem of missing characters in many of the taxa that are known from very incomplete material. At any event, they were unable to derive from the cladogram a satisfactory cladistic classification, and instead proposed to retain the admittedly paraphyletic group 'Plagiaulacida' for the generally primitive forms such as *Plagiaulax* itself, and a possible but by no means very strongly supported group Cimolodonta for many but not all the more advanced forms (Kielan-Jaworowska *et al.* 2004).

Paulchoffia is the most plesiomorphic 'plagiaulacid' (Fig. 5.9(a)). It comes from the Upper Jurassic Guimarota Lignite Mine in Portugal and has retained what is presumed to be the primitive multituberculate dentition of I3/1: C0/0: PM5–4/4–3: M2/2. The lower premolars differ from those of all other multituberculates in their rectangular shape, and horizontal wear facets on the crowns that indicate a crushing rather than shearing function. The molar teeth have a relatively small number of cusps, there being only three or four in each row. The North

American Morrison Formation genus *Ctenacodon* is of a similar age and degree of primitiveness, although it is not apparently closely related to *Paulchoffia*. Indeed, it illustrates well the plesiomorphy of 'Plagiaulacida' because of its mosaic of primitive and advanced features. On the one hand it has retained a primitive molar structure with few cusps and unornamented enamel, but on the other it has more advanced, shearing lower premolars.

The 'advanced' multituberculates consisting of most of the forms once classified as Ptilodontoidea and Taeniolaboidea together constitute the Cimolodonta, which dominated the Late Cretaceous and early Tertiary radiation of the group. Dental characters defining Cimolodonta include loss of the first upper incisor, and reduction of the number of premolars to four uppers and two lowers, coupled with a tendency to concentrate the shearing action on the hindmost ones only. Several of the groups, probably independently, evolved a third row of cusps on the upper molars. Kielan-Jaworowska and Hurum (2001) recognise a number of subgroupings of cimolodonts. One includes the Ptilodontoidea, which are easily recognised by a huge, arch-shaped lower premolar bearing heavy vertical striations (Fig. 5.9(d)). Most, though confusingly not all ptilodontoids have modified the dental enamel to a microprismatic type in which the prisms are small and densely packed, in contrast to the gigantoprismatic enamel of most other groups. Ptilodontoids were predominantly, though not exclusively North American and were one of the groups that survived well into the Tertiary. Another advanced grouping includes a series of largely Asian forms, the djadochtatherioids (Fig. 5.9(c)) plus the cosmopolitan taeniolabidoids. A character shared by all these is restriction of the enamel on the lower incisors to the antero-ventral surface, which gives them a self-sharpening ability analogous to rodent incisors. Other characters are distinctly separate molar cusps and smooth molar enamel. However, all these features do occur convergently in other multituberculate forms that are not clearly recognisable as members of any of these groups.

Functional biology

This inability to resolve the phylogenetic interrelationships of the multituberculates arises in large part because they are a very conservative group, with much minor convergent change but little in the way of clear cut, consistent evolutionary trends evident. Functionally, however, more and more subtle differences have been coming to light in such things as implied diet and mode of locomotion. Two aspects of their biology important for understanding the nature of multituberculates are first that they were small mammals, mostly in the mouse to rat range and even the Tertiary 'giants' were no larger than a good-sized marmot; Krause (1982) estimated that the head plus body length of the ptilodontoids ranged from 6 to 22 cm. Second, multituberculates were very abundant. Typically, something like 75% of the specimens found at Late Cretaceous or Palaeocene mammal localities are multituberculates. On both these points a comparison with modern rodents is compelling, quite apart from any anatomical similarities,

Krause (1982) gave the first detailed analysis of the jaw function of a multituberculate, in this case the North American form *Ptilodus*, mainly by looking at the dental wear facets and the microstriations across them (Fig. 5.10(a)). He showed first that the upper and lower incisors could not have made direct contact, and that the wear on them consisted only of a general abrasion of the tips. Therefore, despite their relatively large size, they could not have been used for gnawing as in modern rodents, but only for collecting and ingesting food items. The premolars underwent a crushing–slicing cycle, with the large, striated, blade-like fourth lower premolar cutting against the occlusal surface of the large, multicusped upper fourth premolar. This was a strictly orthal movement and resulted in horizontal wear of the blade of the lower tooth and the rows of cusps of the upper tooth. The action of the molars was a grinding cycle (Fig. 5.10(b)) in which the lower jaw was protracted, the lower molars brought into occlusion with the upper molars, and then the jaw powerfully retracted. This led to a forceful trituration of food caught between the upper and lower rows of crescentic cusps occupying the molar crowns. Although not possible to be certain of the diet of an animal with this kind of dental action, Krause proposed that it would have been more likely a mixed, omnivorous one than strictly herbivorous. Evidence from the extent of

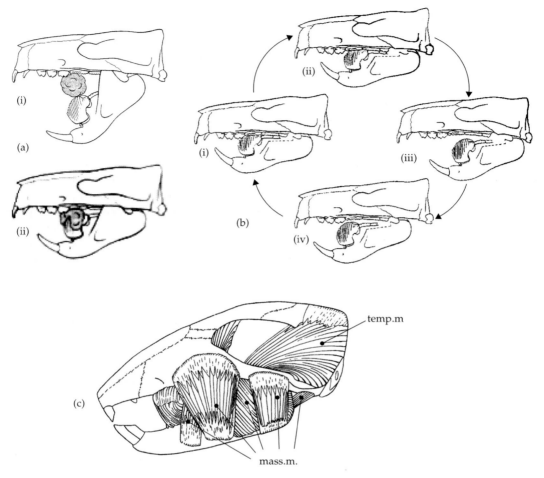

Figure 5.10 Jaw action of multituberculates. (a) Slicing action of premolars and (b) triturating cycle of molar teeth of *Ptilodus* (Krause 1982). (c) Reconstruction of the adductor musculature of a djadochtatherioid (Gambaryon and Kielan-Jaworowska 1995). mass.m, branches of the masseter musculature; temp.m, temporalis muscle.

general abrasion of the teeth, including the more anterior upper premolars and the fourth lower premolar suggests that relatively large, quite tough items were included, such as hard seeds. Furthermore, by comparison with very small modern mammals, a diet of leaves alone would be unlikely to have been nutritionally adequate, and almost certainly a mixture of seeds, tubers, insects, and worms would have been necessary to maintain the relatively high metabolic rates expected in such creatures.

Gambaryan and Kielan-Jaworowska (1995) have reconstructed the jaw musculature of the djadochtatherioids of Asia (Fig. 5.10(c)), which evidently had a similar mode of feeding action to *Ptilodus*. Not surprisingly, the authors demonstrated that the masseter muscle was complex, and that its insertion on the lower jaw extended far anteriorly compared to other mammals, giving it a large vertically directed bite force. The coronoid process of the dentary is well forward, so that the temporalis muscle would have had an exaggerated posterior component during chewing. One characteristic of members of this group, and of the taeniolabidoids is the restriction of the enamel layer to the anteror-ventral surface of the lower incisor. As in rodents and lagomorphs, this gives the tooth a self-sharpening property. As the dentine wears away at a higher rate than the enamel, the latter remains as a sharp, cutting edge at the front of the tooth. The glenoid fossa of the jaw articulation in these forms is extended forwards increasing the maximum anterior shift of the lower jaw. Their upper and lower incisors could therefore come into direct contact for active biting, unlike those of *Ptilodus*. The lower premolar teeth are smaller than those of ptilodontoids, confirming that the premolars had to some extent given way to the incisors in the initial preparation of the food, possibly indicating a diet of seeds that required de-husking prior to mastication.

Several practically complete and many partial postcranial skeletons and isolated bones of multituberculates are known (Fig. 5.11). The skeleton shows a number of similarities to that of modern therian mammals, but at the same time there are several quite different features. This has led to a divergence of views both about the functional anatomy, and about how far the postcranial skeleton does or does not support a relationship between multituberculates and therian mammals. In the forelimb, the scapula is narrow with an outturned anterior spine which, as interpreted by Sereno and McKenna (1995), means there was no development of a supraspinous fossa. However, Kielan-Jaworowska and Gambaryan (1994) regard the antero-medial face of the spine as an incipient supraspinous fossa, onto which the equivalent of the supraspinatus muscle already attached. The coracoid is very reduced as in therians, but remains as a distinctive, pointed process. The glenoid cavity faces more ventrally than laterally, and is only about half the diameter of the bulbous, upturned head of the humerus. Clavicles and interclavicles are present. According to Sereno and McKenna (1995) the elbow joint has a trochlea form similar to that seen in modern therians, and which restricts movements to a hinge action about a single axis. In *Nemegtbaatar*, by contrast, Kielan-Jaworowska and Gambaryan (1994) described the condyles on the end of the humerus for both the radius and the ulna as convexities, and the radial condyle as spherical. From this they infer that the elbow was capable of a wider variety of movements than just strictly hinging about one axis.

The pelvis and hindlimb are also distinct from other mammals in several respects (Krause and Jenkins 1983; Kielan-Jaworowska and Gambaryan 1994). The narrow iliac blade attaches at a relatively very high angle to the line of the sacral vertebrae, while the pubo-ischiadic plate extends deeply below the acetabulum. The acetabulum itself is open dorsally. A pair of epipubic bones is present, although this may be a primitive feature of mammals generally. The articulating head of the femur is on the end of a distinct neck that is set at an angle to the shaft of the bone. A very prominent trochanter major extends well proximal to the head, and a smaller trochanter minor lies on the underside, where the neck and trochanter major meet. There is no sign of a third trochanter. The tibia is unusual in being wider from lateral to medial surfaces than from anterior to posterior, and both tibia and fibula possess a hook-like lateral process on the proximal end of the bone.

There are considerable differences in the interpretation of locomotor function in multituberculates. Jenkins and Krause (1983) argued for an arboreal mode of life in *Ptilodus*, on the basis of the structure of the pes, and the tail (Fig. 5.11(c)). Within the ankle, the joint between the calcaneum and the astragalus permitted extensive flexion-extension. The joint between the distal end of the tibia and the calcaneum consisted of a pair of facets, which permitted the calcaneum to abduct, causing rotation of the distal end of the pes laterally and then posteriorly to the extent that the foot could point almost backwards. They also recognised the possibility of extensive, independent movement of the first digit. Together, these movements would have allowed the foot to be turned backwards and to gain purchase while the animal descended head first down a tree. The tail is very long, possesses well-developed haemal arches on the caudal vertebrae, and there are long neural spines on the sacral vertebrae from which powerful tail muscles could have originated, all of which indicate a prehensile tail.

A second proposed mode of locomotion in multituberculates arises from Sereno and McKenna's (1995) description of a complete, articulated forelimb of the djadochtatherioid *Bulganbaatar*. They argued that its structure indicates a fully therian

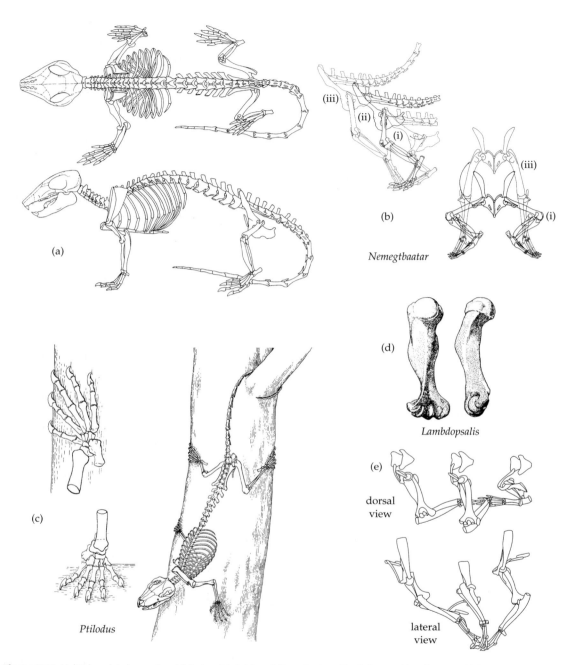

Figure 5.11 Multituberculate locomotion. (a) Postcranial skeleton of *Nemegtbaatar gobiensis*. Presacral length approx. 12.0 cm (Kielan-Jaworowska and Gambaryan 1994). (b) Interpretation of hindlimb function in lateral and anterior views (Kielan-Jaworowska and Gambaryan 1994). (c) *Ptilodus* hindfoot in posteriorly directed and in forwardly directed positions, and whole skeleton descending tree (Jenkins and Krause 1983). (d) Humerus of the presumed burrowing multituberculate *Lambdopsalis bulla*. Length approx. 3.6 cm. (e) Relatively parasagittal action of multituberculate forelimb in dorsal and lateral views, according to Sereno and McKenna (Gambaryan and Kielan-Jaworowska 1997, after Sereno and McKenna 1995).

near parasagittal gait (Fig. 5.11(e)), with the elbow turned back, the forefeet lying under the body, and the whole limb moving in a parasagittal plane. The principal reasons they gave for their interpretation were the reduction of the coracoids, the postero-ventral position of the glenoid and its small size relative to the head of the humerus, and the mobile clavicle–interclavicle joint, all of which are matched in a modern non-cursorial therian such as *Didelphis*. They were not explicit about whether their proposed mode of action of the forelimb is compatible with Jenkins and Krause's (1983) arboreal hypothesis, but presumably it is not.

A third possible mode of locomotion was proposed by Kielan-Jaworowska and Gambaryan (1994; Gambaryan and Kielan-Jaworowska 1997). At least as far as the Asian genera they looked at, such as *Nemegtbaatar*, are concerned (Fig. 5.11(a)), they argued against a parasagittal mode and in favour of a sprawling mode in which both the humerus and femur are held horizontally, at an angle of around 60° to the animal's sagittal plane, so that both the elbow and knee are out to the side. They envisage a form of locomotion similar to, though not exactly matched by modern ricochetal mammals. It consisted of a leaping mode in which the hind limbs acting synchronously create an upward and forward thrust, while the forelimbs, similarly acting together, absorbed the energy of the animal's landing prior to the next leap. According to their account, most of the unique anatomical features of the multituberculate skeleton can be explained by this form of progression. The relatively vertical pelvic girdle, the lateral orientation of the femur, and the form of the musculature as reconstructed by them indicate that thrust was produced only to a limited extent by normal mammalian retraction of the femur. Much of it was generated by adduction, a drawing inwards, of the femora, coupled with extension at the knee and ankle joints, in somewhat frog-like fashion. On landing, the forelimb began in an outstretched position, the humerus partially adducted and the forearm extending antero-laterally. Absorbtion of the kinetic energy of the leap occurred as the distal ends of the humeri were forced upwards. Towards the end of this phase, the elbow joint extended restoring the height of the front of the animal between its forelimbs and preparing it for the next propulsive phase of the hindlimb. Kielan-Jaworowska and Gambaryan's interpretation of locomotion in multituberculates would presumably only apply to rapid, or perhaps predator avoidance locomotion. No doubt these animals were also capable of slow leisurely ambulation, and it seems likely that a more conventional walking mode with the legs acting successively occurred as well.

Yet another mode of locomotion is found in the Late Eocene Asian genus *Lambdopsalis*, which has a number of indications of a fossorial mode of life (Kielan-Jaworowska and Qi 1990; Meng and Miao 1992). The head is broad and flattened, some of the cervical vertebrae fused, and the humerus massive with a huge condyle for articulation with the as yet unknown ulna (Fig. 5.11(d)).

It is thus evident that the multituberculates evolved a range of locomotory adaptations. That there should be arboreal, parasagittal, ricochetal, and fossorial types adds yet further to the view that they were the Mesozoic ecological equivalent of the latter day rodents.

One of the initially unexpected discoveries about multituberculates in recent years has been their possession of true ear ossicles detached from the dentary bone (Fig. 5.12(b)). Meng and Wyss (1995) described the bones of the Chinese form *Lambdopsalis*, showing a close similarity to the corresponding elements of modern mammals. The angular bone forms an ectotympanic for support of the tympanic membrane, and the articular and prearticular are fused to form the malleus. Even the position of the groove for the chorda tympani nerve is the same. Of the modern mammals, the similarity is greatest to the monotremes, because the ectotympanic and therefore the ear drum is oriented horizontally rather than vertically. Hurum *et al.* (1996) confirmed this detailed similarity to modern mammals in the Mongolian *Chulsanbaatar*, although suggesting that the ectotympanic and tympanic membrane were less horizontally inclined than in *Lambdopsalis*. The quality of hearing has been assessed by Meng and Wyss (1995), who suggested that the uncoiled cochlea and the inflated vestibule of multituberculates indicate rather poor reception of high-frequency air-borne sound, and that conduction of low-frequency sound through the bones of the skull was important.

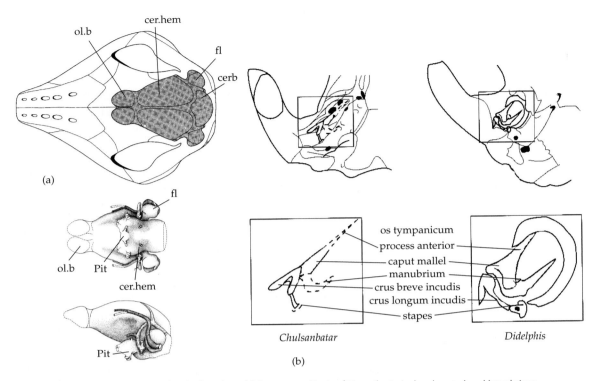

Figure 5.12 Brain and ear ossicles of multituberculates. (a) Reconstructed brain of *Nemegtbaatar* in dorsal, ventral, and lateral views (Kielan-Jaworowska *et al*. 1986). (b) Ear ossicles of *Chulsanbaatar* (left) compared to *Didelphis* (right) *in situ*, and enlarged. (Hurum *et al*. 1996). cerb, cerebellum; cer.hem, cerebral hemisphere; fl, flocculus; ol.b, olfactory bulb; pit, pituitary.

The brain of multituberculates (Fig. 5.12(a)) has been reconstructed from the endocranial casts of a number of genera, particularly *Chulsanbaatar* and *Nemegtbaatar* (Kielan-Jaworowska *et al*. 1986) and *Ptilodus* (Simpson 1937; modified by Krause and Kielan-Jaworowska 1993). All give the same general picture of relatively very large olfactory bulbs and no exposure of midbrain structures dorsally, which suggests that olfaction was a more important sense than vision, and therefore that they had a nocturnal habit. The relative size of the brain of *Ptilodus*, in terms of the Encephalisation Quotient (actual brain weight divided by expected brain weight for a mammal of that body weight) has been estimated by Krause and Kielan-Jaworowska (1993). It lies between a minimum of 0.37 and a maximum of 0.62, depending on which procedure is used, and whether the olfactory bulbs are, or are not included in the estimates of brain mass. Despite the lack of precision, it is clear that the multituberculate brain

was significantly smaller than those of modern mammals.

Multituberculates are one of the extremely few fossil mammal groups from any period where the presence of hair has been positively demonstrated. Skeletal remains of *Lambdopsalis* along with the exceptionally detailed impressions of hair have been found in fossilised coprolites from Late Palaeocene beds of Inner Mongolia (Meng and Wyss 1997). Unfortunately, there are no organic remains of the material, so the attractive prospect of finding out what colour the animal was cannot be satisfied.

Kielan-Jaworowska (1979) speculated that multituberculates were viviparous. She based her argument on the very acute angle at which the two sides of the pelvic girdle meet ventrally, about 40° instead of the 180° or so typical of most mammals. This would allow an egg or neonate with a diameter no greater than about 3.4 mm in *Kryptobaatar* to pass through the birth canal. Relative to body size, no fully developed cleidoic egg that small is known, but it is of the same order of size as a marsupial neonate.

There are few other indications of the details of multituberculate biology. That they occupied an adaptive zone generally comparable to the rodents today seems beyond serious dispute. Their small body size, grinding dentition, and considerable diversity and abundance all point to this. They must have much more disparate in their detailed biology than has yet been shown, with great variation in the details of diet and habitat occupied amongst the 70 known, and presumably many unknown genera.

The final demise of the group right at the end of the Eocene has generally, if uncritically been attributed to competition with advanced therian groups, particularly the rodents, whose explosive radiation commenced during the Eocene. Van Valen and Sloan (1966) suggested that prior to rise of rodents, the Condylarthra and Plesiadapiformes of the Palaeocene had 'weakened' the multituberculates, though there is little evidence for this. A number of suggestions have been made about what might have been the direct cause of the presumed competitive inferiority of multituberculates in the face of placentals, from the vague 'multituberculates remained significantly below the eutherian level of advancement in nearly all areas of their biology' of Hopson (1967) to the specific belief of Kielan-Jaworowska and Gambaryan (1994) that their decline and extinction was a due to their less efficient locomotor system. A hypothesis that the replacement of one higher taxon by anotheris due to competitive interaction is extraordinarily difficult, and usually quite impossible to test (Kemp 1999). At the very least it demands a precise analysis of the pattern of decline of one group and rise of the other. It also requires convincing evidence of adaptations in the two respective groups for utilising an identical resource, over which competition could have occurred. As it happens, Krause (1986) has attempted just such an analysis of this case, showing that there is indeed a significant inverse correlation between both generic diversity, and also abundance of individuals of multituberculates and rodents respectively, through the Late Palaeocene and Eocene of North America.

Basal Holotheria

There is a large group of Mesozoic mammals whose molar teeth have evolved triangulation of the three main cusps. This feature is associated with an exaggeration of the postvallum-prevallid shearing function, that is to say, the front part of a lower tooth shears against the back part of the corresponding upper. In those where it is known, there are also certain minor characters of the petrosal bone. The group embraces very early, primitive forms with no more than the basic triangulated tooth, through several intermediate grades, to the living marsupials and placentals that possess fully expressed tribosphenic molars. Until relatively recently, palaeontologists usually referred to the group as the Theria, confusingly because historically this term was coined for the two groups of living members alone, to distinguish them from the monotremes. With the advent of cladistic analysis (Fig. 5.23) and the associated demand for unambiguous, strictly monophyletic groups, a sequence of essentially node-based taxa have been named to accommodate the sequence of increasingly derived forms. McKenna and Bell (1997) have adopted six such, of which Kielan-Jaworowska *et al.* (2004) recognise four.

1. Holotheria. The group containing all the forms that have triangulated molar cusps, and therefore the clade that contains the common ancestor of *Kuehneotherium*, the living groups, and all its descendants. Kielan-Jaworowska *et al.* (2004) exclude this taxon because they believe the evidence that *Kuehneotherium* is related to the rest is inadequate.
2. Trechnotheria. The clade consisting of the common ancestor of 'Symmetrodonta', the living groups, and all its descendants.
3. Cladotheria. The clade consisting of the common ancestor of 'Eupantotheria', the living groups, and all its descendants.
4. Zatheria. The clade consisting of the common ancestor of *Peramus*, the living groups, and all its ancestors.
5. Tribosphenida. The clade consisting of the common ancestor of Aegialodontidae, the living groups, and all its descendants. Kielan-Jaworowska *et al.* (2004) refers to this clade as Boreosphenida for reasons discussed below.

6. Theria. The common ancestor of the living groups, Marsupialia and Placentalia and all descendants. Kielan-Jaworowska *et al.* (2004) do not use this taxon because there are certain Cretaceous forms that are at the same grade, but whose relationships with marsupials and placentals are unclear: they leave Tribosphenida unresolved.

Kuehneotherium

As well as the abundant *Morganucodon*, a second kind of mammal is found amongst the fragments of bone in the Late Triassic or Early Jurassic fissure deposits of South Wales. *Kuehneotherium* is represented only by rare isolated teeth and a very few fragmentary dentaries (Kermack *et al.* 1968; Parrington 1971). Of course, since the cranial and postcranial bones of the two forms may have been virtually identical, it is possible that a fraction of the abundant skeletal fragments may pertain to *Kuehneotherium*. For all its rarity, *Kuehneotherium* was immediately considered to be very important, for it was believed to be the earliest and most primitive member of the Holotheria. The lower jaw of *Kuehneotherium* (Fig. 5.13(a)) is relatively slender and lacks the distinct angular process of morganucodontids. Postdentary bones have not been found, but the form of the trough on the inner face of the dentary indicates that they were present as in *Morganucodon*, and therefore independent ear ossicles had not evolved. Gill (1974) quotes a lower dental formula of I4: C1: PM6: M4–5, with absorption rather than replacement of the anterior premolars as occurs in *Morganucodon*. The molar teeth are characterised by the triangular arrangement of three main cusps on both the upper and the lower teeth. These are the homologues of identifiable cusps of more advanced holotherian mammals, and can consequently be named in accordance with the nomenclature used for the latter (Fig. 5.13(b) and (c)). The central cusp of the upper molar is the paracone and is much the largest cusp. In front of and slightly labial to it is the anterior accessory cusp or stylocone, while behind, and also labially displaced lies the metacone. The three are connected by a sharp crest, and a shelf-like cingulum lacking cuspules surrounds most of the base of the crown. The lower molars are taller than the uppers, but have a comparable arrangement of cusps, a large protoconid in the middle, and smaller,

lingually displaced paraconid in front and metaconid behind. The lower molars have an additional, small cusp at the back of the tooth which is an incipient talonid, a structure that achieved greater anatomical and functional significance in advanced holotherians. At this stage, however, its main role seems to be as part of the interlocking mechanism of adjacent molars. The cingulum of the lowers is incomplete externally.

The function of the molar teeth was first analysed from the wear facets by Crompton (1971). The upper and lower molars tend to form interlocking reversed triangles (Fig. 5.13(d)). As in morganucodontids, the crests between the main cusps form opposing shearing edges, but the difference is that these edges lie obliquely across the line of the jaw, rather than longitudinally as in *Morganucodon*. According to Crompton, this arrangement improves the efficiency of the system by creating an interlocking effect between the upper and lower rows of crests. There is less of a tendency for food trapped between opposing shearing edges to force those edges apart.

Kuehneotherium held a pivotal role in the development of modern ideas about the evolution of mammals because of its interpretation as the most plesiomorphic holotherian mammal. However, there are some doubts about its phylogenetic position (Kielan-Jaworowska *et al.* 2004). First, there are other mammals, the eutriconodontans, which resemble more advanced holotherians than *Kuehneotherium* in having lost the postdentary trough, and yet which have the non-holotherian linear arrangement of cusps. Second, there are a number of other groups that have independently evolved a similar degree of triangulation of the molar cusps, such as the morganucodontan *Megazostrodon* (Fig. 5.3(d)) and the eutriconodontan *Gobiconodon* (Fig. 5.7(d)). Discovery of more complete material of *Kuehneotherium* will be needed to resolve this issue.

'Symmetrodonta'

Until recently, the symmetrodonts have been known only from isolated teeth, incomplete jaws, and very occasional fragments of the postcranial skeleton of Upper Jurassic and Early Cretaceous age. They have molar teeth similar to those of *Kuehneotherium* in the triangular arrangement of the three main molar cusps and the absence of a talonid in the lowers.

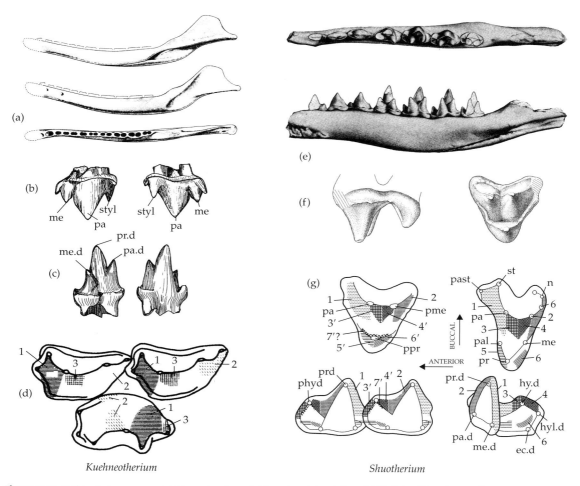

Figure 5.13 *Kuehneotherium praecursoris* (a) Lower jaw in medial, lateral, and dorsal views. (b) Medial and lateral views of upper molar. (c) Medial and lateral views of lower molar. (d) Occlusal diagram of two upper and one lower molars. Length of jaw approx. 1.4 cm (Kemp 1982, from Kermack *et al*. 1968). (e) *Shuotherium dongi* lower jaw in dorsal and medial views. Length of jaw as preserved approx. 1.2 cm (Chow and Rich 1982). (f) Upper molar of *Shuotherium shilongi* in anterior and occlusal views (Wang *et al*. 1998). (g) Postulated occlusal relationships between upper and lower molars of *Shuotherium* (left) compared to a tribosphenidan (right) (Wang *et al*. 1998).

In fact they tend to be even simpler because of the relatively weak development of the cingulum, presumably a secondary reduction. Some have teeth in which the angle between the main cusps is obtuse, such as *Tinodon* from the Jurassic of North America (Fig. 5.14(a)), and the Early Cretaceous *Gobiotherodon* from Mongolia. Given that this pattern of cusps is similar to that of *Kuehneotherium* it is presumably an ancestral feature and the obtuse-angled 'symmetrodontans' are almost certainly paraphyletic (Luo *et al*. 2002).

Others, the spalacotheriids, are the acute-angled 'symmetrodontans', in which the angle between the three main cusps is less than 90°. This group shares certain features with the remainder of the Holotheria, indicating a relationship. The most important is the exaggeration of the postvallum-prevallid shearing action, between the back of the upper molar and the front of the lower, which is the functional reason for the development of the acute angulation of the cusps. *Spalacotherium*, which occurs in Late Jurassic to Early Cretaceous deposits

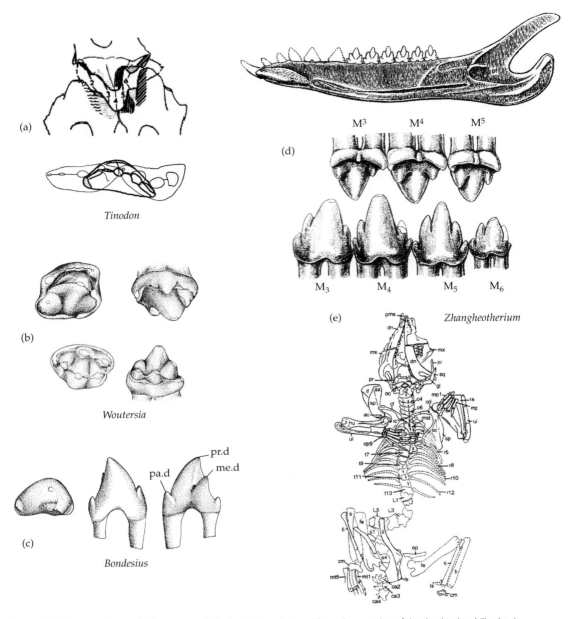

Figure 5.14 Symmetrodontans. (a) Upper molar of *Eurylambda* in occlusion with two lower molars of the closely related *Tinodon*, in bucco-lingual and occlusal views (Crompton and Jenkins 1967). (b) Isolated upper and lower molars of *Woutersia butleri* in occlusal and lingual views (Sigogneau-Russell and Hahn 1995). (c) Isolated right lower molar of *Bondesius ferox* in occlusal, labial, and lingual viens (Bonaparte 1990). (d) Lower jaw of *Zhangheotherium quinquecuspidens* in medial view, with enlarged views of upper and lower molars. Length of jaw approx. 2.5 cm (Hue *et al*. 1997). (e) Skeleton of *Zhangheotherium quinquecuspidens* as preserved. Maximum length approx. 13 cm (Hu *et al*. 1997).

in North America and Europe, is a typical example. Hu *et al.* (1997) described a largely complete spalacotheriid skeleton from the Early Cretaceous Yixian Formation of northeastern China (Zhou *et al.* 2003).

Named *Zhangeotherium* (Fig. 5.14(d) and (e)), it has added enormously to the detailed knowledge of a basal holotherian, and reveals a number of additional characters shared with the later holotherians.

(a)

(b)

Anterior

Lingual

Crusafontia

Henkelotherium

(c)

(d)

Leonardus

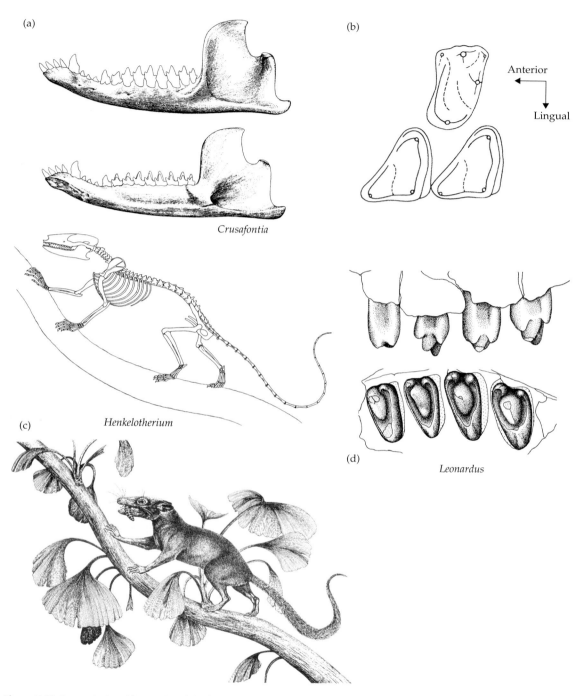

Figure 5.15 Eupantotherians. (a) Lower jaw of the dryolestid *Crusafontia cuencana* in lateral and medial views. Length approx. 2 cm. (b) Occlusal view of one upper and two lower molars of a typical dryolestid (Both Krause 1979 from Clemens 1970). (c) Reconstruction of the skeleton and the whole animal of *Henkelotherium guimarotae*. Presacral length approx. 7 cm (Krebs 1991). (d) Upper molars of the South American dryolestid *Leonardus cuspidatus* in labial and occlusal views. Length approx. 1 cm (Bonaparte 1990).

One is confirmation of the absence of a broad trough for the postdentary bones, thereby distinguishing spalacotheriids from the more primitive *Kuehneotherium*, although, as in the latter, there is no angular process on the dentary. The postcranial skeleton of *Zhangeotherium* has a very modern style of shoulder girdle, with scapula spine and large supraspinatus fossa in front of it.

Sigogneau-Russell and Hahn (1995) described isolated teeth of a very peculiar, and very early possible 'symmetrodontan'. *Woutersia* (Fig. 5.14(b)) is from latest Triassic, Rhaetian deposits in France. It has the obtuse-angled 'symmetrodontan' pattern of main cusps, but also greatly enlarged inner cingulum cusps, a single one on the upper molar, and two on the lower. *Woutersia* may be the earliest 'symmetrodontan' which would be a considerable extension backwards of the time range of the group or, equally likely, an unrelated, specialised holotherian lineage evolved from a *Kuehneotherium*-like ancestor.

At the other end of the timescale, symmetrodontans survived far longer in South America, where isolated molar teeth of *Bondesius* (Fig. 5.14(c)) occur in the Late Cretaceous Los Alamitos Formation of Patagonia (Bonaparte 1990, 1994). The anterior and posterior cusps and the cingulum are all reduced in size, and their relationships are presently obscure.

Shuotherium

In 1982, Chow and Rich described a lower jaw of an extraordinary Upper Jurassic mammal from China, which they named *Shuotherium* (Fig. 5.13(e) to (g)). The jaw is slender, and very reminiscent of that of *Kuehneotherium*, complete with a well-developed trough that would have housed postdentary bones. There are three premolar and four molar teeth, or four and three respectively, according to Kielan-Jaworowska *et al.* (2004), and it is the molars that are so remarkable. Each one has a trigonid of three main cusps of normal holotherian form, but in addition there is a large basin extending from the front of the tooth. This resembles remarkably closely the talonid of the tribosphenic lower molars of advanced holotherians, except that there the development is at the back of the tooth. Wang et al (1998) subsequently found what appears to be an isolated upper molar of *Shuotherium* from the same locality (Fig. 5.13(f)). Not unexpectedly, it possesses a new,

large, lingual cusp, which would have occluded with the basin of the corresponding lower molar. Functionally, the molar teeth of *Shuotherium* had a crushing mechanism between a pseudo-protocone on the upper molar against a pseudo-talonid at the front of the lower, which is analogous to the mechanism in true tribosphenic teeth, but quite independently evolved (Fig. 5.13(g)).

Shuotherium teeth have also been identified in the Upper Jurassic of England (Sigogneau-Russell 1998). Few have doubted the therian affinities of *Shuotherium*, but within that context there has been little agreement in detail because of the combination of uniquely specialised molars with a primitive mandible. Recently, Luo *et al.* (2002) have raised interest in the genus by proposing that it is related to those Gondwanan tribosphenids, and monotremes that they have somewhat controversially combined as Australosphenida, an issue discussed at more length later (page 178).

'Eupantotheria'

Traditionally, the group Eupantotheria has been used for those basal therians which had evolved a definite, but single-cusped, un-basined talonid on the back of the lower molar teeth, but lack a protocone on the uppers. As such, these teeth are structurally intermediate between the simpler *Kuehneotherium* and symmetrodontans on the one hand and the tribosphenids with their fully developed talonid and protocone on the other. There are also several grades within the group. As so constituted, 'eupantotheres' are a diverse paraphyletic group. However, as their precise interrelationships are still uncertain, it remains a useful informal taxon.

The 'eupantotheres' with the least derived dentitions constitute the several families of the Dryolestida (Fig. 5.15(a)). The molar teeth are short from front to back and transversely widened, and the talonid is small (Fig. 5.15(b)). The family Paurodontidae includes the only described complete skeleton of a 'eupantothere', *Henkelotherium* from the Late Jurassic Guimarota mine of Portugal (Krebs 1991). It was a mouse-sized animal (Fig. 5.15(c)) with a remarkably modern postcranial skeleton. The shoulder girdle lacks a procoracoid and interclavicle, has a reduced coracoid, and there is a fully developed supraspinous fossa. In the pelvis, the pubis is excluded from the acetabulum. Krebs (1991) believed that *Henkelotherium* was adapted for an arboreal existence, as indicated by the long, flexible tail, the relative length of the penultimate phalanges, and the recurved claws.

The dryolestidans were abundant and diverse during the Upper Jurassic and Early Cretaceous of Europe and North America. More unexpectedly, they underwent a considerable radiation into several indigenous families during the South American Late Cretaceous (Fig. 5.15(d)), where they form the dominant part of the mammalian fauna of the Campanian-aged Los Alamitos Formation of Patagonia (Bonaparte 1990, 1994).

One of the earliest 'eupantotheres', and also the first to be discovered is *Amphitherium*, from the Middle Jurassic Stonesfield slate of Oxfordshire (Fig. 5.16(d)). As yet it is only known from lower jaws and teeth. What the lower molar teeth show is a relatively large protoconid and smaller metaconid and paraconid cusps, forming a roughly equilateral triangle. The talonid is more developed than in dryolestidans, to the extent of bearing a distinct cusp and extending backwards to overlap the next molar behind.

Vincelestes (Fig. 5.16(a)) is known from several skulls and postcranial remains from the Early Cretaceous of Argentina (Hopson and Rougier 1993). It was a relatively large form, with a short, robustly built skull about 7 cm long. The side wall of the braincase is composed of more or less equal sized anterior lamina of the petrosal and alisphenoid, which more closely resembles the condition in the basal mammal *Morganucodon* than the tribosphenidans. No description of the postcranial skeleton has yet been published (Rougier 1993). The dentition is unusual, with reduction of the dental formula to I4/1: C1/1: PM2/2: M3/3. The upper canine is huge, the first and last postcanines small. The lower molars have low cusps and a small talonid with only a single cusp. The upper molars are a curious shape, being very wide laterally, and narrow medially. The phylogenetic position of *Vincelestes* is not certain because of the specialisation of its dentition, and the lack of cranial material in most other 'eupantotheres' for comparison.

The dentition of the Upper Jurassic *Peramus* (Fig. 5.16(b) and (d)) is very well known, and is important as the 'eupantothere' stage whose teeth are most similar to those of the Tribosphenida (Clemens and Mills 1971). There are eight post-canine teeth, but whether four or five of these should be regarded as premolars is debated, with a growing consensus that the primitive number of premolars is actually five (Novacek 1986*b*; Martin 2002). The lower molars have a talonid that bears two cusps, the hypoconid and hypoconulid, connected by an oblique crest. In view of developments to come, the talonid is described as incipiently basined. The upper molars are triangular and although no protocone has evolved, there is a narrow cingulum on the lingual side of the crown that is the incipient homologue of the protocone. The functional effect of these changes was to enhance the shearing action between the hinder part of the upper tooth and the talonid of the lower tooth, as analysed by Crompton (1971).

Tribosphenida

Aegialodontidae

If the discovery of such an oddity as *Shuotherium* illustrates the unexpectedness of Mesozoic mammal research, then the discovery of a single, worn, and incomplete molar tooth referred to as *Aegialodon* illustrates the rewards of patience (Kermack *et al.* 1965). It emerged from a very large volume of the Early Cretaceous Wealden deposits of Cliff End in Sussex, and its importance lies in the small, but distinctly basined talonid on the back of the tooth. All the advanced holotherian mammals, including the modern marsupials and placentals as well as several extinct Mesozoic taxa possess a new large cusp on the upper molar teeth called the protocone, which occluded with an enlarged talonid on the corresponding lower molar. The talonid is described as basined because crests on all sides enclose it. This new arrangement adds a crushing action to the largely shearing action of more primitive therian teeth such as those of *Amphitherium* and *Peramus,* in which the smaller talonid is still open on the lingual side. Although no upper molar of *Aegialodon* is yet known, a protocone must have been present, for in no other way could the pattern of wear facets on the talonid be explained (Crompton 1971). A couple of decades later, Dashzeveg and Kielan-Jaworowska (1984) described a very similar form from the Early

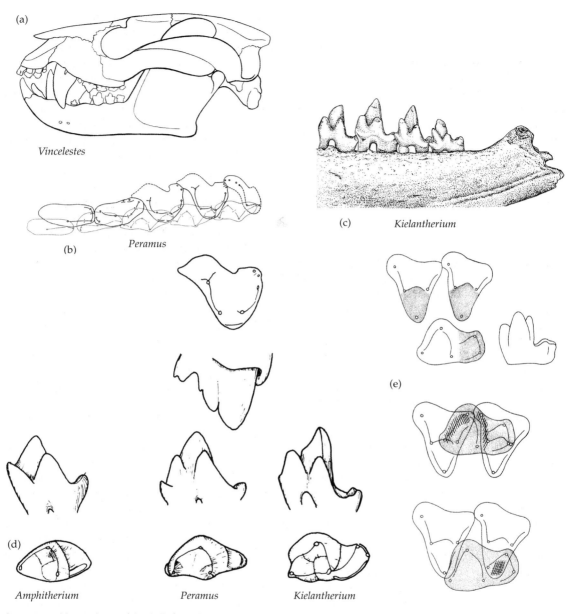

Figure 5.16 (a) Lateral view of the skull of *Vincelestes neuqueniams*. Length approx. 7.5 cm (Hopson and Rougier 1993). (b) Occlusal view of the upper and lower postcanine dentition of *Peramus* in occlusion (Clemens and Mills 1971). (c) Labial view of lower jaw with four molars of *Kielantherium gobiensis*. Length illustrated approx. 0.7 cm (Kielan-Jaworowska and Dashzeveg 1984). (d) Stages in the evolution of lower molars illustrated by a lower molar of *Amphitherium*, upper and lower molars of *Peramus*, and a lower molar of *Kielantherium* (Kemp 1982). (e) Simplified views of a fully evolved tribosphenid upper and lower molars, separated, and at the start and completion of occlusion (Pough *et al.* 2001).

Cretaceous, (Aptian or Albian) of Mongolia. Apart from an isolated molar, *Kielantherium* (Fig. 5.16(c) and (d)) has been described from a partial lower jaw with the posteriormost four teeth in place. While the full dental formula cannot be determined from this specimen, the similarity of all the preserved teeth to one another suggests that they are all molars. Further forwards there are alveoli for four more teeth, all presumably premolars, and Dashzeveg and Kielan-Jaworowska (1984) argue that these are the second to fifth premolars, with the section of the jaw that housed the first premolar missing. Therefore they proposed that the basic tribosphenidan tooth formula consists of five premolars and four molars. The trigonid of the lower molars is still much the dominant part of the tooth, with a very tall protoconid and the metaconid higher than the paraconid. The relatively small talonid bears only two cusps, but unlike *Peramus* is in the form of a completed basin.

There is a possible candidate for an even earlier tribosphenidan than these. *Tribotherium* (Fig. 5.19(c)) is known as isolated mammalian teeth of Berriasian age, right at the base of the Cretaceous, from Morocco (Sigogneau-Russell 1995).

Aegialodon, *Kielantherium*, and possibly *Tribotherium* are placed in the family Aegialodontidae, which accordingly represents the most primitive group of the Tribosphenida, a taxon formally defined by the presence of a protocone on the lingual side of the upper molar working against a basined talonid on the posterior side of the corresponding lower molar. (Luo *et al.* 2001a; Kielan-Jaworowska *et al.* 2004) have replaced this name with a new name, Boreosphenida, on the grounds that they have removed a very small number of certain tribosphenid-like taxa and placed them in a completely different group they name Australosphenida, a move not universally accepted yet, and not very good grounds for abandoning a familiar, long-established name.

The subsequent evolution of the Tribosphenida involved a complex radiation of mammals (Fig. 5.20), whose teeth show further degrees of enlargement of the protocone and talonid, the addition of minor cusps and crests to various parts of the teeth, and modifications to the primitive dental formula. The precise interrelationships of the several lineages are obscure, particularly because many are still only based on isolated teeth. It is clear that the best known groups of tribosphenidans, namely the placentals (eutherians) and marsupials (metatherians) respectively, represent only two of these lineages, albeit the two that eventually survived into the Tertiary to diverge into the mammalian fauna of today. Other lines evolved independently, survived briefly, and disappeared without further trace by the close of the Cretaceous (Clemens and Lillegraven 1986; Kielan-Jaworowska and Cifelli 2001).

Placentalia

There is some discrepancy between the use of the respective names Placentalia and Eutheria. McKenna and Bell (1997) regard them as synonyms and prefer the former. Others, particularly Kielan-Jaworowska *et al.* (2004) discriminate between the two, using Placentalia for the living groups and their fossil relatives, and Eutheria for the more inclusive group consisting of these plus their stem-group. The problem is that the relationships of the various early fossil members to living placentals is far from safely established, and therefore Placentalia in Kielan-Jaworowska et al's sense may be a paraphyletic group. McKenna and Bell's terminology is therefore adopted in the present work.

Placentals are recognisable from the presence of a semi-molariform last premolar, only three molars, and the cusp pattern (e.g. Fig. 5.17(c)). As described in more detail in a later chapter (page 226), the earliest recorded placental is *Eomaia* from China (Fig. 5.17(a) and (b)), which is believed to be Barremian in age and therefore 125–130 Ma (Ji *et al.* 2002). *Murtoilestes* from Russia (Averianov and Skutschas 2001) is believed to be only 4 or 5 younger, occurring as it does around the Barremian–Aptian boundary. Younger Early Cretaceous placentals occur in Aptian-Albian rocks, in the form of *Prokennelestes* (Fig. 5.17(c) and (d)) from Mongolia, and probably *Montanalestes* (Fig. 5.17(e)) from North America (Cifelli 1999).

Once into the Late Cretaceous, placentals started to diversify in both Asia and North America, as discussed in detail in a later chapter. Following 40 years on in the footsteps of the celebrated American Museum expeditions of the 1920s (Colbert 2000), Zofia Kielan-Jaworowska and her colleagues collected numerous skulls and skeletons of a range of

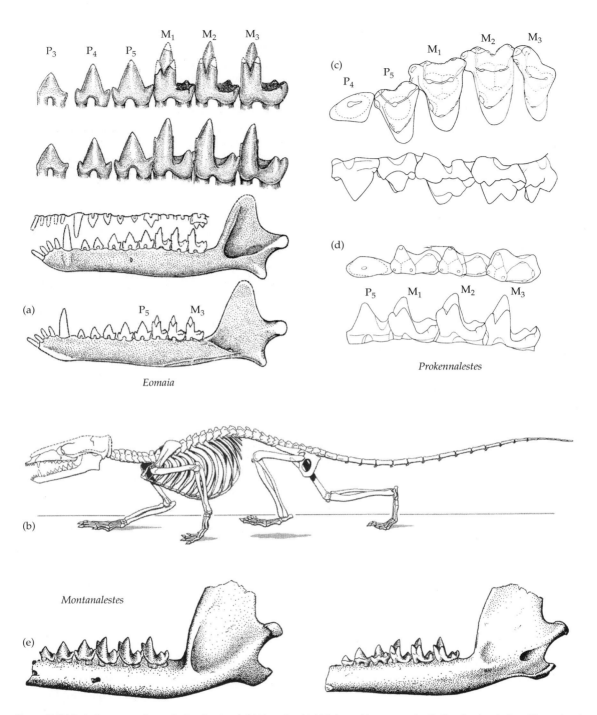

Figure 5.17 Early Cretaceous placentals. *Eomaia scansoria* (a) Lingual and labial views of lower premolars 3–5 and molars 1–3, and internal and external views of lower jaw. Length of jaw approx. 2.3 cm. (b) Reconstruction of skeleton. Presacral length approx. 8 cm (Ji *et al*. 2002). (c) Occlusal and labial views of upper premolars 4–5 and molars 1–3 of *Prokennalestes trofimovi*. Length of illustrated tooth row aprox. 0.7 cm. (d) Occlusal and lingual views of lower last premolar and upper molars of the same (Kielan-Jaworowska and Dashzeveg 1989. (e) External and internal views of lower jaw of *Montanalestes keebleri*. Length of preserved jaw approx. 1.5 cm (Cifelli 1999).

beautifully preserved placentals from Mongolia, dating from the Coniacian through to early Maastrichtian stages. Four genera have been described in a series of papers (Kielan-Jaworowska 1984a; Kielan-Jaworowska et al. 2000). *Kennalestes* and *Asioryctes* are generalised insectivorous mammals in form; *Zalambdalestes* and *Barunlestes* had evolved a specialised postcranial skeleton with long hind legs and apparently a ricochetal mode of locomotion comparable to modern day elephant shrews. Placental diversity also increased in North America, but not until well into the Campanian, around 80 Ma (Fox 1984; Cifelli 1999). Even then, it was not until the last stage of the Cretaceous, the Maastrichtian, that a significantly diverse range of North American placentals had evolved. Mesozoic placental mammals are at present unknown for certain from the southern continents of Gondwana, with one exception: isolated teeth of undoubted placentals, but of unclear phylogenetic relationships, have been described from the latest Cretaceous, Maastrichtian of India (Prasad et al. 1994; Rana and Wilson 2003). The possibility that the Australian Albian-dated form *Ausktribosphenos* and others are placentals is discussed in the context of a possible convergent group of tribosphenidan-grade mammals from the southern continents, Australosphenida.

Marsupialia
In contrast to the placentals, the marsupials are recognised by the possession of four molars, a fully premolariform last premolar, and the labial shelf of the upper molars wide so that the metacone and paracone are set well towards the centre of the crown (e.g. Fig. 5.18(a)). Up to five stylar cusps are developed on this shelf. On this basis, and a number of more minor characters of the molars, the earliest probable marsupial is *Kokopellia* (Fig. 5.18(b)) from Early Cretaceous, Albian aged rocks in Utah (Cifelli 1993a; Cifelli and Muizon 1997). After their initial appearance, marsupials radiated in North America into several groups during the later part of the Cretaceous, heavily dominating the placentals in diversity and abundance.

Until quite recently, Cretaceous marsupials had not been found elsewhere in the world, leading to the conclusion that the group originated in North America and remained restricted to that continent

until after the Cretaceous. However, a Mongolian skeleton of Campanian age (Trofimov and Szalay 1994; Szalay and Trofimov 1996), *Asiatherium*, is undoubtedly that of a marsupial, its dental formula (Fig. 5.18(a)), tooth structure, and cranial features all attesting to that conclusion. Averianov and Kielan-Jaworowska (1999) have argued that the even older, Coniacian-aged *Marsasia* from Uzbekistan is also a marsupial. The material is poor but if confirmed, then it would demonstrate the presence of marsupials in Asia almost as early as in North America. However, the biogeographic situation finally changed dramatically with the recent description (Luo et al. 2003) of *Sinodelphys* (Fig. 5.18(c)), from the same 125 Ma Barremian locality as the earliest placental *Eomaia*. There is a mixture of primitive and derived marsupial characters. Three premolars and four molars are present, and on the lowers the entoconid and hypoconulid are described as approximated, but not fully twinned. The postcranial skeleton, especially the ankle and wrist joints, is very marsupial-like in structure, and apparently adapted for an arboreal existence. Luo et al. (2003) concluded that *Sinodelphys* is a stem member of the Metatheria, which is the group combining marsupials with their possible sister-group the deltatheroidans, as described shortly. The possible biogeographic implication of this identification is that the metatherians arose in Asia, although it still leaves open the possibility that a stem metatherian found its way into North America, and the marsupials as such originated there. This would account for the vastly greater Late Cretaceous diversity in the latter.

The only possible Gondwana record of a Cretaceous marsupial is a single lower molar from Madagascar described by Krause (2001). However, it is very poorly preserved, and Averianov et al. (2003) believe that it is more probably the tooth of a placental.

Deltatheroida
The deltatheroidans (Fig. 5.18(a) and (b)) are a third group of tribosphenidans that are known from skulls and skeletons as well as teeth. They were first collected on the American Museum of Natural History expedition to the Gobi Desert (Gregory and Simpson 1926), and since then a number of partial skulls and

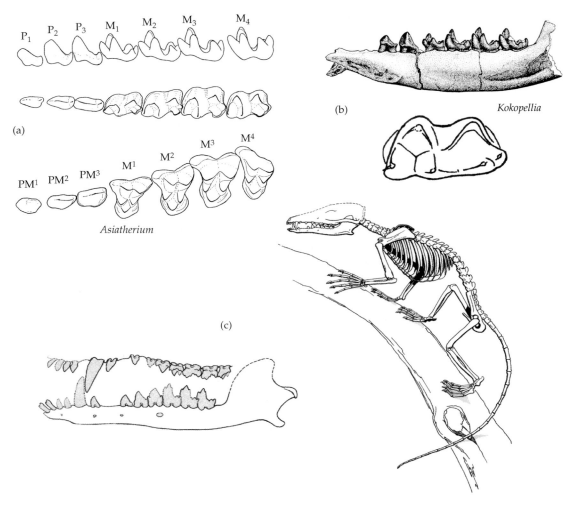

Figure 5.18. Early metatherians. (a) Basic marsupial postcanine dental structure illustrated by *Asiatherium reshetovi*: lowers in lingual and occlusal views, uppers in occlusal views. Length of tooth rows approx. 1 cm (Szalay and Trofimov 1996). (b) Internal view of lower jaw and lower molar of *Kokopellia juddi*. Length as preserved approx. 2 cm (Cifelli 1993a). (c) Lateral view of lower jaw and dentition, and reconstruction of skeleton of *Sinodelphys szalayi*. Presacral length approx. 7 cm (Luo *et al*. 2003).

lower jaws have been found (Kielan-Jaworowska 1975; Rougier *et al*. 1998), dating from the Coniacian to the Campanian of the Late Cretaceous. Fox (1974) has referred some isolated molar teeth from North America to the group, indeed to the same genus, *Deltatheroides*, as one of the Mongolian forms and an upper and lower molar from as early as the Albian of Oklahoma has been tentatively attributed to the Deltatheroida by Kielan-Jaworowska and Cifelli (2001). Deltatheroidans are relatively large

for Cretaceous tribosphenidans, with a skull length of about 4 cm. The cusps of the molar teeth (Fig. 5.19(a) and (b)) are high and sharp, indicating a carnivorous mode of life.

Views on the phylogenetic relationships of deltatheroidans have shifted markedly over the years. When originally described by Gregory and Simpson (1926), they were interpreted as placentals, certainly incipiently carnivorous, and possibly related to the post-Mesozoic creodonts. However, subsequent specimens revealed that there are three premolar teeth and possibly four molar teeth (Butler and Kielan-Jaworowska 1973), which corresponds to the marsupial rather than the placental dental formula, and which led to the group being removed from the Placentalia and placed in the informal category of 'Theria of the metatherian-eutherian grade'. By 1990, Kielan-Jaworowska and Nessov (1990) had come to the conclusion on the basis of the tooth formula, and an incipient tympanic bulla formed from the alisphenoid bone that deltatheroidans are in fact the sister-group of the marsupials, and that the two together should formally constitute the taxon Metatheria. The structure of the molar teeth (Fig. 5.19(b)) does not offer strong support for this conclusion though. In the upper molars, there is a very wide stylar region and a relatively small protocone, and in the lower molars the talonid is much narrower than in typical marsupials. However, these could be specialisations superimposed on a basic marsupial structure for the enhanced shearing action of the teeth of a carnivore. Opposing Kielan-Jaworowska and Nessov's interpretation, Cifelli (1993c) denied a relationship between Deltatheroida and Marsupialia. He suggested that not only were there no shared molar tooth characters, but also that the other similarities may be primitive for tribosphenidans. The most recent stage in the argument is based on some beautifully preserved material of *Deltatheridium* (Fig. 5.19(a)), in which Rougier *et al.* (1998) have shown that the complete dental formula is I4/3: C1/1: PM3/4: M4/4, which confirms a premolar and molar count for marsupials. A second characteristic marsupial feature is the pattern of tooth-replacement, in which only the last premolar is actually replaced after birth; one of the their specimens is a juvenile form that appears to show this was indeed the case. A third is an inflected angular process on the lower jaw, which is characteristic of marsupials. There are also one or two fine details of the anatomy of the braincase supporting the relationship.

Minor tribosphenidan groups
While the Placentalia, Marsupialia, and Deltatheroida are the three best known and diverse Cretaceous groups of tribosphenidan mammals, there are several other groups known only by their particular versions of the tribosphenic tooth whose relationships both to one another, and to the three well-categorised groups are unclear (Fig. 5.20). Taken together, these short-lived, minor branches of the tribosphenidan tree have been informally referred to as 'Theria of the metatherian-eutherian grade', an expression introduced by Patterson (1956). Historically the most important are the 'Trinity Theria', named after the Trinity Formation in Texas, and dating from the Albian stage, at the end of the Early Cretaceous. It was comparison of the 'Trinity Theria' with 'eupantotheres' which enabled Patterson (1956) to establish that the protocone of the tribosphenic upper molar was a new structure, and not the homologue of any of the triangle of cusps of the molars of pre-tribosphenic molars. *Pappotherium* (Fig. 5.19(d)) is the best known, being represented by partial jaws along with isolated upper and presumed but not certainly associated lower teeth. *Holoclemensia* (Fig. 5.19(e)) is similar but known only from upper molars, and *Kermackia* (Fig. 5.19(f)) is based on a lower molar. Early attempts were made to slot the 'Trinity Theria' into existing taxa, notably that of Slaughter (1971), who claimed that *Pappotherium* was a placental and *Holoclemensia* a marsupial. However, all subsequent authors have rejected such a simple phylogeny and accept instead that there were several separate lineages of tribosphenidans, most very short-lived, radiating from a hypothetical metatherian-eutherian grade ancestor during the Early Cretaceous (Kielan-Jaworowska *et al.* 1979b; Clemens and Lillegraven 1986; Cifelli 1993c; Flynn *et al.* 1999).

Other independently evolved molar teeth of no clear affinities are found in the Late Cretaceous, such as *Falpetrus* (Fig. 5.19(h)) from Wyoming and *Bistius* (Fig. 5.19(i)) from Mexico (Clemens and Lillegraven 1986). Clearly, the second half of the Cretaceous Period was a time of considerable evolutionary 'experimentation' as far as advanced, insectivorous mammals were concerned (Fig. 5.20).

Australosphenida and the mystery of the Monotremata

The greatest mystery of all concerning mammalian evolution stretches back for 200 years: the question of what exactly the monotreme mammals are, and

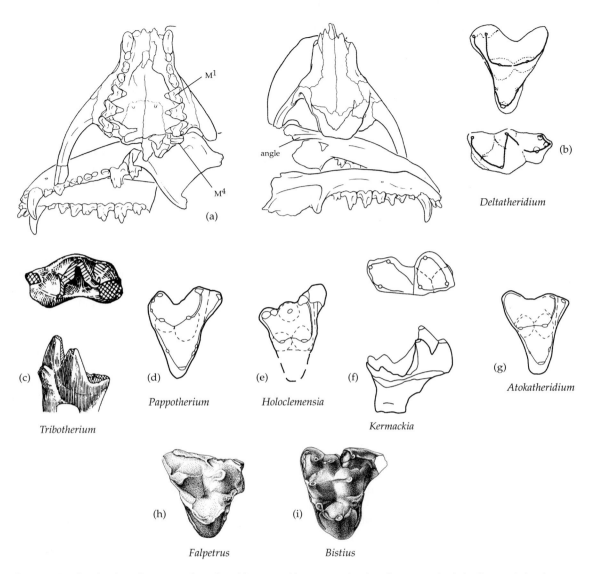

Figure 5.19 Tribosphenidans of uncertain relationships. (a) Upper and lower views of outline of specimen of *Deltatheridium pretrituberculare*, showing four molars, and inflected angular process of the lower jaw. Length of lower jaw approx. 3.6 cm (Rougier *et al.* 1998). (b) Crown views of upper and lower molars of *Deltatheridium*. (c) Lower molar in crown and lingual views of the basal Cretaceous *Tribotherium*. (d) Upper molar of *Pappotherium*. (e) Upper molar of *Holoclemensia*. (f) Lower molar of *Kermackia* in occlusal and lingual views. (g) Upper molar of *Atokatheridium*. (h) Upper molar of the late Cretaceous *Falpetrus*. (i) Upper molar of The late Cretaceous *Bistius* (c) (Sigogneau-Russell 1995, (d-g) Kielan-Jaworowska and Cifelli 2001 (h) and (i) Clemens and Lillegraven 1986).

how they relate phylogenetically to therians (Musser and Archer 1998). The three living genera of monotremes, *Ornithorhynchus* (platypus), *Tachyglossus* (short-beaked echidna), and *Zaglossus* (long-beaked echidna) of Australasia constitute a monophyletic group supported by several characters such as the low, broad skull, reduced dentition, secondary sprawling gait, and hindlimb spur. From time to time, it has been proposed that the monotremes are the living sister group of the marsupials, most

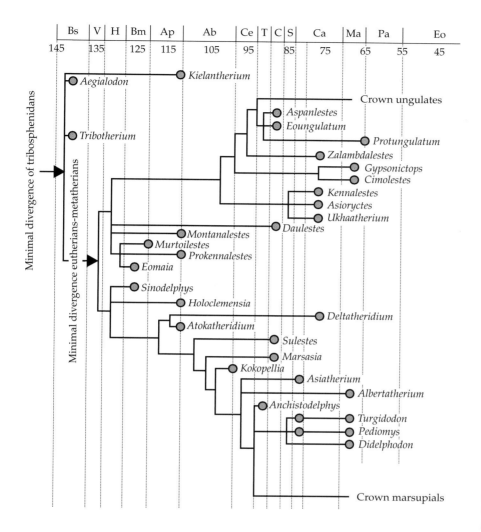

Figure 5.20 Dates of occurrences, and a possible phylogeny of Cretaceous Tribosphenida (Modified from Luo *et al.* 2003).

famously by Gregory (1947) and more recently on the basis of some molecular evidence. However, there is general agreement now that marsupials and placentals are related as crown-group Theria, to the exclusion of Monotremata. As well as numerous cranial and dental characters, an increasing preponderance of molecular evidence supports this view (Killian *et al.* 2001).

The relationship of monotremes to the Mesozoic mammal groups is considerably less clear, and the development of views about this problem has had

an extraordinarily chequered history. At one time it was believed by almost everyone that monotremes had a separate origin from pre-mammalian therapsids, implying convergent evolution of their mammalian characters. The discovery of the Late Triassic mammals of South Wales quickly altered that view because similarities were seen between, on the one hand *Morganucodon* and the monotremes, and on the other *Kuehneotherium* and the living therians. Yet the dental similarities between the two fossil forms was such that there was no longer any reason to doubt that they shared a common mammalian ancestor (Hopson and Crompton 1969; Kermack and Kielan-Jaworowska 1971).

There were two characters quoted in support of a relationship between *Morganucodon* and monotremes. The first was an alleged similarity between the molar teeth. In fact, echidnas completely lack teeth and *Ornithorhynchus* only possesses them as a juvenile, shedding them and replacing them with horny ridges, shortly after emerging from the burrow. These juvenile teeth are three in number and are large, thinly enamelled and have a crown pattern consisting of two or three high, transverse ridges or lophs connecting labial with lingual cusps. They were interpreted as based on a linear arrangement of three cusps, and hence derivable from the 'non-therian' molar teeth of a *Morganucodon*-like ancestor, but specialised by extreme transverse widening. The second supposedly shared character was the structure of the sidewall of the braincase. In living therians this is formed largely from the alisphenoid bone, in monotremes from an anterior lamina of the petrosal bone. In *Morganucodon* both these bones are present but, as in monotremes, the anterior lamina constitutes the sidewall itself. Thus, relying on these two similarities, an early dichotomy into non-therian and therian mammals was envisaged, with monotremes and therians the respective living members of the two.

In due course, however, it was shown that the differences between the two groups in the structure of the sidewall of the braincase were quite trivial. In both cases the bone that forms it is an intramembranous ossification that later in development simply fuses with either the periotic bone behind as in monotremes, or the alisphenoid bone below as in therians. Furthermore, the alleged similarities in tooth structure came to be recognised as far too vague to carry much weight. Compared to this very weak evidence for a monotreme–*Morganucodon* relationship, there are several characters shared by monotremes and therians that can be demonstrated to be more derived than in *Morganucodon* (Kemp 1983). These include the loss of contact of post-dentary bones to the dentary to form ear ossicles, the fusion of the elements of the first vertebra to form a complete atlas ring and of the cervical ribs to the vertebrae, and the enlargement of the cerebellum.

All cladistic analyses now place monotremes closer to living therians than to morganucodontids.

Up to this point, the fossil record of monotremes had not proved helpful. Apart from Pleistocene remains, the only fossil known was *Obdurodon* (Fig. 5.21(a)), a Miocene ornithorhynchid that had retained an adult dentition. Even the beautiful, complete skull found in 1985. (Archer *et al.* 1992; Musser and Archer 1998) added nothing useful to the debate on monotreme relationships, for the teeth (Fig. 5.21(b)) are very similar to the juvenile teeth of *Ornithorhynchus*, and equally difficult to compare with those of other taxa. A palaeontological breakthrough came with the description of a fragment of lower jaw bearing three molar teeth that had been found in the opal mine of Lightning Ridge, New South Wales (Archer *et al.* 1985). *Steropodon* (Fig. 5.21(c)) dates from the Albian stage of the Early Cretaceous. Its molars have a pattern of cusps that can be identified as equivalent to the three trigonid and three talonid cusps of a tribosphenic lower tooth, although modified in a monotreme-like fashion by development of strong transverse ridges (Kielan-Jaworowska *et al.* 1987; Rich *et al.* 2001*b*). Whether the dentition of *Steropodon* was a modified tribosphenic tooth as such, or evolved from a pre-tribosphenic stage was debatable (Kielan-Jaworowska *et al.* 1987). At any event, not only was *Steropodon* the first Australian Mesozoic mammal to be discovered, but also it is one of the best examples of that rare phenomenon of a new fossil intermediate resolving an existing dispute about relationships. Monotremes are highly derived tribosphenidans.

A few more monotremes have been discovered. *Teinolophos* (Fig. 5.21(d)) is also from the Albian of Australia, and the sole specimen to date consists of a single antero-posteriorly compressed molar tooth in a minute partial lower jaw. Rich *et al.* (2001*b*) interpreted it as closely related to *Steropodon*. The first monotreme to be discovered outside Australia is *Monotrematum* (Fig. 5.21(e)). It is of Palaeocene age and occurs in Patagonia. So far only isolated molar teeth have been found, but these resemble quite closely those of *Obdurodon*, indicating that it is an ornithorhynchid (Pascual *et al.* 1992, 2002).

To this point in the story of the origin of tribosphenidans, the timing of the appearance in the fossil record of the different degrees of expression

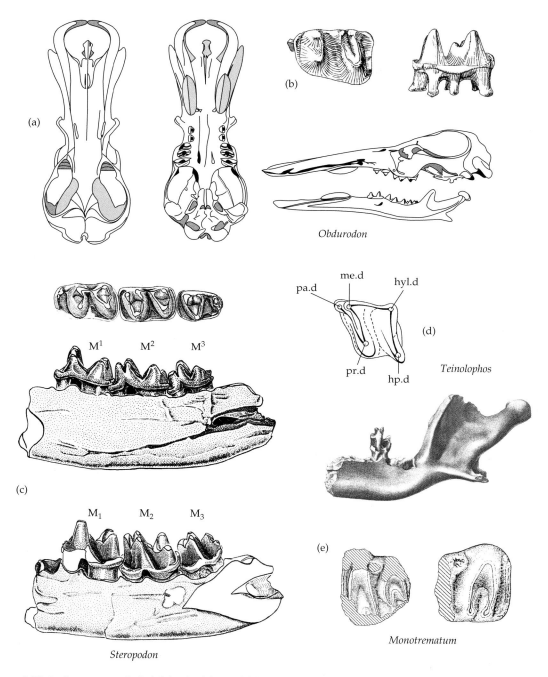

Figure 5.21 Fossil monotremes. Skull of *Obdurodon dicksoni* in (a) dorsal, ventral and lateral views. Skull length approx. 5.2. cm (after Musser and Archer 1998). (b) Lower molar in occlusal and lingual views. (c) Jaw fragment and molars 1–3 of *Steropodon* in occlusal lingual, and lateral view. Length of fragment approx. 2.9 cm (d) Fragment of lower jaw and occlusal view of molar tooth of *Teinolophos trussleri*. Max. length of fragment 3.6 cm (Rich *et al.* 2001. (e) Upper and lower molars in occlusal view, as preserved, of *Monotrematum sudamericanum*. (Pascual *et al.* 2002).

of the tribosphenic molar tooth has agreed with simple evolutionary expectations, from the Middle Jurassic *Amphitherium*, through the Late Jurassic *Peramus*, to the Early Cretaceous aegialodontids and the fully expressed version in the various younger Cretaceous lineages. While not to be read literally as a sequence of ancestor-descendants, these forms do illustrate an approximate morphocline. Even the monotremes fit in well, with the Albian age of the earliest member, *Steropodon*, compatible with the derivation of the group from a basal tribosphenidan ancestor. This simple picture was seriously upset when Flynn *et al.* (1999) described *Ambondro mahabo* from the Middle Jurassic (Bathonian) of Madagascar (Fig. 5.22(a)). The single specimen consists of three teeth in a fragment of the dentary, which the authors interpret as the last premolar and the first two molars. The molars are remarkable for possessing a very large talonid, virtually the same size as the trigonid, in which respect they resemble even more than do aegialodontids the fully expressed tribosphenic molars of advanced members of the Tribosphenida, apart from one primitive respect: there are only two rather then three fully developed cusps on the otherwise completely basined talonid. *Ambondro* extends backwards the known temporal range of the tribosphenic tooth by some 25 Ma, which is equivalent to about 20% of its previous known history. Indeed, such a discovery seemed sufficiently improbable for there to be some doubt about the correctness of the dating, despite the apparently reliable correlation with fossils of undisputed Middle Jurassic age.

Ambondro was also extremely unusual in being a Gondwanan Mesozoic mammal. Indeed, at the time of its description only two others were known. One was the monotreme *Steropodon* already mentioned. The other was *Ausktribosphenos* (Fig. 5.22(b)), an Early Cretaceous, Aptian age specimen consisting of a fragment of lower jaw bearing the posteriormost four postcanine. Rich *et al.* (1997) believed these to represent the last premolar and all three molars and therefore, despite several unusual characters of the teeth, they tentatively proposed that it was a placental mammal. This conclusion was promptly challenged by Kielan-Jaworowska *et al.* (1998) because of the primitive jaw structure, and unique feature of the molars. Rich *et al.* (1999) then went so far as to suggest that it was specifically an erinaceid placental.

These three Cretaceous Gondwana mammals taken together are anomalous. As well as the surprisingly, and in the case of *Ambondro* almost absurdly early dates, the structure of the molar teeth does not conform as closely as might be expected to that of the Laurasian tribosphenidan forms. Luo *et al.* (2001a) addressed the issue by a detailed comparison of the three and pointed out that they share certain unique characters of the molar teeth (Fig. 5.22(d)). The last premolar is triangular but lacks a talonid, there is a cingulum wrapping around the anterior base of the molar, and the trigonid part of the cusp is relatively low. There are also certain minor characteristics of the lower jaw. On the basis of these admittedly quite limited characteristics, and also the Middle Jurassic age of *Ambondro*, they made the radical proposal that the tribosphenic type of molar tooth evolved on two independent occasions. One occurred in a lineage restricted to the southern continents of Gondwana, that they termed the Australosphenida, and it is represented today by the monotremes. The other occurred in a lineage restricted during the Mesozoic to the northern, Laurasian continents, is represented today by the marsupials and placentals, and is referred to by them as the Boreosphenida, although removal of a few fossil jaws plus the monotremes from the rest of the Tribosphenida does not really warrant a change of name for this taxon. The hypothesis of a dual origin of the tribosphenic tooth requires that at least one, and preferably both of the proposed convergent groups can be demonstrated to be related to known pre-tribosphenic mammals. The cladogram of Luo *et al.* (2001a, 2002; Kielan-Jaworowska *et al.* 2004) indicates that both the 'Eutriconodonta' and the basal therian group 'Eupantotheria' are related to the Tribosphenida. Regarding the Australosphenida, Luo *et al.* (2002) argued that the peculiar Chinese *Shuotherium* (Fig. 5.13(e) to (g)), with its pseudotalonid at the front of the lower molar teeth, is related to them. The characters supporting the proposal consist of little more than the form of what may or may not be the last premolar, and Averianov (2002) has strongly criticised it. It might be more convincing if *Shuotherium* was at least from a Gondwanan continent.

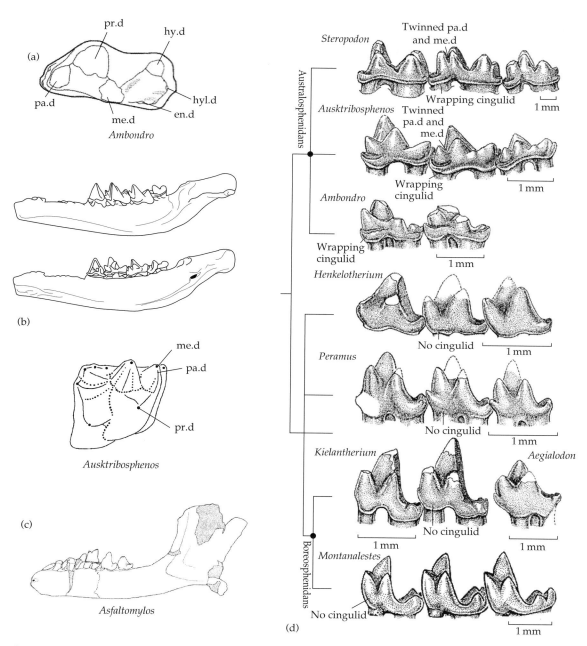

Figure 5.22 Australosphenidans. (a) Occlusal view of lower second molar of *Ambondro mahabo*. (Flynn *et al.* 1999). (b) External and internal views of lower jaw of *Ausktribosphenos nyctos*, with occlusal view of a generalized lower molar (Rich *et al.* 1997). (c) Lateral view of lower jaw of *Asfaltomylos patagonicus*. Length approx. 6.7 cm (Rauhut *et al.* 2002) (d) Characters of australosphenidans compared to tribosphenidans (boreosphenidans) (Luo *et al.* 2002).

Since the first proposal of a taxon Australosphenida, additional Gondwana forms have been discovered, and attributed to the group, most notably *Asfaltomylos* (Fig. 5.21(c)), which extends the geographical range of the group to South America (Rauhut *et al.* 2002). It is from the Middle to Late Jurassic of Argentina, and the only specimen so far consists of a damaged but complete lower jaw showing a possible six premolar teeth and definitely three molars. Rich *et al.* (2001a) described another Early Cretaceous Australian jaw, *Bishops*, which resembles *Ausktribosphenos* and which Luo *et al.* (2002) therefore include in the australosphenidans.

Several authors disagree strenuously with the Australosphenidan concept. Sigogneau-Russell *et al.* (2001) have commented that, while a relationship between *Ausktribosphenos* and *Steropodon* may be well supported, *Ambondro* certainly does not fit at all well, for its tooth is different in a number of respects. Rich *et al.* (2002) strongly argued against a relationship between on the one hand *Teinolophos*, *Steropodon* and the other monotremes, and on the other *Ausktribosphenos* and *Bishops*, disputing the characters supposedly uniting them. They continue to believe that the latter two are related to the true tribosphenidans. Pascual *et al.* (2002) go so far as to revert to an old view that monotreme molar teeth are not even basically tribosphenic, but that they were derived from a completely 'non-therian' ancestry.

Woodburne *et al.* (2003) undertook a cladistic analysis, based on 51 characters, most dental but a few mandibular. They found that the monotremes, including *Steropodon* and *Teinolophos* as basal members, are the sister group of all the therian mammals. Furthermore, the disputed genera *Ambondro*, *Ausktribosphenos*, *Asfaltomylos*, and *Bishops* constitute a monophyletic group that nests within the stem placentals.

So far only incomplete lower jaws and partial lower dentitions of putative australosphenidans have been found. It is therefore too early to be confident that it really is a well-founded taxon, let alone that it evolved independently such an apparently complex dental structure as the tribosphenic tooth, complete with neomorphic protocone and huge expansion of the talonid. More material will doubtless test this nevertheless exciting, and at least

biogeographically satisfying concept. Meanwhile, the solution to the mystery of the monotremes continues to be elusive.

An overview of the interrelationships and evolution

Interrelationships
Several cladistic analyses of the interrelationships of Mesozoic mammals (Fig. 5.23) have been published in recent years and there is good agreement on a number of issues, notably concerning the abandonment of any vestige of the old, simple dichotomy into 'therian' and 'non-therian' mammals. In its place, an increasingly crownward sequence consisting of *Sinoconodon*, morganucodontans, docodontans, eutriconodontans, and Trechnotheria (stem-group plus

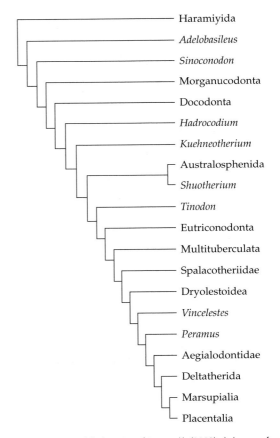

Figure 5.23 Simplified version of Luo *et al.*'s (2002) phylogeny of the major Mesozoic mammal groups.

crown-group therians) is accepted by most as the central phylogenetic framework (Wible and Hopson 1993; Ji *et al.* 1999; Kielan-Jaworowska *et al.* 2004).

There are nevertheless several outstanding problems concerning certain groups. The most intractable of these is the phylogenetic position of the multituberculates, a group whose high diversity and great longevity define its fundamental importance in mammal evolution. The similarity and inferred mode of action of the teeth to those of haramiyidans led to the view that these two groups are indeed related and should be combined in a high-level taxon Allotheria (Hahn 1973; Butler 2000). If true, then the primitive nature and early occurrence of haramiyidans indicate that Allotheria diverged right at the base of the mammal radiation, and that multituberculates evolved a range of derived therian mammalian characters of the skull and postcranial skeleton convergently. A second hypothesis is that multituberculates are the sister-group of the monotremes. This is based on the selected evidence of the braincase (Wible and Hopson 1993) and ear structure (Meng and Wyss 1995), and no cladistic analyses that use all characters supports it. Third, a number of particularly postcranial characters shared by multituberculates and therian mammals led to the hypothesis that multituberculates shared a common ancestor with some or all of the trechnotherians (Hu *et al.* 1997), despite the complete absence of any points of unique similarity between the multituberculate and therian dentitions. The most detailed cladistic analysis to date is that of Kielan-Jaworowska *et al.* (2004), who find slightly more support for the trechnotherian relationship than for the haramiyid relationship, but the difference is barely significant. The problem arises from the multituberculates' combination of highly specialised herbivorous adaptations of the dentition and jaw musculature, superimposed upon a mosaic of primitive and derived mammalian characters. Only new, more plesiomorphic specimens with other combinations of characters such as a less derived dentition are likely to result in resolution of the problem.

There is no serious doubt about the monophyly of the Trechnotheria, characterised by their triangulated molars and features of the lower jaw, braincase and ear region. The one question mark concerns the earliest supposed member, *Kuehneotherium*. Its

lower jaw is very primitive in structure, suggesting it may be a more basal mammal that independently evolved triangulated teeth. Determination of its true relationships will depend on finding cranial material. Aside from the *Kuehneotherium* issue, there is an uncontroversial sequence of stages illustrating the evolution from a very primitive version of the therian tooth in 'symmetrodontans' through a basal 'eupantothere' such as *Amphitherium*, a more progressive 'eupantothere' such as *Peramus*, to the definitive basal tribosphenidans represented by aegialodontids such as *Kielantherium*. However, there is little doubt that these intermediate stage taxa are paraphyletic, their definitions being based only on ancestral stages in tooth and jaw evolution. Continuing to utilise the names 'Symmetrodonta' 'Eupantotheria' and possibly 'Aegialodontidae' is a matter of convenience rather than an issue of conceptual difficulty.

The interrelationships within the Tribosphenida are confusing. There appears to have been a burst of radiation starting in the Early Cretaceous and producing several separate lineages, of which only three are represented by more than isolated teeth or jaw fragments. These are the placentals, marsupials, and deltatheroidans, and there is debate about their interrelationships. A sister-group relationship between the latter two is supported by a number of dental characters, leading to their frequent inclusion in a taxon Metatheria. However, the overall cladistic picture indicates that the placentals are more closely related to the marsupials.

The most recent contentious issue is whether those mammals with tribosphenic molar teeth are a monophyletic group, or whether a separate, independent origin of that form of tooth occurred in the southern continents of Gondwana, within a group named Australosphenida, and which includes amongst others the fossil and living monotremes. As usual, it will take further relevant fossil material to decide one way or the other.

Evolutionary pattern
Given the generally agreed relationships outlined, it is possible to infer a number of interesting aspects of the evolutionary pattern of Mesozoic mammals, the first of which occurs right at the very beginning of the story, when considerable diversification was

immediately manifest. Within a brief window of time, around the end of the Triassic, at least five very distinct groups of mammals occur in the fossil record, namely *Sinoconodon*, morganucodontans, *Kuehneotherium*, haramiyidans, and the possible symmetrodontan *Woutersia*. Furthermore, the phylogenetic position of the docodontans, basal to *Kuehneotherium*, indicates that a sixth distinct form may have been present, even though no certain representatives of it have been found prior to the Middle Jurassic. The age of these early mammals ranges through the Norian and into the Rhaetian, a time period spanning approximately 220–205 Ma. Whether this initial fossil diversity was an artefact due to a significant period during which no fossils of the early stages are preserved, or whether there really was a rapid radiation at the time is unknown. However, the range of small, 'near-mammalian' tritheledontans that occur in the Upper Santa Maria Formation of Rio Grande do Sol, southern Brazil, suggest a possible explanation. They are believed to be Carnian in age, and so slightly older than the mammals. A tritheledontan-grade mammalian ancestor may therefore have dispersed from Gondwana into some part of Laurasia, in which new geographical area it was free to rapidly radiate and fill a set of newly available niches.

The next major event in mammalian history was the Jurassic radiation into a bewildering variety of forms, all small but differing in tooth structure to an extent that indicates a wide range of adaptations to different food resources suitable for small-bodied, endothermic animals. The primitive form of the molar tooth is a single row of three main cusps arranged linearly from front to back of the tooth, surrounded by a cingulum of cuspules at the base of the crown. They were designed for a simple shearing action between crests and presumably functioned mainly in dealing with insect cuticle. From this basic form, several different modes of tooth action evolved, often convergently in different groups. Triangulation of the main cusps developed on a number of occasions. *Kuehneotherium* was the earliest, but the condition also occurred at least to some extent in the morganucodontan *Megazostrodon*, the eutriconodontan *Gobiconodon*, and the symmetrodontans. The greater interlocking between the crests in triangulated teeth may perhaps have given

them a better facility for dealing with relatively larger insect prey that required more extensive mastication before swallowing.

More radical modification of the primitive molar tooth form involved the evolution of a crushing action in addition to the primitive shearing action, was achieved by the development of a lingual extension of the upper tooth, coupled with a labial extension of the lower. One short-lived version is found in the docodontans, which possessed molar teeth virtually as complex in design as the tribosphenic tooth. Presumably it was an adaptation for an omnivorous diet, able to deal with high-energy vegetable matter such as seeds, nuts, tubers and fruits, as well as invertebrates. A much more radical design of crushing teeth was manifested during the Jurassic by the appearance of the multituberculates, with their highly specialised, rodent-like dentition, surely capable of dealing adequately with these plant products. The very high diversity of the group indicates that there was specialisation into many specific food and habitat niches within the group, and several modes of locomotion have also been documented in different forms, including parasagittal, arboreal, ricochetal, and burrowing. The ecological comparison of the group with modern-day rodents remains as compelling as ever.

A third category of dietary specialisation occurred in several of the eutriconodontans, which evolved what was for Mesozoic mammals the relatively large body size of a cat, strongly shearing teeth, and therefore must have been small carnivores.

Initially the Cretaceous Period saw a continuation of the diversity of the non-tribosphenidan groups of mammals, with the multituberculates increasingly dominant. However, the most important evolutionary innovation, towards the end of the Early Cretaceous, was the tribosphenic molar tooth. These are designed to provide two phases of mastication, a generalised puncture crushing when the food is reduced to a pulpy mass, and a precise shearing, in which resilient parts are finely shredded. The origin of the tribosphenic tooth coincided with the beginning of the spectacular Late Cretaceous explosive radiation of flowering plants, and the associated diversification of the insects feeding on them. The several lineages of tribosphenidans, with their many generic and specific variations in the exact form of the tooth, must reflect subtle differences in niche amongst this new, spectacular richness of entomological resources, comparable to what is seen in the insectivorous mammals of today.

Thus the Late Cretaceous became dominated by two mammalian groups, the multituberculates and tribosphenidans. However, the vast preponderance of information about Mesozoic mammal evolution comes from the Laurasian continents of North America, Europe, and Asia, and to a small degree North Africa. It would be misleading to assume that the events outlined above also involved the southern continents. Indeed, evidence is accruing that it would be just plain wrong. Prior to the Late Cretaceous, close to nothing is known at present, apart from the pitifully few australosphenidans. Late Cretaceous mammals occur in the Los Alamitos Formation of Argentina, and are composed entirely of non-tribosphenidan mammals. There are indigenous groups of eutriconodontans, symmetrodontans, dryolestidans eupantotheres, and the strange, hypsodont teeth of the uniquely Gondwanan group Gondwanatheria. Presumably this mammalian fauna evolved in complete isolation from the familiar northern fauna, and no southern tribosphenidan is known for certain prior to the Cenozoic.

The general biology of the Mesozoic mammals

The cladistic relationships of the non-mammalian cynodonts indicates quite clearly that the origin of mammals during the Upper Triassic involved amongst other things fairly extreme miniaturisation. *Sinoconodon* was the largest of the early forms, with a skull length up to 6 cm, which is roughly the size of the skull of the European hedgehog. The postcranial skeleton is unknown, but, continuing the comparison with the hedgehog, it would be expected to have been about 25 cm in presacral length, and to have weighed about 800g. It may also be of significance that *Sinoconodon* retained the primitive condition of continuous growth, rather than the typically mammalian pattern where growth is restricted to a relatively brief juvenile phase. The great majority of the rest of the Mesozoic mammals were considerably smaller, being of the general body size of soricid shrews and murid mice and

rats, with body weights in the range of 200 g down to the 3 g of a minute modern shrew such as *Microsorex hoyi* (Eisenberg 1981). The docodontan *Henkelotherium* is typical. It had a skull length of 4 cm, and a presacral body length of 11 cm, which is roughly equivalent to a body weight of about 20 g. The smallest Mesozoic mammal so far described is the sub-adult specimen of *Hadrocodium*, which has a skull length of only 1.5 cm and an estimated body weight of 2 g (Luo *et al.* 2001*b*). There were also a few relative 'giants'. The largest are the Early Cretaceous eutriconodontans such as *Gobiconodon ostromi*. This had a 10 cm skull and a 35 cm presacral body length, giving it a body size roughly comparable to *Didelphis virginiana*, a mammal whose body weight is around 1.5 kg. A few of the advanced tribosphenidan mammals of the Late Cretaceous also achieved somewhat increased body size. The largest is probably *Zalambdalestes lechei*, but even this was reconstructed by Kielan-Jaworowska (1978) with a skull length of no more than 4.6 cm, and a presacral body length of 14.9 cm.

The hypothetical ancestral mammal, as well as being small, is universally agreed to have been insectivorous, endothermic, and nocturnal. The inference that its primary diet consisted of small, terrestrial invertebrates is based on the sharp-cusped, shearing nature of the postcanine teeth, which were unsuitable for dealing with any significant volume of plant material. The hypothesis that it was endothermic has been considered at length earlier, and it is difficult to imagine that the whole integrated suite of mammalian morphological characters was not associated with insulation and a high metabolic rate. Preservation of the impression of the pelt of the Early Cretaceous placental *Eomaia*, and of hair of a Late Palaeocene multituberculate, powerfully confirms this belief. Nocturnality is more speculative, but is implied by the form of the brain and sense organs. The enlargement of the cerebrum to become the primary control centre in mammals, at the expense of the optic lobes, suggests that the importance of the sense of olfaction found in modern nocturnal insectivores was already present. The increased sensitivity of the sense of hearing, as witnessed by the evolution of the ear ossicle system may also point to nocturnality, although it has to be said that amongst other modern vertebrates, many lizards and birds have equally acute hearing, despite their predominantly diurnal habits. The evolution of tactile vibrissae in mammals is also suggestive of sense organs adapted for night-time use. A second argument for nocturnality is the indirect one, that the nocturnal insectivore habitat was potentially available, and that the great majority of small, sharp-toothed mammals occupy it today. The only radical adaptive shift from the insectivore habitat was towards at least omnivory if not full herbivory. Three taxa independently evolved broader, crushing dentitions suitable for dealing with plant material such as seeds and tubers. The earliest were the haramiyidans, and later the Jurassic docodontans possessed teeth that functioned in a manner that included crushing as well as shearing. The most specialised and diverse of the omnivore/small herbivores were the multituberculates whose superficially rodent-like dentition indicates that they had the sort of cosmopolitan diet found in modern murids. Despite this apparent shift in diet, the members of all three groups nevertheless remained small in body size and presumably nocturnal.

The continuous occupation of this fundamentally small body size, nocturnal habitat by varying taxa of mammals throughout the whole history of the group from the Late Triassic to the present day is not surprising. What is surprising is the virtually complete restriction of mammals to this way of life throughout the Mesozoic. The absence of any mammals at all of larger body size for the extraordinarily long period of time of about 140 million years, the entire length of the Jurassic and Cretaceous Periods together, is not easy to explain. Yet, whilst the post-Mesozoic fauna continued to include large numbers of small insectivorous, omnivorous, and herbivorous mammals, several groups of larger body-sized mammals appeared almost immediately after the end of the Cretaceous, and continued to be a highly conspicuous element of the terrestrial biota ever since. Why were there no Mesozoic equivalents of the foxes, lions, bears, the antelopes, elephants, and kangaroos, the monkeys, anteaters, and sloths? Two alternative explanations have been offered for their absence, neither of which is overwhelmingly convincing. The commoner is that the dinosaurs constituted the larger body-sized terrestrial tetrapods and that by competition for resources they excluded the mammals from evolving into those niches. In this view, the end-Cretaceous extinction of the dinosaurs simply removed their influence, thereby freeing the mammals to fulfil a pre-existing biological potential to evolve increased body-size. The second explanation is that during the Mesozoic phase of their existence, the physiological or structural design of mammals imposed a constraint on how large the body-size could be. In the light of this interpretation, the end-Cretaceous extinction event either created new environmental conditions within which mammals could evolve larger size, or else happened to coincide with an adaptive breakthrough in mammalian design that overcame the constraint.

The competitive exclusion hypothesis. Both the mammals and the dinosaurs originated in the Upper Triassic. The earliest known member of the Mammalia occurs in the Carnian, although only as the single, poorly preserved specimen of *Adelobasileus*. The other early taxa, morganucodontans, haramiyidans, and kuehneotheriids, appeared during the Norian-Rhaetian (Lucas and Hunt 1994). The earliest dinosaurs are also found in deposits of Carnian age, but they too are rare and accompanied by a range of more primitive 'thecodont' archosaurs (Olsen and Sues 1986; Benton 1994). Again like the mammals, several more dinosauran taxa were added during the Norian-Rhaetic prior to the close of the Triassic. Clearly, the timing of the origin and initial radiation of the two groups supports well the hypothesis that competitive exclusion by the dinosaurs was the mechanism by which mammals of large body size were prohibited. It must have been set in place at the very time of origin of the two groups. The pattern of extinctions and originations at the Cretaceous-Tertiary boundary also supports the concept of competitive exclusion. There were members of several families of multituberculates, placentals, and marsupials that survived the K/T mass extinction to be present at the start of the Palaeocene. But, as far as is known, not a single species of dinosaur survived the transition. Within the first few million years of the Palaeocene, a minimum of three and almost certainly rather more of the surviving mammalian lineages had produced species of substantially increased body size.

Against this argument, the great difficulty with the competitive exclusion hypothesis lies in the lack of feasibility that a taxon of endothermic, primarily nocturnal animals like the mammals had a significant enough ecological overlap with large to very large, probably diurnal animals like the dinosaurs. The presumed differences between the two groups in metabolic physiology and mode of food preparation were surely quite as great as between many co-existing groups of organisms sharing comparable body size: mammals and reptiles; carrion birds and mammals; different groups of ungulate mammals. Even accepting the possibility of competitive exclusion to the extent that it could have prevented mammals and dinosaurs from evolving extensively overlapping size ranges, it leaves unanswered the question of why there were no mammals of medium-size. The ancestral size of dinosaurs was large, and though a few modest sized taxa evolved, the vast majority were relatively large animals. The presence of dinosaurs might perhaps account for the absence of Mesozoic mammals above, say, 100 kg body weight, but it certainly does not offer a plausible explanation for the complete absence of the equivalents of foxes, leopards, beavers, medium-sized antelopes, goats, and such like, animals of the order of 1–100 kg body weight. Perhaps the main competitive excluders of medium-sized mammals were juvenile dinosaurs, and that they were active at night as well as day, and that they had the same order of efficiency of feeding mechanisms as mammals. This is not a convincing scenario.

The physiological constraints hypothesis. The alternative hypothesis for the universally small body size of Mesozoic mammals is that their biological design was such that larger body size would not have been viable. The idea that the particular body plan of an organism can constrain its size is commonplace (e.g. Calder 1984). The surface area to volume considerations that limit the size of a unicellular organism dependent on diffusion, the muscular power output to body mass relationship that prevents flying birds from exceeding 10–20 kg, and many more examples are well appreciated.

In the case of size in Mesozoic mammals, there is one obvious possible source of constraint on large body size. Bakker (1971) suggested that the problem lay in the absence of well-developed evaporative ability for cooling. Therefore, the surface area to volume ratio had to remain high to allow adequate heat loss during high levels of muscle activity. It is also perhaps implicit in Crompton *et al.*'s (1978) hypothesis that the origin of endothermy in mammals was an adaptation to nocturnal conditions. At the relatively low body temperature of around 30 °C assumed by this hypothesis, a tendency to overheat during high levels of activity may have been the main thermoregulatory problem in the early forms. It is, however, difficult to accept such a simple thermoregulatory explanation (Lillegraven 1979a). For one thing, many modern mammals do not rely to any significant extent on sweat glands for evaporative cooling, but on the much simpler device of panting. For another, it does not answer the question of why, post-Mesozoic, larger size suddenly became possible, unless by the coincidental evolution of enhanced cooling ability in several separate mammalian lineages.

Comparable objections can be made to any other suggestion that a simple, single biological attribute caused the size restriction. For example, larger mammals must have a relatively more massive skeleton and muscles to support and move the body, a longer lifespan, a lower reproductive rate, and so on. However, in every such case, the invoked physical or biological relationship is neither a bar to large body size in modern mammals, nor offers an explanation for the sudden post-Mesozoic release from the constraint.

It may be more fruitful to consider the Mesozoic mammals in the wider context of their terrestrial ecosystem as a whole. Compared to an equal sized ectothermic tetrapod, an endothermic mammal requires around 10 times as much energy intake per unit time, and therefore requires a sufficiently nutritious and reliable source of food to provide this amount. Consider four possible basic diets. Insects and other small invertebrates offer an adequate diet, but only for mammals of small body size. The discrete occurrence and diminutive size of each individual food item set a limit on the absolute rate of food intake for any individual organism. Within this limit, however, little in the way of extreme specialisation is required. The teeth are not subjected to excessive wear, digestion is simple, and at small body size a secretive, nocturnal, or crepuscular way of life is possible. This is the ancestral mode of life of mammals.

A second possible diet consists of high-energy plant material, seeds, fruits, and storage organs.

Like insects these also tend to occur as small, discrete items requiring individual discovery and collection. Small body size is again a prerequisite for surviving on this diet, but with the complication that teeth adapted to some degree of grinding, and hence subjected to a greater degree of wear must evolve. A number of Mesozoic mammals occupied this role, probably the haramiyids and the docodontans, certainly the multituberculates, and also the Late Cretaceous placental zhelestids.

Predaceous carnivory is a mode of life that readily provides enough food, provided there is a source of suitable sized prey. During the Mesozoic this would have been a problem for a large mammal. There were no medium-sized to large herbivorous mammals, and the dominant herbivores, the dinosaurs, were mostly too large in body size to be prey items for all except very large predators indeed. Of the other terrestrial animals present and forming a potential food source, some tended to be too large such as crocodiles, and others such as lepidosaurs too small to provide an adequate diet for a middle to large mammal carnivore. Therefore, no large carnivorous mammals evolved. As it happens, the largest Mesozoic mammals that did exist were the predaceous members of the Eutriconodonta, and their body size of up to 1.5 kg is consistent with a diet of amphibians, small lizard, and other mammals.

The fourth basic diet to consider is leaf-eating, browsing herbivory, the diet followed by the majority of large herbivorous mammals today. This mode of nourishment creates special problems because of the relatively low energy content of foliage, and its low rate of assimilation due to the need for a gut fermentation flora. Large body size is necessary so that the basal metabolic rate is relatively less, the storage capacity of the gut is larger, and the potential foraging area is greater. Small mammals cannot collect or assimilate this kind of food at a rate that is high enough to satisfy their relatively higher metabolic demands, a problem exacerbated in a female committed to lactation.

From these comments, simple as they are, it follows that the fundamental point was the inability of Mesozoic mammals to adapt to the large herbivore habitat because of the absence of, or inability to acquire food at a high enough rate for the endothermic metabolism. A possible reason for this is a den-

tition unable to cope with the abrasion involved. As it happens, the first appearance of the form of enamel found in modern chewing mammals was in the Early Palaeocene herbivorous mammals. Koenigswald *et al.* (1987) showed that the arrangement of the enamel prisms into what are termed Hunter–Schreger bands was present in certain condylarths. The apparent function of this organisation of the enamel is to reduce the tendency of teeth to split when actively masticating plant food. Since similar enamel is found in most large herbivore mammal taxa thereafter, it must have evolved convergently several times, implying that it was a morphological adaptation in response to a common ecological opportunity in different lineages. Precisely what that opportunity consisted of is unclear.

To conclude, there are two seriously proposed physiological constraints that might have been responsible for the persistent small body size of Mesozoic mammals: over-heating in the absence of adequate cooling adaptations, and under-nourishment in the absence of adequate folivorous adaptations.

End of the era: the K/T mass extinction and its aftermath

Sixty-five million years ago, an episode of mass-extinction affected the world's biota to the extent that some 65–75% of the species disappeared in, geologically speaking, an instant. Such events have occurred every few tens of millions of years throughout the history of life and this one was not even the largest. That accolade goes to the end-Permian mass extinction which was experienced by the therapsid ancestors of mammals 250 Ma while even the Late Triassic mass extinction of 210 Ma, around the time of the origin of mammals and dinosaurs, was at least of comparable magnitude (page 87). Nevertheless, the end-Cretaceous, or K/T event was the mass-extinction most significant in the story of the origin and evolution of mammals. It marks the transition from the ecologically very limited Mesozoic mammal radiation to the Tertiary radiation that culminated in today's mammalian fauna with its huge variation of body size and habitat. Whether the mass extinction acted by removing the dinosaur competitors of mammals, by creating environmental circumstance more favourable for large mammals, or whether it was simply a coincidence, has been exercising palaeobiologists for many years. The answer, as discussed in the previous section, is not known.

Even the basic cause of the K/T mass extinction continues to be debated at length, despite the frequent and confident assertions in public to the effect that the question has been answered. That a large bolide, variously described as a meteor or a comet, struck the Earth in the vicinity of Chicxulub, on the Yucatan Peninsular of southern Mexico at this time is completely agreed upon. However, whether it was the sole cause of the mass-extinction, or at most a *coup de grâce* applied to an already environmentally stressed world remains a matter of dispute (Archibald 1996; Kaufman and Hart 1996; Hallam and Wignall 1997). The evidence for the single catastrophe hypothesis is mainly the geochemical and petrological indications that there was indeed such an impact at that time, of a magnitude that could not fail to have had a huge, deleterious effect on the biota. The worldwide enriched iridium layer at precisely that time, the shocked quartz grains and microtectites, and the discovery of the submerged Chicxulub Crater all point in that direction. There is also a sudden fall in the proportion of the stable isotope ^{13}C present in calcareous marine fossils at the time, which indicates a large fall in planktonic photosynthetic activity. This would be compatible with a period of darkness following the impact.

The evidence that prior to the impact the environment was already undergoing severe alterations consists of various signals that indicate other kinds of environmental stress not in any obvious way associated with the postulated impact. A slight rise in the proportion of the isotope ^{18}O at the K/T boundary indicates a fall in sea temperature, but this appears to have been the culmination of a cooling trend throughout the last few million years of the Cretaceous. Vast volcanic activity was occurring around the time, forming the central Indian Deccan Traps. A major regression of the sea, followed soon afterwards by a transgression, occurred as shown by the position of ancient shorelines.

A mass extinction is by definition a biological process of loss of species, and the most important

observation is therefore the rate and pattern of extinction at the time. The vast majority of the evidence of the effect on the biota comes from the marine record, with its far more extensive exposures and generally much higher temporal resolution and precision of dating than the continental fossil record. Yet even the marine fossil record has insufficient resolution to determine whether the K/T mass-extinction was a single, catastrophic event, or whether the loss of species was a continuous or possibly stepped process over tens or even hundreds of thousands of years. In the context of studying how the K/T mass extinction affected mammalian evolution, what is needed is a fossil record spanning the Cretaceous–Tertiary boundary with sufficient resolution to indicate changes in the environment, in the flora, and in the various tetrapod groups including mammals. Gratifyingly, there is one such a record. The western interior of North America from New Mexico in the south to the Yukon in northwest Canada contains numerous exposures of the K/T boundary, some richly fossiliferous.

At the time, an epicontinental sea lay between what is now western and eastern North America, and the fossil deposits are in the lowland areas forming its western margin. In 1990, Johnson and Hickey published their very detailed study of the megafloral fossil record across the K/T boundary of Montana and adjacent areas. Below the boundary, they found continuous taxonomic turnover. This included a shift towards a higher percentage of plants with smooth margins, a characteristic of increasingly warm conditions. There was also a fall in the average leaf size, which is associated with a wetter environment. Many plants characteristic of warm conditions were present, such as palms, the dicotyledonous Laurales and *Dryophillum tennesseensis*, and araucariaceous conifers. However, exactly at the K/T boundary, as indicated by the iridium anomaly, there was a sudden and profound change in the flora. Only 21% of the species survived into the base of the Palaeozoic and most of these had been very rare in the earlier rocks. Very few of the Late Cretaceous dicotyledonous angiosperms survived. The change in flora is accompanied by evidence of a small change in the physical environment. Lignite and coal deposits appear in the rocks, which, with other lithological features, indicate a shift from an area of poorly drained soils and meandering river channels to conditions of standing ponds and swamps. The conclusion of Johnson and Hickey (1990) is that the abruptness and extensiveness of extinction of the plants were too great to be accounted for solely by climatic change. Rather, they strongly support the idea of a sudden, catastrophic cause. Palynological studies of fossil pollen grains and spores in the western interior are more equivocal. That of Nichols and Fleming (1990) supports the impact hypothesis of a sudden extinction event co-incidental with the iridium anomaly. Furthermore, the discovery of the celebrated K/T 'fern-spike', a large rise in the percentage of fern spores in palynological samples of the time, points to a suddenly imposed environmental shift. However, another palynological study, by Sweet et al. (1993), indicates that the floral extinctions actually began significantly earlier than the K/T boundary in far northern latitudes, which is consistent not with an impact but with a gradual climatic deterioration that affected higher latitudes tens of thousands of years earlier than lower ones. Taking these studies together, there is a good case to be made from the evidence of the fossil flora for an impact induced catastrophic extinction in North America that was superimposed on an environment that was altering more gradually anyway, as indeed most environments are most of the time. What very limited evidence is currently available suggests that the flora of other parts of the world did not suffer the abrupt mass-extinction evident in North America (Archibald 1996). Studies across the K/T boundary in New Zealand (Johnson 1993), Northern Russia (Golovneva 1994), and Antarctica (Askin 1990; Elliot et al. 1994) all fail to produce evidence for a single large extinction, and are more compatible with gradual environmental change.

The described pattern of taxonomic change suffered by the various elements of the fauna, including the dinosaurs and the mammals, is no less ambiguous than that of the flora. Even more so than in the case of the plants, knowledge of the extinction of tetrapods across the K/T boundary is at present virtually completely restricted to the North American western interior sequence, and notably the fossil vertebrates of the Late Cretaceous Hell Creek Formation and the overlying earliest Palaeocene Tullock formation in Montana. Regarding the dinosaurs, there is still no agreement on whether there was a gradual decline and that the K/T boundary only marks the end of that process, or whether there was a catastrophic extinction at that time. Sheehan et al. (1991) made a very detailed collection and found no statistically significant decline in number of dinosaur families through the Hell Creek Formation, which remained at eight throughout. They inferred from this that the dinosaur extinction was a single, large, and rapid event. However, Hurlbert and Archibald (1995) pointed out that the taxonomic level of family is too insensitive to detect a decline, since a large proportion of their contained species could have disappeared without whole families becoming extinct. It seems that the fossil record even here is simply inadequate to distinguish between a gradual decline and a catastrophic mass extinction of the group. Indeed, it cannot even indicate whether any dinosaurs survived until the actual K/T boundary. The highest stratigraphic level of any specimen recorded by Sheehan et al. (1991) was 60 cm below the boundary, a result equally explicable as the true absence of later dinosaurs, or as an artefact due to the increased rarity of fossil preservation in the later part of the Hell Creek Formation.

Archibald and Bryant (1990; Archibald 1991, 1996) made a detailed study of the fossil collections of vertebrates across the K/T boundary in northeastern Montana. After allowing for occurrences in places other than this area, and for pseudoextinction, where a species is believed to have evolved into a new species rather than become extinct, they documented species level survival percentages for all the major vertebrate taxa. Even after making these adjustments, there is still the possibility of large errors because of the relatively low resolution of this fossil record. Certain groups disappeared completely, notably the dinosaurs but also the freshwater sharks. Only 30% of the squamate species survived, but amphibians (100%), turtles (88%), and crocodiles (80%) were barely affected. As far as the mammals are concerned, about 44% of the latest Cretaceous species also occur in the earliest Palaeocene, but different taxonomic groups were very differently affected. The hardest hit were the 11 species of marsupial, of which only one, *Alphadon marshi* survived. On the other hand, all six of the placental species survived, *Gypsonictops illuminatus*, four species of the palaeoryctid genus *Cimolestes*, and another palaeoryctid *Batodon tenuis*. Finally, 5 out of the 10 multituberculate species survived. Archibald (1996) offered a possible explanation for the much more severe effect on the marsupials. The re-establishment of the land bridge across the Bering Straits that resulted from the regression of the sea allowed the immigration of several species of mammals, at the very start of the Palaeocene. These were mainly primitive 'condylarths', which are related to the later ungulate groups of mammals. At this stage, however, the condylarths such as *Purgatorius* were small, lacked extensive herbivorous adaptations, and may have had a similar diet to the marsupials. Thus he proposed that competition by the invaders affected marsupials more than the other two mammal groups.

To conclude, what is known of the environmental and ecological setting within which the mammals underwent their transition from the Mesozoic to the Tertiary is, to say the least, sketchy and geographically very limited. In North America there was a gradual environmental shift during the Maastrichtian, the last stage of the Cretaceous, involving an increase in temperature and humidity. At the K/T boundary this process reversed, mean temperature falling and conditions becoming dryer. Simultaneously, there was a dramatic shift in the vegetation, characterised by a reduction in angiosperms and an increase in ferns for a short time. These latter changes may well have been associated with the bolide impact. One other important geographical feature at about this time was the regression of the sea level, causing a large reduction in coastal areas within North America, and also opening up a land connection across the Bering Straits to the Asian land mass (Smith *et al.* 1994).

For all the detailed effort put into study of the K/T boundary and the extinction patterns of the components of the biota in North America, it ought to be borne in mind that this represents only one area of the world, and it is impossible to generalise from it. For one thing, mid-western North America lies close to the site of the bolide impact, so presumably would have been disproportionately affected by that event. Second, it was only one environmental type, namely a mid-latitude, low-lying area immediately adjacent to a large epicontinental sea. Third, once into the Palaeocene it was rapidly colonised by the new taxa of mammals that are assumed to have immigrated into the area from Asia, and which must represent lineages of mammals that had survived the K/T transition elsewhere in the world, possibly under very different circumstances. In fact, several mammalian lineages are found in the Palaeocene of other parts of the world that have no initial representation in North America. In China, Early Palaeocene mammals of the Shanghuan Formation include both endemic forms, and primitive members of groups that appear later in North America, such as the herbivorous pantodonts and tillodonts (Wang *et al.* 1998; Lucas 2001). In South America, Palaeocene faunas containing diverse marsupials, and indigenous placental groups leave wide open the question of where and when they originated, and how they relate to the North American extinction (Marshall *et al.* 1997). As to what happened to mammals at the K/T boundary of Australia, Africa, and Antarctica, there is simply no evidence at all.

At any event, the new, exciting Tertiary world opened with a range of surviving Cretaceous groups of mammals, and so far as is known no dinosaurs. Increase in diversity was rapid, with the known standing diversity in North America rising from 15 species at the K/T boundary to over 60 a mere 5 million years later (Wing *et al.* 1995; Alroy 1999*b*). This part of the story is pursued in more detail in the following two chapters.

CHAPTER 6

Living and fossil marsupials

There are about 265 living species of marsupial mammals, the majority in Australasia, about 60 in South America, and a handful in Central and North America (Macdonald 2001). They are distinguishable from the placental mammals by many characters, but most profoundly by their mode of reproduction. Compared to the placentals, there is only a relatively brief intrauterine period, during which the embryo exchanges nutrients and gases with the mother via a simple, non-invasive yolk sac placenta. There is no development of the complex, highly invasive chorio-allantoic placenta found in the placentals with the partial exception of the bandicoots in which there is a small, short-lived, but true chorio-allantoic placenta. The marsupial neonate is born at a very immature stage, and most of the total maternal provision comes via lactation. In the majority of cases the young are carried in a pouch, although there are exceptions to this. Whether pouched or not, the young attach themselves continuously to the teat for an extended period of time. There has been much discussion about whether the marsupial mode of reproduction is ancestral to that of the placental mammals, or whether it represents an independent, parallel acquisition of viviparity. Lillegraven (1979), Lillegraven *et al.* (1987), and Szalay (1994), for example, regarded the marsupial mode as primitive and inefficient compared to the placental mode, and that it was failure of the marsupials to evolve a mechanism to prevent immunological rejection of the embryo by the mother that prevented any extension of the gestation period. Placentals, they argued, solved the problem by evolving the trophoblast layer of embryonic cells that performs the function of preventing the maternal antibodies from damaging the embryo. Conversely, several authors such as Parker (1977) have argued that the marsupial mode is an alternative, but equally well-adapted strategy of

reproduction to that of placentals. It is one of low investment but low risk, and is therefore suitable for a more unpredictable environment. Tyndale-Biscoe and Renfree (1987) suggested that primitive marsupials and placentals had quite similar reproduction, with relatively immature neonates and a relatively long lactation period. Subsequent specialisation in the two groups went in different directions. The marsupials extended the lactation period, which became more complex and controlled than that of placentals. The composition of the milk changes much more as the young grows, in accordance with its changing nutritional requirements. In contrast, the placentals extended the gestation period instead. Renfree (1993) quotes evidence that the total energetic cost to the mother of rearing young is similar in the two groups, but that this is spread over a longer period in the marsupials, perhaps reflecting a continuous rather than a seasonal food supply. From this, she concludes that the typical placental is better adapted for reproduction during a climatically controlled short feeding season.

The marsupial dentition, as befits such a disparate group of widely varying adaptive types, is very variable, but can be related to an ancestral condition including the dental formula I5/4 : C1/1 : P3/3 : M4/4. Tooth replacement is limited to the third premolar. The molar teeth (Fig. 6.1(a)), at least in their unmodified condition, are readily distinguished from those of placental and other tribosphenidan groups. The upper molar teeth have the metacone and paracone set towards the centre of the tooth, and the stylar shelf is large, with up to five stylar cusps present on its occlusal surface. The lower molars have a lingually placed paraconid on the trigonid part, while on the talonid the hypoconulid and entoconid cusps are close together, or 'twinned', and well separated from the hypoconid.

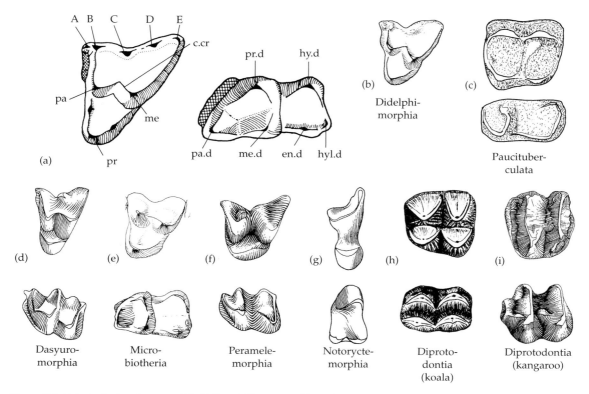

Figure 6.1 Molar teeth of modern marsupials. (a) The primitive marsupial form of upper and lower molars in occlusal view (Reig *et al.* 1987). (b) Didelphimorphia (*Monodelphis*). (c) Paucituberculata (Caenolestes) (d) Dasyuromorphia (*Sminthopsis*). (e) Microbiotheria *(Dromiciops).* (f) Peramelemorphia (*Perameles*). (g) Notoryctemorphia (*Notoryctes*). (h) Diprotodontia (*Phascolarctos*). (i) Diprotodontia (*Simosthenurus*). c.cr, centrocrista; en.d, entoconid; hy.d, hypoconid; hyl.d, hypoconulid; m, metacone; me.d, metaconid; pa, paracone; pa.d, paraconid; pr, protocone; pr.d, protoconid.

There are several diagnostic characters of the marsupial skull, including the foramen of the lacrimal lying external to the orbit, an alisphenoid component to the auditory bulla, a pair of palatal fenestrae, and an inflected angular process of the dentary. Unique characters of the postcranial skeleton are few but include the presence of a prepubic bone, and details of the structure of the ankle joint.

Characters of the soft tissues, which may well be primitive for tribosphenidan mammals, include the relatively small size of the brain, and the structure of the urinogenital tract in which the uterus and vagina are paired.

Living marsupials and their interrelationships

The several detailed classifications of marsupials of recent years agree with one another as far as the recognition of seven major taxa of living marsupials is concerned (Aplin and Archer 1987; Marshall *et al.* 1990; Szalay 1994; McKenna and Bell 1997), although there is little agreement about their interrelationships to one another. The seven groups are usually designated as orders. Three are American and four Australian, differing in numerous characters, including molar tooth form (Fig. 6.1(b)–(i)).

Didelphimorphia

There are 16 genera containing about 65 living species in the single family of American opossums, Didelphidae, living mostly in South America, but with several species ranging into Central America and Mexico. *Didelphis virginiana* has extended its range throughout most of the USA to Canada, as a result of its association with human habitation. Size varies from a head plus body length of 7 cm, and weight of 10 g up to the 50 cm length and 5 kg of the largest specimens of Virginia opossum. They are a structurally conservative group with many primitive characters, and few morphological features are clear synapomorphies defining the group. The primitive marsupial dentition is retained, and the incisors are unmodified. The most distinctive character of the upper molar teeth is their dilambdodont condition, in which the crest connecting the metacone to the paracone tends to be W-shaped (Fig. 6.1(b)). Whether this is a primitive, or a convergently derived character shared with other marsupial groups is discussed later. The didelphimorph ankle joint has certain minor distinctive features described by Szalay (1994).

The majority of didelphids are arboreal, with an opposable first digit and often a prehensile tail. The diet is omnivorous and very adaptable, to which feature Szalay (1994) attributed the lack of dental specialisation. Among the more specialised members, the yapok *Chironectes*, and *Lutreolina* are the only marsupials that are even semi-aquatic. They have webbed feet and a sealable pouch. The caluromyine woolly opossums are specialised fruit and nectar eaters, hanging by their prehensile tails in the branches of the high forest canopy.

Paucituberculata
The Caenolestidae or shrew-opossums are the sole living family of this South American order. There are only three genera, comprising five species, and they live primarily in the Andes. All are small, with body weights ranging from 15–40 g, and they possess an elongated snout for the mainly insectivorous diet. The fur is particularly thick to provide the insulation necessary for their high-altitude habitat.

The caenolestid dentition (Figs 6.1(c) and 6.7(c)) is modified by reduction of the number of lower incisors and greatly increasing the length and procumbency of the first pair, a condition described as diprotodont. They use them to pierce their prey. The upper molars are also very distinctive, having evolved a roughly square shape in crown view.

Microbiotheria
Dromiciops gliroides (Fig. 6.10(a)), the monito del monte, is the sole living species of this order. It is a small, mouse-like form, weighing up to 30 g and confined to the cool, humid forests of Chile and Argentina. Its diet is primarily insects.

The single most characteristic feature of this group is the complex structure of the tympanic bulla, housing the middle ear structures. It is a composite structure involving the alisphenoid and the petrosal bones, which completely encloses the ectotympanic. There are also characteristic features of the molar teeth, notably a reduction of the stylar shelf of the uppers, and elongation of the talonid of the lowers (Fig. 6.1(e)).

Dasyuromorphia
These are the Australian carnivores, and consist of 19 genera containing 73 species. Amongst their diagnostic characters there is a reduction of the incisor teeth to four uppers and three lowers, and possession of lower molars in which the talonid has been reduced in size relative to the trigonid (Fig. 6.1(d)). The braincase and the ankle joint also have distinctive osteological features.

The great majority of dasyuromorphs are members of the family Dasyuridae, the quolls, Tasmanian devil, antechinuses, dunnarts, and marsupial mice. They occupy a wide range of habitats, from rain forest through woodland and savannah to very arid regions. All are basically carnivores or insectivores, although the Tasmanian devil has large teeth and a short powerful jaw which it uses for dealing with carrion as well. Two other families, both monospecific, complete the major living subgroups of Dasyuromorphia. The termite-eating numbat, *Myrmecobius fasciatus* has a somewhat elongated snout and long tongue for collecting the termites that it digs up with its strong front claws. The Tasmanian tiger or wolf, *Thylacinus cynocephalus*, can be counted as the largest recent dasyuromorph species, although there have been no confirmed reports of a living specimen since 1936 when one died in Hobart Zoo. There have nevertheless been many unconfirmed claims of sightings of living thylacines or their dung (Paddle 2000).

Notoryctemorphia
Two species of the genus *Notoryctes* constitute the living marsupial moles. They occur in the sandy central deserts of Australia, consuming insects, notably burrowing beetle larvae, and small lizards. The body is profoundly adapted for a fossorial existence. The forelimbs are very short and powerful,

bearing two huge claws on digits three and four for digging. The hindlimbs have three large claws for pushing soil and sand backwards as the animal moves through the substrate. The snout is pointed and bears a protective plate, the eyes are vestigial, and external ears and vibrissae are absent. The molar teeth (Fig. 6.1(g)) are highly modified, with fusion of the metacone and paracone of the uppers and loss of the talonid of the lowers. These teeth are thus a version of the zalambdodont tooth, which evolved convergently in a number of other marsupial and placental groups.

Peramelemorphia
There are 18 species in the seven living genera of bandicoots and bilbies that belong to this order. They are relatively small insectivores and omnivores, up to the size of a rabbit, that are distributed throughout Australia and New Guinea, in a wide variety of habitats. Peramelemorphs are one of the two syndactylous groups of marsupials, in which the second and third digits of the hindfoot are enclosed in a single sheath. The function of the syndactyl foot in bandicoots is apparently grooming of the fur, rather than locomotor specialisation (Hall 1987).

The most remarkable character of the peramelemorphs is the presence of a chorio-allantoic placenta. During the course of development of the embryo, a normal marsupial type of yolk sac placenta is supplemented by a placental type of chorio-allantoic placenta. The effect of this is to accelerate the development of the embryo so that the neonate emerges from the birth canal to enter the pouch at a significantly younger age. However, it is no more developmentally advanced at parturition than are the neonates of other marsupials.

The peramelemorph dentition is characterised by laterally compressed incisors. The molars tend to be square-shaped, and the talonid of the lowers is as large or larger than the trigonid (Fig. 6.1(f)).

Diprotodontia
There are 40 living genera containing about 128 living species in this primarily herbivorous group, ranging from the minute, nectivorous 7 g pygmy possum and 10 g honey possum to the large kangaroos weighing close to 100 kg. The various diprotodont families of phalangers, possums, gliders, koalas, wombats, and

kangaroos of Australasia combine two distinctive, though individually not unique characters. The dentition is diprotodont, with great enlargement of the first pair of lower incisors to form a base against which the upper incisors can work for browsing and grazing. The second is the syndactylous hindfoot, in which the second and third digits are combined in a single tissue sheath to form a functionally single digit. There are also certain characteristic features of the basic postcanine dentition of diprotodonts, notably crenulated molars, and details of the molar cusp pattern (Fig. 6.1(h) ands (i)), which is based on a bilophodont arrangement. However, there is a wide variety of detailed tooth form in this ecologically disparate taxon.

Interrelationships

For many years, two conflicting derived characters dominated the question of the evolutionary interrelationships of the marsupials. Diprotodonty (Fig. 6.2(a)), the enlargement of one of the pairs of lower incisors, occurs in both the South American caenolestids and the Australian diprotodonts, suggesting a relationship between these two. But syndactyly (Fig. 6.2(b)), the coupling of the second and third digits of the hindfoot, occurs in the peramelemorphs and diprotodonts, suggesting a different, mutually incompatible sister-group relationship. The biogeographic coincidence of the two Australian orders might be seen to have supported the Syndactyla hypothesis, except that discussions at the time were also coloured by the common opinion that the Australian thylacine and the South American fossil borhyaenids were related, for they possess very similar adaptations for a dog-like, carnivorous mode of life; if this relationship was true, then biogeographic evidence could not count for much. How the remaining marsupial groups fitted into either scheme, beyond the assumption that dasyuromorphs and didelphimorphs were more or less ancestral, was obscure.

Kirsch's (1977) application of the serological technique to marsupials produced one of the earliest results of molecular systematics. He demonstrated that all the Australian taxa constituted one monophyletic group, and the American taxa a second, and therefore that the diprotodont dental condition

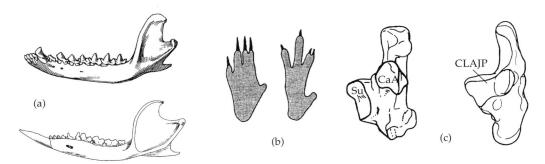

Figure 6.2 Marsupial characters. (a) Comparison of the polyprotodont (above) and diprotodont (below) conditions. (b) Syndactyly of the hindfoot of two marsupials. (c) Left calcaneum of an ameridelphian showing the separate lower ankle joint pattern (SLAJP) and an australidelphian showing the continuous lower ankle joint pattern (CLAJP).

of the diprotodonts and caenolestids, respectively, must be convergent. However, Kirsch had not included the microbiothere *Dromiciops* in his study, and not long afterwards Szalay (1982) challenged the simple biogeographically supported dichotomy, when he showed that the structure of the lower ankle joint of *Dromiciops* was modified in the same way as in the Australian marsupials (Fig. 6.2(c)). There is one continuous joint surface of the calcaneum for the astragalus, instead of the two separate joints found in the remaining American groups. Szalay (1982) therefore placed *Dromiciops* with the Australian forms, separate from the American groups, and named the two taxa Australidelphia and Ameridelphia, respectively. Certain details of the dental morphology are also shared by microbiotheres and the Australian radiation (Aplin and Archer 1987; Marshall *et al.* 1990), and most of the subsequent molecular evidence strongly supports the relationship (Kirsch *et al.* 1991; Springer *et al.* 1998; Palma and Spotorno 1999), as also does the detailed morphological analysis of Horovitz and Sánchez-Villagra (2003), which used a very large database of 230 characters. It is now universally accepted.

The interrelationships of the five australidelphian orders among themselves are poorly understood and consequently very controversial, for the morphological evidence alone is extremely inconsistent. Concerning the relationship of *Dromiciops*, Marshall *et al.*'s (1990) cladogram placed it as the sister group of all the other four together, Szalay (1994) placed it as the sister group of the Dasyuromorphia alone, and

Horovitz and Sánchez-Villagra (2003) placed it as the sister group of Diprotodontia Fig. 6.3(a)) alone.

Szalay (1994) recognised a monophyletic group Syndactyla for peramelemorphs and diprotodonts, which none of the other authors do. The marsupial mole, *Notoryctes*, has been associated variously with diprotodonts (Marshall *et al.* 1990), or peramelemorphs (Szalay 1994).

The phylogenetic relationships of the American marsupials, as inferred from morphology, have been no clearer. Even after the microbiothere *Dromiciops* is excluded, there remains the matter of whether the other two South American orders, Didelphimorphia (didelphids) and Paucituberculata (caenolestids) constitute a monophyletic group Ameridelphia. Based on morphology, there are three alternative views. One is that Ameridelphia is indeed monophyletic (e.g. Marshall *et al.* 1990). The only morphological character that supports this is the pairing of the spermatozoa in the male reproductive tract, and even here there are differences in the details suggesting possible convergence (Temple-Smith 1987). No characters of the dentition, cranial, or postcranial skeleton clearly link them. The second view is that the didelphids are a monophyletic group, but that it is the sister group of the rest of the marsupials, which is to say the caenolestids plus the Australidelphia. Characters of the Didelphidae so defined are limited to details of the ankle joint and the absence of a third trochanter of the femur (Marshall *et al.* 1990; Szalay 1994). The third possible view is that Didelphidae is a paraphyletic grouping that contains the independent ancestry of both the caenolestids and the australidelphians. Szalay and Sargis (2001) argued this from the postcranial skeletal structure, suggesting that the didelphid carpus and tarsus bone patterns are ancestral to the basic australidelphian patterns.

There have been two comprehensive cladistic studies based on morphological characters. Springer *et al.* (1997a), before going on to discuss the molecular evidence, analysed 102 morphological characters by PAUP and found that the most parsimonious tree generated did not even recognise Ameridelphia or Australidelphia as monophyletic. The dasyuromorphs came out as the sister group of all other living marsupials; didelphids as the sister group of the caenolestids plus the remaining australidelphians; *Dromiciops* as the sister group of the diprotodonts. However, most of these groups are weakly supported, and rather than elucidating relationships, their cladogram illustrates the considerable limitations of morphology alone to resolve marsupial interrelationships. Horovitz and Sánchez-Villagra's (2003) analysis of 230 morphological characters is far the most detailed to date (Fig. 6.3(a)), and it also embeds *Dromiciops* deeply within Australidelphia as the sister group of diprotodonts. However several of their groups are very weakly supported, and it must be concluded that as far as morphology is concerned, there is still merit in Aplin and Archer's (1987) refusal to accept any supraordinal taxa, but simply to list the respective australidelphian and ameridelphian orders.

Turning to the more recently acquired molecular evidence of relationships, this too has been marked by inconsistency, although as more sequences are becoming available, agreement seems to be emerging. One of the first was a hybridisation DNA study by Kirsch *et al.* (1991), which supported Australidelphia, and again found Dromiciops with diprotodonts. Springer *et al.* (1998) published a study using a variety of rRNA, cytochrome *b*, nuclear IRBP, and protamine P1 sequences, and they were able to propose a complete resolution of the interrelationships of the living marsupial orders. Their cladogram supported quite strongly the caenolestid plus australidelphian clade, with didelphids basal to it. The most unexpected point was that the peramelemorphs come out as the sister

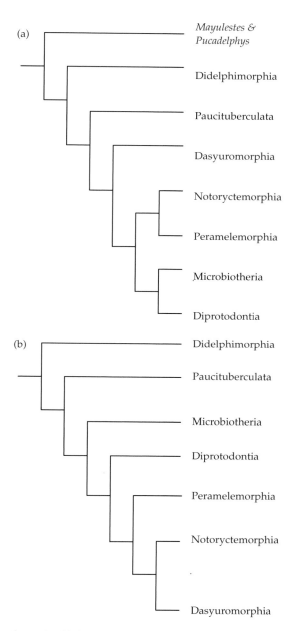

Figure 6.3 (a) The morphological based cladogram of Horovitz and Sánchez-Villagra (2003). (b) The molecular cladogram of Amrine-Madsen *et al.* (2003).

group of the rest of the Australidelphia. *Dromiciops* emerges as the next most basal australidelphian group. Finally, *Notoryctes* is the sister group of the dasyuromorphs.

Some of these hypothesised relationships have been supported by subsequent workers. Springer *et al.*'s placement of the peramelemorphs and of *Notoryctes* are corroborated by the results of Krajewski *et al.* (2000), although this study was actually focussed on the interrelationships of the dasyuromorphs, and did not include *Dromiciops*. Phillips *et al.* (2001) also concluded that the peramelemorphs are the sister group of the rest of the Australidelphia, although again *Dromiciops* was not included in the study. Other studies, however, have contradicted this consensus. Palma and Spotorno (1999) analysed 12S rRNA of several marsupials and came to the unexpected conclusion that peramelemorphs are related to the caenolestids, although they were unable to resolve the relationships of the two together with the rest of the marsupials. Amrine-Madsen *et al.* (2003) have applied the largest data set so far, consisting of five genes and a total of 6.4 kilobases, and included Bayesian as well as Maximum Likelihood methods of analysis (Fig. 6.3(b)). They strongly support the monophyly of Australidelphia, and, though less strongly, the paucituberculates (caenolestids) as its sister group. They found they were unable to resolve the position of *Dromiciops* confidently, although there is some preference for placing it as the sister group of all the Australian groups.

Thus, it has to be concluded that molecules have not yet resolved the interrelationships of marsupials much more than morphology, beyond confirming the monophyly of Australidelphia including *Dromiciops*.

The fossil record of marsupials

The taxonomy and biogeography of living marsupials generates a number of fascinating issues that can only hope to be answered by reference to the fossil record. The most obvious one is how the group as a whole arrived at such a disjunct distribution, being found today only in Australasia and America, two continental land masses now separated by a large, inhospitable distance. Second is how the marsupial mammal radiation paralleled the placental mammal radiation in many, but not in all ways, and what this may imply about the constraints and potentials of mammalian adaptive evolution. Third is the effect of secondary contact of hitherto isolated marsupial and placental faunas, a phenomenon well illustrated by the history of the group and often, but arguably claimed to be an example of competitive interactions between clades.

Origin and Cretaceous radiation

As briefly discussed earlier, the marsupial clade was part of the radiation of tribosphenidan mammals that commenced towards the end of the Early Cretaceous. The earliest possible fossil marsupial by a long way is the 125 Ma Chinese *Sinodelphys* (Fig. 5.18(c)), although it appears in Luo *et al.*'s (2003) cladogram as the sister group of the Marsupials and Deltatheroidans combined. That is, as a stem-metatherian. Otherwise, the earliest recorded fossil marsupials, *Kokopellia* and other similar genera from Utah in North America (Cifelli 1993a; Cifelli and Muizon 1997), are dated about 100 Ma, close to the Albian/Cenomanian boundary. The dentition of *Kokopellia* (Fig. 5.18(b)) has many, but not all the characters of the rest of the primitive marsupials, indicating that it is the most plesiomorphic form. It lacks the characteristic development of stylar cusps other than the stylocone on the upper molars, and the twinning of the entoconid and hypoconulid cusps of the talonid of the lower molars.

A small number of specimens of more advanced forms of marsupials have been found in rocks of the succeeding Cenomanian stage, such as the stagodontid *Pariadens* (Cifelli and Eaton 1987), and these were followed during the course of the remainder of the Late Cretaceous of North America by a radiation of four families, known almost exclusively by their teeth and incomplete jaws.

Didelphidae

This family is most abundantly represented by *Alphadon* (Fig. 6.4(a)), which is often used as a model for the 'archetype' marsupial (Fig. 6.1(a)). Its dentition is generalised, as is still to be seen in the living members of the family. The dental formula is the basic marsupial one of 5/4 : 1/1 : 3/3 : 4/4, apart from an occasional specialised genus. The upper molars tend to develop the dilambdodont condition of a V-shaped centrocrista between the paracone and metacone, and the complete set of five stylar cusps are usually present. The lower molars

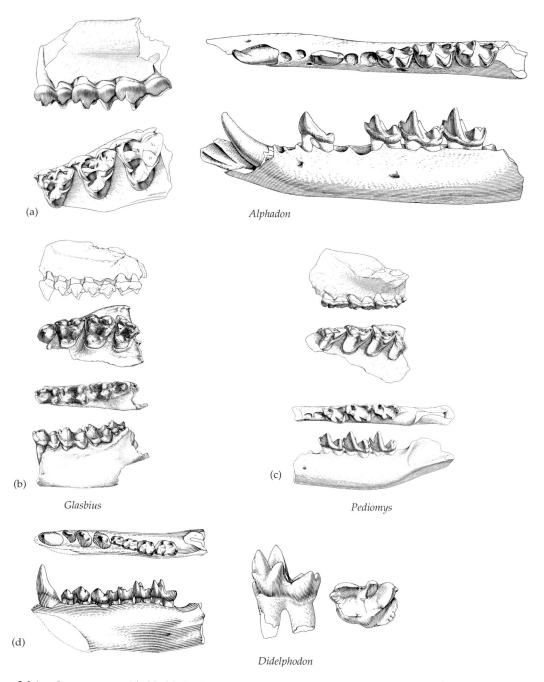

(a)

Alphadon

(b)

Glasbius

(c)

Pediomys

(d)

Didelphodon

Figure 6.4 Late Cretaceous marsupials. (a) *Alphadon rhaister* upper molars 1–3 in labial and occlusal views; *Alphadon lulli* lower dentition in occlusal and labial views. (b) *Glasbius intricatus* upper last premolar and molars 1–3 in labial and occlusal views; lower molars 1–4 in occlusal and labial views. (c) *Pediomys elegans* upper molars 1–4 in labial and occlusal views; *Pediomys hatcherii* lower molars 2–4 in occlusal and labial views. (d) *Didelphodon vorax* lower dentition in occlusal and lingual views; lingual and occlusal views of an isolated 4th molar (Clemens 1979).

are characterised by the lingual shift of the hypoconulid to lie close to the entoconid, the 'twinned' condition.

Pediomyidae

Forms such as *Pediomys* and *Aquiladelphis* have more specialised molar teeth (Fig. 6.4(c)), in which the upper stylar shelf is reduced and the trigon basin of the upper and talonid basin of the lower are relatively enlarged at the expense of other parts of the teeth, enhancing somewhat the crushing function at the expense of the shearing.

Stagodontidae

In contrast to the pediomyids, the stagodontids tended to accentuate the shearing action of the molars (Fig. 6.4(d)), indicating a more carnivorous diet. This was achieved by a definite carnassial notch in the protocristid crest between protoconid and metaconid cusps of the lower molars. The Stagodontidae are best known from *Didelphodon*, one of the largest of the Mesozoic tribosphenidans, with individual molars about 6 mm in length.

Glasbiidae

This family is based solely on the genus *Glasbius*, which Clemens (1966, 1979) originally thought to be a specialised didelphid. It has low, bunodont molars (Fig. 6.4(b)) suitable for a more simple crushing action, suggesting a partially herbivorous diet. They most closely resemble the teeth of the caroloameghiniids, a South American Palaeocene family (page 204).

Asian Cretaceous marsupials

For many years, Cretaceous marsupial fossils were believed to occur only in North and South America. However, the picture has changed radically on two counts. With one possible, but doubtful exception, the earliest South American rocks yielding marsupial fossils are now believed to be Early Palaeocene rather than Late Cretaceous in age. On the other hand, definite Late Cretaceous marsupials do occur in Asia. Trofimov and Szalay (1994; Szalay and Trofimov 1996) described a virtually complete specimen from the Campanian of Mongolia as *Asiatherium* (Fig. 5.18(a)). They regard it as the most plesiomorphic marsupial known, primitive in practically all its anatomy, and placed in its own order

Asiadelphia. It does, however, possess the twinned entoconid and hypoconulid and other dental characters indicating that it is a more derived marsupial than *Kokopellia* (Cifelli and Muizon 1997). Even earlier, though far more incomplete material from Uzbekistan has also been attributed to the Asiadelphia by Averianov and Kielan-Jaworowska (1999). *Marsasia* is dated as Coniacian which makes it about 88 Ma, but unfortunately it is represented so far only by several toothless fragments of lower jaw and one with the fourth molar in place.

Although not as old as the Northern American *Kokopellia*, these specimens dispel any simple view that marsupials originated in North America and were restricted to that continent during the whole of the remainder of the Cretaceous Period. Krause's (2001) identification of a Late Cretaceous tooth from Madagascar as a marsupial is very doubtful (Averianov *et al.* 2003), but is a timely reminder that Late Cretaceous mammalian faunas are extremely poorly represented outside North America and Mongolia.

Ameridelphia: the South American radiation

In stark contrast to the situation in North America, South America was home to a major radiation of marsupials from the Palaeocene onwards, which included all the insectivorous, small omnivorous, and carnivorous mammals until the arrival of certain placental mammals in the later part of the Cenozoic. It is not yet certain exactly when the South American marsupial fossil record begins. Gayet *et al.* (2001) have recently reported on a Late Cretaceous site, dated as Middle Maastrichtian, from Pajcha Pata in southern Bolivia. Amongst the vertebrate microfossils collected, there are two very poorly preserved tribosphenic teeth, so incomplete that it is impossible to say to which taxa they belong, not even whether or not they are marsupials. They may prove to be australosphenidans, even monotremes, or perhaps further specimens will reveal that they are indeed marsupials, in which case they will be critical for understanding the earliest phase of marsupial history in South America.

Apart from the possibility of Pajcha Pata, Laguna Umayo in Peru is the oldest marsupial bearing locality in South America. For a long time it was believed to be Late Cretaceous in age, but this is disputed and it is more probably Early Palaeocene (Marshall *et al.* 1997). In any event, the fossils themselves are sparse and fragmentary, there being a possible tooth of the didelphid *Peradectes* and a possible member of the Pediomyidae (Sigé 1972). Taken at face value, these identifications are very significant because both taxa otherwise only occur in the latest Cretaceous, Maastrichtian, of North America. This suggests that a faunal interchange occurred, presumably from north to south, at a time very close to the end-Cretaceous, although the temporal resolution is not good enough to say whether just before, at, or just after that event. A richer and younger site in the same formation as Laguna Umayo has been described at Chulpas, and here there are reported to be as many as eight marsupial taxa, including a polydolopoid (*Sillustania*), a primitive caroloameghiniid (*Chulpasia*), caenolestids, and didelphids. All but the last of these groups are only known from South America (Crochet and Sigé 1996; Muizon and Cifelli 2001). The Chulpas fauna is tentatively dated close to the K/T boundary and presumably represents the beginning of the indigenous radiation that followed the initial immigration of marsupials from North America.

Far the richest Palaeocene localities on the continent occur in the Santa Lucia Formation at Tiu Pampa in Bolivia, which is either Early (Muizon 1998) or possibly early Late (Marshall *et al.* 1997) Palaeocene in age, and at Itaboraí in Brazil, where the fissure-fill fauna is slightly younger. Both sites have yielded abundant and diverse marsupial faunas, including six or seven indigenous South American families between them. There are also representatives of the modern marsupial family Didelphidae, and the genus *Khasia* which is a member of the biogeographically challenging Microbiotheria to which the modern *Dromiciops* belongs. Other, less-rich Palaeocene mammal localities occur in the San Jorge Basin of Patagonia, some of which are Early Palaeocene (Marshall *et al.* 1997).

There is inevitably a great deal of disagreement about the detailed phylogenetic interrelationships and thus taxonomy of ameridelphians, particularly as many of them are still represented only by teeth and jaws. However, the majority of authors accept three lineages (Aplin and Archer 1987; Marshall *et al.* 1990; McKenna and Bell 1997). Two of these have living members and have already been noted, although they both have a much wider range of fossil forms. Very generally speaking, the didelphimorphs retain primitive, sharp-cusped tribosphenic teeth; the sparassodonts have carnivorous, carnassial teeth; and the paucituberculates have broader, more bunodont teeth implying an omnivorous, somewhat rodent-like habit.

Didelphimorphia

Didelphimorphs consist of small- to medium-sized opossums. They are generally primitive in nature, mostly retaining the full marsupial dental formula, and possessing simple incisors and basic tribosphenic molars. The most controversial character is the dilambdodont condition, which is to say the V-shaped centrocrista, the crest on the upper molars that connects the metacone to the paracone (Fig. 6.1(b)). This has been taken by some authors to indicate that the didelphimorphs are a monophyletic group (Aplin and Archer 1987; Muizon and Cifelli 2001). However, a very similar dilambdodont condition is also found in primitive Australian dasyuromorphs and certain other primitive marsupials (Godthelp *et al.* 1999). As discussed later, this suggests the alternative view that dilambdodonty is primitive for both the ameridelphian and the australidelphian marsupials, and indeed that the didelphimorphs may be ancestral to both and therefore paraphyletic. There are actually certain possibly unique characters of the didelphimorph ankle joint (Szalay 1994), although this evidence is very limited because of the few adequately preserved postcranial skeletons.

The Tiupampan form *Szalinia* (Fig. 6.5(a)) is the most primitive South American didelphimorph because its molar teeth are intermediate in structure between the basal North American marsupial *Kokopellia*, and more typical didelphimorphs from Tiu Pampa such as *Andinodelphys* and *Pucadelphys* (Muizon and Cifelli 2001). For instance, the centrocrista of *Szalinia* is only incipiently V-shaped. *Pucadelphys* is in fact much the best-known Tiupampan marsupial; several well-preserved skulls and skeletons have been found (Marshall *et al.* 1995). It is a typical didelphid in size and general appearance, with a skull length about 3 cm

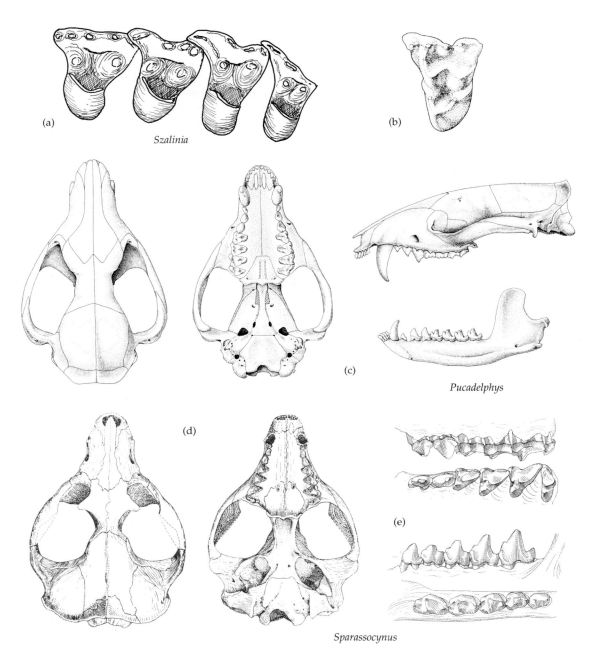

Figure 6.5 Didelphimorphia. (a) Upper molars of *Szalinia gracilis* in occlusal view (redrawn after Muizon and Cifelli 2001). (b) Upper second molar of *Pucadelphys andinus* in occlusal view (Marshall *et al*. 1995). (c) Skull of *Pucadelphys andinus* in dorsal, ventral, and lateral views. Skull length approx. 3 cm (Marshall *et al*. 1995). (d) Skull of *Sparassocynus derivatus* in dorsal and lateral views (Reig *et al*. 1987). (e) *Sparassocynus* sp. upper dentition in labial and occlusal views, and lower dentition in labial and occlusal views (Reig *et al*. 1987).

(Fig. 6.5(c)). The dentition exhibits the characteristics of the family Didelphidae, including a fully expressed V-shaped centrocrista. *Pucadelphys* was part of an extensive radiation of small didelphid opossums well under way during the Palaeocene; there are four or five Tiupampan genera, and Itaboraí has yielded nine (Marshall *et al.* 1997). During the Eocene and Oligocene there was a marked decline in didelphid diversity although this is due at least in part to the relatively few fossil-bearing localities yet unearthed for this period of time (Flynn and Wyss 1998). From the mid-Miocene onwards, the known diversity increased, reaching its maximum at the present day.

Didelphids have also been reported from Antarctica, in the Eocene La Meseta Formation of Seymour Island (Reguero *et al.* 2002). Furthermore, they are the only marsupials to have appeared in the northern continents and Africa during the Cenozoic. The very generalised *Peradectes*, and various closely related genera, occur sporadically until the Miocene in deposits in North America, Europe, Asia, and Africa, although never diversifying, or contributing a particularly significant part of the mammalian faunas of those areas. One noteworthy European occurrence is in the Eocene Lagerstätten of Messel in Germany of completely preserved specimens (Koenigswald and Storsch 1992). Shortly afterwards they disappeared from the Old World (Zeigler 1999) and didelphids remained exclusive to South America until the secondary invasion of Central and North America during the Plio-Pleistocene, when they participated in the northern dispersion during the Great American Biotic Interchange (page 285).

The Sparassocynidae is the only other family of didelphimorphs usually recognised, and it consists solely of the genus *Sparassocynus*, a Miocene and Pliocene taxon (Reig *et al.* 1987). While the molar teeth are basically similar to those of typical didelphids, including the V-shaped centrocrista, they have been modified for more specialised carnivory (Fig. 6.5(d) and (e)). Both the protocone and the talonid are reduced and carnassial crests have evolved. Sparassocynids possess a curiously specialised ear region with a highly expanded tympanic bulla and a very large ectotympanic sinus above the middle ear, of unknown functional

significance. The palatal vacuities are lost. The relationships of *Sparassocynus* have long been debated, including the suggestion at one time that they are related to borhyaenoids. However, they are now generally regarded as aberrant didelphimorphs with certain features converging on those of the borhyaenids (Reig *et al.* 1987), and worthy of their own family status.

Sparassodontia (Borhyaenoidea)

This is an entirely extinct order containing the specialised carnivorous marsupials of South America, some of which achieved the size and extremes of adaptation found in the larger placental felids. The molar teeth are specialised for shearing. In the upper molars, the centrocrista is straight, in contrast to the dilambdodont V-shape of the didelphimorphs. The protocone and the stylar region are reduced, while the metacone and postmetacrista expanded, which tends to create a longitudinal, carnassial cutting edge. In the lower molars, the posterior part of the tooth, metaconid and talonid, are reduced. There are also well-defined sparassodont cranial characters, such as the absence of a palatal vacuity, and postcranial skeletal characters such as reduced epipubic bones, and details of the ankle structure (Marshall 1978).

Discovering the origin and relationships of this group has been made difficult by the convergent development of carnivorous adaptations of the teeth in other predominantly carnivorous marsupial taxa. At one time it was believed that sparassodonts were related to the thylacines of Australia, despite the fairly obvious superficiality of the similarities in tooth and limb structure. Another, more carefully considered view was that sparassodonts were the sister group of the Cretaceous North American stagodontids (page 198), but this was necessarily based solely on details of molar morphology, since no skull or postcranial skeleton of a stagodontid has yet been found. One of the most interesting possibilities is the view of Marshall and Kielan-Jaworowska (1992) that sparassodonts are the most basal group of marsupials, with plesiomorphic similarities to the deltatheroidans. As described earlier (page 171), Deltatheroida is a Late Cretaceous group of Asian and North American tribosphenidans that may be the sister group of all marsupials. Others disagree.

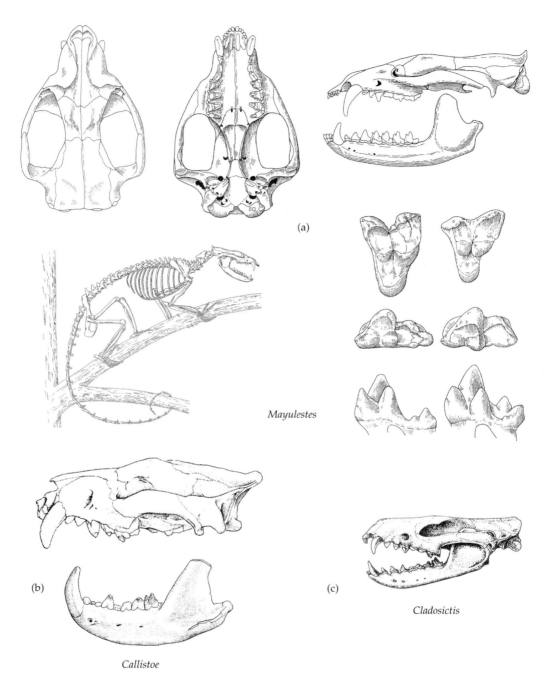

Figure 6.6 Sparassodontians (borhyaenoids). (a) *Mayulestes* skull in dorsal, ventral, and lateral views, postcranial skeleton, upper molar in occlusal view and lower molars in occlusal and lingual views (Muizon 1998). (b) Skull and lower jaw of *Callistoe vincei*. Length of skull approx. 24 cm (Babot *et al.* 2002). (c) *Cladosictis* skull. Length approx. 13 cm (Savage and Long 1986, after Sinclair). (d) *Thylacosmilus* skull. Length approx. 20 cm (Carroll 1988, after Riggs). (e) Skeleton and reconstruction of *Prothylacinus patagonius*. Body length approx. 65 cm (Carroll 1988, after Sinclair, and Marshall 1980).

(d)

(e)

Thylacosmilus

Prothylacinus

Figure 6.6 (*continued*).

Muizon *et al.* (1997) compared the primitive Tiupampan sparassodont *Mayulestes* with the didelphimorphs *Pucadelphys* and *Andinodelphys* from the same locality. They found a number of similarities suggesting a relationship, such as a medial process of the squamosal, and an enlarged, slightly procumbent first upper incisor and staggered second lower incisor. Limited though this evidence is, it is rather more convincing because it does take cranial anatomy into account as well as dental. In fact, this is a reversion to another long-held view, that sparassodonts are simply specialised carnivores derived from a basal didelphimorph stock (Marshall 1978).

Mayulestes (Fig. 6.6(a)) is the oldest known sparassodont and is represented by a skull and partial skeleton (Muizon 1998). It is placed in a separate family Mayulestidae, along with the contemporary *Allqokirus* that is only known from isolated teeth. *Mayulestes* has retained the primitive marsupial dental formula of I5/4 : C1/1 : P3/3 : M4/4, and the structure of the molars is also of the basic marsupial form. However, the skull does show the diagnostic sparassodont characters of a contribution

of the squamosal to the roof of the alisphenoid sinus, and the loss of the prootic canal. The postcranial skeleton is indicative of a mammal with an agile, scansorial habit; the fore and hindlimbs had considerable three-dimensional mobility and the tail was probably prehensile. Muizon (1998) believed that the mode of life would have been comparable to that of a small pine marten, and suggested that this was the primitive habit, not only of sparassodonts, but of marsupials generally.

Members of the more progressive, and far more abundant and diverse family Borhyaenidae appear in the slightly younger Itaboraí fauna, for example *Patene*. The Early Eocene *Callistoe* (Fig. 6.6(b)) is related to the sabre-toothed *Thylacosmilus*, as indicated by its already very enlarged canines. The borhyaenids subsequently went on to radiate as middle-sized and larger carnivores of the South American Tertiary (Marshall 1978), a role shared amongst the mammals only with a few of the largest didelphids, and modest-sized procyonid placentals that entered the continent in the Late Miocene. In fact, the largest predators were amongst the flightless phorusrhacid birds, with some species approaching

3 m in height, and bearing a huge head with ferocious, raptorial beak. The maximum borhyaenid diversity was reached in the Miocene, with some two dozen genera described (McKenna and Bell 1997). *Prothylacinus* was about 80 cm in body length (Fig. 6.6(e)), *Borhyaena* jaguar sized, and *Proborhyaena* said to be larger than a grizzly bear. At the other extreme *Cladosictis* (Fig. 6.6(c)), for example, was no larger than an otter. From their Miocene peak onwards, borhyaenids declined in both diversity and maximum body size, perhaps in part due to the radiation of the phorusrhacid birds during this period of time. By the time the new carnivorous placental immigrants arrived in South America in the late Pliocene, borhyaenids had already disappeared, with the one exception of *Thylacosmilus* (Marshall and Cifelli 1990).

The genus *Thylacosmilus* (Fig. 6.6(d)), the marsupial sabre-tooths, were the most extraordinary borhyaenids of all, with their remarkable convergence on the placental sabre-toothed cats. The upper canine is enormous and, unlike the placental equivalent, it was continuously growing. When the jaws were closed it was protected by an equally large ventral extension of the jaw. At the other end, the jaw articulation was modified to permit the necessary gape of 90° and more. *Thylacosmilus* lived through the Miocene and Pliocene and were the last borhyaenids to survive, lasting until about the time of the entry of placental carnivores from North America.

Paucituberculata

The third of the ameridelphian orders includes the living caenolestids or shrew-opossums, but there is a very much broader range of extinct forms, constituting about 12 families. Because of the width of the radiation, there are not many characters that universally define the group and there is indeed considerable disagreement about the membership. The main character is the distinctive molar teeth, which are relatively broad and bunodont. Two of the stylar cusps of the upper molars, identified as cusps B and D, have greatly enlarged to increase the crushing action of the outer part of the crown. A second trend in the group is the development of procumbent lower incisors and canines. The cladistic analysis of Sánchez-Villagra (2001) is based mainly on the

dentition, and it supports the monophyly of two of the paucituberculate families, Caenolestidae and Argyrolagidae. However, no other paucituberculate families were included.

Reflecting the difficulties of understanding the interrelationships of the paucituberculates, there is divided opinion about which are the most basal members. Several authors regard the superfamily Caroloameghinioidea as occupying this position, with the ludicrously named Palaeocene Tiupampan species *Roberthoffsteteria nationalgeographica* as the earliest South American member (Aplin and Archer 1987; Marshall 1987). The North American Late Cretaceous genus *Glasbius* has very similar molars, is usually included in the same superfamily, and therefore offers an important link between the faunas of these two continents around the end of the Cretaceous. Other authors, however, do not even accept that the caroloameghinioids belong within Paucituberculata, but argue that they are basal to the marsupials as a whole, as discussed by Marshall *et al.* (1990). *Caroloameghina* (Fig. 6.7(a)) is known from complete jaws and dentitions, and exhibits well the molar characters of its group. In the uppers, the cusps are bunodont and bulbous, and the stylar shelf large with equal-sized, enlarged stylar cusps B and D. In the lowers, the talonid is short, broad, and of about the same height as the trigonid. These adaptations were presumably for an omnivorous, and perhaps specialised frugivorous diet in a small mammal (Marshall *et al.* 1983), a style that was to be inherited or repeated in the rest of the polydolopoids. Whatever their true relationships, the caroloameghiniids were a short-lived group, being unknown after the Early Eocene.

The Palaeocene polydolopoids are the earliest undisputed members of the Paucituberculata. *Epidolops* (Fig. 6.7(b)) has been identified in the Itaboraí fauna, and *Polydolops* in the San Jorge Formation. They are diagnosed by reduction of the incisors to 3/2, procumbent lower canines, and development of quadrilateral upper molar teeth associated with evolution of a new cusp, the hypocone. Polydolopoids radiated during the first half of the Eocene but had dwindled practically to extinction by the Early Oligocene. During this time, members of the family Polydolopidae also existed in western Antarctica. *Antarctodolops* was found some

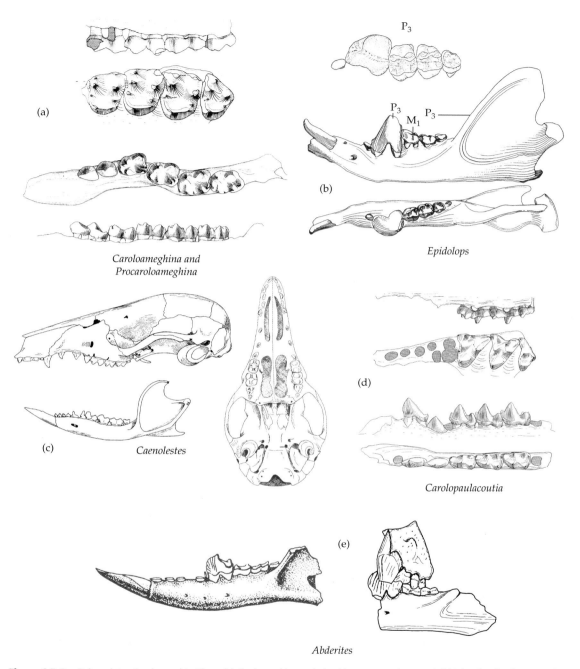

Figure 6.7 Paucituberculates. Caroloameghinoidea—(a) *Caroloameghina tenuis* dentition: upper molars 1–4 in labial and occlusal views and *Procaroloameghina pricei* lower P2–M5 in occlusal and lingual views (Reig *et al*. 1987). (b) Polydolopoidea—*Epidolops* dentition: upper molars in occlusal view, lower jaw in lateral and dorsal views (Goin and Candela 1996, Marshall 1980). Caenolestoidea—(c) *Caenolestes obscurus* skull in lateral and ventral views (Gregory 1910); (d) *Carolopaulacoutia (=Sternbergia) itaboraiensis* dentition: last three upper molars in labial and occlusal views and lower P_2–M_4 in labial and occlusal views (Reig *et al*. 1987); (e) *Abderites* lower jaw, length approx. 5.5 cm, and detail of occluding upper and lower enlarged last premolar (Romer 1966; Dumont *et al.* 2000).

time ago in Eocene deposits of Seymour Island (Woodburne and Zinsmeister 1982, 1984). More recently, specimens identified as *Polydolops*, and also *Perrodelphys* which is a member of another polydolopoid family Prepidolopidae, have been found (Reguero *et al.* 2002).

The Caenolestoidea is the superfamily that contains the only living paucituberculates. Like the polydolopoids this group also developed quadrituberculate upper molar teeth, but in a different way, namely by enlargement of the metaconule rather than development of a new cusp, the hypocone. A second dental character is the procumbent nature of the first or second lower incisor, accompanied by reduction of the first two premolars. With one controversial exception, caenolestoids did not appear in the fossil record until the Oligocene, much later than the other South American superfamilies. The possible but doubtful exception is the Itaboraí form *Carolopaulacoutoia* (=*Sternbergia*), as interpreted by Szalay (1994). However, it is known only from its fairly generalised dentition (Fig. 6.7(d)), which lacks the characters of caenolestoids just noted. The family Caenolestidae (Fig. 6.1(c) and 6.7(c)), the shrew-opossums of the modern South American fauna, retained the basic caenolestoid dentition, but used it for a generally insectivorous habit.

During the Miocene, two caenolestoid families evolved teeth adapted for a more specialised diet. The Abderitidae (Fig. 6.7(e)) possessed a pair of heavily ridged shearing teeth very similar to those of multituberculates and of several of the Australian diprotodonts such as *Phalanger*. In the case of the abderitids, these specialised cutting teeth are the last upper premolar and the first lower molar teeth. Judging from the nature of the wear facets, and by comparison with the similar system in *Phalanger*, Dumont *et al.* (2000) inferred that the function of the shearing system was to deal with both relatively hard food such as nuts, straw, and particularly tough insects, and also more pliant material that tends to stretch and bend rather than sever. From this they concluded that abderitids were adapted for a broader, more cosmopolitan diet than caenolestids. The Palaeotheniidae constitute the second specialised Miocene family. They also had a modified last upper premolar and first lower molar forming a specialised shearing system. However,

they differ from abderitids in the details of the tooth anatomy, and in having molar teeth also specialised for a shearing function (Bown and Fleagle 1993).

The final superfamily of the Paucituberculata is Argyrolagoidea, which includes two highly specialised families whose closeness of relationship is doubtful, but which do share certain characters related to a rodent-like, small herbivore diet. The argyrolagids (Fig. 6.8(a) to (c)) were superficially similar to small hopping rodents with greatly elongated hindlegs (Simpson 1970). The snout is elongated, possibly indicating a mobile proboscis, and there is a prominently procumbent pair of lower incisors which worked against two or three pairs of sharp-edged upper incisors. These are followed by a rodent-like diastema. The molar teeth are high-crowned, hypselodont, and according to Sánchez-Villagra and Kay (1997) they were adapted for particularly abrasive plant material, possibly seeds in the main. Argyrolagids have been found from the Late Oligocene through the Plio-Pleistocene (Sánchez-Villagra *et al.* 2000).

The genus *Groeberia* (Fig. 6.8(e)) is an extremely peculiar little mammal from the Late Eocene/Early Oligocene of Argentina, whose very marsupial status has been doubted in the past, although the palatal vacuities, inflected angular process of the dentary, and four molar teeth seem to settle the question. It shares little apart from some superficial aspects of the dentition with argyrolagids, or even with paucituberculates generally, and Pascual *et al.* (1994) regard it as a member of a quite separate order of presumably ameridelphian marsupials, which they term Groeberida.

Groeberia was a small animal with the dentition and jaw musculature highly modified for dealing with hard food. The rostrum is extremely short and deep, and there is a bony platform, unique amongst mammals, extending posteriorly from the fused symphysis of the two lower jaws, and so providing a solid floor to the oral cavity. There are two upper incisors on either side, the first one being enormous, and described as gliriform in that the roots extend within the skull as far as the orbit. The enamel is restricted to the front edge as in rodents, giving it a self-sharpening ability. The single lower incisor is similarly constructed. Unlike the rodents, there is no diastema and the incisors are followed

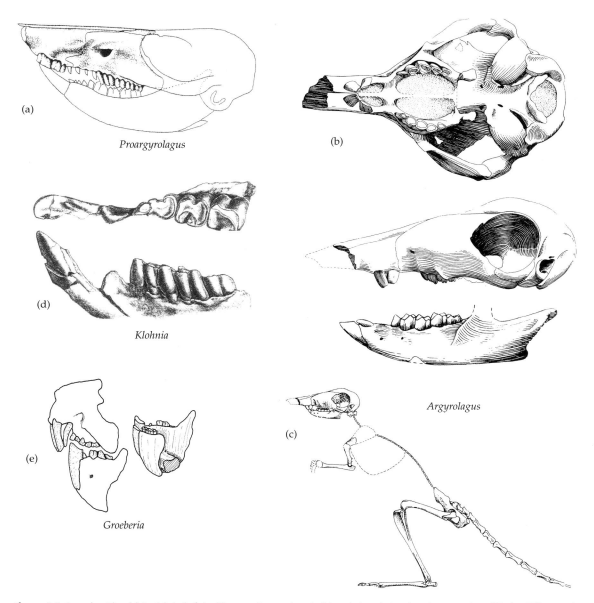

Figure 6.8 Argyrolagoidea. (a) Partial skull of the Oligocene *Proargyrolagus bolivianus* in lateral view. Length approx. 3 cm (Sánchez-Villagra and Kay 1997). (b) Skull of *Argyrolagus scaglii* in ventral and lateral views. Skull length approx. 5 cm (Rich 1991 from Simpson 1970). (c) Postcranial skeleton of *A. scaglii* (Simpson 1970). (d) Jaw fragment of *Klohnia* in occlusal and lateral views (Flynn and Wyss 1999) (e) Anterior fragment of snout and lower jaw in lateral view and lower jaw in postero-lateral view of *Groeberia minoprioi* (redrawn after Pascual *et al*. 1994).

immediately by high-crowned premolars, three upper and two lower. There are then four low-crowned, squarish molars with flat occlusal surfaces covered in thick enamel. A powerful adductor musculature had evolved to operate the dentition, as indicated by a deep extension of the zygomatic arch below the orbit. Pascual *et al*. (1994), in describing new specimens, concluded that the diet was probably nuts and other hard food, and that in addition to the gnawing and breaking function of the incisors, a particularly well-developed tongue may have been involved, working in association with the symphyseal floor. Nothing is known of the rest of the skeleton.

A second groeberiid, *Klohnia* (Fig. 6.8(d)), has been described from the late Eocene/Early Oligocene by Flynn and Wyss (1999), and it too is poorly known, with little more than partial dentitions preserved. A third possible member, *Patagonia* (Pascual and Carlini 1987), was originally described as a member of yet another family of specialised, rodent-like caenolestoids. It is based on a single partial lower jaw with a gliriform incisor. Flynn and Wyss (1999) made a preliminary analysis of the interrelationships and concluded that *Groeberia*, *Klohnia*, and *Patagonia* do indeed constitute a valid family, and that Groeberiidae is indeed the sister group of Argyrolagidae.

Australidelphia: the Australian radiation

As discussed earlier, the Australasian marsupials form a monophyletic group, provided that the living microbiotherian *Dromiciops* and its fossil relatives are admitted. Unlike the South American radiation, in Australia the marsupials included middle- to large-sized herbivores as well as the carnivores. Indeed, apart from the minute number of monotremes, numerous bats, and a few dozen species of murid rodents that entered the continent in the later part of the Cenozoic from Asia, the entire pre-human mammalian fauna was marsupial. Also in contrast to South America, the fossil record of Tertiary Australian marsupials during the critical early stages is extremely poor (Vickers-Rich et al. 1991; Long et al. 2002). Nothing at all is known of the group during the Palaeocene, and the earliest record is the very modest Tingamarra fauna of Queensland, which is dated approximately 55 Ma, and so is latest Palaeocene or earliest Eocene. This is followed by another relatively huge gap; the next window on to the radiation is not until Late Oligocene times, around 25 Ma, with several fossil-yielding localities, including the start of the great tropical rain forest Riversleigh system, with its abundance of beautifully preserved fossils continuing through the Miocene

(Archer et al. 1991). Other important, though less rich fossil-bearing areas from the Late Oligocene include the Etadunna Formation in the Lake Eyre Basin of South Australia, and the Namba Formation which is also in South Australia. With one exception, all the Oligocene and Miocene marsupials are accommodated in one or another of the four extant Australian orders. At a lower taxonomic level, 19 of the total of 23 families of australidelphians listed by McKenna and Bell (1997) occur in the Oligocene–Miocene, and of the 10 extant families only 4 are as yet unknown as fossils from that time. Thus, by Riversleigh times a taxonomically richer version of what is essentially the modern marsupial fauna existed; nevertheless there are a few surprises and mysteries.

The Tingamarra marsupials

The Early Eocene or possibly Late Palaeocene fossils from the Tingamarra Local Fauna, near Murgon in southeastern Queensland, are the oldest marsupial mammals from Australasia (Godthelp et al. 1992). Unfortunately, only a few very fragmentary jaws and isolated teeth are preserved, and none of them are clearly attributable to particular modern groups. *Djarthia* (Fig. 6.9(a)) was described by Godthelp et al. (1999) on the basis of several jaws and teeth. Its upper molars are dilambdodont, having a V-shaped centrocrista, and there are four well-developed stylar cusps. The proportions of the talonid and trigonid of the lower molars are of standard marsupial pattern. From these dental characters, it is impossible to tell whether *Djarthia* is a didelphid or a basal australidelphian, yet this is a fundamentally important issue for understanding Australian marsupial history. If *Djarthia* is a didelphid, then it means that both ameridelphians and australidelphians occurred together there. If it is a basal australidelphian, then the Australian radiation can be accounted for as a consequence of a single didelphid-like immigrant into the continent prior to the Eocene.

A second Tingamarra form is equally tantalising. Archer et al. (1993) described isolated teeth of *Thylacotinga* (Fig. 6.9(b)) as a possible peramelemorph. However, this view has been revised, and Long et al. (2002) regard it as a possible member of the South American and Antarctic group Polydolopoidea. The molar teeth have swollen, bunodont cusps. The stylar shelf of the upper molar is

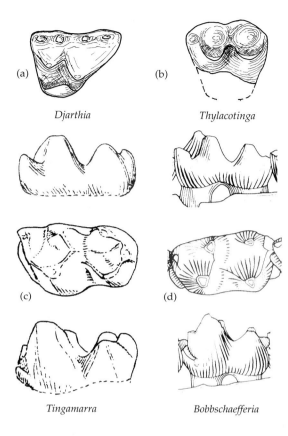

Figure 6.9 Tingamarra mammals. (a) Upper molar of the didelphimorph-like *Djarthia murgonensis* in occlusal view (redrawn after Godthelp *et al*. 1999). (b) Isolated, incomplete upper molar of *Thylacotinga bartholomaii* (redrawn after Long *et al*. 2002). (c) Lower molar of *Tingamarra porterorum* in lingual, occlusal, and labial views, with (d) lower molar of *Bobbschaefferia flumensis* also in lingual, occlusal, and labial views for comparison (Woodburne and Case 1996).

reduced in area and the stylar cusps are equal in size to the main cusps, creating a broad, crushing tooth. Yet other Tingamarran teeth have been attributed to the Microbiotheria (Archer *et al.* 1999) but this is also very tentative indeed.

The most enigmatic Tingamarran fossil of all is a single lower molar tooth named *Tingamarra* (Fig. 6.9(c)). Godthelp *et al.* (1992) claimed that it is not a marsupial because it lacks the twinning of the entoconid and hypoconulid and other features that are so characteristic of marsupials. Instead, they attributed it to the primitive placental group

Condylarthra which, if true, would make it the only early Cenozoic placental known from Australia. Woodburne and Case (1996) challenged their interpenetration and proposed instead that *Tingamarra* is actually a marsupial, possibly a member of the ameridelphian family Protodidelphidae. Members of this family, of which the most familiar is *Bobbschaefferia* (Fig. 6.9(d)), have only been found otherwise in the Palaeocene Itaboraí deposits of South America, and therefore, on this view, *Tingamarra* would be an ameridelphian group common to both continents. An alternative possibility is that *Tingamarra* is a specialised australidelphian convergent upon such forms. Only better specimens are likely to resolve the question of its true identity. Thus the tantalising nature of these fragmentary Tingamarra mammals can be appreciated. At one extreme, they could all be basal members of the radiation of Australidelphia, and to varying degrees exhibiting a dental morphology convergent on other marsupial groups. At the other extreme, there could be representatives of as many as three ameridelphian taxa, didelphids, polydolopoids, and protodidelphids, and of microbiotheres that are also common to both continents. This conclusion would indicate that at the time of the Palaeocene–Eocene boundary there was virtually a single marsupial fauna throughout Gondwana. Only more diagnostic material will help to resolve the issue, which will be returned to later.

Microbiotheria
Once Szalay's (1982) comparative study of ankle structure, and subsequent molecular and morphological analyses confirmed that the surviving South American microbiothere *Dromiciops* (Fig. 6.1(e) and 6.10(a)) is related to the Australian rather than the South American marsupials, the biogeographic and stratigraphic history of the group became very important. Characterised by the elongation of the talonid of the lower molars, and reduction of the stylar shelf of the upper molars, the earliest record is *Khasia* (Fig. 6.10(b)), molar teeth of which have been found in the Palaeocene Tiu Pampa locality of Argentina (Muizon 1991). Various other genera of microbiotheres occur through the South American fossil record, for instance *Microbiotherium* itself (Fig. 6.10(c)) in the Eocene and Miocene. They are

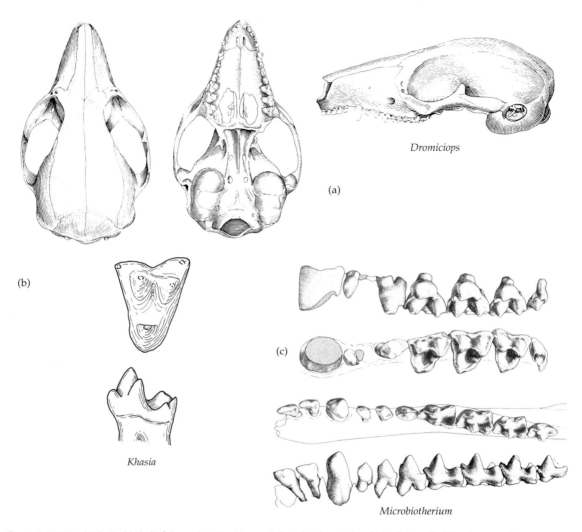

Dromiciops

(a)

(b)

(c)

Khasia

Microbiotherium

Figure 6.10 Microbiotheria. (a) Skull of the modern *Dromiciops australis* in dorsal, ventral, and lateral views. Skull length approx. 3 cm (Reig *et al*. 1987). (b) The Tiupampan *Khasia cordillerensis*: upper molar in occlusal and lower molar in lingual view (redrawn after Marshall and Muizon 1992). (c) Dentition of the Eocene/Miocene *Microbiotherium tehuellcum*: upper in labial and occlusal views, lowers in occlusal and labial views (Reig *et al*. 1987).

also represented in Antarctica, where teeth and jaws of *Marambiotherium* occur in the Eocene deposits of Seymour Island on the Antarctic Peninsular (Goin and Carlini 1995; Goin *et al*. 1999). In the light of this occurrence, the presence of microbiotheres in Australia would be more plausible although, as already mentioned, suggestions that they are represented in the Tingamarra Local Fauna have not yet been confirmed.

Yalkaparidontia

The only Australian order of marsupials that has not survived to the present day is represented by the single genus *Yalkaparidon* (Fig. 6.11(a) and (b)) from the Miocene Riversleigh deposits (Archer *et al*. 1988). It was quite a small animal with a skull length of 5 cm and an extremely unusual dentition. The dental formula is reduced to I3/1 : C1/0 : P3/3 : M3/3. The first upper incisor and the lower incisor are huge, sharp

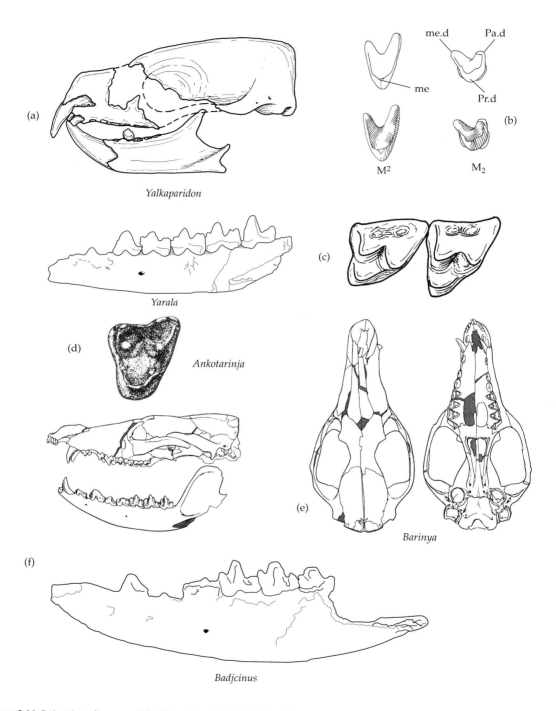

Figure 6.11 Extinct Australian marsupials. *Yalkaparidon coheni*: (a) Skull in side view, and (b) occlusal views of upper (left) and lower (right) molars (redrawn after Long *et al*. 2002). (c) The peramelemorphian *Yarala*: fragment of lower jaw with 3rd premolar and molars (after Long *et al*. 2002), and two upper molars in occlusal view (Muirhead and Filan 1995). (d) The basal dasyuromorph *Ankotarinja* upper molar in occlusal view (Savage and Long 1986). (e) Skull of the earliest dasyurid *Barinya wangali* in lateral, dorsal, and ventral lateral views (Wroe 1999). (f) Lateral view of the lower jaw of the basal thylacine *Badjcinus*. Length of fragment approx. 6 cm (after Muirhead and Wroe 1998).

and open-rooted so they were continuously growing. The molars have evolved a highly specialised form, described as zalambdodont, by losing the protocone from the uppers and the talonid from the lowers, and developing a V-shaped crest between the high, sharp remaining cusps. The resulting occlusal surface is crescentic in shape, and resembles the teeth of several placental groups such as tenrecs and golden moles. There is also a superficial similarity to the zalambdodont-like molars of the marsupial mole, although no other evidence points to a relationship of yalkaparidonts to the latter.

Judging from the molars, the diet consisted of soft invertebrates, but the role in feeding of the gross incisors is hard to decipher. Archer *et al.* (1991) suggested that the food consisted of invertebrates that required puncturing by the front teeth, but were then easy to masticate with the simple transverse crests of the molars. Earthworms and beetle grubs would be candidates. The postcranial skeleton is as yet unknown, so it cannot be said whether *Yalkaparidon* had the digging adaptations that would be expected in an animal with such a food preference.

Extinct Peramelemorphia
Some molecular evidence points to the possibility of the bandicoots and bilbies being the sister group of the rest of the Australidelphia, although this view does not gain any support from their fossil record. Apart from the possible but very doubtful peramelemorph teeth from Tingamarra, the earliest and most primitive known member is *Yarala* (Fig. 6.11(c)) from Riversleigh (Muirhead 2000). During the Late Oligocene and Miocene, peramelemorphs were actually the most abundant marsupial group. There were around a dozen species, all members of the family Yaralidae, and they constituted the majority of the small, mouse-sized omnivore/carnivore guild of mammals (Archer *et al.* 1991). The living genera of the modern family Peramelidae did not start to appear until the Pliocene (Rich 1991). So far, peramelemorphs are not represented in the fossil record of either South America or Antarctica.

Extinct Dasyuromorphia
There are certain mouse-sized Riversleigh specimens such as *Keeura* and *Ankotarinja* (Fig. 6.11(c))

that were considered to be primitive members of the Dasyuromorphia (Wroe 1996), and indeed they may well be correctly assigned to that group. However, they are known only from jaw fragments and isolated teeth, and Archer (1982) and Godthelp *et al.* (1999) have claimed that the teeth of these genera cannot actually be distinguished from didelphid teeth, as is the case for the Tingamarran *Djarthia*. Didelphid and basal dasyuromorph molars both exhibit the dilambdodont condition superimposed upon a basic marsupial tooth structure, including a complete set of five well-developed stylar cusps. Therefore, the possibility exists that these Late Oligocene/Miocene taxa are surviving members of a presumably paraphyletic Didelphimorphia, from which the dasyuromorphs had arisen. More material is required to be sure one way or the other.

The earliest indisputable dasyuromorph, indeed member of the surviving family Dasyuridae is also from Riversleigh. *Barinya* (Fig. 6.11(e)) is well known from a complete skull which has a number of characters unique to the living dasyurids, and several jaws (Wroe 1999). As a group the Oligocene–Miocene dasyurids were less diverse than they are today, and it has been suggested that this is related to competition from the relatively more abundant bandicoots and thylacines.

The Thylacinidae in contrast were more diverse then, being represented by at least six genera in the Late Oligocene to Early Miocene, ranging in size from that of a cat to that of a large dog (Muirhead 1997). *Badjcinus* (Fig. 6.11(f)) is the most plesiomorphic form, with a number of primitive features of molar morphology compared to other thylacinids, such as the wide-angled rather than straight centrocrista and the relatively unreduced metaconid (Muirhead and Wroe 1998). By the Late Miocene onwards diversity of thylacinids had declined and only the modern genus *Thylacinus* existed, and even it occurred only as a single species at any one time. The family had effectively been replaced by the larger of the dasyurids, such as the quolls *Dasyurus*, and the Tasmanian devils *Sarcophilus*.

The third living family of dasyuromorphs, the numbats or Myrmecobiidae, are as yet unknown as fossils, prior to the Pleistocene occurrence of the living species.

Extinct Notoryctemorphia

At present the sole fossil relatives of the marsupial mole are teeth and isolated skeletal elements from the Riversleigh deposits. The molar teeth are not as extremely modified as those of *Notoryctes* (Fig. 6.1(g)), and when a description is published it may indicate how the notoryctid-like version of the zalambdodont condition evolved from a basic marsupial tooth (Archer *et al.* 1991).

Extinct Diprotodontia

As the principal herbivorous Order of Australian marsupials, diprotodonts are, as expected, the most abundant and diverse component of the fossil, as well as of the living, fauna. In fact, the family level diversity of the group at their first appearance in the Late Oligocene and through the Early Miocene exceeds that of the present day. Of the total of 16 diprotodont families recognised by McKenna and Bell (1997), only two are not known during this period, the monospecific Tarsipedidae (honey possum), and dispecific Acrobatidae (feather-tailed glider and feather-tailed possum). On the other hand, eight of the mid-Cenozoic families have subsequently disappeared. These extinct families include some very peculiar and highly specialised diprotodonts.

Wynyardiidae. Members of this family only occur in the Late Oligocene/Early Miocene, although they were fairly abundant during this time. Two complete skeletons of *Muramura* (Fig. 6.12(a)) from the Etadunna Formation of the Lake Eyre Basin have been found (Pledge 1987), so it is a particularly well-known group. *Muramura* was dog-sized and its unspecialised syndactyl hindfeet indicate a simple, terrestrial quadrupedal locomotion. The molars had a primitive degree of development of transverse lophs, and the stylar cusps on the uppers are still distinguishable. As far as the details of the wynyardiid braincase are concerned, Aplin (1987) believed it to be the most plesiomorphic of all diprotodonts. Therefore, on dental, cranial, and postcranial structure, the wynardiids appear to be close to the base of the diprotodont radiation.

Diprotodontidae. The diprotodontids (Fig. 6.12(i) and (j)) include the largest marsupials ever to evolve, ranging in size from that of a sheep to giants larger than hippos. The Late Pleistocene *Diprotodon optatum* was almost 3 m long, stood 2.6 m high at the shoulder, and weighed as much as 2.78 tonnes (Wroe *et al.* 2003). The diprotodontid dentition was modified for browsing on relatively soft vegetation, with enlarged, bilophodont molar teeth. Unlike the other major herbivore family, the macropodids, diprotodontids remained fully quadripedal. The group finally became extinct in the Late Pleistocene, as part of the general megafaunal mass extinction of that time.

Palorchestidae. This family consisted of forms quite similar to diprotodontids and they were at one time actually included in the latter family. Also like the diprotodontids, they survived until the Late Pleistocene, when the best-known genus *Palorchestes* was the size of a horse. However, palorchestids do differ from diprotodontids in details of the dentition, and also in the structure of the ear region, in which they more closely resemble the wombats. Very large, powerful claws tended to develop on the forelimbs suggesting a diet of tubers and roots, and in the later forms such as *Palorchestes* the snout was elongated in a manner that suggests that it bore a mobile proboscis or short trunk.

Thylacoleonidae. Famously the most extraordinary diprotodonts of all were the marsupial lions, which are also one of the comparatively rare examples of the evolution of a carnivore from a specialised herbivorous ancestor. The Late Oligocene–Early Miocene thylacoleonid *Priscileo* was a relatively small, cat-sized animal (Rauscher 1987), and *Wakaleo* (Murray *et al.* 1987) had a 15-cm skull (Fig. 6.12(f)). By the Plio-Pleistocene, *Thylacoleo* (Fig. 6.12(g)) had evolved into the size and general form of a leopard. The dentition is modified for extreme carnivory (Nedin 1991), or perhaps a hyaena-like scavenging that included dealing with large bones. The first upper and sole lower incisors are enlarged and caniniform. The upper third premolar (Fig. 6.12(h)) is an enormous shearing blade that worked against the almost equally large lower third premolar plus the first molar. The rest of the molars are reduced in both number and relative size.

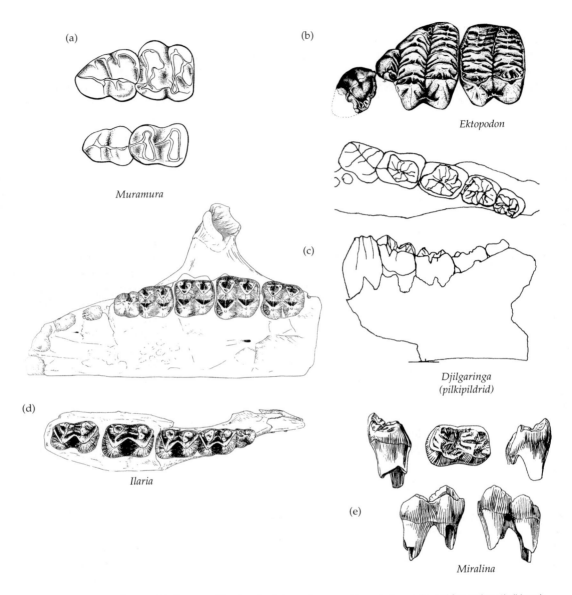

(a) *Muramura*

(b) *Ektopodon*

(c) *Djilgaringa (pilkipildrid)*

(d) *Ilaria*

(e) *Miralina*

Figure 6.12 Extinct diprotodontians. (a) *Muramura williamsi:* occlusal views of upper and lower last premolar and first molars. Skull length approx. 13 cm (Pledge 1987). (b) *Ektopodon stirtoni* upper last premolar and first two molars in occlusal view. Length of fragment approx. 1.8 cm (Woodburne 1987). (c) The pilkipildrid *Djilgaringa gillespiei* lower dentition in occlusal and buccal views. Length of fragment approx. 1.9 cm (Archer *et al*. 1987). (d) *Ilaria illumidens* upper and lower cheek teeth. Length of upper dentition approx. 1.0 cm (Rich 1991). (e) *Miralina doylei*, lower right molar (Woodburne *et al*. 1987). (f) Skull of the basal thylacoleonid *Wakaleo vanderleuri*. Skull length approx. 18 cm (Murray *et al*. 1987). (g) Skull of *Thylacoleo carnifex*. Skull length approx. 27 cm (from Rich 1991, after Lydekker 1887). (h) Comparison of occlusal views of the upper dention of *Wakaleo* (Miocene), *Thylacoleo crassidentatus* (Pliocene) and *Thylacoleo carnifex (Pleistocene)* showing enlargement of PM3. (i) Upper and lower cheek dentition of *Diprotodon* (Murray 1991). (j) Reconstruction of the skeleton of *Diprotodon optatum*. Length approx. 3 m (Murray 1991).

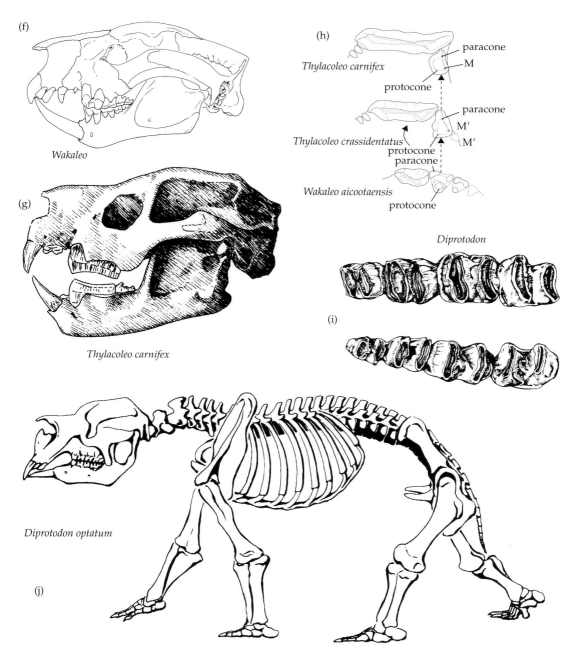

(f)

Wakaleo

(g)

Thylacoleo carnifex

(h)

Thylacoleo carnifex

paracone

M

protocone

paracone

M′

Thylacoleo crassidentatus

M′

protocone

paracone

Wakaleo aicootaensis

protocone

protocone

Diprotodon

(i)

Diprotodon optatum

(j)

Figure 6.12 *(continued)*.

Ilariidae. This exclusively Middle Miocene group of medium to large herbivores is poorly known (Tedford and Woodburne 1987). It has some superficial resemblances to the Koala, notably the essentially selenodont upper molar structure (Fig. 6.12(d)). However, the lower molars in particular are very different from those of Koalas, having retained a recognisably tribosphenic form.

Ektopodontidae. Known almost exclusively by its extraordinary teeth (Fig. 6.12(b)), these small animals have at one time or another been taken for multituberculates, or for monotremes. The crown of the relatively huge molar teeth, both uppers and lowers, consists of two transverse rows of up to about nine short, longitudinal crests. Their diet was presumably soft plant material, perhaps fruit, but nothing is really known about their habits or habitat. There is no good argument for a relationship of ektopodontids with any particular diprotodont family, and nor is there anything close to an analogous living tooth structure. They survived right through from the Late Oligocene/Early Miocene into the Early Pleistocene (Woodburne 1987).

Pilkipildridae. This family is also very poorly known, being represented only by a few teeth and jaw fragments from Oligocene to Middle Miocene deposits (Fig. 6.12(c)). They were small animals and their dentition combines an enlarged lower-third premolar and low-crowned, basined molars (Archer *et al.* 1987). The diet was perhaps hard food such as seeds and tough fruit, and a relationship with the petaurids has been tentatively proposed.

Miralinidae. The final extinct family to be considered is yet another Miocene group known only from teeth and jaw fragments, from South Australia (Woodburne *et al.* 1987). In this case there are large, sectorial third premolars, and the molars bear transverse rows of weakly developed cusps (Fig. 6.12(e)). A possible relationship with ektopodontids, whereby miralinids are interpreted as having a primitive version of the former's teeth, has been suggested.

Living families. Of the remaining eight families of diprotodonts usually recognised, two as mentioned have no pre-Holocene fossil record: Tarsipedidae and Acrobatidae. The other six occur from the Oligocene–Miocene to the present day. These are Vombatidae (wombats), Phalangeridae (cuscuses and brush-tailed possums), Burramyidae (pygmy possums), Macropodidae (kangaroos), Petauridae (gliders and ring-tailed possums), and Phascolarctidae (koalas).

An overview of marsupial evolution

The history of the marsupial mammals, like that of any other large taxonomic group, involved a complex interplay between biogeographic and ecological opportunities on the one hand and the evolutionary potential to adapt to a particular variety of habitats and niches on the other. Some of these factors can be determined from the fossil record that has just been described, and others from what is known of the biology of the living members.

Time and place of origin
The marsupials are one of the several lineages of tribosphenidan mammals that appeared in the Early Cretaceous, most of which were short-lived. Identifying either the time or the place of origin of a taxon is notoriously difficult because of the stratigraphic and the biogeographic incompleteness of the fossil record, and because diagnostic morphological characters do not necessarily evolve at the same time as lineages separate. In most cases concerning mammals, dates of divergence based on molecular sequences are consistently much earlier than the latest dates indicated by the fossils, and until recently this case was no exception. The oldest fossil identifiable as a stem group marsupial is about 125 Ma, from the Barremian of China. Both Kumar and Hedges (1998) and Penny *et al.* (1999), using different molecular data, placed the marsupial–placental divergence at about 170 Ma, which is Middle Jurassic, and would imply a long 'ghost-lineage' of missing fossils. However, a more recent analysis by Springer *et al.* (2003) using a far larger molecular database, estimated the divergence date between 102 and 131 Ma, which is perfectly compatible with the fossil date.

From the information that is currently available, the simplest hypothesis is that marsupials arose from a stem-member that had dispersed from an ultimate Asian origin into North America in the Early Cretaceous. This would account for the existence of the stem marsupial *Sinodelphys* very early on in Asia, but the vastly greater diversity of Marsupialia in North America during the Late Cretaceous, compared to the one or two sporadic occurrences recorded in Asia for the whole of the Late Cretaceous. Unfortunately, Early Cretaceous fossil mammals, which might help clarify this question, are exceedingly scarce in North America (Cifelli and Davis 2003).

At any event, during the Late Cretaceous the North American marsupials radiated into 4 families, and about 19 genera, all of them relatively small forms, but variously adapted, some for insectivory, some for a rather more carnivorous habit, and some for a more cosmopolitan or omnivorous diet. Incidentally, it should be remembered in this context that the Tribosphenida as a whole were less abundant and diverse than the contemporary multituberculate mammals. It needs also to be borne in mind that so very little is known about mammals of the Cretaceous in the Gondwanan continents, including Africa, that a major surprise is always possible. The reported, though disputed presence of a Late Cretaceous marsupial in Madagascar is a reminder of this.

The K/T mass extinction event has been described. As far as the marsupial mammals are concerned, the thriving North American radiation was all but eliminated, and the only marsupials represented on that continent in the Palaeocene were the didelphid *Peradectes*, and one or two closely related genera. Possibly, as discussed earlier in the chapter, the mass extinction of the marsupials was due to competition from similarly adapted placental immigrants from Asia. Whatever the reason, didelphids did nevertheless persist and indeed radiated in a small way through the Eocene and possibly into the Miocene. Members of the family also appeared sporadically in Europe, Africa and Asia during this time, presumably by dispersal from North America. Although important as small carnivorous members of certain Old World mammalian faunas, they never achieved significant taxonomic diversity, and have not been recorded in these regions since the Miocene (McKenna and Bell 1997).

South America

The next major point in known marsupial history occurred around the time of the K/T boundary. Having all but disappeared from North America, they left the Laurasian and African world virtually exclusively to the placentals. However, at about the same time, marsupials first appear in the South American fossil record. No marsupial, or placental, mammals have been found in undisputed Cretaceous rocks anywhere in South America. The mammalian fauna that was present, as most richly represented by the Campanian or Maastrichtian Los Alamitos Formation of Argentina, consisted solely of non-tribosphenidan Mesozoic mammal groups that are more typical of Jurassic and Early Cretaceous deposits elsewhere, and which are represented here mostly by subgroups endemic to South America. Whether the very earliest South American marsupials are strictly latest Cretaceous or Early Palaeocene is not certain, but what is evident is that the fauna included possible members of two groups characteristic of the Late Cretaceous of North America, peradectids and pediomyids. Add to these the very slightly later appearance in South America of possible relatives of the specialised North American genus *Glasbius*, and the conclusion seems inescapable that the South American fauna was derived by immigration from North America close to 65 Ma. During the Late Cretaceous, the Caribbean Archipelago was forming, and provided a potential dispersal route by island hopping. Dispersal of several other groups of terrestrial vertebrates, including dinosaurs, had indeed been in progress during the latest Cretaceous (Rage 1988; Gayet *et al.* 1992).

Within a very short space of time, less than five million years, a truly indigenous South American marsupial fauna had evolved from the presumed immigrant ancestors. There was a radiation of primarily insectivorous and omnivorous didelphids, that retained the primitive tooth form. Simultaneously, the medium and large carnivore mammal niches started to be filled by the sparassodonts (borhyaenoids), a group which was also probably derived from the immigrant didelphid ancestor. The paucituberculates may be related to the original *Glasbius*-like immigrant, and they evolved into a series of small- and medium-sized omnivore, and specialist grain, fruit, and tuber-eating herbivores. However, the South American marsupial radiation was limited by them having to share their continent with certain placental mammal groups, which had also arrived at the start of the Palaeocene. These contributed the medium- to large-sized ungulate herbivores, and also certain specialists, such as anteaters and sloths, that constitute the Xenarthra. Later in the Cenozoic, marsupials had to contend with the giant flightless predatory birds, such as the Miocene *Phorusrhacus* that stood over 2 m high, with a huge raptorial beak, and the introduction during the Oligocene of the placental primates and rodents.

This general picture of South American marsupial diversity remained intact throughout the succeeding 60 million years (Marshall and Cifelli 1990). A considerable degree of endemicity occurred, as would be expected in a large, single continent, with faunal differences reflecting regions with different climates (Pascual and Ortiz 1990; Flynn and Wyss 1998). Although there were shifts in the relative fortunes of particular groups, there were no extinctions of higher taxa until the Plio-Pleistocene, when a combination of large climatic changes, and the secondary connection of North and South America, resulted in the events referred to as the Great American Biotic Interchange that is described in more detail in Chapter 7. Around this time, several South American marsupial taxa disappeared, notably the last of the carnivorous sparassodonts (borhyaenids), while a handful of didelphid species dispersed northwards into Central and North America, where they are still to be found.

Antarctica

In principle, the palaeo-biogeography of American marsupials offers no problems, and the same is true of Antarctica (Woodburne and Case 1996). The only mammalian fossils so far described are from the Middle or Late Eocene of Seymour Island, on the Antarctic Peninsula immediately south of South America. At the time, the oceanic current system, and the prevailing high levels of CO_2 (Pearson and Palmer 2000) were responsible for a warm to cool temperate forested environment, dominated by angiosperms. The mammals all belong to taxa that are also found in the Palaeocene of South America. They include marsupial polydolopoids, didelphimorphs, and a microbiothere, and there are also placental ungulates, xenarthrans, and bats; there is even a possible phorusrhacid bird (Reguero *et al.* 2002). There was still a connection between the South American and Antarctic land masses, and clearly the Antarctic Eocene marsupial fauna was continuous with that of South America, with no more difference than would be expected in different

areas of the same large land mass. However, Drake's Passage between the two continents opened up in the Late Eocene, about 36 Ma, with disastrous consequences. A circum-Antarctic current was established that prevented the warming influence of the more southern oceans upon the Antarctic climate and therefore caused an increased temperature gradient to build up between warm tropics and the cold, southern polar region. The event also coincided with a reduction in CO_2 level, presaging the start of a worldwide cooling. Before long the permanent ice sheets characteristic of Antarctica built up, and the last remnants of its Eocene biota disappeared.

Australia

The origin and early history of marsupials in Australia is shrouded in ignorance. To start with, absolutely nothing is known about any fossil mammals of eastern Antarctica, the region of Gondwana to which Australia was connected during the Mesozoic. Yet eastern Antarctica was separated from western Antarctica, and ultimately therefore from South America, by the Transantarctic Mountain chain, which may have posed a significant barrier to dispersal of South American groups of marsupials, and therefore may have been a critical influence on the origin and nature of the Australian fauna. The exact timing of the separation of Australia from Antarctica by the opening up of the Southern Ocean is not certain (Woodburne and Case 1996). Rifting between the two had commenced by 80 Ma and the last remaining physical connection, the Tasman Rise foundered around 64 Ma. However, it was not until about 52 Ma that a complete seaway developed (Fig. 6.13). Therefore, while an easy dispersal route may have ended at the start of the Cenozoic, a potential sweepstake dispersal route probably existed throughout the Palaeocene and even into the Eocene. To add to the ignorance about eastern Antarctica, there is no information at all about Australian mammals prior to the latest Palaeocene or earliest Eocene Tingamarra Fauna.

If the marsupials found in Australasia were a monophyletic group unrepresented elsewhere, then their origin could be explained as the consequence of a single entry of a 'didelphimorph-like' ancestor into the continent, via a sweepstake route

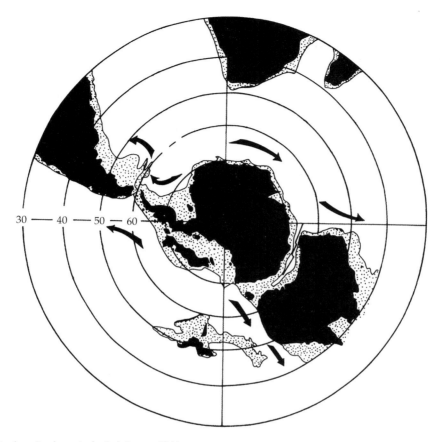

Figure 6.13 Southern Gondwana in the Early Eocene, 53 Ma.

from Antarctica, followed by radiation into the five indigenous orders. No such simple picture as this is possible, however, because of the relationships of the microbiotheres. These are taxonomically australidelphians, yet occur in South America, from the Palaeocene to the Recent, and in the Eocene of western Antarctica. They may possibly occur in the Tingamarra fauna, but this has yet to be substantiated.

There are a number of possible solutions to what is termed the '*Dromiciops* problem'. One is that the whole australidelphian radiation, including the origin of the microbiotheres, did indeed occur in Australia after the immigration of a single ancestor, but that microbiotheres dispersed from Australia into Antarctica and South America. This sequence of events would have to have been completed by the Early Palaeocene for the microbiothere *Khasia*

to appear in the Tiupampan fossils of that age, which means hardly any time at all after the initial entry of marsupials from North America into the southern continents in the first place. Meanwhile, the microbiotheres must have gone extinct in Australia, if not by the Eocene, then certainly before the great Oligocene–Miocene faunas of Riversleigh. This view is consistent with Godthelp *et al.*'s (1999) description of the Tingamarran genus *Djarthia*, whose teeth they cannot distinguish from primitive members of either Australidelphia or Ameridelphia. Its upper molars are dilambdodont, and they propose that this is the condition of the common ancestor of both these groups; *Djarthia* is therefore seen as a conservative descendant of the original single immigrant. However, their interpretation still requires the initial dispersal event to have occurred by the very start of the Palaeocene, so that a microbiothere could have reversed back along the sweepstake route into Antarctica and South America by Tiupampan times.

The second possible explanation for the Australian marsupial fauna is that the initial radiation of the taxon Australidelphia did not occur in Australia, but in other parts of Gondwana, namely Antarctica and/or South America. Under this scenario, if the microbiotheres are the sister group of the rest of the australidelphians (Fig. 6.36), as weakly supported by Amrine-Madsen et al. (2003), then the divergence between them must have occurred during the earliest Palaeocene, and only a single dispersal into Australia by the ancestor of the Australian radiation need have occurred. This event could have taken place any time in the Palaeocene, prior to the age of the first Australian fossil marsupials. However, if the microbiotheres are taxonomically nested within the Australidelphia, there would have to have been a dispersal into Australia of the respective ancestors of each of the lineages more basal than the microbiotheres. For example Horovitz and Sánchez-Villagra's (2003) cladogram (Fig. 6.3(a)), requires three such dispersals: a dasyuromorphian, a notoryctemorphian–peramelemorphian, and a diprotodontian. Woodburne and Case (1996) have proposed a solution along these lines. They argue that the dilambdodont condition evolved independently in didelphids and basal australidelphians. They then point to the dentition of *Andinodelphys* from the Tiupampan Palaeocene of South America, which they claim has plesiomorphic australidelphian features, including the pre-dilambdodont condition of a linear centrocrista, and prominent stylar cusps on the upper molars. They infer therefore that *Andinodelphys* represents the basal stock of all the Australidelphia, and that the origin of the group occurred in Gondwana, outside Australia. Their view also implies that some divergence had already occurred by Tiupampan times, because microbiotheres are distinct by then. The authors thus propose that representatives of the Australian groups dispersed via eastern Antarctica and into Australia by a sweepstake route, prior to the complete separation of Australia and Antarctica. Acceptance of this hypothesis hinges primarily on their interpretation of *Andinodelphys*. Godthelp et al. (1999) do not accept its australidelphian affinities but regard it as a didelphid although, as already pointed out, they cannot actually distinguish basal didelphids from basal dasyuromorphs on teeth alone.

The third possible explanation for the biogeography of the Gondwanan marsupials is that Australia, Antarctica, and South America shared a single, taxonomically broad marsupial fauna in the Early Palaeocene. Following the isolation of Australia, differential extinction accounted for its subsequent fauna. For this hypothesis to be true, there would have to have been ameridelphians such as polydolopoids and true didelphimorphs in Australia at one time, as well as the australidelphians. Possible support for this pan-Gondwana hypothesis comes from interpretation of the Tingamarran fauna as including representatives of the didelphimorphs, polydolopoids, and possibly also microbiotheres and protodidelphids. Not one of these identifications is indisputable, but all are possible. Under this scenario, it would be expected that the Gondwanan placentals would also be represented in the Palaeocene of Australia, particularly those known to have existed in the Eocene of Antarctica, namely xenarthrans, and the ungulate orders of litopterns and astropotheres (Reguero et al. 2002).

It is impossible at present to choose between these alternatives, but at least a solution is realistically likely to be found. The first alternative would be supported by the discovery of very early Palaeocene faunas in Australia of australidelphians including microbiotheres; the second by Palaeocene faunas in Antarctica yielding other australidelphian groups as well as microbiotheres; the third by the undoubted presence in Australia of polydolopoids, didelphimorphs, other ameridelphian groups, and also Gondwanan placentals. In short, Early Palaeocene mammalian faunas of Australia, and Antarctica, if and when discovered, may be expected to reveal the critical evidence.

Whatever the detailed history of their arrival in Australia, the subsequent radiation of marsupials in that continent was profoundly affected by the absence of placental ungulates, the mammals that occupied the medium-sized and large herbivore roles everywhere else in the world, including South America and also the Eocene of Antarctica. Among lots of smaller and medium-sized species, the Diprotodontia also included genera of diprotodontids, kangaroos, and wombats that adopted various roles as large herbivores. To date absolutely nothing is known of the radiation of these and the other australidelphians before the Late Oligocene, about 25 Ma. At this time, Australia was largely occupied by rain forests, tropical and temperate, in which lived a rich marsupial fauna containing members of most of the modern families, and about as many again no longer extant. Therefore, whatever the situation today, it is clear that the australidelphian marsupials were primarily adapted for non-arid, forested conditions, an environment that prevailed for another 10 million years. However, by the Late Miocene, drier conditions had spread bringing with them a great extension of grasslands in place of forests. This was caused in part by the general

worldwide cooling of the later Miocene, and in part by the northward drift of Australia, eventually colliding with Asia and causing the rise of the New Guinean mountains which cast a rain shadow (Archer *et al.* 1991). One consequence was a great increase in the diversity of grazing diprotodonts. These were mostly the macropodine kangaroos, with their pattern of tooth replacement modified so that new teeth continued to replace old, worn teeth, and wombats with high-crowned, open-rooted molars that continuously grew.

From this point, despite further reduction of forests and desertification of the centre of the continent, the fauna remained unchanged in essence until the arrival of humans 50,000–60,000 years ago, and the simultaneous commencement of the extinction of the larger marsupial species, as discussed at the end of the next chapter.

CHAPTER 7

Living and fossil placentals

Living placentals and their interrelationships

The vast majority of living and fossil mammals are placentals. Today there are about 4,400 species, which are traditionally organised into 18 Orders (Table 2.4), with an extra one if the Pinnipedia are separated from the Carnivora, and a twentieth if the recently extinct Malagasy order Bibymalagasia is recognised as such. There have been many attempts to discover supraordinal groupings from amongst these Orders based on morphological characters, though few proposals have been universally accepted. It is only with the advent of increasingly large sets of molecular sequence data in the last few years that a reasonably robust resolution looks imminent, although these contemporary analyses are remarkably and controversially at odds with the traditional ones.

Novacek *et al.* (1988) summarised the then current situation regarding supraordinal classification of placentals, a time at which morphology was still dominant but molecular data was at the threshold of significance (Fig. 7.1(a)). They accepted a basal group Edentata that combined the Xenarthra of the New World with the Pholidota of the Old, based on a few cranial characters, loss of the anterior teeth, and reduction of the enamel of the remaining ones. This left the rest of the living placentals as a monophyletic group Epitheria, sharing such apparently minor characters as the shape of the stapes bone in the ear. They found very little resolution within the Epitheria, and concluded that there was a polychotomy of no less than nine lineages arranged as a 'star' phylogeny. No remnant of the previously recognised taxon Ferungulata, created by Simpson (1945) for the Carnivora plus the ungulate orders Artiodactyla, Perissodactyla, Proboscidea, Hyracoidea, Sirenia, and Tubulidentata remained. On the other hand,

three supra ordinal taxa of earlier authors did survive. One was Gregory's (1910) Archonta, consisting of generally conservative forms and by now composed of the Primates, Dermoptera, Scandentia, and Chiroptera, but excluding the Lipotyphla. The second was Glires, originating with Linnaeus (1758) and widely accepted ever since, for the Rodentia and Lagomorpha; Novacek *et al.* (1988) tentatively placed the Macroscelidea as the sister-group of the Glires. The third supraordinal taxon recognised was, like Glires, well-established if not universally accepted. The Proboscidea and Sirenia constitute a group Tethytheria and the addition of its supposed sister-group Hyracoidea creates the Paenungulata.

Even for the supraordinal taxa that were proposed, morphological characters supporting them are few in number, fine in level of detail, and frequently challenged. Rose and Emry (1993), for example, rejected the Edentata, and Pettigrew *et al.* (1989) dismembered the Archonta by separating the Microchiroptera (echo-locating bats) and tree shrews on the one hand, from the Macrochiroptera (fruit bats), Primates, and Dermoptera on the other. Few authors have accepted uncritically the relationship of macroscelideans to rodents plus lagomorphs, and Fischer and Tassy (1993) continued to argue for a relationship of hyracoids with perissodactyls, rather than with the elephants and sirenians as Paenungulata. In evolutionary terms, this general paucity and triviality of supraordinal diagnostic characters points to one of two conclusions. Possibly the living placental orders all diverged at a low taxonomic level, from within a radiation of primitive, insectivorous forms that differed from one another in little more than such things as the course of various minor foramina, nerves and blood vessels in the skull, or the fine details of the structure of the

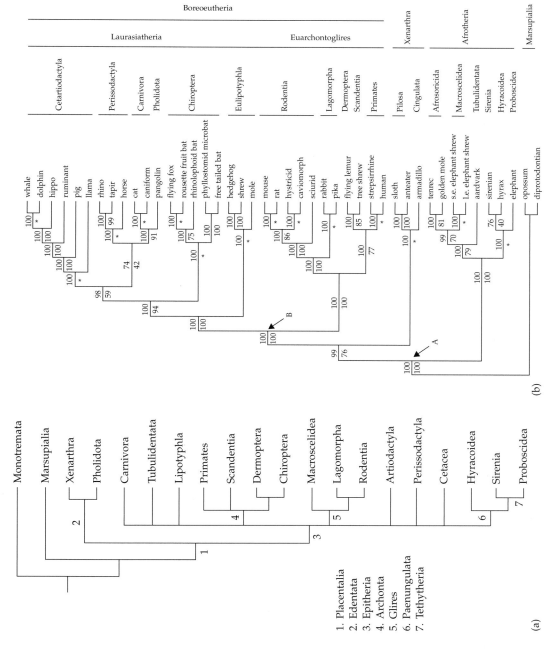

Figure 7.1 Phylogeny of the living orders of placental mammals. (a) Novacek *et al.*'s (1988) morphological-based cladogram. (b) Murphy *et al.*'s (2001b) molecular-based cladogram. Numbers above the branches are percentage Bayesian posterior probabilities; numbers below the branches are maximum likelihood bootstrap support values.

middle ear. Alternatively, the basal members of related orders may have shared significant derived characters, but these have subsequently been lost or transformed out of recognition as the individual orders evolved into their modern representatives. Further purely morphological analysis of living mammals is unlikely to lead to a recognition of which of these two phylogenetic patterns is true, let alone to resolve the interrelationships. Two centuries of study is surely approaching saturation. In principle the fossil record can improve the taxonomic resolution by providing character combinations not found amongst the living groups, and has indeed contributed much to the debates on supraordinal relationships, as discussed later. Unfortunately, palaeontological evidence will always be hampered by the bugbear of missing data: there are too many cases of incongruence amongst dental and skeletal characters for fossils alone to be regarded as reliable trackers of phylogeny.

As with the marsupials discussed in Chapter 6, analysis of molecular sequence data is proving the likeliest source of information for solving the riddle of placental interrelationships, and some remarkable but by no means fanciful hypotheses have now emerged, almost to the extent of providing a tree that is fully resolved at the ordinal level. In one of the first contributions, de Jong *et al.* (1993) analysed the eye lens protein αA-crystallin and found that the macroscelidean *Elephantulus rufescens* had three amino acid replacements otherwise unique to the paenungulates (elephants, sirenians, and hyraxes) and the aardvark, suggesting a monophyletic group for these very different orders, whose sole non-molecular common feature is that their history is mainly African. Within a few years, this relationship had become strongly supported by DNA sequences, both mitochondrial and nuclear, and even extended to include the Tenrecida (Springer *et al.* 1997b; Stanhope *et al.* 1998). These are the tenrecs of Madagascar and the related otter shrews of mainland southern Africa, which together had been regarded as a subgroup of lipotyphlan insectivores. One final taxon of African mammals has been added to what became named the Afrotheria. *Chrysochlorids* are the golden moles of southern Africa and, like the tenrecs, they too had been firmly believed to be lipotyphlans. The study of

Stanhope *et al.* (1998) placed it firmly in the afrotherian clade. The Afrotheria concept goes a long way towards understanding the previously poorly understood relationships of the Tubulidentata and Macroscelidea. However, it contradicts all morphological based classifications of Lipotyphla, by removing the tenrecs and golden moles from their position nested deeply within the group, and related to the shrews and moles (Butler 1988).

Meanwhile, another radical rearrangement was emerging. The relationships of the cetaceans had been the subject of a long dispute, even to the extent of whether they had evolved from a carnivorous, or an herbivorous ancestor. On the basis of the fossil evidence, the view prevailed that they had evolved from a secondarily carnivorous group of primitive, 'condylarth' ungulates known as mesonychids. This consensus was actually disturbed long ago, when, in a very early essay into molecular systematics, Boyden and Gemeroy's (1950) immunological method suggested a relationship between whales and artiodactyls. Little notice was taken at the time. The recent era was marked by Graur and Higgins (1994), who analysed the molecular evidence then available, mainly protein sequences, and also concluded that cetaceans were most closely related to the Artiodactyla. Furthermore, their study pointed to the egregious conclusion that whales nested within the artiodactyls, as the sister-group of the hippos. Since then, a very large quantity of DNA sequence data bearing on the question has been analysed, with results overwhelming confirming the hippo-whale clade (Waddell *et al.* 1999; Gatesy and O'Leary 2001). At first, morphological evidence continued to fail to support even the overall group Cetartiodactyla, as it had come to be named (O'Leary 1999), although the similarities that do exist between whales and hippos in particular, such as the reduction of hair, development of subcutaneous fat, and ability to suckle the young underwater, were not lost on the proponents of the new grouping. Far from being the convergences hitherto supposed, they can be taken as evidence of a common ancestor between the two that was already adapted to a semiaquatic existence. Modern whales are so extremely derived that the key characters that would allow the recognition of their relationships, notably limb structure and dental structure, are effectively absent. However, increasingly convincing fossil evidence of the foot structure of primitive, still quadrupedal cetaceans does now offer good support for Cetartiodactyla, rather than for the mesonychid relationship, although not specifically supporting a relationship with hippos. Moreover, Naylor and Adams (2001) found that if dental characters are omitted from the analysis, the rest of the morphology supports the molecular-based relationship, indicating that most of the evidence for mesonychid affinities of whales may be due to convergence of the tooth structure in primitive members.

Most of the remaining modern mammalian orders had at one time or another been associated in one of two supraordinal groupings. The Ferungulata was used by Simpson (1945), for example, and included the Carnivora plus the main ungulate groups of Artiodactyla, Perissodactyla, Tubulidentata, and the three paenungulate orders Sirenia, Proboscidea, and Hyracoidea. As the molecular evidence now shows, the paenungulates along with the Tubulidentata fall into the Afrotheria supraordinal group. Meanwhile molecular evidence has been accumulating that not only supports the rump of the Ferungulata consisting of Carnivora, Perissodactyla and what is now Cetartiodactyla, but also adds two other orders never previously associated with it. Pumo *et al.* (1998) sequenced the complete mitochondrial DNA of a fruit bat and demonstrated a relationship between bats and ferungulates. This surprising conclusion, subsequently confirmed by Cao *et al.* (2000) denies the long-standing association of bats with Primates and Dermoptera. Waddell *et al.* (1999) reported an analysis based on seven genes that supported the ferungulate and bat relationship, but also added the pangolin to this group. On this evidence, therefore, the Pholidota can no longer be associated with the Xenarthra of South America as a supraordinal group Edentata.

The final association to emerge during this flurry of analysis of ever-increasing lengths of sequence data concerns the fate of the traditional group Archonta. This quite widely approved morphological-based taxon had consisted of the Primates, Scandentia (tree shrews), Dermoptera (colugos), and Chiroptera (bats). The Chiroptera have been removed, as discussed, but the molecular evidence supports a relationship between the remaining

three. It also indicates, to much surprise, that the well supported group Glires, consisting of the Rodentia and Lagomorpha (Huchon *et al.* 2002), is also a member of this supra-order.

The old placental order 'Insectivora' has a long and very chequered history. Originally it incorporated all the placentals, living and fossil, that retained such primitive characters as a full dentition, the basic or at least little modified form of the tribosphenic molar tooth, five plantigrade digits, and a small brain. As such it had virtually no phylogenetic meaning and over the years, even on morphological grounds, more and more subgroups were removed and elevated to ordinal status in their own right. Thus the Scandentia (tree shrews) and Macroscelidea (elephant shrews) were long since rejected, and the diminished order became referred to as the Lipotyphla. With this recent removal of the Tenrecida and Chrysochlorida on molecular grounds, only four families Soricidae (shrews), Talpidae (moles), Solenodontidae (west Indian shrews) and Erinaceidae (hedgehogs) remain, and the order is now renamed Eulipotyphla. There has even been some doubt about the monophyly of the Eulipotyphla based on molecular sequences, and one study (Cao *et al.* 2000), using mitochondrial gene sequences, found the hedgehogs to be the basal group of all placentals, and the moles to be associated with the expanded ferungulate plus bat group. Other studies, however, kept the Eulipotyphla intact (Waddell *et al.*1999).

This series of molecular analyses of ever-increasing numbers of base pairs in an ever-increasing variety of genes culminated in 2001, when two papers containing complete phylogenies of all the major groups of placentals appeared simultaneously in the journal *Nature*. Madsen *et al.*'s (2001) study was based on two molecular data sets, respectively 5,708 and 2,947 base pairs in length, taken from both mitochondrial and nuclear genes. Murphy *et al.*'s (2001*a*) study used sections of 18 genes totalling 9,779 base pairs. Despite the independence of the two research groups, their results are in complete agreement over the recognition and composition of the four supraordinal groupings, towards which the previous five years or so of work had been increasingly pointing.

- **Xenarthra**. The armadillos, sloths, and anteaters of South America, a group which, unlike the other three, has been long accepted on morphological evidence: they share a unique kind of accessory articulations between the vertebrae, from which the taxon name comes, and the absence of enamel on the teeth.

- **Afrotheria**. Proboscidea, Sirenia, and Hyracoidea, together constituting the long accepted Paenungulata of morphological based schemes, along with Tenrecida (tenrecs and otter shrews), Tubulidentata (aardvarks), Macroscelidea (elephant shrews), and Chrysochlorida (golden moles).
- **Euarchontoglires**. Primates, Dermoptera (colugos), and Scandentia (tree shrews) left over from the old Archonta, combined with Rodentia and Lagomorpha (rabbits and pikas), which constitute Glires.
- **Laurasiatheria**. Eulipotyphla (hedgehogs, moles, and shrews), Chiroptera (bats), Pholidota (pangolins), Carnivora, Perissodactyla (odd-toed ungulates), and Cetartiodactyla (even-toed ungulates and whales).

Murphy *et al.* (2001*a*) and Madsen *et al.* (2001) differed over the interrelationship between these four super-orders, and neither could produce a well-supported hypothesis. However, they have now combined their respective data sets, and applied the powerful Bayesian methods in addition to Maximum Likelihood methods to what has become a 16.4 kilobase sequence (Murphy *et al.* 2001*b*). The result is a strongly supported inference that the Euarchontoglires and Laurasiatheria are sister groups of what they term Boreoeutheria. Boreoeutheria is the sister-group of Xenarthra, and Afrotheria is the basal group of living placentals.

The only other molecular analyses involving very large data sets are those using the ever-increasing range of taxa whose complete mitochondrial genome has been sequenced (Arnason *et al.* 2002; Lin *et al.* 2002). These studies relatively weakly support Xenarthra, Afrotheria, and most of Laurasiatheria, but rodents and erinaceids tend to come out in unexpected positions. The latter family, hedgehogs, are found to be basal to all the placentals in Arnason *et al.*'s tree, and in one of Lin *et al.*'s versions. Corneli (2003) concluded that supraordinal relationships as they appear in trees based on mitochondrial genomes are very poorly supported because the branches are too short. On the other

hand, he found that complete mitochondrial sequences do resolve the relationships of families within orders very well.

Murphy *et al.* (2001*b*) have generated the most complete tree of the relationships of the subgroups within the super-orders, most of which have good statistical support both from this work and from the further analysis by Springer *et al.* (2003), and is the basis of the molecular-based cladogram in Fig. 7.1(b).

There are important implications for the biogeographic dimensions of the early diversification of placental mammals inherent in this classification, which are taken up later, after the fossil evidence for the emergence of the recent orders has been reviewed.

Cretaceous fossils

Like the marsupials, the placental (eutherian) lineage is one of the several groups of tribosphenidan mammals that first appeared in the Early Cretaceous fossil record. Primitive placental mammals can be recognised by several dental characters. There are only three molars in the upper and the lower jaws, and the last of the premolars, particularly the upper one, tends to be more complex than the others, and is described as submolariform. The upper molar teeth have a narrow stylar shelf so that the metacone and paracone are well towards the buccal side of the crown. These characters are associated with a series of skull characters.

Eomaia (Fig. 5.17(a) and (b)) is the earliest described placental (Ji *et al.* 2002). This astonishingly well-preserved skeleton of a little 25 g mammal even has impressions and carbonaceous traces of its pelt. It is from the Barremian part of the Yixian Formation of China, which gives it an age of about 125 Ma. The dental formula conforms to the presumed primitive placental one of I5/4: C1/1: PM5/5: M 3/3, and the last premolar shows little molarisation. The postcranial skeleton indicates that *Eomaia* was a climbing and perhaps permanently arboreal animal. A few other Early Cretaceous placental mammals have been found that are only slightly younger than *Eomaia*, although none anywhere near such complete specimens. The isolated teeth of the Russian *Murtoilestes* date from around the end of the Barremian some 120 Ma (Averianov

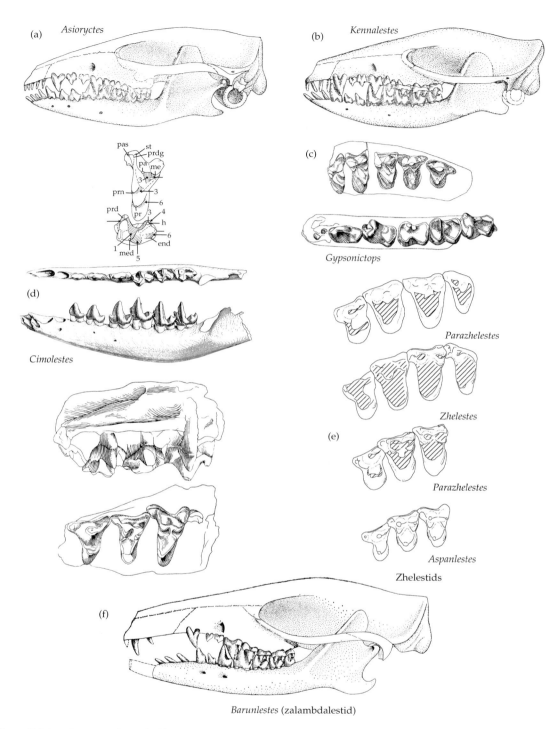

(a) *Asioryctes*

(b) *Kennalestes*

pas
st
prdg
pa
me
prn
prd
pr
med

(c)

Gypsonictops

(d)

Cimolestes

Parazhelestes

Zhelestes

(e)

Parazhelestes

Aspanlestes

Zhelestids

(f)

Barunlestes (zalambdalestid)

Figure 7.2 Late Cretaceous placentals. (a) *Asioryctes* skull in lateral view, and occlusal views of an upper and a lower molar. Skull length approx. 5.4 cm (Kielan-Jaworowska *et al*. 2000, after Kielan-Jaworowska 1975). (b) *Kennalestes* skull in lateral view. (c) Upper and lower dentition of *Gypsonictops* in occlusal view. Approx. length of upper jaw fragment 1.3 cm (Kielan-Jaworowska *et al*. 1979, after Lillegraven 1969). (d) *Cimolestes* lower jaw in occlusal and labial views, and upper molars in lingual and occlusal views. Length of fragment of mandible approx. 3 cm (Kielan-Jaworowska *et al*. 1979, after Lillegraven 1969). (e). Upper molars in occlusal view of four zhelestids, *Parazhelestes robustus, Zhelestes temirkazyx, Parazhelestes minor,* and *Aspanlestes aptap*. Approx. width of teeth between 1.5 and 2.0 cm (Nessov *et al*. 1998 figure 20). (f) *Barunlestes*. Skull lenght approx. 8.3 cm (Kielan-Jaworowska *et al*. 2000, after Kielan-Jaworowska 1975).

and Skutschas 2001) and the Mongolian *Prokennalestes* (Fig. 5.17(c) and (d)) is Aptian or Albian in age, making it perhaps 10 Ma younger still (Kielan-Jaworowska and Dashzeveg 1989).

Although it is generally supposed from their early appearance in Asia that this was the area of origin of placentals, Cifelli (1999) described an associated pair of dentaries from the Cloverley Formation of Montana, which is also dated as Aptian-Albian. *Montanalestes* (Fig. 5.17(e)), as this specimen is named, has a submolariform last premolar followed by only three molars. However, the teeth differ in several respects from *Prokennalestes* (Fig. 5.17(d)), and until material of upper molars is found, a placental attribution of *Montanalestes* remains a biogeographically intriguing, but not totally established possibility. Another possible, though even more dubious, very early placental has been described from isolated molar teeth (Fig. 5.19(c)) from Morocco and named *Tribotherium* (Sigogneau-Russell 1995). It is dated as ?Berriasian, which is the very base of the Cretaceous. The teeth do have some placental characters such as the narrow labial shelf of the upper molar, but lack others such as a third talonid cusp on the lowers. Were *Tribotherium* actually to prove to be a stem placental, it would be important because it would put the fossil date of divergence of Placentalia from Marsupialia back to at least the start of the Cretaceous around 135 Ma. However, the information available is presently far too sparse to make such an assertion with confidence.

During the Late Cretaceous, a radiation of placentals occurred. All remained small in body size, and fairly conservative in dental and, where known, cranial structure to such an extent that there is limited agreement on their classification. Relatively small differences have been used for ordinal and even higher level separation, but had they all been living today most if not all the Late Cretaceous placentals might well have been incorporated into no higher a taxon than a superfamily. At one time it was in fact customary to include them in a single suborder Proteutheria, as a constituent of the order Insectivora (e.g. Romer 1966; Kielan-Jaworowska *et al.* 1979a). Inevitably, the taxon Proteutheria has long since disappeared from usage due to its manifestly paraphyletic nature, based as it necessarily had to be solely on ancestral dental characters.

Kennalestes and *Asioryctes* from Asia, and *Gypsonictops* and *Cimolestes* from North America are among the best-known representatives of the Cretaceous placentals, and the level of taxonomic uncertainty surrounding them is well illustrated by a very brief, recent history of the classification of these four forms.

• In the definitive review of the time, Kielan-Jaworowska *et al.* (1979a) placed *Gypsonictops* and possibly *Kennalestes* in a super-family Leptictoidea. *Asioryctes* and *Cimolestes* were placed together in a family Palaeoryctidae that was part of another super-family, Palaeoryctoidea.

• In their comprehensive classification of all mammals, McKenna and Bell (1997) placed *Asioryctes* into a group basal to most of the placental mammals, *Kennalestes* and *Gypsonictops* together in the same family, Gypsonictopidae, of a super-order Leptictida, and *Cimolestes* in an entirely separate Order Cimolesta.

• Most recently, Kielan-Jaworowska *et al.* (2004) have associated the two Mongolian genera *Asioryctes* and *Kennalestes* in separate families of the same super-order Asioryctitheria. *Gypsonictops* is classified in a different super-order Insectivora as a leptictidan, and *Cimolestes* in yet a third super-order Ferae.

Many other less comprehensive proposals of relationships concerning these forms have been made over the years, and there have also been a number of suggestions of relationships between specified Cretaceous placentals and particular groups of living mammals, considered later.

Irrespective of the disputatious details of possible interrelationships, there are five main groups of Late Cretaceous placentals, the primitive form *Asioryctes* and its possible relatives, the leptictidans, the zalambdalestids, the sharper-toothed palaeoryctids, and the incipiently bunodont-toothed zhelestids.

Asioryctida

Asioryctes has been described from about 10 skulls and several fragmentary skeletons from the Coniacian and Campanian of Mongolia and elsewhere in Asia (Fig. 7.2(a)). It has retained several primitive placental characters that are modified in all other known Late Cretaceous forms, including the five upper and four lower incisors that occur in *Eomaia* and *Prokennalestes*, although it does posses the more derived number of four premolars. In the skull there is a large lachrymal and a deep jugal bone. In the postcranial skeleton *Asioryctes* is uniquely primitive in its incompletely co-ossified atlas ring, and in the absence of superposition of the astragalus on the calcaneum in the ankle that characterises all other placentals. Instead, the two bones lie side by side, so the astragalus has full contact with the ground (Kielan-Jaworowska 1977; Novacek 1980).

The molar teeth are unusual in being exceptionally transversely widened and narrow from front to back, so that the paracone and metacone of the uppers lie very close together. The trigonid of the lower molars is shorter than the talonid.

The only closely related form is another Mongolian genus *Ukhaatherium*, of which there are also several skulls and skeletons (Novacek *et al.* 1997). It differs from *Asioryctes* in having less extremely narrowed molar teeth, in which feature it is presumably more primitive. An epipubic bone has been found in *Ukhaatherium*, indicating that this is a primitive feature of placentals as well as marsupials.

Leptictida

The North American *Gypsonictops* is only known from incomplete jaws and isolated teeth (Fig. 7.2(c)), although these are relatively common in the Campanian and Maastrichtian. The dentition is often taken as close to the unspecialised condition for Late Cretaceous placentals. The number of incisors is unknown, but the primitive condition of five premolars is probably present, the posteriormost of which tends to be well molarised in form. The molar teeth are uncompressed from front to back, and have well separated metacone and paracone.

Kennalestes (Fig. 7.2(b)) is from the Coniacian to Campanian of Asia, and is a great deal better known than *Gypsonictops*, on the basis of complete skulls although not, as yet, postcranial skeletons. Like, *Asioryctes*, the skull is primitive in most respects, and therefore it is not at all certain that *Kennalestes* should actually be included in the Leptictida (Novacek 1986a; Kielan-Jaworowska *et al.* 2004): However, the premolar and molar teeth are very similar to those of *Gypsonictops*. One derived character of *Kennalestes*, unknown for the American form, is the reduction of

the number of incisor teeth to four uppers and three lowers, from the ancestral 5/4 condition.

Leptictidans continued into the Palaeocene and Eocene, when they are represented by complete skull and skeletal material described by Novacek (1986a), who considered them to be relatives of the Lipotyphla on the basis of several cranial characters.

Zalambdalestidae

The earliest known zalambdalestid specimens are the jaws and teeth of *Kulbeckia*, from the Coniacian of Uzbekistan, which is dated at 85–90 Ma (Archibald *et al.* 2001; Archibald and Averianov 2003), although the more familiar members are the two Mongolian genera, *Zalambdalestes* and *Barunlestes* (Fig. 7.2(f)), which are Campanian in age. There are no known North American representatives.

The family includes the largest of the Cretaceous placentals, with a skull length up to around 5 cm. There are only three incisors, and the first lower one is greatly enlarged and strongly procumbent. The lower molars have developed a relatively large, strongly basined talonid. The postcranial skeleton resembles superficially that of an elephant shrew, with elongated hind legs indicating a saltational ability. Kielan-Jaworowska *et al.* (1979a) speculate that they lived in a rocky habitat, and the procumbent lower incisor could extract insects from crevices.

As far as the relationships of the group are concerned, they have often been associated with the anagalids, which occur in the Asian Palaeocene. More recently, some authors have claimed a relationship with the Glires, the supraordinal taxon that includes rodents and lagomorphs (Archibald *et al.* 2001; Fostowicz-Frelik and Kielan-Jaworowska 2002). The main character supporting this view is the similarity of the zalambdalestid lower incisor to that of the Glires. Both are enlarged, procumbent, and have the enamel restricted to the labial side.

Palaeoryctida

The North American genus *Cimolestes* (Fig. 7.2(d)) is usually accepted as a basal palaeoryctidan, an otherwise largely Palaeocene group of small, but incipiently more specialist carnivores (Lillegraven 1969). They are characterised by the development of very high, sharp cusps and crests on the molar teeth. The uppers are transversely expanded and the paracone is markedly taller than the metacone. In the lowers, there is a very marked difference in height between the trigonid and the talonid. This tendency to develop transversely oriented shearing edges suggests that the teeth were designed for coping with relatively hard insects, or small vertebrate prey.

Zhelestidae
The family Zhelestidae (Fig. 7.2(e)) consists of a series of small mammals that have incipiently ungulate molar teeth (Nessov *et al.* 1998). The majority of the numerous zhelestid genera recognised, for example, *Zhelestes* and *Aspanlestes*, are from the Turonian-Coniacian of Central Asia, approximately 88 Ma (Averianov 2000). Younger, Maastrichtian-age genera occur in Western Europe (Gheerbrant and Astibia 1994), and also in North America where they extend from the Late Cretaceous into the Palaeocene (Nessov *et al.* 1998).

There are five upper premolars. The crowns of the upper molars are broad, and subrectangular in shape. This was achieved by antero-posterior expansion of the protocone region of the tooth, and a wide separation of that cusp from the paracone and metacone. In the lower molars, the trigonid is relative lower, and so is more nearly equal in height to the talonid, while the talonid is expanded to about the same antero-posterior length as the trigonid. Zhelestids were presumably tending to omnivory in habit, their teeth adapted for high-energy plant food as well as insects.

Although none are yet known from more than teeth and jaw fragments, zhelestids have assumed great significance as possible basal members of a single radiation of ungulate mammals, culminating ultimately in such orders as artiodactyls, perissodactyls, elephants, and even whales. Indeed, Nessov *et al.* (1998) concluded from their cladistic analysis of dentitions that Zhelestidae is paraphyletic, and contains basal relatives of all these later forms.

It was thought for a long time that there are true ungulate mammals in the latest Cretaceous of North America, and in particular the renowned Bug Creek *Protungulatum* (Fig. 7.6(a)). In fact, while its 'condylarth', and therefore 'ungulate' credentials are not in much doubt, and it is more advanced than zhelestids, *Protoungulatum* is now believed to be Early Palaeocene in age (Archibald and Lofgren 1990).

Palaeocene fossils: the archaic placentals

The mass extinction event at the end of the Cretaceous Period that marked the demise of almost all the North American marsupials had far less effect on the placentals; indeed quite the opposite. The handful of Maastrichtian placental genera on that continent increased dramatically in the Early Palaeocene (Alroy 1999*b*), the result of a combination of survival, evolutionary radiation, and immigration. The picture is less clear elsewhere for want of fossil beds that span the K/T boundary. Rich Early Palaeocene placental faunas occur in the Shanghuan Formation of China (Lucas 2001), and at Tiu Pampa in Bolivia, the latter in association with the marsupials found there. In Europe, mammals are unknown until the Middle to Late Palaeocene of Hainin in Belgium and Cernay in France (Agustí and Antón 2002). The African Palaeocene is even more poorly represented, being largely restricted to the modest fauna of the Ouarzazate Basin of Morocco (Gheerbrant 1992, 1995).

Cretaceous survivors
Many of the new genera were members of pre-existing Late Cretaceous groups, notably leptictidans and palaeoryctidans. Small, insectivorous leptictidans occur abundantly in North American Early Palaeocene mammal faunas, and persisted through the Eocene and into the Oligocene. The skull of the Late Eocene *Leptictis*, which has been described in full detail by Novacek (1986*a*), illustrates the conservative structure. The group is also found in Europe, where again they survived into the Oligocene. There are superbly well-preserved, complete skeletons of *Leptictidium* in the Eocene Lagerstätten at Messel in Germany (Koenigswald *et al.* 1992). It was a relatively large form (Fig. 7.3(a)), up to 90 cm in presacral length, and possessed very long hind legs and short fore legs, indicating a saltational habit resembling modern jerboas. There is one family present in the Chinese Palaeocene, Didymoconidae, that is regarded by McKenna and Bell (1997) as a constituent of Leptictida.

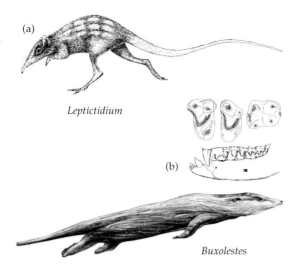

(a)

Leptictidium

(b)

Buxolestes

Figure 7.3 Two Early Eocene mammals from the Messel Oil Shales. (a) The leptictidan *Leptictidium* (Koenigswald *et al.* 1992). Presacral length approx. 30 cm. (b) The pantolestidan *Buxolestes*, with occlusal view of upper molars and lateral view of dentition. Total length of body approx. 80 cm (Savage and Long 1986).

Palaeoryctidans were also common in the North American Palaeocene and Eocene, and present in Europe. None have yet been found in China, although there are teeth of a family of primitive insectivores, Micropternodontidae, which Butler (1988) regarded as derived from palaeoryctids. Others dispute this relationship (e.g. McKenna and Bell 1997). The shrew-like *Eoryctes*, described from a complete Early Eocene skull (Thewissen and Gingerich 1989), is a typical micropternodontid.

Of the other Cretaceous placental groups, a few zhelestid teeth have been found in the Early Palaeocene of North America, but neither asioryctidans nor zalambdalestids appear to have survived.

In addition to pre-existing groups, there were several new lineages that were related to, but taxonomically distinct from the Cretaceous groups. The majority of these still consisted of relatively small, basically insectivorous or omnivorous mammals. However, within as little as 2 or 3 million years after the start of the Palaeocene, several groups of radically new kinds had arisen, and entered habitats not previously occupied by mammals during any of their previous 145 Ma of existence. There

were middle to quite large-sized herbivorous groups, and dog-sized to small bear-sized carnivores.

Anagalida

The anagalidans were an Asian group, appearing in considerable diversity in the Early Palaeocene of China. They were small herbivores or omnivores, with somewhat enlarged, procumbent lower incisors (Fig. 7.4(a)). The premolar and molar teeth are broad, low-crowned, and hypsodont. Their evolutionary affinities are not clear, but several authors accept the relationship with the Asian Cretaceous zalambdalestids that was proposed by Szalay and McKenna (1971).

Mixodonta

The mixodonts are also an exclusively Asian group of small omnivorous mammals, also possibly related to zalambdalestids. They are notable for their fully gliriform incisor teeth, and for this reason are widely regarded as related to the origin of the super-order Glires (rodents and lagomorphs). Li and Ting (1993) go so far as to claim that different families of mixodonts are relatives of lagomorphs and rodents respectively. A mimotonid, such as the Early Palaeocene *Mimotona* (Fig. 7.4(b)), has the dental formula, 2/2: 0/0: 3/3: 3/3, along with several cranial characteristics of lagomorphs. In contrast, the early eurymylid *Heomys* (Fig. 7.4(c)) has a rodent-like dentition including a single upper incisor and only two lower premolars. There are also characters of the postcranial skeleton that resemble those of rodents.

Pantolestida

The pantolestidans are almost, if not exclusively North American and European in occurrence, appearing in the Middle Palaeocene and surviving into the Oligocene (Koenigswald 1980). Otter-like in general size and proportions, the canines were well developed and while the molars of the earlier forms had sharp cutting edges, they tended to evolve a lower, broader form coupled with powerful adductor musculature suggestive of mollusc eating. The Eocene *Buxolestes* (Fig. 7.3(b)) from Messel has fish remains preserved in its stomach (Koenigswald *et al.* 1992). The relationships of pantolestidans are unclear, though it is generally assumed that they are a highly derived lineage related to palaeoryctidans.

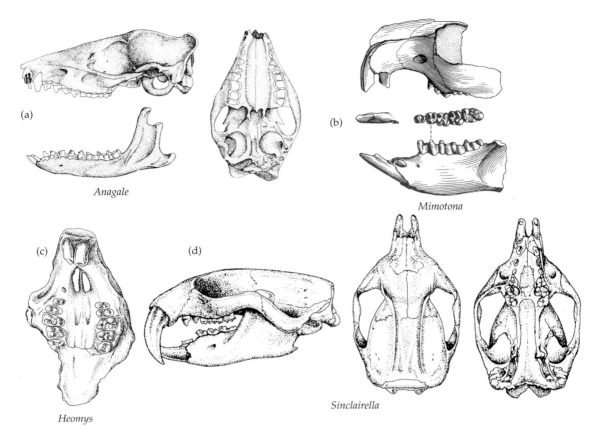

Figure 7.4 New groups of small Palaeocene placentals. (a) *Anagale* skull and lower jaw. Length of skull approx. 5.5 cm (Lucas 2001. (b) *Mimotona*. Skull and jaw fragments in lateral view, with occlusal view of lower dentition. Length approx. 2.0 cm (Li and Ting 1993). (c) Palatal view of front part of the skull of the eurymylid *Heomys*. Length approx. 1.7 cm (Carroll 1988, after Li and Ting). (d) Skull of the apatemyid *Sinclairella* in three views. Length of skull approx. 60 cm (Savage and Long 1986, after Scott and Jenson).

Apatemyida

The apatemyidans such as *Sinclairella* (Fig. 7.4(d)) are another Early Palaeocene North American and European group of obscure relationships, again possibly highly derived relatives of palaeoryctidans. They have a very unusual dentition, in which the single lower incisor is a huge, procumbent tooth, with what is described as a spoon-shaped blade meeting an enlarged first upper incisor. Immediately behind the lower incisor there is a large, blade-like premolar. The molar teeth tended to evolve into a flattened form with bunodont cusps. The group survived into the Late Oligocene.

Plesiadapiformes

At one time, the genus *Purgatorius* (Fig. 7.5(a)) was regarded as a highly significant fossil, because it was believed to be a Late Cretaceous primate. However, this early date was based on a single tooth from the Hell Creek Formation of Montana, which spans the Cretaceous–Palaeocene boundary, and the specimen has since been reinterpreted as Early Palaeocene, and therefore contemporaneous with more complete specimens from Wyoming. Furthermore, *Purgatorius* is the earliest member of a diverse group, Plesiadapiformes, which many authors no longer believe to be primates. That they

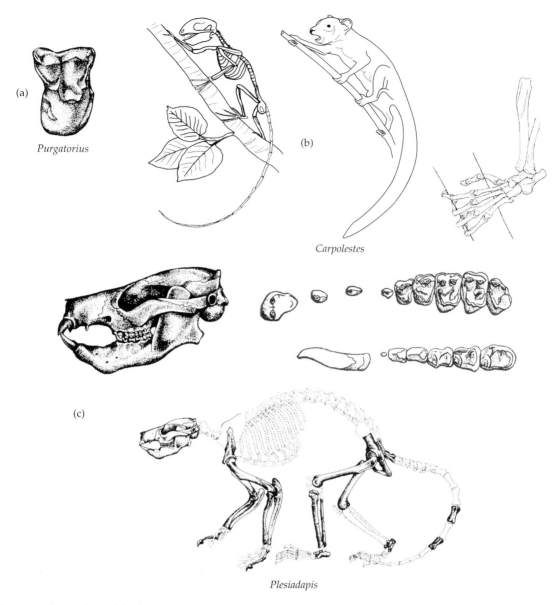

Figure 7.5 Plesiadapiformes. (a) Isolated upper molar of *Purgatorius* in occlusal view (Savage and Long 1986). (b) *Carpolestes* skeleton, restoration, and hind foot to show grasping structure. Presacral length of skeleton approx. 12 cm (Bloch and Boyer 2002). (c) *Plesiadapis* skull, occlusal views of upper and lower dentitions, and skeleton. Length of skull approx. (Savage and Long 1986 and Romer 1966 after Simpson).

are part of the super-order Euarchontoglires is not doubted, but within it they could be stem-group Primates (Bloch and Silcox 2001; Bloch and Boyer 2003*a,b*), basal members of the Dermoptera (Beard 1993), or have no sister-group relationship with anyone of the other euarchontogliran orders (Kirk *et al.* 2003).

Plesiadapiforms were a relatively abundant, diverse part of the Palaeocene radiation of North America and Europe. The dentition (Fig 7.5(c)) is superficially rodent-like, with enlarged incisors and reduced canines. The molars are rectangular and the cusps blunt, indicating an omnivorous, or perhaps frugivorous diet. Most plesiadapiforms were small, such as *Carpolestes* (Fig. 7.5(b)) whose head and body length is about 15 cm (Bloch and Boyer 2003a), and skeleton adapted for an arboreal existence. Others, notably *Plesiadapis*, were relatively large in body size, weighing up to 5 kg (Fig. 7.5(c)). The skeleton of this particular genus has been variously interpreted as either terrestrially adapted on the basis of the length of the limbs, or alternatively arboreally adapted as suggested by the well-developed claws and the structure of the ankle joint (Szalay and Delson 1979).

'Condylarthra'

The concept of 'Condylarthra' as a group of basal ungulate herbivores has been very important in the development of ideas about Palaeocene placental evolution. It contains a variety of small, primitive members, but more importantly a number of lineages of specialised and significantly larger herbivores, and secondary carnivores are included. Furthermore, the ancestral roots of all the later, specialised ungulate orders of placentals have been assumed to lie within the group. Thus the 'condylarths' are profoundly paraphyletic, and the interrelationships of this mixture of ancestral and derived condylarths are practically impossible to disentangle. Most agree that, while the term has long since outlived any formal usefulness, it is still virtually impossible to do without it: hence its retention but with quotation marks, or alternatively use of the synonymous expression 'archaic ungulates'. The characteristics of 'condylarths' are simply the ancestral features of ungulates generally (Archibald 1998; Nessov *et al.* 1998). The molar have evolved a crushing action in place of the primitive shearing function. They are low-crowned, with bunodont cusps. The uppers are rectangular in occlusal view with a fourth cusp, the hypocone, developed. The trigonid of the lower molar is shortened from front to back and bears a prominent metaconid cusp. The second feature of 'condylarths' concerns the terminal phalanges of the digits. They are elongated and are not grooved or fissured, and are therefore at least incipiently hoof-like rather than claw-like.

The distinction between 'condylarth' molars and those of the presumed ancestral grade seen in zhelestids are not marked, and nor are there clear differences between the most progressive 'condylarths' and stem members of some of the ungulate placental orders. Nevertheless, accepting 'condylarths' as a grade of primitive ungulates, several groups can be distinguished at least some of which are probably monophyletic (Archibald 1988; Prothero *et al.* 1998). The great majority are from the Palaeocene of North America, where more than half the Early and Middle Palaeocene mammal species are 'condylarths'. Many are herbivorous and include species of large body size. Others have a modified dentition suitable for an omnivorous, or a carnivorous diet, and amongst the latter are the first placental mammals to enter the medium to large body-sized carnivorous habitat.

Protoungulatum. *Protungulatum* (Fig. 7.6(a)) is the earliest and most primitive 'condylarth', a genus that occurs in North America at the base of the Early Palaeocene. Apart from a few cranial and postcranial fragments, it is only represented by teeth and jaws, which have the ancestral 'condylarth' form. The dental formula is I3/3: C1/1: PM4/4: M3/3, there is no diastema behind the incisors, and the premolar teeth show only a slight tendency to become more molariform. The cusps of the molars are relatively tall, and the upper molars are transversely wide. The hypocone is barely developed.

Arctocyonidae. Several families of 'condylarths' appear in the Early Palaeocene, distinguished mostly by details of the dentition. The most primitive are the Arctocynonidae, which occur in North America and Europe. *Protoungulatum* is sometimes included as the most basal member, although it shares only ancestral characters with them. Arctocyonid molar teeth are relatively low-crowned, and there is a fully developed hypocone cusp on the uppers, indicating that the crushing function had fully evolved. A virtually complete skull of *Arctocyon* (Fig. 7.6(b)) from Europe shows lower canines, a deep sagittal crest, and a low position of the mandibular articulation. All these characters indicate an ability to take significant live prey, even though there was no tendency to develop any form of carnassial, or specialised shearing teeth. Thus, arctocyonids seem to

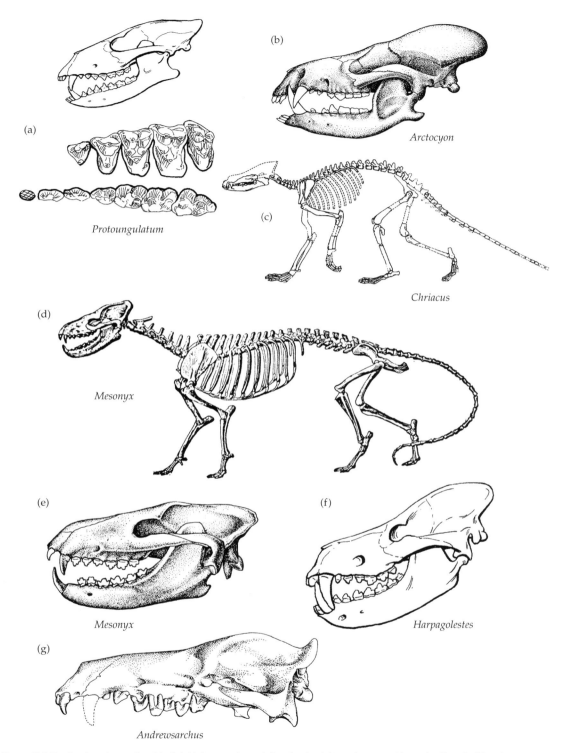

(a)

Protoungulatum

(b)

Arctocyon

(c)

Chriacus

(d)

Mesonyx

(e)

Mesonyx

(f)

Harpagolestes

(g)

Andrewsarchus

Figure 7.6 Basal and carnivorous 'condylarths'. (a) *Protoungulatum* skull and occlusal views of upper and lower dentition. Skull length approx. 6 cm (Archibald 1998, after Szalay and Sloan and Van Valen). (b) *Arctocyon* skull in lateral view. Length approx. 26 cm (Romer 1966, after Russell). (c) Postcranial skeleton of *Chriacus*. Presacral length approx. 50 cm (Rose 1987). (d) Skeleton and (e) skull of the mesonychid *Mesonyx*. Skull length approx. 28 cm (Romer 1966 after Scott). (f) Skull of the mesonychid *Harpagolestes*. Length approx. 40 cm. (g) Skull of the giant mesonychid *Andrewsarchus*. Skull length approx. 83 cm (Carroll 1988, after Osborn).

have been generalist omnivores. In body size, they ranged up to that of a small bear, and probably had a comparably cosmopolitan diet. *Chriacus* (Fig. 7.6(c)) was a very lightly built, agile form.

Carnivorous 'condylarths'. Arctocyonids were common in the Early Palaeocene and structurally represent the base of the 'condylarth' radiation, from which various more specialised groups diverged. Some became more adapted for carnivory, with a tendency to evolve molar teeth with higher, sharper cusps, and shearing edges, in place of the bunodont cusps of the arctocyonid-grade (Szalay 1969). These constituted the very first radiation of relatively large, predaceous mammals. Three groups of 'condylarths' exhibiting an increasing expression of this trend are usually combined into a taxon Cete (Archibald 1998; Prothero *et al.* 1988). The triisodontids are the most primitive members of this group, and include one of the very earliest 'condylarths'. Isolated 5 mm long molars of *Ragnarok* occur at the base of the Palaeocene 65 Ma. Carnivorous characteristics of the dentition are only incipiently developed. The later group Hapalodectidae are more progressive in this respect, having higher, sharper cusps. *Hapalodectes* was a small, rat-sized animal, not found until the Eocene in North America, but already present in the Late Palaeocene of Asia. The Mesonychidae express to the greatest extent the tendency towards specialised carnivory. The molar teeth are laterally compressed, and shearing edges have developed on the front of the lower teeth, which acted against edges on the back of the uppers. This condition is described as prevallid-postvallum shearing, a system that is only analogous to the postvallid-prevallum shearing system of the carnivorous teeth that evolved in the later specialist carnivore groups Creodonta and Carnivora. Some *Mesonyx* such as *Mesonyx* itself (Fig. 7.6(d) and (e)) were the size of a wolf and had lightly built, digitigrade limbs. Others, such as *Harpagolestes* (Fig. 7.6(f)) were even larger, heavily built with powerful jaws and teeth and were probably scavengers rather than hunters. The mesonychids are also known from Asia, where they occur in the Early Palaeocene Shanghuan and the Late Palaeocene Nonshanian faunas (Wang *et al.* 1998; Lucas 2001). One of the last forms, *Andrewsarchus* (Fig. 7.6(g)), is from the Late Eocene of Mongolia, and is often claimed to be the largest terrestrial carnivorous mammal ever discovered. Its skull was over 80 cm in length.

Herbivorous 'condylarths'. Other lineages of 'condylarths' evolved increasingly specialised herbivorous adaptations from an arctocyonid-like ancestry. To varying degrees the premolar teeth enlarged and became more molar like, and the molars evolved broader, six-cusped occlusal surfaces. The limbs tended to become digitigrade, with well developed hooves on each of the digits. Four main families are recognised (Archibald 1998) and all of them first occur in the Early Palaeocene of China as well as North America (Wang *et al.* 1998; Lucas 2001).

The hyopsodontids (Fig. 7.7(a)) were small, rabbit-sized animals with short limbs. The canines were very small, and the last premolar is slightly enlarged. The paraconule and metaconule cusps of the upper molars had enlarged to give the six-cusped condition. *Hyopsodus* extended into the Eocene and is therefore one of the last surviving 'condylarths'.

The family Mioclaenidae (Fig. 7.7(b)) were also relatively small, incipiently specialist herbivores. The premolars were more molarised, and the occlusal surfaces of the molar teeth were simplified by having less distinct cusps, suggesting a more continuous grinding action of tougher food. Mioclaenids, or at least close relatives of this family occur in the Early Palaeocene Tiupampan fauna of the Santa Lucia Formation of Bolivia, where they have been implicated in the origin of the indigenous South American ungulate orders. Gheerbrant *et al.* (2001) have described 'condylarth' teeth of Early Eocene age from Morocco in North Africa, one of which, *Abdounodus* (Fig. 7.7(b)), is comparable to mioclaenids.

Periptychids varied from small, squirrel-sized mammals to *Ectoconus* (Fig. 7.7(c)), which was the size of a sheep. In this family, the premolars were more or less fully molarised, and crescentic crests had developed from the cusps of these and the molars (Fig. 7.7(d)). Archibald (1998) interpreted the tooth structure as an adaptation for dealing with tough, fibrous vegetation that was first shredded by the swollen premolars and then pulverised by the molars. The skeleton of periptychids is rather heavily built and the limbs short. The feet too are short,

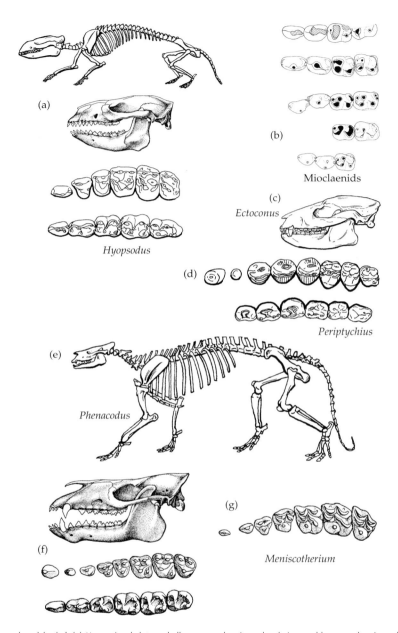

Figure 7.7 Herbivorous 'condylarths'. (a) *Hyopsodus* skeleton, skull, upper molars in occlusal view and lower molars in occlusal view. Length of skull approx. 6 cm (Archibald 1998 after Gazin and Romer 1966 after Matthew). (b) Comparison of mioclaenid lower teeth in occlusal view. From the top *Abdounodus* from North Aftica; *Promioclaenus* from North America; *Tiuclaenus* from South America; *Molinodus* from South America; *Pucanodus* from South America (Gheerbrant *et al.* 2001). (c) Skull of the periptychid *Ectoconus*. Length approx. 16 cm (Archibald 1998, figure 20.2 after Gregory 1951). (d) Occlusal views of upper and lower dentition of *Periptychius* (Archibald 1998 after Matthews). (e) *Phenacodus* skeleton. Presacral length approx. 1.5 m (Archibald 1998, after Gregory). (f) Skull and occlusal views of upper and lower dentition of *Phenacodus*. Skull length approx. 22 cm (Romer 1966, after Cope and Matthew). (g) *Meniscotherium* upper teeth in occlusal view (Romer 1966, after Matthew).

with five widely spaced digits bearing small hooves, and probably associated with a semi-plantigrade stance.

The Phenacodontidae were the most specialised herbivorous family of all. They appeared in the Middle Palaeocene of North America and survived into the Eocene. *Phenacodus* (Fig 7.7(e) and (f) also occurs in the Early Eocene of Europe, and there is a possible specimen of similar age from Morocco (Gheerbrant *et al.* 2001), but the group is not represented for certain in Asia (Thewissen 1990). Taxonomically, the phenacodontids are not usually included with the more conservative herbivorous families, but as a separate group of more progressive ungulates, possibly related to the Tethytheria (Prothero *et al.* 1988; Janis *et al.* 1998*a*), a question returned to later. Phenacodontids show the extreme of 'condylarth' specialisation for herbivory of both the dentition and the postcranial skeleton. The posterior premolars are the most molariform of all. The molars are very low-crowned with the cusps developed into crescents or lophs of various patterns in the different genera. The limbs are well-designed for cursorial locomotion: elongated, slender, and bearing reduced first and fifth digits. *Phenacodus* (Fig. 7.7(e) and (f)), with a presacral body length up to 2 m, had the least specialised version of phenacodontid teeth; others such as the 60 cm long *Meniscotherium* (Fig. 7.7(g)) reached the zenith of 'condylarth' herbivorous dentition, with its fully expressed pattern of crescents and lophs on the crowns.

The phenacolophids are a poorly known, but phylogenetically very important group of Asian 'condylarths' because they have been variously implicated in the origins of the embrithopod, tethythere, and perissodactyl ungulate orders (McKenna and Manning 1977; Prothero *et al.*1988). They are known from little more than teeth of Mid-Late Palaeocene age, the molars of which are dilophodont, having two transverse lophs or crests.

Eocene 'condylarths' from Morocco have been mentioned. There are also Palaeocene 'condylarths' from the Ouarzazate Basin of Morocco (Gheerbrant 1995), but the affinities of these specimens are unclear. Given the molecular evidence for the origin of the African ungulate orders as part of a geographically isolated radiation of Afrotheria, it will be very important to discover whether these African 'condylarths' are members of groups existing elsewhere, or are actually basal members of Afrotheria, related perhaps to the Proboscidea and Hyracoidea.

Taeniodonta

The taeniodonts were omnivorous mammals occurring exclusively in the North American Palaeocene and Eocene (Schoch 1986; Lucas *et al.* 1998). In body size they vary from 5 kg to over 100 kg. The most primitive member is the Early Palaeocene *Onychodectes* (Fig. 7.8(a)), in which there is a moderately large canine followed by widely spaced, interlocking premolars. The molars are high crowned with low, bunodont cusps and a roughly rectangular occlusal surface. The postcranial skeleton is that of a fairly generalised, non-cursorial mammal perhaps able to climb well, and Lucas *et al.* (1998) suggest that it had a mode of life comparable to that of the Virginia Opossum. Later taeniodonts, notably *Stylinodon* (Fig. 7.8(b)), were larger and had evolved a much shorter, powerfully built skull. The canines and upper incisors were greatly enlarged, open-rooted, and the enamel only covered the anterior faces, giving them a chiselling function. The anterior premolars had evolved into cutting blades while the posterior teeth were simplified crushing pegs of dentine. The whole skeleton had become heavily built, and the powerful forelimbs bore large, flattened claws, presumably adapted for digging out and consuming roots and tuberous parts of plants.

The relationships of taeniodonts are obscure. There is no evidence at all to relate them to 'condylarths', but rather to one of the more conservative primitive insectivorous taxa. McKenna and Bell (1997) classify them with palaeoryctids. A more precise possibility is that they are related to the otter-like pantolestids, which are themselves believed to have palaeoryctid affinities.

Pantodonta

The pantodonts (Fig. 7.8(c)) include the very first of the large, herbivorous placental mammals to evolve after the Cretaceous, although others were much smaller, less than 10 kg. The earliest and most primitive forms are from the Early Palaeocene of China, where *Bemalambda* and *Hypsilolambda* occur in the Shanghuan Formation (Wang *et al.* 1998). *Bemalambda*

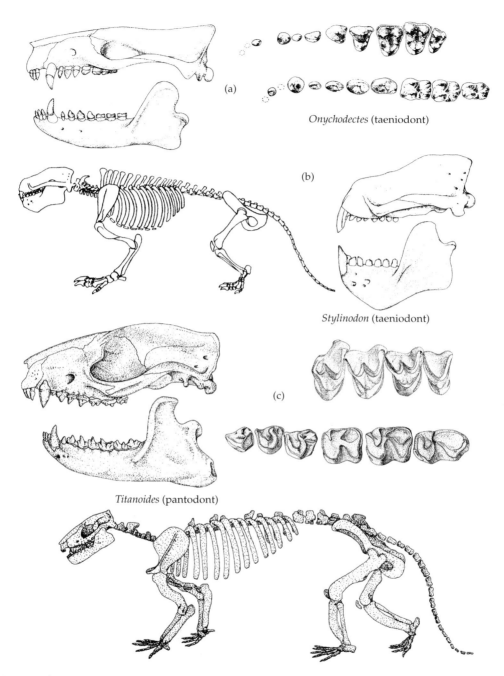

Onychodectes (taeniodont)

Stylinodon (taeniodont)

Titanoides (pantodont)

Figure 7.8 Taeniodonts and pantodonts. (a) The taeniodont *Onychodectes* skull and occlusal views of upper and lower dentitions. Skull length approx. 10 cm (Lucas *et al*. 1998). (b) The taeniodont *Stylinodon* skeleton and skull. Skull length approx. 25 cm (Lucas *et al*. 1998). (c) The pantodont *Titanoides* skull, occlusal views of upper and lower dentitions, and skeleton. Length of skull approx. 15 cm (Lucas 1998 after Simons).

was the size of a large dog, with a skull about 20 cm in length. In North America, the group makes its appearance in the middle of the Palaeocene, and it survived into the Middle Eocene as represented by *Coryphodon*. Teeth of pantodonts also occur in the Palaeocene Tiupampan fauna of South America (Muizon and Marshall, 1987, 1992) and the Eocene of Antarctica (Reguero *et al.* 2002).

The dentition of pantodonts is complete, with modest canines and no diastema. The most characteristic feature is the development of lophs on the postcanine teeth. In the more primitive bemalambdids of Asia, the lophs are V-shaped, but in all other forms, collectively classified as Eupantodonta, the molar tooth lophs have evolved a W-shape (Lucas 1993). The postcranial skeleton varies from that of the lightly built, possibly arboreal genus *Archaeolambda*, which is known from a complete Late Palaeocene skeleton from China, to the massive, graviportal *Barylambda* that is often likened to a giant ground sloth, complete with a massive pelvis and heavy tail that suggest a habit of browsing on high vegetation.

Despite their large body size and herbivorous habit, there are no dental or postcranial features suggesting that pantodonts are related to the 'condylarths' as was once believed. McKenna (1975) followed more recently by Lucas (1993, 1998) proposed that, like the taeniodonts, pantodonts are possibly related to the palaeoryctid group of small Cretaceous insectivorous mammals, on the basis of features of the dentition shared by the primitive pantodonts and a form such as *Didelphodus*.

Tillodonta

The tillodonts are another group of mainly Palaeocene herbivores, whose most primitive members such as *Lofochaius* occur in the Early Palaeocene of China, where they survived until the Late Eocene (Lucas 2001). They make their appearance in North America in the latest Palaeocene, where they underwent a brief radiation in the Early to Middle Eocene before becoming extinct. Early Eocene tillodonts are also found in Europe, but so far none have turned up in South America (Lucas and Schoch 1998*a*).

Tillodonts varied from small, with a skull length of only 5 cm, to large mammals such as the bear-like *Trogosus* (Fig. 7.9(a)), which has a 35 cm skull

and an estimated body weight of about 150 kg. The tillodont dentition (Fig. 7.9(b)) is highly distinctive, particularly due to the enlarged, chisel-like second incisors in which the enamel is restricted to the anterior edge. The other incisors, canines and anterior premolars tended to be lost, leaving only the square, hypsodont molars with a dilambdodont pattern of crests, which rapidly wore down. The teeth are carried in powerfully built jaws, and the body as a whole was heavily built, with well-developed and recurved claws. All the indications are of a ground grubbing mammal feeding on roots and tubers.

At one time tillodonts were regarded as derivatives of 'condylarths', but there are no clear ungulate characters. The similarity of the dilambdodont molars of tillodonts and pantodonts has suggested to several authors that the two are sister groups (Chow and Wang 1979; Lucas 1993), though the evidence is not very convincing: the similarities may be convergent in two groups independently derived from palaeoryctid-like ancestors. The biogeography of the group has been investigated by Schoch and Lucas (1982; Schoch 1986), who argue that it probably had an Asian origin. The similarity between the Chinese *Meiostylodon* and the North American *Esthonyx* can be explained by a dispersal into North America by the Late Palaeocene. They also suggested that the relatively early demise of the group in North America was due to competition from the similarly adapted taeniodonts.

Dinocerata

The dinoceratans include the largest of all the archaic Palaeocene–Eocene herbivores, ranging in estimated body weight from 175 kg to 4.5 tonnes. The earliest record of this short-lived group is *Prodinoceros* from the latest Palaeocene of both China and North America (Lucas and Schoch 1998), and none survived beyond the Middle Eocene. Although massively built, *Prodinoceros* (Fig. 7.9(c)) was relatively small for the group and lacked the bony protuberances of the skull so characteristic of the later members, such as the huge mid-Eocene *Uintatherium* (Fig. 7.9(d)), which occurred in North America and Asia, although not in Europe. The dentition of dinoceratans shows a tendency to reduce and eventually lose the upper incisors, but to retain and enlarge the upper canine. The premolars

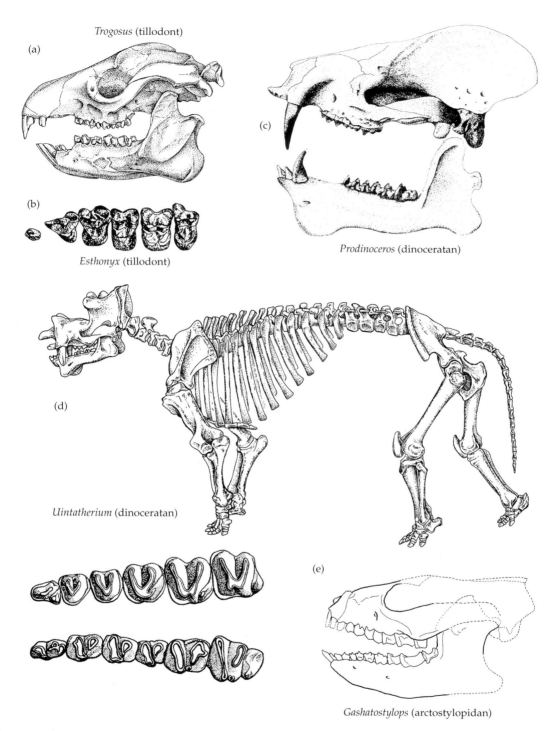

Figure 7.9 Tillodonts, dinoceratans, and arctostylopidans. (a) Skull of the tillodont *Trogosus*. Length approx. 30 cm (Carroll 1988 after Gazin). (b) Upper dentition of the tillodont *Esthonyx* in occlusal view (Savage and Long 1996 after Gazin). (c) Skull of the dinoceratan *Prodinoceros*. Length of skull approx. 45 cm (Lucas and Schoch 1998b after Flerov). (d) *Uintatherium* skeleton and occlusal views of upper and lower dentition. Height at shoulder approx. 1.6 m (Carroll 1988 after Gregory 1957, and Romer 1966). (e) The arctostylopidan *Gashatostylops* skull in lateral view. Length of restored skull approx. 4 cm (Cifelli and Schaff 1998).

are molarised and the cheek teeth have a well-developed and characteristic bilophodont form.

Until quite recently (Prothero *et al.* 1988), dinoceratans were believed to be ungulates related to the paenungulate group of elephants, hyraxes, and sirenians, a view based on a number of ambiguous postcranial characters. Lucas (1993) has, however, argued the case for a sister-group relationship with the Pyrotheria, which is an exclusively South American, Eocene order. He notes a number of detailed similarities in the molar teeth.

Dinoceratans, particularly the more advanced uintatheres, must have been browsing forms with a rhinoceros-like mode of life. The loss of the upper incisors suggests that a mobile proboscis was present. Why such apparently uniquely adapted forms should have gone extinct as early as they did is mysterious; they had no obvious direct competitors.

Arctostylopida
Arctostylopidans are small, rabbit-sized, and exclusively Late Palaeocene mammals. All are Asian apart from the single North American genus, *Arctostylops* (Fig. 7.9(e)). Little more than dentitions are known for the group, which are characterised by the even size of all the teeth, including reduced canines and molarised premolars. Lophs are developed on the molars, which develop as longitudinal shearing edges. They were, perhaps, comparable in feeding habits to the modern day hyraxes. Several authors had placed arctostylopidans as a basal group of the South American ungulate order Notoungulata, but Cifelli *et al.* (1989; Cifelli and Schaff 1998) have argued that the dental similarities are superficial and probably convergent. Thus their affinities are at present unclear.

Meridiungulata: South American Ungulates
One of the more remarkable features of the whole placental fossil record is the revelation of a great radiation of advanced ungulates in South America, from the Early Palaeocene right through until the Plio-Pleistocene, 60 million years later. Five orders are recognised, mostly restricted to that continent although their presence in the Eocene of Antarctica (Reguero *et al.* 2002), and the dispersal of a few into North America towards the end of their existence are

recorded. All have the fundamental ungulate characteristics of large, grinding, lophodont premolar and molar teeth, and fully hoofed feet. McKenna (1975) formalised the commonly held view that all five orders constitute a monophyletic group, by creating a super-order Meridiungulata, and there has been fairly general acceptance of the concept, despite the lack of clearly defined characters supporting it. However, the issue has become complicated by Muizon and Cifelli's (2000) cladistic analysis of dental characters that supports a relationship between the Early Palaeocene Tiupampan 'condylarths' such as *Pucanodus*, the North American mioclaenid 'condylarths' (Fig.7.10(a)), the South American didolodontids (Fig. 7.10(b)), which are advanced 'condylarths', and the Meridiungulate order Litopterna (Fig. 7.10(c) and (d)). The implication to be drawn is that a mioclaenid entered South America from the North around the start of the Palaeocene, and radiated into the somewhat more progressive didolodontids, and the fully ungulate litopterns. None of the other four Meridiungulate orders can be shown to belong to this same group, although neither is there strong evidence that they do not. They all have teeth too specialised to reveal their relationships clearly, and therefore any of them could have originated from a different 'condylarth' grade ancestor. Indeed, on the basis of certain details of tooth structure, Schoch and Lucas (1985; Lucas 1993) proposed that the two of the Meridiungulate orders, Pyrotheria (Fig. 7.10(g)), and Xenungulata (Fig. 7.10(h)), are related to the Dinocerata of the northern continents (Fig. 7.9(d)), and referred to this grouping as the Uintatheriomorpha. They then claimed that these uintatheriomorphs are related to the anagalidans (Fig. 7.4(a)), a group of small omnivores/herbivores from the Asian Palaeocene. If true, the relationship would imply that there was a second, independent immigration into South America from Laurasia of a progenitor of ungulate taxa, this one not even a 'condylarth'.

For the time being, the Meridiungulate hypothesis that all five orders of South American ungulates form a monophyletic group that could have been derived from a single dispersal of a mioclaenid-like ancestral 'condylarth' cannot be satisfactorily refuted and therefore remains the simplest explanation for

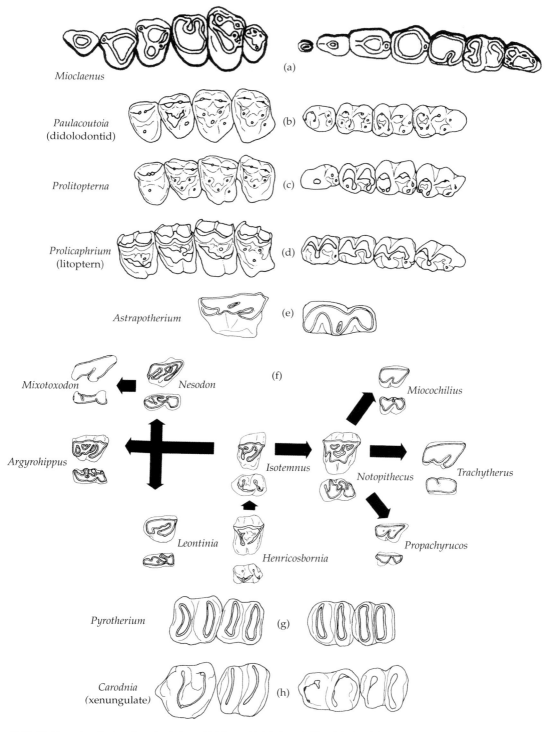

Figure 7.10 Meridungulate dentitions in occlusal views. (a) Mioclaenid *(Mioclaenus)*. (b) Didolodontid *(Paulacoutoia)*. (c) Plesiomorphic litoptern *(Prolitopterna)*. (d) Litoptern *(Prolicaphrium)*. (e) Upper and lower molar of *Astrapotherium* in occlusal view. (f) Molar tooth diversity in notoungulates. (g) Upper and lower molars of *Pyrotherium*. (h) Upper and lower molars of the xenungulate *Carodnia* (a) after Archibald 1998 after Matthew 1937. Others Cifelli 1993).

Mioclaenus (a)

Paulacoutoia (didolodontid) (b)

Prolitopterna (c)

Prolicaphrium (litoptern) (d)

Astrapotherium (e)

(f)

Mixotoxodon   *Nesodon*

Miocochilius

Argyrohippus   *Isotemnus*   *Notopithecus*   *Trachytherus*

Leontinia   *Henricosbornia*   *Propachyrucos*

Pyrotherium (g)

Carodnia (xenungulate) (h)

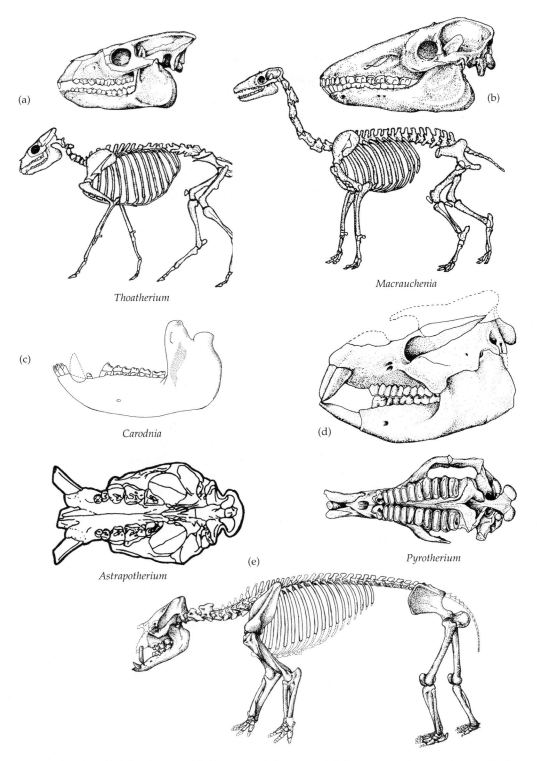

Figure 7.11 Meridungulates. (a) Skull and skeleton of the litoptern *Thoatherium*. Length of skull approx. 15 cm (Savage and Long 1986, after Scott). (b) Skull and skeleton of the litoptern *Macrauchenia*. Length of skull approx. 45 cm (Savage and Long 1986, after Scott and Burmeister). (c) Lower jaw in lateral view of the xenungulate *Carodnia* (Carroll 1988, after Paula Courto). (d) Skull of *Pyrotherium* in lateral and ventral views (Savage and Long 1986, after Loomis). (e) *Astrapotherium* skull in ventral view and skeleton. Length of skull approx. 50 cm (Savage and Long 1986, after Simpson).

their presence. The five orders may be described briefly as follows.

Litopterna. The litopterns were relatively conservative as far as the dentition is concerned, and the teeth of the earliest members of the group (Fig 7.10(c)) resemble those of 'condylarths', particularly the South American group Didolodontidae (Fig. 7.10(b)). Some authors have even included the latter in the Litopterna. The crowns are generally low, and the development of lophs weak. In contrast, the limbs were the most highly evolved of the ungulates, with a strong tendency to reduce the side toes leaving the third digit dominant. Of the two groups of litopterns, the proterotherioideans were remarkable for their horse-like build and the Miocene member *Thoatherium* (Fig. 7.11(a)) possessed a single-toed, extraordinarily equid-like condition of its limbs. The other group, the macraucheniolideans are invariably described as camel-like, with their elongated neck and limbs, and broad foot consisting of three almost equal toes. The group is also represented in the Eocene of Antarctica by teeth of *Victorlemoinea*. The Plio-Pleistocene *Macrauchenia* (Fig. 7.11(b)) has its nostril opening high up on its skull, which led to the belief that there was a proboscis present, although it might alternatively have been associated with a swampy habitat, or keeping the brain cool by evaporation. This genus also has the most derived molar teeth of all litopterns, having become high-crowned and lophodont.

Notoungulata. The notoungulates were far the most diverse of the meridiungulate orders, with more than 100 genera in 13 families evolving in the course of the Cenozoic (Fig. 7.10(f)). At one time, it was believed that there were Palaeocene notoungulates in China and North America, which were placed in a primitive family Arctostylopidae (page 242). However, the dental similarities on which the proposed relationship was based are evidently convergent and the arctostylopids are properly placed in an entirely unrelated placental order of their own (Cifelli *et al.* 1989; Cifelli and Schaff 1998). This leaves notoungulates as exclusively South American, with the single exception of the Pleistocene *Mixotoxodon*, which briefly extended its range into central America.

Notoungulates are categorised by a generally broad, flat skull, and unique features of the auditory region. The molar teeth tend to be strongly lophodont, and full hypsodonty is commonly developed. The most basal group, the Notioprogonia, were medium-sized, heavily built, and digitigrade animals (Fig. 7.12(a)). In the course of the Tertiary three groups evolved from notioprogonian-like ancestors. The toxodonts included the largest of all South American ungulates, with the late surviving *Toxodon* (Fig. 7.12(e)) the size and build of a large rhinoceros. Its incisor teeth were chisel-shaped and the molars hypsodont, indicating a low-level browsing and grazing habit. The typotheres (Fig. 7.12(b) and (c)) were rather like giant rodents, with loss of the anterior dentition except for broad, open-rooted, chisel shaped incisors, and again hypsodont molars. The third group, the hegetotheres (Fig. 7.12(d)), resembled the typotheres in dentition, but tended to develop elongated hind legs, giving them a rabbit-like appearance.

Astrapotheria. The astropotheres were very large ungulates, with the best-known genus, the Oligocene *Astrapotherium* (Fig. 7.10(e) and 7.11(e)), some 3 m in length. The upper incisors of astropotheres are absent and presumably there was a horny pad against which the lower incisors bit. The upper canines are huge and tusk like while the lowers are less enlarged but still prominent. The last two molar teeth evolved enormous size. The nostrils have shifted to the top of the skull, suggesting the possibility of a proboscis as in the litoptern *Macrauchenia*. The postcranial skeleton was extremely peculiar. The forelegs were very stout and strongly built, but the hindlegs relatively slender. Romer (1966) suggested the possibility of an amphibious mode, perhaps comparable to hippos. Astropotheres survived into the Late Miocene, and are one of the two ungulate orders represented by teeth in the Eocene of Antarctica.

Pyrotheria. The pyrotheres (Fig. 7.10(g) and 7.11(d)) were also very large and elephant-like in general form (MacFadden and Frailey 1984). The similarity is enhanced by the tusk-like incisors, two upper and one lower on either side. The six cheek teeth resembled those of primitive proboscideans in the transversely expanded, bilophodont condition. Compared to other South American ungulate orders, pyrotheres were short-lived, appearing in the early Eocene and disappearing during the Oligocene.

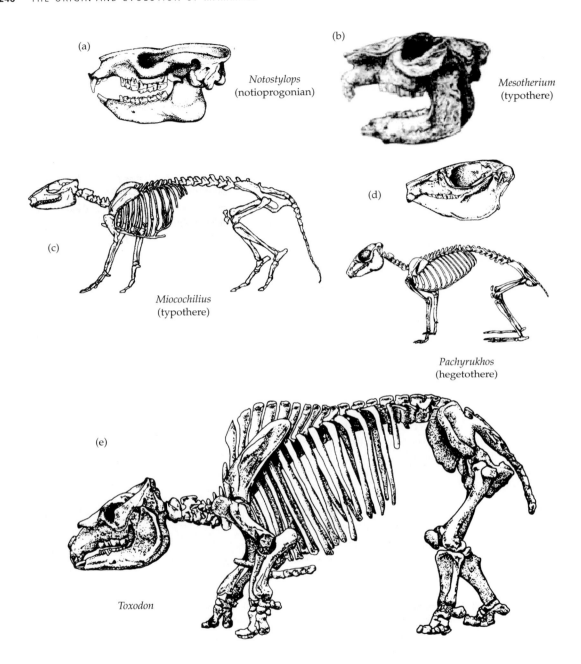

Figure 7.12 Meridungulates: notoungulates. (a) Skull of the notioprogonian *Notostylops*. Length approx. 14 cm (Savage and Long 1986, after Simpson). (b) Skull of the Pleistocene typothere *Mesotherium*. Length approx. 27 cm (Savage and Long 1986, after Rovereto). (c) Skeleton of the Miocene typothere *Miocochilius*. Presacral length approx. 70 cm (Savage and Long 1986, after Stirton). (d) Skull and skeleton of the hegetothere *Pachyrukhos*. Skull length approx. 7 cm (Savage and Long 1986 after Scott). (e) Skeleton of *Toxodon*. Presacral length approx. 2.7 m (Savage and Long 1986 after Piveteau).

Xenungulata. Only two Late Palaeocene genera are presently included in the xenungulates. The best known, *Carodnia* (Fig. 7.10(h) and 7.11(c)), has at least superficial similarities to the pyrotheres (Cifelli 1993*b*). It was large, with chisel-like incisors, and basically bilophodont cheek teeth.

Palaeanodonta
The palaeanodonts are a Late Palaeocene and Eocene group of North American mammals. They were small to moderate in size and had robustly built skeletons, with heavily built limbs and strong claws, clearly adapted for a fossorial life. The teeth were reduced in number, size, and enamel, trends most highly expressed in Eocene members such as *Metacheiromys* (Fig. 7.13(b)). All these features suggest that the palaeanodonts were an early placental group adapted for ant-eating, though whether the resemblances to xenarthrans and pholidotans indicate relationship or convergence has been much debated. Rose and Emry (1993) suggested that, while they may be related to pholidotans, there is certainly no good evidence for a relationship with xenarthrans. As far as the origin of palaeanodonts is concerned, Rose and Lucas (2000) described a relatively complete skeleton of *Escavadon* (Fig. 7.13(a)), a 0.5–1 kg specimen from the Middle Palaeocene of North America, and concluded that it is the most primitive palaeanodont known. It has a number of postcranial similarities to other members of the group, but the dentition is much more primitive compared to a typical palaeanodont. In the latter respect, the authors note certain similarities to the Palaeocene otter-like pantolestids.

There is also a very early Chinese form, *Ernanodon* (Fig. 7.13(c)), which was regarded as a palaeanodont by Radinsky and Ting (1984). However, the similarities to any other placental anteaters are believed by Rose and Emry (1993) to be no more than superficial convergences for a fossorial mode of life.

Carnivorous mammals: Creodonta and Carnivora
The earliest mammals to have increased in body size and adapted for carnivory were amongst the 'Condylarthra' discussed above, culminating in the mesonychids. In addition to these, members of two more or less exclusively carnivorous placental orders also made their appearance during the Palaeocene,

although members of neither actually achieved large size until later. The two orders are the Creodonta, stigmatised as archaic because they did not survive beyond the Miocene, and the Carnivora, which eventually radiated to become the dominant terrestrial carnivores of today. Most accept that the two are related as a monophyletic group termed Ferae, although almost equally as often admit that the characters supporting the relationship are not very impressive (Flynn *et al.* 1988; Wyss and Flynn 1993). They include restriction of the specialised, shearing function of carnassial teeth to a limited part of the postcanine dentition, although the actual teeth involved differ from group to group. In the Carnivora it is invariably the upper PM4 and lower M1 that are the major carnassials. In creodonts there tends to be some shearing along the entire molar row, but with emphasis on either the upper M1 and lower M2 as generally in oxyaenids, or the upper M2 and lower M3 as generally in the hyaenodontids. These differences indicate convergent evolution of the specialised carnassial teeth. Other shared characters of carnivorans and creodonts are a bony lamina, called the osseous tentorium, dividing the cerebellar from the cerebral hemisphere regions of the cranial cavity, certain details of the tympanic bulla and the course of the internal carotid artery, and aspects of the structure of the ankle joint.

The possibility of a sister group relationship between Carnivora and Creodonta became less likely with the description of teeth of Early Palaeocene carnivoran mammals from Canada by Fox and Youzwyshyn (1994). *Pristinictis* (Fig. 7.14(d)) is described as a primitive member of the basal group Viverravidae, and there is a similar form, *Pappictidops*, from the Early Palaeocene Shanghuan Formation of China. *Ravenictis* is more primitive than any other known member of the Carnivora. These specimens lack any dental features shared uniquely with the Creodonta, and therefore the Carnivora must have evolved from a form even more primitive in its molar structure than creodonts. This conclusion accords with that of Wyss and Flynn (1993), who pointed out a number of braincase similarities between primitive members of the Carnivora, the living insectivorous Eulipotyphla, and the early Cenozoic insectivorous Leptictida (Novacek 1986*a*).

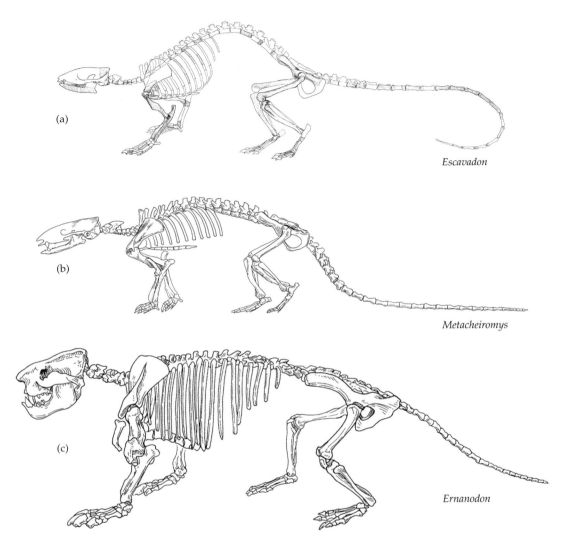

Escavadon

Metacheiromys

Ernanodon

Figure 7.13 Early Tertiary diggers. (a) *Escavadon*, a primitive palaeanodont. Presacral length approx. 30 cm (Rose and Lucas 2000). (b) The Middle Eocene palaeonodont *Metacheiromys*. Rose and Emry 1993 after Simpson. Presacral length approx. 25 cm. (c) The Late Palaeocene Chinese *Ernanodon*. Presacral length approx. 50 m (Rose and Emry 1993 after Ding).

No equivalent new evidence about the relationships of the creodonts has yet emerged, and it is not even possible to be sure whether they are a monophyletic group, because the two constituent families, Oxyaenidae and Hyaenodontidae, have no known unique similarities (Janis, *et al.*1998*b*; Gunnell 1998). The oxyaenids(Fig. 7.14(a) and (b)) are the earlier of the two to appear in the fossil record, in the form of the small, cat-sized *Tytthaena*

from the Middle Palaeocene of North America. It was not until into the Eocene that any large bodied oxyaenids, such as the wolverine-sized *Oxyaena* (Fig. 7.14(a)), evolved, and by the middle of that period they were extinct in North America, although lingering until the Late Eocene in Europe and Asia. In contrast, hyaenodontids made their appearance not in North America, but in Europe at the end of the Palaeocene (Agustí and Antón 2002).

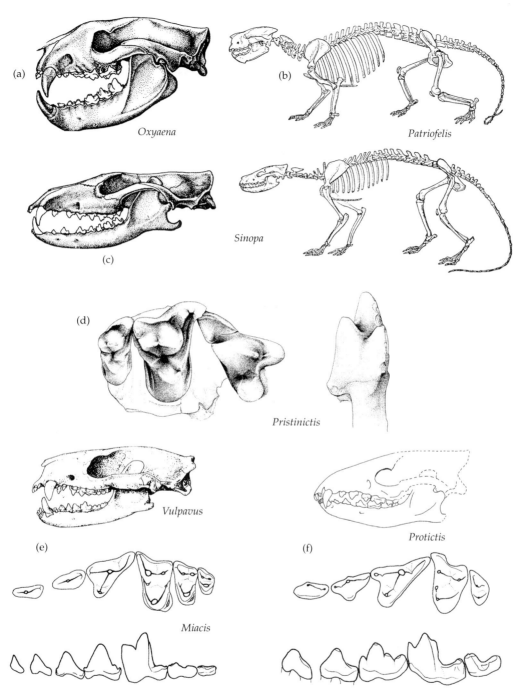

Figure 7.14 Creodonts and early carnivorores. (a) Skull of the oxyaenid creodont *Oxyaena*. Length approx. 20 cm (Romer 1966, after Wortman and Matthew). (b) Skeleton of the oxyaenid creodont *Patriofelis* (Gunnell 1998, after Osborn). (c) Skull and skeleton of the hyaenodont creodont *Sinopa*. Skull length approx. 15 cm (Romer 1966, after Wortman and Matthew, and Gunnell 1998 after Matthew). (d) *Pristinictis*. Upper molars in occlusal view and lingual view of an incomplete lower molar (Fox and Youzwyshn 1994). (e) Skull of the basal caniform miacid *Vulpavus*. Length approx. 10 cm (Savage and Long 1986). Upper dentition in occlusal view and lower dentition in labial view of *Miacis*. (Flynn 1998, figure 6.3). (f) Skull of the viverravid *Protictis*. Length approx. 9 cm (Carroll 1988 after MacIntyre 1996) and upper dentition in occlusal view and lower dentition in labial view of *Protictis* (Flynn 1998, figure 6.3).

Hyaenodontids such as *Sinopa* (Fig. 7.14(c)) tended to be more agile, with slender jaws, and a range of body sizes comparable to the modern-day carnivorans. In feeding specialisations too, they radiated into a variety of highly active predators at one end and presumed bone-crushing scavengers at the other. The hyaenodontids also survived for far longer than the oxyaenids, not finally disappearing until the Miocene of Europe, Asia and Africa.

As already mentioned, the Carnivora actually predated the creodonts, for the earliest members were *Pappictidops*, a viverravid from the Early Palaeocene Shanghuan Formation of China (Wang *et al*. 1998; Lucas 2001) and the Early Palaeocene Canadian genera *Pristinictis* and *Ravenictis*. At one time, all the early, primitive carnivorans were included in a group 'Miacoidea', which was assumed to be basal to the whole order. However, recent cladistic analysis indicates that different 'miacoids' are related to different lineages of the Carnivora (Flynn 1998; Janis *et al*. 1998*b*). The family Viverravidae (Fig. 7.14(f)), including the Early Palaeocene members mentioned, are related to the Feliformes on the basis of reduction of the number of molars to two, and an elongated skull. The family Miacidae (Fig. 7.14(e)), which do not appear until the very end of the Palaeocene, are primitive Caniformes as indicated, for example, by the relatively shorter skull, and the loss of contact between the calcaneum and fibula in the ankle. Surprisingly therefore, the two respective main lineages of modern Carnivora can be traced by fossils back to the Early Palaeocene.

All the primitive, Palaeocene members were relatively small carnivores, varying only from the size of a weasel to that of a small cat, and it was not until the rise of the modern families such as the canids, felids, and ursids later in the Cenozoic that any members of the order achieved the large body size that certain mesonychids and creodonts had done far earlier.

The origin and radiation of the modern orders

The post-Palaeocene fossil record of placental mammals and its associated literature is enormous, and far beyond the scope of this work to deal with in more than the barest outline. The times of first occurrence, nature of the early members and their possible relationships, and palaeogeographic distribution are of the most immediate relevance for understanding the Cenozoic mammalian evolution. It is also, of course, irresistible to deal briefly with the more unfamiliar, often quite bizarre kinds of extinct mammals. Several modern orders appear in the Palaeocene fossil record, undoubtedly the Carnivora in the Early Palaeocene, and the Xenarthra, Rodentia, Eulipotyphla, and possibly Perissodactyla in the Late Palaeocene. However, even in these cases, their main radiation did not commence until the Eocene.

The molecular evidence for four super-orders of modern placental mammals has been discussed earlier in the chapter, and is taken as the basis for the following account. As it happens, this radical new phylogenetic hypothesis is surprisingly little in conflict with conventional interpretations of the fossil record, apart from two aspects. One is the extensive taxonomic redistribution of the primitive 'insectivorous' orders; the other is the recognition that the 'ungulate' orders arose within two independent radiations. Otherwise, the molecular story contributes towards the formulation of new answers to several long-standing questions that the fossils alone have been unable to provide.

Xenarthra

Only the three living sub groups Cingulata (armadillos), Pilosa (sloths), and Vermilingua (anteaters) constitute the Xenarthra (Delsuc *et al*. 2001), now that a relationship with either the old world Pholidota, or the North America Palaeocene Palaeanodonta has been discarded (Rose and Emry 1993). The earliest fossil record consists of Late Palaeocene armadillo scutes of *Prostegotherium* in the Itaboraí fissure deposits of Brazil. Armadillos diverged throughout the Cenozoic of South America, and a few genera reached North America in the Late Pliocene faunal interchange. The Glyptodonta were the most spectacular cingulate xenarthrans, with their bony helmet, hugely exaggerated carapace of interlocking scutes, and tail massively protected by rings of bone. The cheek teeth, all that remain of the dentition, are tri-lobed, continuously growing, and form a grinding surface, suitable for relatively soft vegetation. Appearing in the Middle Eocene, glyptodonts radiated widely for the remainder of the Cenozoic. The largest was the Pleistocene genus *Glyptotherium*, which reached over 3 m in body length. They also dispersed into Central and North America during the Plio-Pleistocene (Gillette and Ray 1981).

The Phyllophaga are familiar today only as arboreal forms, but most of the Cenozoic members of the group were ground sloths. They date from the Late Eocene, and by the Miocene had become an abundant part of the South American mammalian fauna. An undetermined form has also been reported from the Eocene of Antarctica (Vicaíno and Scillato Yané 1995). The mylodontids were fairly large animals, bear-like in size and build and with the forelimbs shorter than the hind limbs, indicating their terrestrial rather than arboreal habit. Early megatheriids were relatively smaller and possessed equal-length fore and hindlimbs suggesting a tree-living habit. However, later members of this group, such as *Megatherium*, achieved enormous body size, as large as a rhinoceros and, as has been reconstructed many times, were no doubt capable of an erect stance for feeding off high branches, with the help of a massive balancing tail. The Megalonychidae include genera that dispersed into the Caribbean Islands in the Plio-Pleistocene, many becoming dwarf forms.

The Vermilingua contains the South American anteaters, which evolved a very long snout and complete loss of the teeth. They were late to appear, not occurring on that continent until the Middle Miocene, and their fossil record remains generally poor. One of the enduring puzzles of mammalian palaeontology is the European *Eurotamandua*, which consists of a beautifully preserved specimen discovered in the Messel oil-shale (Storch and Richter 1992a). It has the elongated, edentulous snout, and robust skeleton with powerful digging limbs of an anteater of some description. It also possesses extra articulations on two of its vertebrae, and a second scapular spine, both of which characters suggest it is a xenarthran (Storch 1981). Gaudin and Branham (1998) performed a detailed cladistic analysis and found that the most parsimonious position for *Eurotamandua* is as the sister-group of Pilosa, which

is the xenarthran group composed of sloths plus anteaters. On the other hand, Rose and Emry (1993) believe that the characters shared by *Eurotamandua* and xenarthran anteaters are convergent, and that it has certain similarities instead to the pangolins, and also (Rose 1999) to the palaeanodonts of North America. The question of the relationships of *Eurotamandua* is therefore not resolved. If it is a xenarthran, then it is a unique representative of the indigenous South American Eocene mammals that mysteriously found its way into Europe by a very obscure route, and so far as is known, unaccompanied by any other taxon.

Although no other mammals, fossil or living can be demonstrated to be xenarthrans, the possibility must be entertained that the South American groups of large placental herbivores, the Meridiungulata, are members of the super-order. It would be no more absurd than having to accept a relationship of elephants with golden moles and tenrecs, within Afrotheria, and would be consistent with their biogeography. Probably only molecular evidence would permit this hypothesis to be tested, but actually it is not inconceivable that ancient DNA techniques applied to Late Pleistocene specimens of toxodont notoungulates might provide it.

Afrotheria

The wealth of molecular data supporting the supraordinal group Afrotheria is consistent with the palaeobiogeography of the constituent groups (Fig. 7.15). The earliest known fossils of all the living afrotherian orders occur in Africa, with one exception, the Sirenia, which is readily explicable by their marine habitat. There are also two extinct orders that have long been associated with the paenungulates, and which, if this is correct, are also therefore afrotherians, an interpretation recently confirmed by Asher *et al.* (2003). These are the large, ungulate-like embrithopods, which are primarily African but actually first appear in Europe, and the semiaquatic Desmostylia whose exclusively northern Pacific distribution is anomalous.

The fossil record of the relevant period of time necessary for understanding the origin and early radiation of mammals in Africa is very poor indeed (Cooke 1978). Mammal-bearing Palaeocene deposits

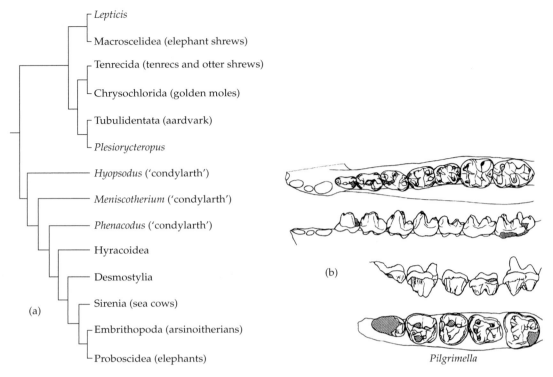

Figure 7.15 (a) Asher *et al.*'s 2003 cladogram of Afrotheria based on an Adams consensus of molecular and morphological data. (b) Dentition of the anthacobunid *Pilgrimella*: lower teeth in occlusal and buccal views, upper teeth in buccal and occlusal views. Length of lower jaw fragement approx. 10 cm. (Wells and Gingerich 1983)

are limited to a few low-yielding sites in North Africa (Gheerbrant 1995; Gheerbrant *et al.* 1996), and the Early Eocene is barely any better represented. It is not until the great Qatrini Formation of the Fayum Depression in Egypt, which dates from Late Eocene through Early Oligocene (Rasmussen *et al.* 1992), that abundant collections of good quality fossil mammals have been made. The next oldest mammals come from the much sparser Late Oligocene fauna at Chilga in Ethiopia (Kappelman *et al.* 2003). What these and other localities indicate is that throughout the Oligocene afrotherians dominated the African mammalian fauna, with only a few of the northern groups, notably primates, present. The orders of large mammals that are so conspicuous in Africa today, artiodactyls, perissodactyls, and carnivorans, made little impact until the Miocene, from which time a good fossil record comprehensively illustrates the later Cenozoic history of mammals in Africa.

During the Late Eocene and Oligocene, the two main herbivorous mammalian taxa in Africa were the Hyracoidea and Proboscidea. Until recently opinion diverged about their mutual relationship (Prothero 1993). One group argued that the Hyracoidea were more closely related to Perissodactyla, citing a range of morphological details of the braincase, teeth, and foot structure in support (Fischer 1989; Prothero and Schoch 1989). Another group argued for a relationship between Hyracoidea and Proboscidea, along with the Sirenia, the three forming a taxon Paenungulata (Shoshani 1986; Novacek *et al.* 1988). Tassy and Shoshani (1988) noted the incongruent evidence and were unable to decide between these possibilities. However, molecular evidence strongly supports the paenungulate group. The morphological characters that define Paenungulata, as listed by Archibald (1998), are an astragalus with flattened navicular facet, 19 or more thoracic vertebrae, and the third molar tooth larger than the second in both upper and lower tooth rows. It is on the basis of this limited morphological evidence, that the two extinct orders, Embrithopoda and Desmostylia are regarded as members too, although hardly surprisingly there is no agreement about their exact respective relationships, either to one another or to other paenungulates.

From time to time certain Asian fossils have been tentatively attributed to the Paenungulata, to the extent of implying an Asian rather than an African origin for the group as a whole (Beard 1998; Fischer and Tassy 1993). *Minchenella* is represented by lower jaws from the Late Palaeocene of China. Its molar teeth are 'condylarth'-like but there are minor features somewhat reminiscent of those of the primitive proboscidean *Moeritherium*. Anthracobunids (Fig. 7.15(b)), which may be related to *Minchenella*, are an Early to Middle Eocene family centred on Pakistan (Wells and Gingerich 1982). Depending on fine details of molar morphology (Tassy and Shoshani 1988), anthracobunids have been interpreted by various authors, respectively, as basal members of Proboscidea, or Desmostylia, or Tethytheria (Proboscidea plus Sirenia).

The interrelationships of the four other modern orders of placentals that have been shown to be members of Afrotheria are not yet well resolved, and elucidation is little helped by the very poor fossil record of all four. Two of them, Chrysochlorida, the golden moles, and Tenrecida, the tenrecs and otter shrews, are probably sister-groups and included in a super-ordinal taxon Afrosoricida (Scally *et al.* 2002; Asher *et al.* 2003). The other two are Macroscelidea, the elephant shrews, and the solitary species of Tubulidentata, the aardvark.

Hyracoidea
The hyracoids are represented today by only about a dozen species in three closely related genera, all of quite small body size. Surprisingly therefore, they were at one time the most abundant medium and large-sized herbivores in Africa, playing this role long before the perissodactyls and artiodactyls arrived on the continent (Rasmussen 1989). Hyracoids also dispersed into much of Asia. The earliest record is from the Late Eocene and Oligocene

of Fayum, where they ranged in size from *Thyrohyrax*, which was about the size of a modern hyrax, through *Megalohyrax* (Fig. 7.16(a)) with a 40 cm long skull (Thewissen and Simons 2001), to the giant *Titanohyrax* which was as large as a small rhinoceros, and bigger than any contemporary proboscideans. The morphology of the cheek teeth was also enormously variable, from simple bunodont to fully lophodont and selenodont molars. As the major, principally forest dwelling group of medium to large-sized browsers, there were species equivalent to the pigs antelopes, and small rhinos of today (Rasmussen and Simons 2000).

By the Miocene, hyracoids had declined in diversity, an event presumably connected with the arrival in Africa of several artiodactyl and perissodactyl families. By this time, a second radiation had occurred, the pliohyracines, which spread from Africa throughout Europe and Asia. In these northern areas they underwent a modest increase in diversity, evolving high-crowned, hypsodont cheek teeth for feeding on more abrasive plant material. Most of the hyracoids had become extinct by the Pliocene, although one highly specialised lineage, procaviids, survived as the living members of the order.

Proboscidea
As mentioned, anthracobunids (Fig. 7.15(b)) are represented by dentitions of mid-Eocene age from India and Pakistan, which have one or two proboscidean characters (Wells and Gingerich 1983), and accordingly they have been accepted by several authors as the most basal members of the group (e.g. Fischer and Tassy 1993; Shoshani et al 1996). However, earlier-dated, undisputed proboscideans have since been discovered in North Africa. These are from the earliest Eocene in age and are part of the Ouled Abdoun Basin fauna of Morocco. Gheerbrant *et al.* (1996; 1998) described isolated upper teeth of *Phosphatherium* (Fig. 7.17(a)). They were from a relatively small animal, with a body weight estimated from tooth size of 10–15 kg. The molar teeth have the true lophodont condition of crests running transversely and uninterrupted by conules, that characterises proboscideans, and in this form there are two such lophs, bilophodonty, which is the primitive proboscidean condition. *Daouitherium* (Fig. 7.17(b)) is a very much larger

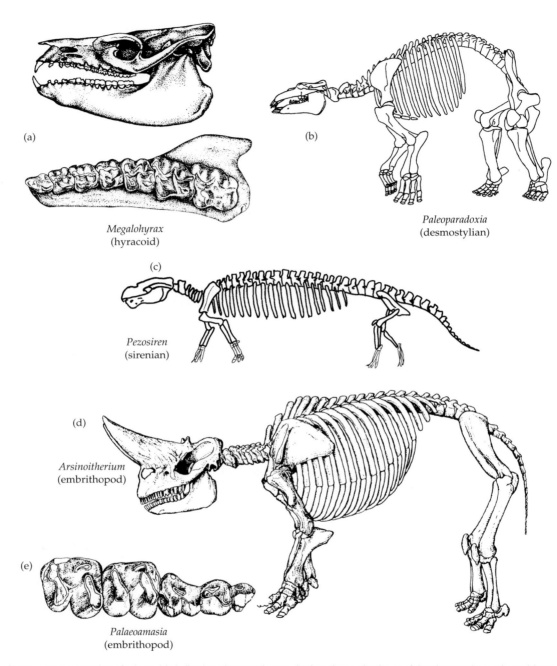

Megalohyrax
(hyracoid)

Paleoparadoxia
(desmostylian)

Pezosiren
(sirenian)

Arsinoitherium
(embrithopod)

Palaeoamasia
(embrithopod)

Figure 7.16 Paenungulate afrotheria. (a) Skull in lateral view and upper cheek teeth in occlusal view of the Oligocene hyracoid *Megalohyrax*. Length of skull approx. 30 cm (Romer 1966, and Carroll 1988, after Andrews). (b) The Miocene desmostylian *Paleoparadoxia*. Length approx. 2.2 m (Domning 2002). (c) The Eocene sirenian *Pezosiren*. Length approx. 2.1 m (redrawn after Domning 2001). (d) Skeleton of the embrithopod *Arsinoitherium*. Length approx. 1.8 m (Savage and Long 1986, after Andrews). (e) Upper cheek teeth in occlusal view of the Asian Oligocene embrithopod *Palaeoamasia* (Savage and Long 1986, after Sen and Heintz).

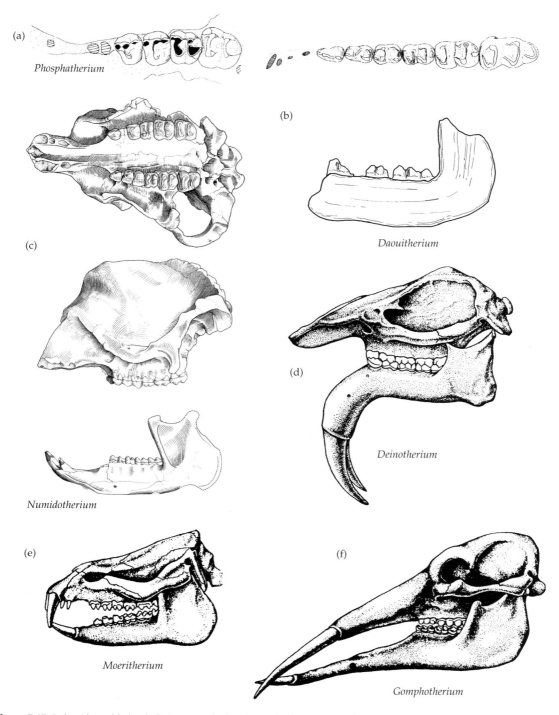

Figure 7.17 Proboscideans. (a) *Phosphatherium* upper cheek teeth in occlusal view. Length of fragment approx. 6 cm (Gheerbrant *et al.* 1996). (b) *Daouitherium* lower dentition in occlusal view, and lower jaw in lateral view. Length of jaw fragment approx. 14 cm (Gheerbrant *et al.* 2002). (c) Skull and lower jaw of *Numidotherium*. Length of lower jaw approx. 45 cm (Mahboubi *et al.* 1986). (d) *Deinotherium*. Skull length approx. 45 cm. (e) *Moeritherium*. Skull length approx. 33 cm. (f) *Gomphotherium*. Skull length approx. 75 cm (*d–f* from Romer 1966).

animal, known from the lower jaw and dentition, plus a possible upper premolar (Gheerbrant *et al.* 2002). As far as can be judged, the dentition is similar to that of *Phosphatherium*, and both resemble a slightly younger, but much more completely preserved Algerian proboscidean. This is *Numidotherium* (Fig. 7.17(c)), which is represented by skull and postcranial skeleton specimens (Mahboubi *et al.* 1984, 1986; Court 1995). It is small for a proboscidean, with a height of about 1 m, but has the relatively large head with pneumatised bones, nostrils high up on the head, and graviportal limbs characteristic of the order. One of the lower incisors and the canine were lost, and the second upper incisors are the largest, and therefore can be interpreted as incipient tusks. The molar teeth are large and, like the Moroccan genera, bilophodont.

The next oldest fossil proboscideans occur in the Fayum and other Late Eocene localities of North Africa, of which *Moeritherium* (Fig. 7.17(e)) is the best known Like *Numidotherium*, it too has enlarged second incisors and bilophodont molar teeth. It was the size of a large pig and the long body, reduced tail and short, stout limbs suggest that it was a semi-aquatic, hippo-like animal. It has actually been proposed that it is not a proboscidean at all, but a relative of the Sirenia, or even the extinct Desmostylia, both of which are partially or fully aquatic mammals (Coppens and Beden 1978). *Barytherium* is a second, less well-known proboscidean from the Fayum. It was far larger, as much as 4 tonnes in weight, and standing 3 m at the shoulder.

Palaeomastodon is a third member of the Fayum fauna. It possessed well-developed tusks, no trace of canines, very large molars with three complex lophs and thin enamel, and an extremely elongated symphysis connecting the relatively short lower jaws. With one exception, a hypothetical Oligocene ancestral form resembling *Palaeomastodon* is believed to have given rise to the subsequent proboscidean radiation in Africa (Tassy 1996; Shoshani *et al.* 1996). The exception are the deinotheres (Fig. 7.17(d)), large animals that are first known in the Late Oligocene of Ethiopia (Kappelman *et al.* 2003). They possess primitive molar teeth with only two or three simple lophs, and thin enamel. The tusks, uniquely, consist of the pair of lower incisors extending ventrally, due to a downward reflection of the symphyseal region of the lower jaws.

Deinotheres probably evolved directly from a primitive, *Moeritherium*-like ancestor.

During the Late Eocene and through the Oligocene, proboscideans were restricted to Afro-Arabia. In the Middle Miocene, tectonic movement of the Arabian peninsula created a connection across the Tethys Sea between Africa and Eurasia, and at least three proboscidean taxa crossed what has been termed the 'Gomphothere Land Bridge' (Agustí and Antón 2002). The elephantoid *Gomphotherium* (Fig. 7.17(f)), the mammutid *Zygolophodon*, and the deinotheriid *Deinotherium* (Fig. 7.17(d)) all appeared in various parts of Eurasia in the mid Miocene. The former two also reached North America at this time, from which moment the elephants radiated to become a significant part of the large herbivorous mammalian fauna of that continent. Gomphotheres also briefly existed in South America, after the formation of the Panamanian land bridge in the Plio-Pleistocene.

Sirenia

The first fossil remains of the dugongs and manatees come from the Early Eocene of Europe, and by the Late Eocene this fully aquatic, herbivorous group had a virtually worldwide distribution, including the shores of Africa, Asia, North America, and more doubtfully South America (Fischer and Tassy 1993; Domning 2002*b*). *Prorastomus* from Jamaica is the most primitive fossil skull of a sirenian (Savage *et al.* 1994). It has, surprisingly, five rather than the usual four premolars, which is presumably a secondary reversion rather than retention of the tooth formula of the Cretaceous ancestors of placentals. The molars are bilophodont and therefore comparable to those of early proboscideans. It lacked the characteristic ventral deflection of the rostrum found in all more progressive sirenians. The postcranial skeleton of *Prorastomus* is unknown, but that of a similar, slightly younger Jamaican form, the Middle Eocene *Pezosiren* (Fig. 7.16(c)), is represented by an almost complete set of postcranial bones (Domning 2001). It was about 2 m in length and had a relatively uncompressed neck, unlike later sirenians. Several characters of the postcranial skeleton indicate that *Pezosiren* was capable of terrestrial locomotion, such as tall neural spines on the anterior dorsal vertebrae for head support, a strong connection between the sacral vertebrae and ilium, and well built though relatively short limbs. These coexist with other characters that are aquatic adaptations, including a dorsal position of the nostrils and a long row of very dense ribs.

Desmostylia

The desmostylians are a group of semiaquatic mammals as large as hippos. They were limited taxonomically to about half a dozen genera, temporally to the Late Oligocene and Miocene, and geographically to the east and west margins of the northern Pacific Ocean (Domning *et al.* 1986; Domning 2002*a*). Complete skeletons of *Paleoparadoxia* (Fig. 7.16(b)) demonstrate well-developed fore and hind limbs, with broad, presumably paddle-like hands and feet, suggesting an amphibious, seal-like existence. The incisors and canine teeth are well developed, and there is a long diastema. The cheek teeth resemble those of the early proboscideans, particularly *Moeritherium*. Clementz *et al.* (2003) analysed stable isotope contents of the bones of *Desmostylia*, and concluded that they could forage on sea grass, and also on land during terrestrial excursions.

That desmostylians are paenungulates is not seriously disputed; a relationship with the tethythere grouping of proboscideans and sirenians is usually preferred (Fischer and Tassy 1993; Asher *et al.* 2003).

Embrithopoda

The embrithopods are best known as the celebrated Late Eocene *Arsinoitherium* (Fig. 7.16(d)) from the Fayum. It was a huge, rhinoceros-sized animal with the skull adorned with a pair of massive horns. Neither tusks nor enlarged canines are present, and the cheek teeth are bilophodont and hypsodont in form (Court 1992). Teeth of earlier and more primitive, hornless forms (Fig. 7.16(e)) occur in Romania and central Turkey (Maas *et al.* 1998).

As with the desmostylians, there is more or less complete agreement that the Embrithopoda are part of the paenungulate radiation, but not about precisely where they fit. Tassy and Shoshani (1988) regarded them as the most basal paenungulate group. Court (1990), Fischer and Tassy (1993) and Asher *et al.* (2003) have them closer to the Proboscidea than to the Sirenia, within the Tethytheria.

Macroscelidea

Fossil elephant shrews occur only in Africa, the earliest record being teeth of *Chambius* from the Middle Eocene of Tunisia, followed by the Late Eocene *Herodotius* from the Fayum (Simons *et al.* 1991; Butler 1995). They have four-cusped, bunodont molars lacking lophs, similar to what is seen in later macroscelideans.

Comparison of the dental characters of these primitive forms has been made with hyopsodontid 'condylarths' by a number of recent authors (Simons 1991 *et al*; Butler 1995). Most recently, Tabuce *et al.* (2001) have also supported the 'condylarth' origin of the Macroscelidea, and proposed that the dentition of the primitive members is close to the ancestral state for paenungulates. However, since the non-afrotherian Perissodactyla appear in their cladogram as relatives, convergent evolution of dental characters must have occurred at least at some level, which considerably blurs the picture.

Tubulidentata

Represented today by *Orycteropus afer*, the single species of aardvark, the fossil record of this group is scarcely any richer. The four or five known genera are all placed in the same family by McKenna and Bell (1997) and they date from the Oligocene of Europe and Early Miocene of Africa.

There are subfossil bones of a very peculiar mammal in Madagascar, named *Plesiorycteropus*, preserved along with giant lemurs, elephant birds, etc. in Quaternary deposits. Representatives of most of the bones are present, except for the facial region, jaws, and dentition. It was a heavily built, digging animal, probably with a long narrow snout and lacking teeth. Most commentators have regarded *Plesiorycteropus* as a tubulidentate. However, McPhee (1994) failed to find any characters uniquely shared with *Orycteropus*, and his cladistic analysis did not group them exclusively together. He concluded that *Plesiorycteropus* is sufficiently distinct from any other placentals as to warrant its own order, Bibymalagasia. In the most recent analysis, Asher *et al.* (2003) found that *Plesiorycteropus* is at least a member of Afrotheria.

Afrosoricida

Seiffert and Simons (2000) have described lower jaw fragments of a Late Eocene form from the Fayum fauna, which they claim may be related to chrysochloridans and/or tenrecs, or possibly to the West Indian eulipotyphlan family Solenodontidae.

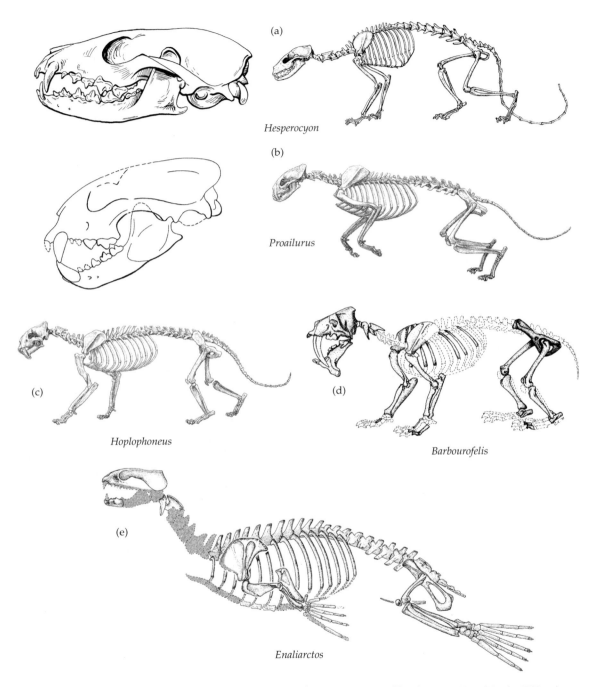

Hesperocyon

Proailurus

Hoplophoneus

Barbourofelis

Enaliarctos

Figure 7.18 Carnivora. (a) Skull and skeleton of the Middle Eocene canid *Hesperocyon*. Presacral length approx. 45 cm (Munthe 1998 and Savage and Long 1986). (b) Skull and skeleton of the Early Miocene felid *Proailurus* (Carroll 1988 from Radinsky 1982, and Turner and Anton 1997). (c) The nimravid *Hoplophoneus*. Height at shoulder 48 cm (Turner and Anton 1997). (d) The nimravid *Barbourofelis* (Martin 1998). (e) The Oligoncene pinniped *Enaliarctos*. Presacral length approx. 1.25 m (Berta and Ray 1990).

The salient feature of *Widanelfasia* is the presence of incipiently zalambdodont molar teeth. The trigonid cusps, especially the protoconid are very high and form a narrowly acute triangle, and the talonid is much lower. The fully zalambdodont molar of solenodontids is presumed to be convergent, but that of the tenrecs and the rather specialised version found in chrysochloridans may well be homologous. It is therefore possible, though by no means positively demonstrable yet, that *Widanelfasia* is a stem afrosoricid. Seiffert and Simons (2000) also point to the upper teeth of an Early Eocene primitive 'insectivore' from Tunisia, *Chambilestes*, described by Gheerbrant and Hartenberger (1999) and note that they have a structure compatible with the lower teeth of their Fayum mammal.

Chrysochlorida. The golden moles have a very poor fossil record. It is entirely restricted to Africa, dates from no earlier than the Early Miocene, and consists of specimens that can be placed in the same family as the nine living genera.

Tenrecida. The tenrecs of Madagascar along with the otter shrews of sub-Saharan Africa have the poorest fossil record of all the afrotherian orders, consisting solely of zalambdodont teeth of an Early Miocene African form *Parageogale* (Butler 1984), prior to subfossils of the Pleistocene.

Laurasiatheria

Carnivora

The very early appearance of members of the order Carnivora has already been described, and within the Early Palaeocene differentiation of the two primary subgroups of the order had apparently also occurred, for one of them, Feliformes, is represented by the viverravids of that time (Fig. 7.14(f)). The miacids (Fig. 7.14(e)), which appear in the Late Palaeocene, are basal members of the other group, Caniformes (Flynn 1998; Janis *et al.* 1998b). Both of these primitive groups, consisting of relatively small animals, radiated through the first half of the Eocene, during which period it was the creodonts that provided the larger carnivorous mammals.

The family Nimravidae were the earliest group of modern-style Feliformes, a group specialised for pure carnivory and convergent in many features on the true cats, the Felidae (Martin 1998a). They appeared in the Late Eocene of North America and Eurasia, and radiated primarily during the Oligocene; the last survivor, *Barbourofelis* (Fig. 7.18(d)), disappeared no later than the Late Miocene of North America. Nimravids, like felids, had short, powerful skulls, highly developed carnassial teeth and reduction of the more posterior molars. Their postcranial skeleton indicates that they were long-legged, cursorial predators with feline-like claws. Even more markedly than in the felids, nimravids had a tendency to evolve sabre-toothed forms. The early, leopard-sized *Hoplophoneus* (Fig. 7.18(c)) already had evolved enlarged canines. The trend is most dramatically expressed in the lion-sized *Barbourofelis* (Turner and Antón 1997).

The first true cat, the felid *Proailurus* (Fig. 7.18(b)), is well known from the Oligocene and Early Miocene of Europe. It was the size of a modern ocelot and its relatively short, robust limbs and retractile claws indicate an accomplished arboreal animal (Agustí and Antón 2002). The major radiation of felids commenced in the Late Miocene (Martin 1998b), including their entry into North America, where they rapidly replaced the nimravids as the top carnivores. A number of genera independently evolved into sabre-toothed cats, most famously *Smilodon*, huge numbers of which are preserved in the Rancho La Brea asphalt pits of Los Angeles and which survived until the end-Pleistocene about 10,000 years ago.

The other feliform families, Viverridae (civets), Herpestidae (mongooses), and Hyaenidae also commenced their radiations during the Miocene, and apart from a single Pliocene hyaenid, were restricted to Eurasia and Africa. This is in inexplicable contrast to the Felidae, which were abundant in North America.

The second branch of the Carnivora are the Caniformes, all the families of which do occur in the fossil record of North America, as well as Eurasia and Africa. As mentioned, the earliest caniniforms are believed to be the Late Palaeocene miacids of North America and Europe. The modern caniform families appeared at various times after this, in a complex biogeographic pattern. Canids were a predominantly North American family (Munthe 1998a), where the small-dog sized *Hesperocyon* (Fig. 7.18(a)) of the Middle Eocene was the earliest, and the family was not represented in Europe until well into the Miocene, nor Asia and Africa until the Pliocene. The Ursidae (bears) also first occur in North America (Hunt 1998), in the Late Eocene, as the very small, gracile *Parictis* that had a skull only 7 cm long. Like the canids, this family too does not appear in the Eurasian or African record until the Miocene. In contrast, the other caniform families Amphicyonidae (the extinct beardogs), Procyonidae (raccoons), and Mustelidae (weasels, otters etc.) all occur by the Late Eocene or Early Oligocene in both the Old World and the New.

The Pinnipedia are a monophyletic group related to the caniforms. Both the morphological evidence (Wyss and Flynn 1993; Janis *et al.* 1998b) and molecular sequence data (Corneli 2003) indicate that they are most closely related to the ursids. They occur from the Oligocene. *Enaliarctos* (Fig. 7.18(e)) is a Late Oligocene Californian genus known from more or less complete skeletons. It was relatively small, about 1 m in presacral length, and both fore and hind feet had evolved into flippers, although they are sufficiently well developed to imply a reasonable degree of manoeuvrability on land (Berta and Ray 1990). The molar teeth are relatively unmodified carnassials.

Fossil members of the three constituent families, which are identified by various cranial characters, are all represented in the fossil record by the Early to Middle Miocene, Otariidae (eared-seals) and Obodenidae (walruses) on the North Pacific coasts, Phocidae (true seals) on the North Atlantic coasts (Berta and Sumich 1999; Berta 2002).

Perissodactyla

The earliest perissodactyls (Fig. 7.19(a)) date from the beginning of the Eocene of North America and Europe, and within the Early Eocene representatives of all the three main lineages are already present (Prothero and Schoch 1989). The most primitive perissodactyls include famously *Hyracotherium* ('Eohippus'), although the taxonomic situation is confusing because specimens attributable respectively to true horses, Equoidea, and to

the primitive European group, Palaeotheriidae, have been placed in this genus (Hooker 1994; Froehlich 2002). Basal members of both these groups consisted of animals no larger than a small dog, with a primitive version of the perissodactyl molar tooth structure consisting of two transverse lophs connected externally by an ectoloph. Early Eocene representatives of the other two major lineages, recognisable by details of their dentitions, were the titanothere *Eotitanops* and the moromorph *Homogalax*.

For many years the 'condylarth' grade phenacodontids (Fig. 7.7(e) to (g)) have been accepted as the origin of the perissodactyls, on the basis of similarities in the bilophodont tooth structure (Prothero and Schoch 1989). However, McKenna *et al.* (1989) described the skull and dentition of *Radinskya* (Fig. 7.19), and demonstrated that the molar teeth are similar to those of primitive perissodactyls like *Hyracotherium* and *Homogalax*. However, *Radinskya* also resembles the group of Chinese 'condylarth' grade phenacolophids, usually regarded as basal embrithopods. This proposed relationship of phenacolophids and perissodactyls excludes the phenacodontids from consideration, and also implies that the perissodactyls arose in Asia and dispersed to Europe and North America around the end of the Palaeocene (Beard 1998). The interrelationships between this order, other ungulate orders, and 'condylarths' continue to be argued over without agreement (e.g. Thewissen and Domning 1992; Fischer and Tassy 1993; Holbrook 2001).

The subsequent Eocene radiation of perissodactyls was spectacular, and the mere six surviving genera today represent a small fraction of their erstwhile diversity (Prothero and Schoch 1989). Several lineages of brontotheres (Titanotheriomorpha) evolved into very large, rhinoceros-like mammals (Fig. 7.19(b)), with the skull adorned by a pair of bony bosses or protuberances at the front of the head, supposedly used more for intraspecific combat than for protection (Mader 1997). The molar teeth remained low-crowned and suitable for browsing on relatively soft vegetation. Restricted almost exclusively to North America and Asia, brontotheres soon disappeared from the fossil record, at the start of the Oligocene.

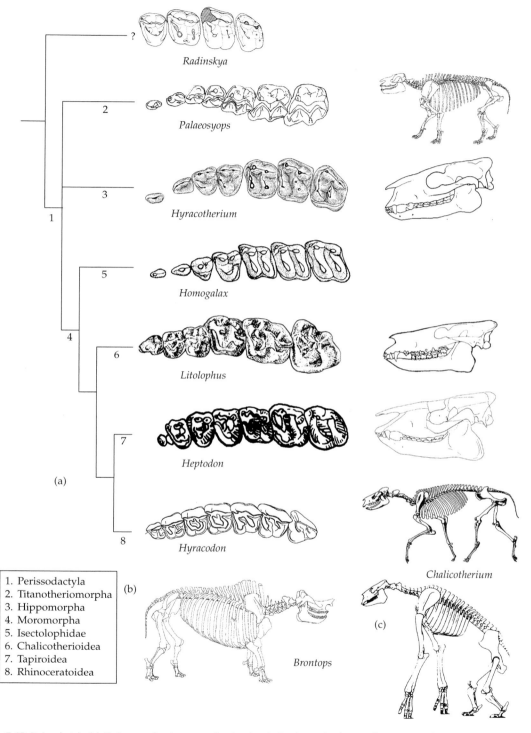

Figure 7.19 Perissodactyla. (a) Cladogram of main groups of perissodactyls showing occlusal views of upper teeth (from various sources). (b) Skeleton of the titanotheriomorph *Brontops*. (Moder 1998, from Osborn 1929) Height at shoulder approx. 2.5 m. (c) Skeleton of the Miocene chalicothere *Chalicotherium* (Coombs 1989 from Zapfe).

The following labels appear within the figure:

- ? Radinskya
- 2 Palaeosyops
- 3 Hyracotherium
- 1
- 5 Homogalax
- 4
- 6 Litolophus
- 7 Heptodon
- (a)
- 8 Hyracodon

1. Perissodactyla
2. Titanotheriomorpha
3. Hippomorpha
4. Moromorpha
5. Isectolophidae
6. Chalicotherioidea
7. Tapiroidea
8. Rhinoceratoidea

(b) Brontops

(c) Chalicotherium

The Equoidea also radiated in the Eocene, particularly in Europe where *Palaeotherium* was a common, superficially tapir-like browsing animal. However, a second major radiation followed in the Miocene, centred in North America but with representatives throughout Europe, and dominated by grazing forms with hypsodont teeth culminating in the modern genus *Equus* (McFadden 1992, 1998).

The third major lineage of perissodactyls are the Moromorpha, which is the most diverse. The chalicotheres were another group of bizarre perissodactyls (Coombs 1998). Although present in the Eocene, their major radiation occurred in the Miocene, when large body size evolved, along with retractable claws in place of hooves. *Moropus* was like a very large horse with its forelimbs somewhat longer than the hindlimbs. This trend reached its culmination in the European and African *Chalicotherium* Fig. 7.19(c)), in which the forelimbs were about 50% longer than the hindlimbs and the animal must have been capable of standing on its hindlegs to pull down branches from quite large trees. Chalicotheres survived into the Pleistocene of Asia and Africa and indeed there have been occasional, unsubstantiated reports from Kenya of sightings of a living chalicothere (Savage and Long 1986).

Rhinocerotoids were far more diverse from the Eocene through the Miocene than they are now. In addition to the familiar horned rhinoceroses of today, they included small, agile, dog-sized browsers, semiaquatic hippo-like grazers, tapir-like forms, and gigantic forms capable of browsing on high trees (Prothero and Schoch 1989). *Hyrachyus* is the earliest and most primitive rhinocerotoid, abundant throughout North America and Eurasia during the Early Eocene. The indricotheres were a purely Oligocene Asian group of huge mammals, amongst which is included *Paraceratherium*, variously referred to in the past as *Indricotherium* and *Baluchitherium*. It was a strictly quadrupedal, graviportal animal but with a height reaching over 6 m, a 1.3 m long skull, and a body weight estimated at around 30 tonnes, it was capable of browsing on leaves and twigs well above the reach of any other mammals of its day. By the Late Oligocene and through the Miocene, the true rhinoceroses radiated and became an abundant part of the

ungulate population of all the northern continents and Africa (Heissig 1989).

The final group of perissodactyls to mention are the tapirs, the sister group of the rhinocerotoids. As with perissodactyls as a whole, the paucity of modern species, no more than four in this case, is a small echo of a once much more diverse group (Schoch 1989; Holbrook 2001). The greatest diversity of tapiroids occurred during the Eocene, when there was an abundance of genera across Eurasia and North America. They remained conservative, browsing forms and their only remarkable feature was a tendency to evolve a proboscis, or short trunk.

Cetartiodactyla

Despite the difference between living whales and the even-toed ungulates, there is overwhelming molecular evidence that the former are taxonomically nested within the latter. As it happens, the fossil evidence is also tending to point to this conclusion. The first known cetartiodactyls are the Early Eocene 'Dichobunidae', almost certainly a paraphyletic group including relatives of more than one of the subsequent lineages (Gentry and Hooker 1988; Stucky 1997). *Diacodexis* (Fig. 7.20(c)) is the most completely preserved of these. It was a rabbit-sized mammal occurring in Eurasia and North America. Despite its small size, the long, slender limbs, and digitigrade stance indicate a cursorially adapted form and the single most characteristic artiodactyl feature (Fig. 7.20(b)), an astragalus with pulley-shaped articulating surfaces for the tibia above and the rest of the foot below was present (Rose 1982, 1987). This arrangement effectively defines what is termed the paraxonic foot, in which the main axis runs between digits three and four. The dentition, and indeed the skull generally of *Diacodexis*, was little advanced from the 'condylarth' grade and gives little away about the origin of the artiodactyls. An earlier proposal by Van Valen (1971) that they were related to the arctocyonid 'condylarths' has received little support (Rose 1987; Prothero *et al.* 1988; Thewissen and Domning 1992).

From this Early Eocene start, the artiodactyls radiated into around 20 families (Fig. 7.20(a)), including the ten that are usually recognised

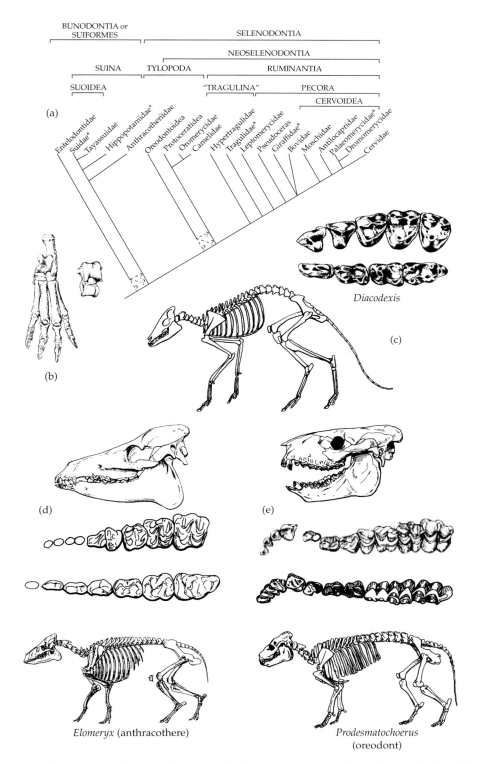

Figure. 7.20 Terrestrial artiodactyls. (a) Phylogeny of the atiodactyl families .Asterixes indicate families exclusive to the old world (Janis *et al*. 1998). (b) Basic hind foot and astragalus structure of artiodactyls (Carroll 1988). (c) Upper and lower dentitions in occlusal view, and skeleton of *Diacodexis*. Presacral length approx. 25 cm (Savage and Long 1986 and Stucky 1998). (d) Skull, upper, and lower teeth in occlusal view, and skeleton of the anthracothere *Elomeryx*. Shoulder height approx. 1.7 m (Kron and Manning 1998). (e) Skull, upper and lower dentition in occlusal view, and skeleton of the oreodont *Prodesmatochoerus*. Height at shoulder approx. 40 cm (Lander 1998).

amongst the modern mammalian fauna (Janis *et al.* 1997(*c*)). It has always been a relatively conservative order in terms of size and body form, not producing any really bizarre forms such as are found in the perissodactyls. On the other hand, artiodactyls were eventually to prove far and away the most diverse of all the ungulate orders, creating extraordinary difficulties for phylogenetic analysis. Homoplasy of morphological characters is rife, and disagreements at the lower taxonomic levels widespread.

The more basal forms are included in the Suiformes, or Bunodontia, in which the molar teeth are bunodont, with low, rounded cusps appropriate for browsing and omnivory. The anthracotheriids (Fig. 7.20(d)) were the earliest family to radiate, in the Middle Eocene of Europe and Asia. By the Late Eocene, North American representatives are found (Kron and Manning 1997), and by the Oligocene they had arrived in Africa (Black 1978). Anthracotheriids varied in size from that of a small pig to that of a large hippopotamus, and being relatively heavily built and short-limbed, it is usually assumed that they had a semiaquatic, hippo-like existence. Indeed their decline in importance during the Miocene and onwards has sometimes been attributed to the Miocene origin of a second suiform family, the Hippopotamidae. However, this family never entered North America, even though anthracotheriids disappeared there too. The third suiform family, Suidae, are first found in the Oligocene of Eurasia, and like the hippos went on to radiate from the Miocene onwards exclusively in the Old World. In contrast, the fourth group, Tayassuidae, or peccaries, were exclusively New World in distribution (Wright 1998). These generally pig-like forms appeared and radiated widely during the Miocene, and entered South America during the Plio-Pleistocene.

All other artiodactyls are included in a taxon Selenodontia, on the basis of the evolution of selenodont cheek teeth, whose crescentic-shaped crests contributed to a structure capable of dealing with tougher vegetation. The least modified of the selenodonts are the Tylopoda, a primarily North American group. It consists of the Middle Eocene and Miocene Oreodontidae (Fig. 7.20(e)), which were pig-like forms (Lander 1998), and the

Camelidae. The latter were a very important part of the large herbivore fauna during the Miocene, and only finally disappeared from that continent in the Late Pleistocene (Honey *et al.* 1998). The familiar Eurasian and African camels of today are the descendants of Late Miocene immigrants (Agustí and Antón 2002).

The Ruminantia constitute by far the majority of artiodactyls, most of which are Old World in distribution. Fossil ruminants are recognisable by the loss of the upper incisors, and fusion of the cuboid and navicular bones of the forefoot. The earliest, and most primitive members are represented by families in North America and Eurasia by the Upper Eocene, and in Africa by the Early Miocene. A complex evolutionary radiation and dispersal pattern of the ruminants followed in all three areas (Gentry and Hooker 1988; Janis and Scott 1988; Janis *et al.* 1998*c*), culminating in the dominant position amongst ungulate mammal faunas that they hold today. The more progressive members, the pecorans were always predominantly Old World forms, and include the Giraffidae, and particularly the Bovidae that are the most diverse artiodactyls of Africa. The Cervidae were always more cosmopolitan, and are the dominant Eurasian ruminants. Only one or two rare fossils of either cervids or bovids have been discovered in North America before the Late Pleistocene. In fact, only one pecoran family at all, Antilocapridae, was ever very significant in North America. They were quite abundant during the Miocene, but disappeared during the Plio-Pleistocene with the sole exception of *Antilocapra*, the living pronghorn antelope (Janis and Manning 1998).

The very strong molecular evidence for a sister-group relationship between whales and hippos was discussed earlier in the chapter. Until recently, the fossil evidence did not support this, and on the basis of dental and braincase characters in particular, cetaceans were regarded as related to, indeed descended from mesonychid 'condylarths'. The fossil record of cetaceans is actually remarkably good, owing to their marine habitat, and a diversity of primitive, Eocene whales have been described (Williams 1998; Thewissen and Williams 2002). They are mainly from the Indian subcontinent and Egypt, both of which bordered the Tethys Sea at the time, which suggests that it was on its shores that Cetacea originated (Thewissen 1998; Gatesy and O'Leary 2001). These early whales are referred to as the Archaeoceti, which is probably a paraphyletic group because it is based on ancestral characters. The dentition is relatively differentiated, and includes molar teeth recognisable as modified ungulate in form. Hindlimbs are still present to various degrees, which casts light on both the relationships, and the mode of evolution of these most highly specialised of mammals.

The earliest specimen yet collected is of *Himalayacetus*, from the 53 Ma Early Eocene of northern India (Bajpai and Gingerich 1998). Unfortunately it consists only of a fragment of lower jaw with the second two molar teeth in place, and nothing at all is known of its postcranial skeleton. The 5 Ma younger pakicetids are reasonably well known and they possess the least modified and therefore the most terrestrially adapted of whale skeletons. *Pakicetus* (Fig. 7.21(c)) is identifiable as a cetacean from the structure of the braincase and ear region, and various features of the cheek teeth such as reduction of the paraconid and metaconid, and presence of only a single talonid cusp of the lower molars. Most spectacular, however, is the postcranial skeleton, which was still fully adapted for terrestrial life (Thewissen *et al.* 2001). *Pakicetus* was about the size of a wolf, with full length, slender limbs of terrestrial ungulate proportions and unmodified pectoral and pelvic girdles. The vertebrae are variable in size, and the zygapophyses indicate a stiff thoracic and lumbar column, as in modern cursorial ungulates. This structure of the skeleton, and the absence of fish-eating adaptations of the dentition, or a specialised aquatic ear suggest that *Pakicetus* was no more adapted for an aquatic habitat than many terrestrial ungulates.

Other archaeocetes had become adapted for an amphibious mode of life. The Middle Eocene *Ambulocetus* (Fig. 7.21(d)) had reduced limbs, but they were still large enough to support the body weight on land, and the pelvis was attached to the vertebral column (Thewissen *et al.* 1994, 1996). However, the feet were large and could probably have paddled effectively, and though there was a normal tail so could not have been a tail fluke, the structure of the vertebral column indicates that it was capable of extensive dorso-ventral flexion and extension.

The protocetid genera *Artiocetus* (Fig. 7.21(a)) and *Rodhocetus* (Fig. 7.21(e)) had comparable sized limbs to *Ambulocoetus*, but larger fore and hind feet, which were probably webbed (Gingerich *et al.* 2001). Like *Ambulocetus*, they would still have been able to move on land, but very clumsily and presumably more infrequently, after the style of a modern seal.

Basilosaurus (Fig. 7.21(f)) is a Middle Eocene Egyptian whale, in which a minute hindlimb is still present (Gingerich *et al.* 1990). The individual bones are recognisable, although several of them have fused. The tiny pelvis, like that of modern whales, has lost its attachment to the vertebral column. The forelimb is also small, and could only have functioned in the control of undulatory swimming. Thus, *Basilosaurus* must have possessed a tail-fluke, and the modern mode of cetacean locomotion had finally evolved.

The subsequent fossil history of cetaceans is relatively uncontroversial (Uhen 1998). The radiation from the Middle Eocene onwards of the basilosaurid whales includes the origin of the two modern groups, Mysticeti (baleen whales) and Odontoceti (toothed whales). Both appear as fossils in the Oligocene, and share numerous characters indicating their monophyly.

The limbs of the primitive archaeocete whales possess the artiodactyl structure (Fig. 7.21(b)). Even the highly reduced hindlimb of *Basilosaurus* can be seen to be paraxonic in structure, although it is too modified to reveal whether it had evolved from a fully artiodactyl form, or is consistent with an origin from the more primitive, mesonychid ankle (Gingerich *et al.* 1990). However, the much less reduced, and still functional ankle joints of the other genera have the trochleated facet of the astragalus for the navicular, and the unique arrangement of the astragalus–calcaneum joint that unambiguously define Artiodactyla (Rose 2001). Furthermore, the principal axes of the feet are clearly distinguishable. In the forefoot, the third digit is the largest, indicating a mesaxonic type of foot. In the hindfoot, the third and fourth digits are equally large, which is the paraxonic

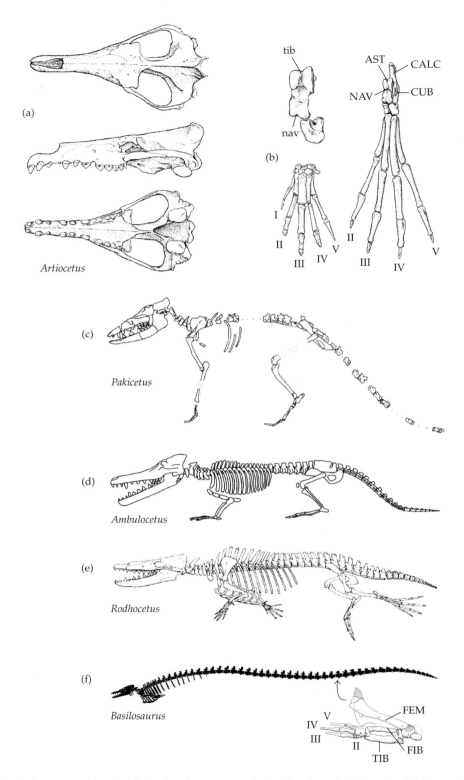

Figure 7.21 Limbed cetaceans. (a) Skull of *Artiocetus*. Length approx. 57 cm (Gingerich *et al.* 2001). (b) Astragalus of *Artiocetus* and complete manus and pes of *Roodhocetus* (Gingerich *et al.* 2001). (c) Skeleton of *Pakicetus*. Estimated presacral length 1.1 m (Thewissen *et al.* 2001). (d) Skeleton of *Ambulocetus*. Presacral length approx. 2 m (Thewissen *et al.* 1996). (e) Restoration of the skeleton of *Rodhocetus*. Maximum length approx. 2.75 m (Gingerich *et al.* 2001). (f) Skeleton of *Basilosaurus*, with enlarged view of hind limb (Gingerich *et al.* 1990). AST, astragalus; CALC, calcaneum; CUB, cuboid; FEM, femur; FIB, fibula; NAV, navicular; nav, articulation surface for navicular; TIB, tibia; tib, articulation surface for tibia.

condition. This particular combination is the ancestral artiodactyl arrangement as seen, for example, in *Diacodexis* (Fig. 7.20(b)).

The foot structure of these primitive whales therefore supports a relationship between Cetacea and Artiodactyla, but is not adequate to determine whether a monophyletic Artiodactyla as a whole, or just the Hippopotamidae is the sister-group of the whales. However, Naylor and Adams's (2001) investigation of the different cladistic relationships generated by different categories of characters included the limb features. Their conclusion was that a cladogram based on all the evidence except the dentition points to a sister-group relationship of whales and hippos. However, one using only dental characters indicates a relationship of whales with mesonychid 'condylarths'. The authors believe that non-independence of the different dental characters, combined with convergence of tooth structure

between these two groups, creates a misleading indication of the relationships.

Pholidota

The earliest undisputed relative of the living pangolins are some astonishing specimens from the Eocene oil shales of Messel in Germany. *Eomanis* (Fig. 7.22(a)) is represented by complete skeletons of a 50 cm long, essentially modern-type pholidotan. Even the overlapping horny scales are preserved. It lacks teeth, the jaw is long and slender, and the limbs powerfully built for digging. There are also gut contents preserved, and they consist mainly of plant material, which is very surprising indeed since the full complement of supposedly ant-eating adaptations are present. Storch and Richter (1992b) ingeniously suggested that the ant-eating habit evolved initially in a lineage of folivores that started to catch leaf-carrying ants with the help of an elongated tongue.

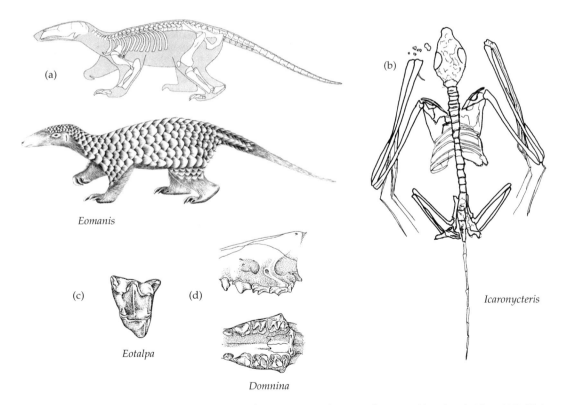

(a)

Eomanis

(b)

Icaronycteris

(c) *Eotalpa*

(d)

Domnina

Figure 7.22 *Early laurasiatherians.* (a) Skeleton and restoration of the Eocene pangolin *Eomanis* from Messel (Storch and Richer 1992). (b) The early Eocene bat *Icaronycteris* (redrawn from Jepsen 1996). (c) Isolated molar tooth of the Eocene mole *Eotalpa* (Savage and Long 1986). (d) Partial skull and dentition of *Domnina*, a Middle Eocene shrew.

Fossil pholidotans occur sporadically through the rest of the Cenozoic in Eurasia and Africa, corresponding to their modern distribution, and there is a single North American genus, *Patriomanis*, dating from the Oligocene. The possibility that the Pholidota are related to the Palaeocene and Eocene palaeanodonts of North America (Rose and Emry 1993) was mentioned earlier in the chapter.

Chiroptera

Early Eocene bats are almost identical in structure to modern forms. The skeletons of *Icaronycteris* (Fig. 7.22(c)) from North America and *Archaeonycteris* from Europe show that full evolution of the flight mechanism had occurred by that time, and the only primitive features they possessed that are absent from modern bats are such details as retention of a claw on the second finger. Isolated teeth attributed to chiropterans have actually been described from a slightly earlier time, the Late Palaeocene of Europe, but unaccompanied by any cranial or postcranial material (Russell *et al.* 1973). Unlike the case of the whales, the chiropteran fossil record consequently reveals absolutely nothing at all concerning intermediate stages in the evolution of their highly specialised locomotion.

Bats also achieved a Gondwanan distribution very early in their history, for they were present in Australia in the Late Palaeocene or Early Eocene Tingamarra Fauna, as the sole certain representatives of placental mammals on that continent prior to the immigration of Pliocene rodents (Hand *et al.* 1994).

The date of the first appearance of Megachiroptera (fruit bats) in the fossil record is unclear. There are isolated teeth of a possible pteropodid from the Late Eocene of Thailand, and a disputed Oligocene form *Archaeopteropus* from Italy (Schutt and Simmons 1998), but otherwise they are unknown prior to the Miocene, when they occurred throughout the Old World.

Simmons and Geisler (1998; Simmons 2000) undertook a morphological cladistic analysis, and concluded that the Eocene bats are basal members of a monophyletic Microchiroptera, containing all the echolocating families of modern bats. The Megachiroptera are their sister-group. This is at odds with molecular evidence suggesting that the Megachiroptera nest within a non-monophyletic Microchiroptera. Teeling *et al.*'s (2000, 2002; Murphy *et al.* 2001*b*) phylogenetic analysis of four nuclear and three mitochondrial genes concluded that the megabats are the sister-group of the rhinolophoid microbats alone amongst the microbats. Apart from reducing Microchiroptera to a paraphyletic group, their conclusion carries the implication that echolocation either evolved more than once in microbats, or, as seems more plausible, was lost in the ancestor of the mainly fruit-eating megabats.

Eulipotyphla

The successive stages in dismemberment of the classic 'Insectivora' were outlined earlier. As a result, finally, of molecular evidence, the remaining living members, referred to as Eulipotyphla, are reduced to the Soricidae (shrews), Erinaceidae (hedgehogs), Talpidae (moles), and Solenodontidae (the extinct solenodons of the West Indies). It is a difficult group to define by unambiguous derived dental, cranial, and skeletal characters (Butler 1988; MacPhee and Novacek 1993). The earliest fossils that can confidently be placed into the modern families occur relatively late. *Litolestes* is an erinaceid from the Late, and possibly earlier Palaeocene of North America; *Domnina* (Fig. 7.22(d)) is a Middle Eocene soricid also from North America; *Eotalpa* (Fig. 7.22(b)) is a Late Eocene European talpid. The solenodontids are unknown prior to *Solendon* itself in the Pleistocene of Cuba.

However, various Late Cretaceous and early Cenozoic fossil genera have been regarded by different authors as erinaceomorphs, soricomorphs, or stem eulipotyphlans. Discussion of relationship to, if not actual membership of the order mainly centres on two primitive, insectivorous groups that appeared in the Late Cretaceous and radiated in the Palaeocene. McKenna *et al.* (1984) proposed that the palaeoryctidans, notably the Late Cretaceous *Batodon*, are primitive soricomorphs, although others rejected this view. Alternatively, Novacek (1986*a*; MacPhee and Novacek 1993) have argued for a relationship between the leptictidans and the Eulipotyphla.

Euarchontoglires

The fossil record has long supported, or at least been seen to be consistent with a relationship between Primates, Dermoptera (colugos), and Scandentia (tree shrews), which formed part of the old supraordinal taxon Archonta. However, the anatomy of neither fossil nor living forms has ever offered support for either excluding the bats, or including the rodents and lagomorphs with them, as is now indicated by molecular evidence.

Primates

The fossil record of primates (Fig. 7.23) is notoriously very poor because they are largely arboreal, forest-dwelling animals, and the early members at least were all small in size (Taveré *et al.* 2002). On the other hand, the degree of effort devoted to finding them and the level of attention given to their interpretation tends to mitigate their sparseness. The Plesiadapiformes (page 232) used to be regarded as the most primitive primates, indicating a very early, immediately post-Cretaceous origin for the order, followed by a radiation through the Palaeocene resulting in at least five plesiadapiform families, distributed in North America and Europe. However, the lack of such fundamental primate characters as a postorbital bar, shortened snout, or opposable hallux and pollex eventually led most authors to exclude Plesiadapiformes from the order, interpreting them instead as basal archontans, with the possible relationship to Dermoptera mentioned later (p272). The issue has re-emerged recently with the description by Block and Boyer (2003*a*,*b*) of a more or less complete skeleton of the carpolestid plesiadapiform *Carpolestes* (Fig. 7.5(b)), showing it does in fact possess opposable, nail-bearing first digits. Their phylogenetic analysis has carpolestids as the sister-group of what they term the Euprimates, and they conclude that plesiadapiforms ought to be included in the order Primates after all. Kirk *et al.* (2003) are unconvinced and believe that the postcranial similarities are convergences in two separate lineages of arboreal euarchontan mammals.

Living primates fall into two informal categories. The 'prosimians' are the primitive primates and include lemurs, galagos, lorises, and tarsiers. They retain a series of ancestral characters, including a small brain, a long snout, and a postorbital bar that is not expanded to enclose the back of the orbit. The incisors are rounded rather than spatulate and

the symphysis of the mandible is unfused. The anthropoids are advanced primates and consist of the monkeys and apes. 'Prosimii' is a paraphyletic group because it is based entirely on ancestral characters; using derived characters demonstrates that some prosimians are basal to the anthropoids. Two monophyletic groups are now recognised. The lemurs, galagos, and lorises are the Strepsirhini, characterised by the presence of a grooming toothcomb formed from the compressed lower incisors and canines. The tarsiers and anthropoids are the Haplorhini, characterised, amongst other features, by loss of the moist rhinarium on the nose and cleft upper lip, loss of the tapetum lucidum, and presence of a haemochorial placenta.

'Prosimian' grade primates first appear in the earliest Eocene, and there followed immediately an Eocene radiation throughout the northern continents, that produced several distinct lineages. The relationships of the Eocene to the modern 'prosimian' groups are far from clear, but most authors accept that both the strepsirhines and the haplorhines were represented, and that the latter had already divided into the tarsier and the anthropoid lineages (Covert 2002). All were relatively small, and arboreal, and variously had insectivorous and frugivorous diets.

Adapiformes are first represented by dentitions and partial skulls of *Cantius* (Fig. 7.23b), from the earliest Eocene of Europe and North America. They had unspecialised incisors and non-shearing molars, suggesting a primarily frugivorous diet. Later forms were generally quite large for Eocene primates. *Notharctus*, for example, had an estimated weight of 7 kg (Gebo 2002) and proportions resembling a lemur. A number of characters suggest that adapiforms are strepsirhines (Kay *et al.* 1997; Ross *et al.* 1998). These include the lemuriform-like structure of the ear region, with a ring-shaped ectotympanic bone within the tympanic cavity, and similarities in the bones of the hand and the foot.

The Omomyoidea are the second group of primates to occur in the Early Eocene of Europe and North America. *Teilhardina* was only about 70 g in estimated body weight, and no omomyoid was larger than a squirrel. Their unspecialised dentition suggests that insects formed an important part of the diet, and at least some members of the family

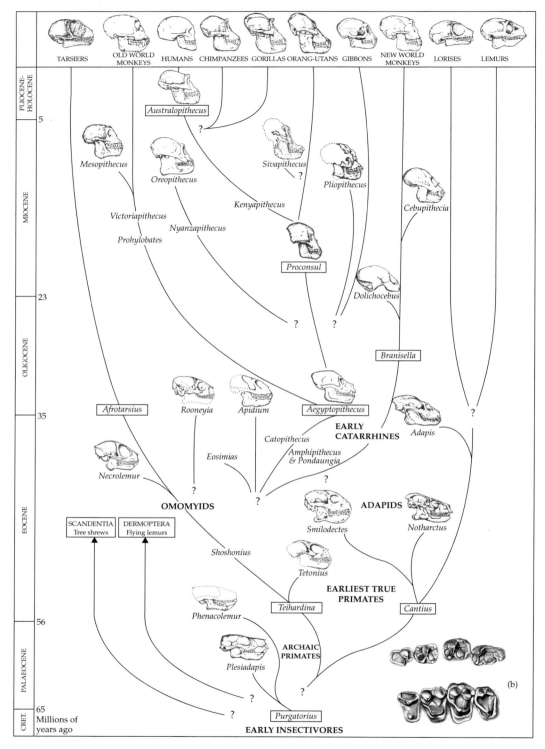

Figure 7.23 (a) Phylogeny of primates (modified after Simons 1992). (b) Occlusal views of the lower and upper cheek teeth of *Cantius* (Gebo 2002)

have postcranial adaptations for arboreal leaping, such as an elongated calcaneum and opposable first digits. Several authors have argued that omomyoids are basal members of the Haplorhini, on the basis of the relatively large orbits, loss of the first premolars, and details of tooth structure and the ear region of the skull (Ross *et al.* 1998; Gunnell and Rose 2002). Some go so far as to relate them to the tarsiers in particular; certainly the short snout, large, forwardly directed eyes, and highly agile skeleton give a very strong impression of a tarsier, as reviewed by Martin (1993).

The modern groups of Strepsirhini have a poor fossil record, even for primates. There is no certain lemuriform at all prior to the Late Pleistocene of Madagascar, with the dubious and controversial exception of the Oligocene *Bugtilemur* from Pakistan (Marrivaux *et al.* 2001) The other strepsirhine branch, the lorisiformes, have been recorded in the late Middle Eocene of the Fayum region of Egypt, but only as isolated teeth: *Karanisia* is a possible loris, and *Saharagalago* a possible galago (Seiffert *et al.* 2003).

For the living haplorhines, the earliest member of the modern tarsiiform family Tarsiidae has been identified from teeth attributed to the modern genus *Tarsius*, found in Middle Eocene deposits of China (Beard *et al.* 1994).

The diversity of the adapiforms, omomyoids, lorisiforms, and tarsiiforms show that the Eocene was a time of considerable radiation of primitive primates, producing a significant fauna of small, arboreal mammals. This observation might have been of only marginal interest had the radiation not also included basal members of the anthropoids, the monkeys, apes, and humans. What at the time was in effect another 'prosimian' grade of Eocene primates appears in the same Middle Eocene deposits of China that yielded *Tarsius* (Beard *et al.* 1996; Dagosto *et al.* 1996). *Eosimias* is known only from dentitions and some probably associated limb bones, and would indeed warrant little attention except for its probable position as the most basal member of the Anthropoidea (Ross *et al.* 1998; Ducrocq 2001). It was a small animal, the lower jaw length being about 2.6 cm. Characters that support its anthropoid affinity are mostly dental as no cranial and very little postcranial material has yet been found. The incisors are reduced in height, spatulate, and vertical

rather than slightly procumbent, and the talonid of the last lower molar is much reduced. Isolated teeth of equally ancient basal anthropoids occur in the Glib region of Algeria (Godinot and Mahboubi 1992), indicating that from the start anthropoids occurred in Africa as well as Asia.

The exact relationship of the Anthropoidea to the other Eocene primates remains a matter of considerable debate, and virtually every possibility has been proposed by one author or another (Martin 1993). Dagosto (2002) notes that hypotheses linking them to, respectively, adapiforms, omomyoids, tarsiids, omomyoids plus tarsiids, or to no known group have all been proposed by one author or another within the last decade or so.

From the Late Eocene onwards, basal anthropoids radiated in Africa and Asia (Beard 2002a). Several genera are found in the Late Eocene and early Oligocene of the Fayum deposits, notably *Apidium*, which is represented by partial skulls (Simons 1995). *Siamopithecus* from Thailand (Ducrocq 1999) is an Asian representative although so far known only from the dentition. The interrelationships amongst these several lineages of Eocene anthropoids are very unclear at present (Simons 1992; Ross *et al.* 1998; Ducrocq 2001), and it is not even established whether the respective taxa from the two continents are sister-groups, or part of a contiguous fauna.

This uncertainty about exact relationships is also true of the two modern groups of anthropoids. The oldest of the New World monkeys, the Platyrrini, is *Branisella*, represented by Late Oligocene teeth and fragmentary jaws from Bolivia. The oldest actual skull so far described is that of *Chilicebus*, which is Early Miocene in age and a member of the existing family Cebidae (Flynn *et al.* 1995). No African or Asian platyrrini have been found, although Simons (1997; Takei *et al.* 2000) noted some dental similarities between *Branisella* and the Fayum *Proteopithecus*. They proposed a relationship between the two which, if true, offers taxonomic support for the view that the origin of the New World monkeys was a dispersal event directly from Africa during the Oligocene. However, not all accept the proposed relationship, so the question is not closed, and an American or Asian origin for Platyrrini remains a possibility (Dagosto 2002).

The other modern anthropoid group are the Catarrhini, the Old World monkeys and the apes. Several primitive members occur in the Late Eocene levels of the Fayum deposits, such as *Catopithecus*, for which there are several skulls preserved (Simons and Rasmussen 1996). It combines derived catarrhine characters of the dentition, including a reduction from three to two premolars and fused symphysis, with primitive features of the skull such as the relatively small brain and elongated snout. *Aegyptopithecus* is the most familiar basal catarrhine, coming from the slightly younger Early Oligocene level in the Fayum sequence. It was the size of a typical monkey, well adapted for arboreal life, and the rounded cusps of the molar teeth indicate a diet of fruit. These basal catarrhines are conveniently grouped as the Propliopithecoidea, despite presumably being a paraphyletic group defined only by primitive characters, and containing the ancestors of the later anthropoid taxa (Rasmussen 2002). As with the platyrrhines, there is no agreement about precisely what the relationship is between the known basal anthropoids and the catarrhines.

There is a considerable gap in the fossil record between the Early Oligocene basal catarrhines such as *Aegyptopithecus*, and the first occurrences of the derived lineages. The Old World monkeys, Cercopithecoidea, are unknown until the Early Miocene of Africa, when the Kenyan *Victoriapithecus* is found. The first of the apes, Hominoidea, may be slightly earlier if the tooth-bearing maxillary fragment referred to as *Kamoyapithecus* (Leakey *et al.* 1995) is correctly identified and correctly dated as Late Oligocene. Otherwise, hominoids too commence their radiation in the Early Miocene of Africa in the form of the well-known genus *Proconsul* (Harrison 2002). By the Middle Miocene, radiations of hominoids were also well under way in Asia and Europe. The third derived catarrhine group are the Pliopithecoidea, an exclusively Middle and Late Miocene group restricted to Europe and Asia. During their brief radiation, a range of size and adaptations comparable to that of the Old World monkeys in Africa evolved (Begun 2002).

The story of the origin and evolution of the Hominidae, the australopithecines and hominines of Africa and their subsequent worldwide dispersal has been told so many times that it does not need repeating here. As far as mammalian radiation is concerned, the hominids were a very minor group of trivial significance, prior to the ecological effect of one species, *Homo sapiens*, since the Late Pleistocene a few thousand years ago.

Dermoptera

The sole living genus of the order Dermoptera is the colugo or flying lemur *Cynocephalus*, although despite this paucity, there is fossil evidence for a very long, if not very diverse history of the group. The dentition is particularly distinctive. There is a gap between the front of the upper tooth rows, and the two pairs of lower incisors have a unique, comb-like construction, consisting of many separate 'tines'. The cheek teeth are wide and have a curiously wrinkled enamel that creates a series of sharp edges after a certain amount of wear. The postcranial skeleton is also uniquely designed for supporting a gliding membrane stretched between all four limbs. On the basis of these characters, a fully evolved fossil dermopteran, *Dermotherium*, is recognisable in the Lower Eocene of Thailand (Ducrocq *et al.* 1992). There are much earlier specimens sometimes regarded as members of the order, namely the plagiomenids that occurred from the Early Palaeocene to Middle Eocene of North America. The lower dentition of a form such as *Plagiomene* certainly resembles that of *Cynocephalus* (Rose and Simons 1977), although others doubt the relationship (MacPhee *et al.* 1989).

Dermoptera have been linked by several authors to the plesiadapiforms, the group of mainly Palaeocene forms sometimes regarded as basal primates. Kay *et al.* (1990) described similarities in the ear region, and Szalay and Lucas (1993) detected homologies in the postcranial skeleton of the two groups. Other authors have, however, disputed the association of any, let alone all the plesiadapiforms with the Dermoptera, regarding the similarities as merely superficial, or primitive for the orders (Wible 1993; Bloch and Silcox 2001).

Scandentia

The tree shrews, at various times confidently classified as Insectivora, and as basal, virtually ancestral primates, are a distinct mammalian order, but one undoubtedly related at a supraordinal level to primates and dermopterans (Luckett 1980). The depauperate modern fauna of but five or six genera is matched by an extremely sparse fossil record. Only the Eocene *Eodendrogale* from Asia predates the one or two Late Miocene genera known, also from Asia and including the modern genus *Tupaia*.

Glires: Rodentia

With over 2000 living species in about 28 families, rodents are the most diverse mammalian order of all. The majority of the families appeared as part of a bush-like explosive radiation during the Late Eocene. Traditionally, the North American ischyomyids of the late Palaeocene have been regarded as the basal rodents. They have the standard rodent dentition of only two gliriform incisors, with a double layer of enamel on the front surface, and fully molariform fourth premolars. *Paramys* (Fig. 7.24(a)) is the classic example, a squirrel-sized animal, with the primitive arrangement of the masseter jaw closing musculature, described as protrogomorphous. The family radiated throughout the Eocene of North America, Eurasia, and Africa, and is probably the stem group of two of the three great super-families, Sciuromorpha and Myomorpha. However, it is not so simple because there is

another primitive rodent group, represented by the Chinese Eocene *Cocomys* (Fig. 7.24(b)), which has certain similarities to the dentition of the third super-family, Hystricomorpha (Flynn *et al.* 1986; Jaeger 1988).

Several other Asian Palaeocene genera have also been considered closely related to, if not actually members of the Rodentia. The Early Palaeocene eurymylids, such as *Heomys* (Fig. 7.4(c)), are one of the families of Mixodonta. They have a single pair of gliriform incisors, and postcanine teeth very similar to those of *Cocomys*, and have therefore been proposed as basal rodents (Li and Ting 1993). The Late Palaeocene *Tribosphenomys* (Meng *et al.* 1994; Meng and Wyss 2001) is even more rodent-like, while McKenna and Meng (2001) described the front part of a skull and dentition of *Sinomylus*, which has a single pair of incisors, but is primitive in retaining the second premolar. They interpret it as basal to both rodents and eurymylids.

One group of rodents arrived in South America, to be first recorded there in the Early Oligocene Tinguiririca Fauna of central Chile (Wyss *et al.* 1993). These are the Hystricognatha, and once there they radiated to fill not only standard rodent niches, but also a number of niches that are occupied by quite different placental groups elsewhere

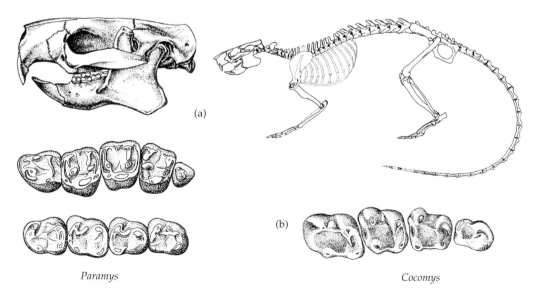

Paramys   *Cocomys*

Figure 7.24 Rodents. (a) *Paramys* skull, skeleton, and upper and lower teeth in occlusal view (Savage and Long 1986, after Wood). (b) Occlusal view of lower teeth of the Chinese Palaeocene *Cocomys* (Carroll 1988 after Dawson *et al.*).

in the world. These include medium to large bodied semiaquatic forms, some of which still exist as capybaras. Sánchez-Villagra *et al.* (2003) described a skeleton of the 700 kg, hippo-sized *Phoberomys* from the Late Miocene of Venezuela, the largest rodent of all. Others evolved relatively long, cursorial limbs, as represented by the living Patagonian hare, for example. The closest living relatives of the hystricognaths are certain African families, Ctenodactyla (gundis) and Anomaluridae (spring hares), all of which constitute a group Hystricomorpha. This suggests that the South American radiation originated with dispersal from West Africa across the Atlantic, some time during the Oligocene.

Glires: Lagomorpha
The second branch of the Glires that includes living members can be recognised by the presence of two pairs of incisors, and the single rather than double layer of enamel that covers them. The mastication mechanism involves lateral rather than strictly backwards and forwards jaw movements. On the basis of these, and other cranial and postcranial characters, the earliest undoubted lagomorphs do not appear in the fossil record until the Middle Eocene of North America. Like the rodents, however, there are possible Palaeocene relatives from Asia, this time in the form of the mimotonid Mixodonta, such as *Mimotona* (Fig. 7.4(b)). In addition to the lagomorph dental formula, there are several features of the skull supporting the relationship (Li and Ting 1993).

Overview of placental evolution

Given the relative completeness of the placental fossil record, coupled with the biological and taxonomic informativeness of many of the individual specimens, theirs is possibly the single most important part of the entire fossil record for illustrating macroevolutionary patterns and inferring evolutionary processes (Kemp 1999). From the pioneering work of Simpson (1944, 1953) through that of numerous subsequent workers such as Philip Gingerich, Elizabeth Vrba, Christine Janis, John Alroy, and so on, many of the concepts currently important in macroevolutionary theory originated

in the study of extinct placental mammals. Adding another dimension, the ever increasing volume of molecular evidence bearing on phylogenetic relationships has variously contradicted, corroborated, or expanded the palaeontological perspective. Combining the two disciplines, the evolutionary story of the placentals, a story of great diversity and disparity set against a background of shifting continents, fluctuating climates, and changing biotas, is becoming ever clearer.

The Cretaceous: origin and radiation

The earliest known stem placental *Eomaia*, and stem marsupial *Sinodelphys* occurred 125 Ma in China. As tabulated by Bromham *et al.* (1999), molecular-based estimates of the time of divergence of placentals from marsupials are extremely variable, from as unbelievably high as 218 Ma to as patently incorrectly low as 104 Ma, and subsequent estimates continue to fall within this range. Cao *et al.* (2000), using mitochondrial DNA, estimated the date as 160 Ma. Springer *et al.* (2003), using a large database of nuclear and mitochondrial, sequences concluded only that the date lies between 102 and 131 Ma, depending on such variables as which constraints are used in the model. Therefore the dating of *Eomaia* is consistent with at least several of the molecular-based estimates, especially this last quoted, which is the most extensive and methodologically sophisticated to date.

It is a different matter for the origin and diversification of the modern placental orders, however. As far as the fossil record is concerned, it reveals the history of the placentals throughout the Late Cretaceous to have consisted of a modest radiation of small, mostly insectivorous forms. The only specialised kinds found either possessed broader, lower-cusped molar teeth for a more omnivorous diet, or sharper crested ones to deal with larger invertebrate, and small vertebrate prey. However, it is possible that what is revealed by the Cretaceous fossil record is a very limited and distorted view of what was really happening, because there is a very considerable inconsistency between the fossil-based and the molecular-based estimates of the dates of divergences of the modern lineages. Not a single fossil of a modern placental order can be shown indisputably to occur prior to the end of the Cretaceous. Even within the Early Palaeocene, between 65 and 60 Ma, there are representatives only of Carnivora, and possibly stem Primates. Otherwise, all of the modern orders made their actual appearance in the fossil record in a narrow window of time, either side of the Palaeocene–Eocene boundary (Archibald and Deutschman 2001). Thus a great stir was caused when Kumar and Hedges (1998) published a comprehensive timescale for the divergences of placental orders based on a molecular clock. From an analysis of 658 nuclear genes, and a calibration of the clock based on the fossil date of divergence of birds from mammals, they estimated that at least five lineages of modern placentals had diverged more than 100 Ma, and most of the orders had differentiated before the end of the Cretaceous at 65 Ma.

All subsequent molecular-based studies, undertaken by several groups of workers using different methods and molecular databases, have supported early divergence times. Waddell et al. (1999), on the basis of mitochondrial DNA data, estimated that divergence times of some orders could have been as early as 150 Ma. In another study, also based on mitochondrial DNA, Cao et al. (2000) used several presumed reliable fossil calibration points as the basis of a clock. They calculated a divergence of Afrotheria from Xenarthra 102 Ma, and the differentiation of the Laurasiatherian orders from one another about 77 Ma. Scally et al. (2001) found the mean estimate for divergence of Afrotheria from the northern groups to be 105 Ma, and that the basal splits within both took place in excess of 85 Ma. Eizirik et al. (2001) estimated possibly somewhat younger dates, finding that Afrotheria diverged from the rest of the placentals in the period 76–102 Ma, and xenarthrans from the northern groups 72–104 Ma. Divergence times of orders within the northern supra-ordinal groups were all in the range of 65–95 Ma. Springer et al. (2003) have conducted the most comprehensive analysis, using a large, diverse data set of nuclear and mitochondrial genes for 42 placentals representing all the living orders (Fig. 7.25). Their results generated divergence dates for the Afrotheria of about 107 Ma, and for Xenarthra of about 102 Ma. The two northern super-orders, Laurasiatheria and Euarchontoglires,

separated about 94 Ma. All the individual orders within these super-orders were differentiated from one another during the Late Cretaceous, between 82 and 77 Ma, apart from the Afrotheria. In the latter case, the Tenrecida separated from the chrysochlorida about 65 Ma, right on the K–T boundary, and the three Paenungulata orders (sirenians, hyraxes, and elephants) diverged from one another about 60 Ma.

This marked and consistent lack of agreement between the pictures based respectively on the fossils and the molecular evidence has led to considerable controversy over attempts at a resolution (Foote et al. 1999; Archibald and Deutschman 2001). There are four possible explanations.

Incompleteness of the fossil record. In order to test this, Foote et al. (1999) created a model based on reasonable expectations of preservation rates of Cretaceous mammals, coupled with the assumption that if they were present, members of the living orders would be morphologically recognisable. The outcome of the study was that if the ordinal-level divergences really had occurred in the Late Cretaceous, then in order to explain the absence of fossils of members of these modern groups, the mean preservation rate would have had to be a whole order of magnitude less than had been assumed likely. This is taken by the authors to be an unrealistically low rate, and therefore they conclude that simple incompleteness of the fossil record cannot be the explanation for the discrepancy between fossils and molecules. Hedges and Kumar (1999) questioned whether the preservation rate assumed in the model to be reasonable should have had such a high value. The early representatives of the living orders might have been very rare animals. However, other studies based on similar kinds of models of the preservation process by Alroy (1999b), and Archibald and Deutschman (2001) have reached the same conclusion, that the surge in origination of ordinal lineages of placentals in the early Cenozoic is a real phenomenon.

Incorrectness of the molecular clock. By supporting the fossil record, these latter studies explicitly conclude that the problem lies with the inappropriateness of assuming a clock-like rate of molecular evolution,

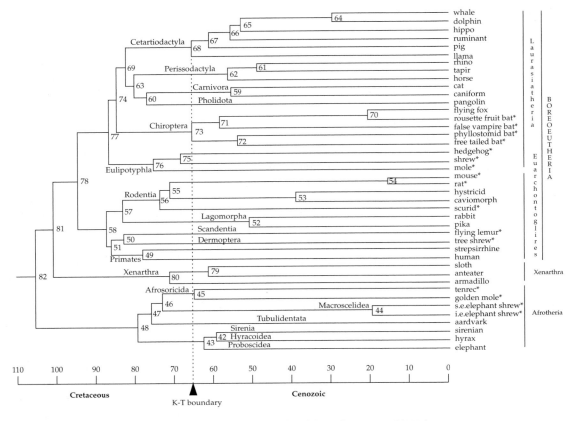

Figure 7.25 Divergence of major supraorders of placentals. Molecular based dates of Springer *et al.* (2003).

which is indeed the palaeontologist's commonest response. Benton (1999), in a typical example of such an argument, suggested that the high rate of speciation and morphological evolution during a phase of rapid lineage splitting, may be associated with an increased rate of selection at the molecular level. This would lead to greater molecular differences between groups, and therefore earlier estimated dates of divergence if it was falsely assumed that molecular evolution had been clock-like.

The most recent molecular studies use methods that not do not depend on a single, simple molecular clock model, for instance the Bayesian approach, used by Springer *et al.* (2003). Their study also compared the divergence dates generated by different categories of the molecular data, namely only nuclear genes, only mitochondrial genes, only exons, or only third codon positions, and found relatively little difference amongst them. It would be

very surprising for all these categories to have exactly the same pattern of variability of evolutionary rate under the influence of varying intensities of molecular-level selection. The authors also calculated that forcing the molecular differences between the groups onto a tree constructed on the fossil-based divergence dates implies extremely variable rates of molecular evolution amongst the orders and subordinal groups. The highest rates, between certain diverging orders, would have to be up to almost 700 times the lowest rates found within certain subordinal groups. This level of rate variation seems far-fetched, even for the most committed of molecular selectionist views.

The Garden of Eden hypothesis. An explanation for the discrepancy between the fossil and molecular dates, which allows that both the fossil and the molecular data are correct, was discussed by Foote *et al.* (1999).

The Late Cretaceous mammal record is extremely sparse, with the sole exceptions of North America and Mongolia, and the Garden of Eden hypothesis is that the origins and divergences of the modern placental orders occurred in a part of the world not at present adequately sampled for Cretaceous mammal fossils. The possibilities are the Gondwana continents of Australia, South America, Antarctica, and Africa, in all of which, there is extremely little in the way of Cretaceous fossil mammals. On this view, the early Cenozoic explosive appearance of the modern orders was due to their immigration into the northern continents of Laurasia.

The Garden of Eden hypothesis does have some arguments in its favour, principally that the molecular dates of the major divergences within placental mammals coincide approximately with the various phases of the break up of Gondwana. Murphy *et al.* (2001*b*) suggested that while the basal placentals (eutherian ancestors) occurred in the northern continents, as indicated by the Early Cretaceous fossil record, the Placentalia originated in Gondwana (Fig. 7.26(a)). The Afrotheria diverged when Africa separated from the rest of Gondwana about 105 Ma (Smith *et al.* 1994). Around 5–10 Ma later, according to the molecular dates, the Boreoeutheria differentiated, presumably as a consequence of their ancestor entering North America from South America. At this time, there may have been several small islands that would have offered a route northwards.

A second possibility is that the boreoeutherians originated in India, which separated from Africa at about the same time as did South America, becoming another island continent. It is not certain when India eventually collided with Asia (Hallam 1994). Most have believed this to be Late Palaeocene (Krause and Maas 1990; Prothero 1994), but other evidence, including geomorphological and palaeontological, suggests that at least faunal interchange may have been possible a good deal earlier, around the end of the Cretaceous (Jaeger *et al.* 1989). This hypothesis is consistent with the difficulty of resolving the base of the supraordinal tree, which is due to the very close dates of divergence between Afrotheria, Xenarthra, and Boreoeutheria.

The only palaeontological support explicitly claimed for the Garden of Eden hypothesis comes from Rich *et al.* (1999; Woodburne *et al.* 2003), who believe that the Early Cretaceous, 100 Ma Australian mammal jaw *Ausktribosphenos* is an erinaceid-like member of the Eulipotyphla, and therefore marks the area of origin of placentals. Only more material will confirm or refute this presently highly disputed identification.

The Long Fuse hypothesis. The fourth possible explanation for the incongruent dates has been proposed most actively by Archibald and Deutschman (2001). The Long Fuse hypothesis is that most of the Cenozoic orders of placentals did diverge in the Late Cretaceous; that they are in fact represented by known fossils from that time; but that these have been unrecognised because they lack the diagnostic characters of the modern representatives. Rather, they remained small, unspecialised mammals with a basically insectivorous, or at most omnivorous dentition. On this view, the post-Cretaceous explosion of recognisable members of the modern orders was actually their evolution from existing, primitive, stem members. This, the hypothesis implies, occurred in many orders during a small window of time, and was possibly in response to the extinction of the dinosaurs and therefore the release of the habitats and niches from which mammals had hitherto been competitively excluded. In this light, there are several candidates for Cretaceous stem-membership of modern orders. The Leptictida have often been included with the Eulipotyphla (Novacek 1986*a*), and the Zalambdalestidae interpreted as stem Glires (Archibald *et al.* 2001). The Zhelestidae possess incipiently ungulate-like molar teeth and could conceivably include stem members of any of the numerous Cenozoic ungulate groups (Archibald 1999). The Palaeoryctida have from time to time been allied to the Carnivora, and also some of the extinct Cenozoic orders such as Creodonta, Taeniodonta, Tillodonta, and Pantodonta. The evidence for all of these putative relationships is certainly not at all strong, often being based almost completely on minor dental similarities, although of course this in itself is a prediction of the very hypothesis.

Springer *et al.*'s (2003) molecular study already referred to (Fig. 7.25) is consistent with the Long Fuse hypothesis, insofar as it places most of the

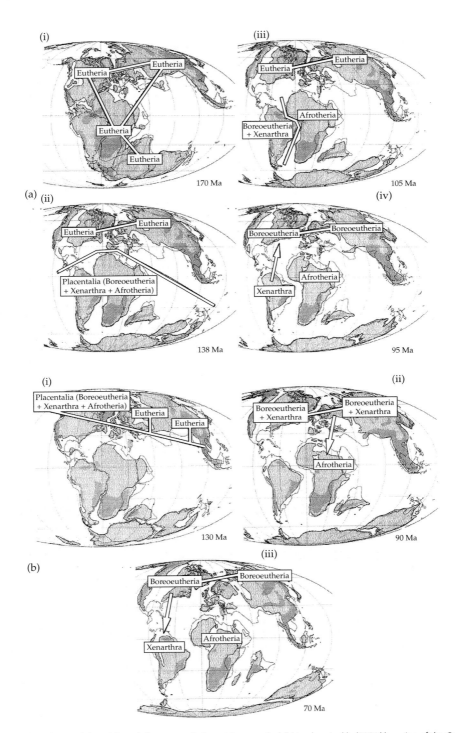

Figure 7.26 Two hypotheses of the origin and divergence of placental mammals. (a) Murphy *et al.'s* (2001*b*) version of the Garden Eden hypothesis. (b) Archibald's version of the long-fuse hypothesis. (Archibald 2003).

interordinal divergences in the Late Cretaceous, but most of the intraordinal splits at or after the end of the Cretaceous. Only Xenarthra, Eulipotyphla, Rodentia, and Primates have any intraordinal divergences between their constituent groups dated to the Late Cretaceous, these varying between 77 and 70 Ma. Intraordinal divergences within Cetartiodactyla and Chiroptera commenced at 65 Ma, the time of the K–T boundary, and those within Carnivora, Perissodactyla, and Lagomorpha 10–15 Ma after that event.

Another argument in favour of what amounts to a decoupling of lineage splitting from the evolution of modern diagnostic characters is that, even to this day, three of the now recognised super-orders, Afrotheria, Laurasiatheria, and Euarchontoglires include small, insectivorous members, which presumably have retained the general ancestral structure of the respective stem-members. The very fact that at one time tenrecs, shrews, and tree shrews, were all included in a single order Insectivora attests to the likelihood that the Cretaceous members of the superorders would not expect to be readily distinguished at a high taxonomic level.

Archibald (2003) has related the long fuse model to the molecular dates (Fig. 7.26(b)). He agrees that the placental stem group of fossil eutherians was a Laurasian radiation, but furthermore that the origin of the crown Placentalia also occurred in the north. The existence of the Afrotheria and the Xenarthra is explained by dispersals to those two continents respectively, in the Late Cretaceous. The argument is supported to the extent that relatives of the Afrotheria may occur in Asia. If 'condylarths' such as *Minchinella*, or anthracobunids such as *Pilgrimella* are indeed related to afrotherian orders, then the origin of the Afrotheria may well be sought in Asia. However, these phylogenetic associations are far from agreed upon. Furthermore, there are no serious suggestions as yet concerning the identity of possible stem Xenarthans in the northern continents.

The Palaeocene: explosive radiation

Irrespective of whether the majority of placental orders arose in the Cretaceous, or not until the early Cenozoic, there was certainly an explosive

appearance of new mammalian taxa right from the start of the Palaeocene. Once the effects of the presumed end-Cretaceous bolide impact had waned, warm, greenhouse conditions pertained, with gentle climatic gradients (Prothero 1997). Under these circumstances, and in the absence of dinosaurs, a mammalian fauna that was essentially modern in ecological structure though not taxonomic constitution was rapidly established. Alroy (1999*b*) plotted the species numbers for successive time bins, and after allowing for differing sample sizes, found an increase from about 23 in the latest Cretaceous to about 60 a mere 5 million years later. Equally significant, while the majority of the new kinds of Early Palaeocene mammals were still small-bodied insectivores and omnivores, there were several new groups with medium-sized (5–40 kg) and large (>40 kg) members. The most abundant were the 'condylarths' of various kinds, which, in addition to primitive ungulate herbivores, included the first ecologically carnivorous placental mammals in the form of actocyonids and mesonychids. Other new placental orders were the specialised herbivorous taeniodonts, tillodonts and pantodonts, all of which were possibly derived from Cretaceous palaeorycti-dans. The arboreal plesiadapiforms, possibly stem primates, and the earliest Carnivora were mainly small, but included some cat-sized animals.

By far the best Early Palaeocene (65–60 Ma) mammal record is from North America, followed by that of China, but it is not clear exactly how the new taxa related to one another biogeographically. Some, such as the taeniodonts and the plesiadapiforms, are assumed to have evolved in North America because of their absence elsewhere other than, later, Europe. Others were exclusively Asian, such as mixodonts and anagalidans. However, some Early Palaeocene taxa occurred in both continents, but with inconsistent dates. The Chinese pantodonts are slightly earlier and more primitive than the American ones, but exactly the reverse is true of the Carnivora.

The East of Eden hypothesis (Beard 1998) is that the centre of origin and early radiation of placental orders was the Asian landmass, and that on successive occasions, throughout the Palaeocene and Early Eocene, climatic conditions permitted the northwards expansion of mammal faunas. At these times, taxa crossed the temporarily dry Bering Straits to enter North America from its northwestern corner. While the evidence is too sparse and ambiguous to support the East of Eden hypothesis for the first half of the Palaeocene, there is strong evidence that a succession of at least three, closely spaced waves of immigration by this route subsequently took place, leading to the spread of several new placental groups by earliest Eocene times. Between about 57 and 54 Ma, a period spanning the Palaeocene–Eocene boundary, several taxa known from earlier in the Palaeocene of Asia first occur in the North American fossil record (Beard 2002b; Bowen et al. 2002). These included the dinoceratans, arctostylopidans, tillodonts, and hyaenodontid creodonts amongst the archaic groups. It also includes members of the modern orders Perissodactyla, Artiodactyla, Rodentia, and Primates. The palaeoclimatic evidence is also compatible with the East of Eden hypothesis. A brief, intense period of global warming occurred at this time. It is indicated by a shift in $^{13}C/^{12}C$ ratio that is possibly associated with the release by volcanic activity of methane and CO_2, causing a greenhouse warming effect (Pearson and Palmer 2000). According to Kent et al. (2003), the event may have been triggered by a cometary impact. The average world temperature reached its maximum Cenozoic level, and the associated climate would be expected to have allowed the expansion of mammalian faunas into higher northern latitudes, and to open up a dispersal route from the Asian to the North American continents via the temporarily dry Bering Straits. The East of Eden hypothesis also predicts that there would have been at least some dispersal in the opposite direction, from North America to Asia. There are, in fact, no clear examples of this except possibly the Carnivora, and nor is obvious why it was in Asia rather than elsewhere that new groups were originating in the first place. Beard (1998) suggested that both phenomena might be related simply to the much larger area of the Asian landmass compared to the North American.

The Palaeocene mammalian fauna of Europe is sufficiently similar to that of contemporary North America that they are presumed to have been interconnected through a northern Atlantic route. With the onset of the warm conditions of the Palaeocene,

dispersal through what is now northern Canada and Greenland is perfectly plausible, and there is neither systematic nor palaeobiogeographical evidence that any Asian taxa arrived directly from the east.

Of the Palaeocene mammals in the southern continents, vastly less is known. Early and Middle Palaeocene mammals are virtually unknown in Africa, and therefore there is no palaeontological confirmation of the taxonomic inference discussed earlier, that the Afrotheria were diversifying at that time, in isolation of the rest of the world. A little more is revealed about the early history of placentals in South America. Pantodonts and 'condylarths' were present in the Early Palaeocene, and presumably had arrived from the north along with the marsupials, at or soon after the end of the Cretaceous. Xenarthran fossils are absent prior to the Late Palaeocene, when they occur alongside the first of the South American ungulates. The latter, Meridiungulata, may have originated monophyletically from an immigrant North American 'condylarth'. However, it is also possible that some of them evolved from an immigrant dinoceratan that also entered during the Palaeocene.

The Indian subcontinent is another Gondwanan landmass that may, for all that is known, have had an important role in the story. During most of the Cretaceous it was an isolated plate, drifting northwards from Africa, and eventually colliding with the Asian continent. The earliest time at which mammals might have been able to disperse from India to the main Asian landmass is unclear. It may have been around the Palaeocene–Eocene boundary, or possibly significantly earlier, even close to the Cretaceous–Palaeocene boundary. Any placental groups that are first recorded in the Asian Palaeocene could conceivably have originated in isolation in India, and, as mentioned above, this could have included either or both the Laurasiatheria and Euarchontoglires. Unfortunately, there are as yet no suitable Palaeocene fossils from the Indian subcontinent to test this idea.

The Eocene: flourishing

The very warm conditions arising at the end of the Palaeocene continued into the Eocene, and are associated with the time of maximum placental diversity on Earth (Janis 1993). Fully tropical rain forests extended as far as latitude 30°, and subtropical, angiosperm dominated forests to 60°. At even higher latitudes there were forests containing broadleaf deciduous trees, such as that revealed by the rich fossil localities of Ellesmere Island at 75° north, and the cool-temperate forests preserved on Seymour Island on the Antarctic Peninsula (Wing and Sues 1992; Prothero 1994). In the context of such favourable environmental conditions of warm temperatures, very high primary productivity, and a large habitable area, the existing mammalian groups radiated extensively, and a number of new taxa appeared.

The fauna of North America, Europe, and Asia was quite similar, although with distinct endemism at lower taxonomic levels, indicating that interdispersal between these northern areas during the Early and mid-Eocene was relatively free (Ting 1998; Agustí and Antón 2002). Land connections existed between northeastern America and Europe, one via Greenland and Norway, and a second further south, via Greenland, Iceland, the Faroes, and Scotland. The dry Bering Strait connected North America with Asia (Prothero 1994). In this single, extensive landmass the archaic Palaeocene groups continued to radiate. Tillodonts and large pantodonts, and various of the 'condylarth' lineages flourished. The creodonts, particularly the hyaenodontids, were the prominent carnivorous mammals, for the Carnivora were still relatively small, though abundant animals.

The newly introduced modern orders were also rapidly diversifying. Perissodactyls in particular soon became an important group of medium to large-sized herbivores, perhaps because their hindgut fermentation system was better adapted for the high fibre content of many Eocene forest plants (Janis 1989). They diversified into the early representatives of the modern families of horses, rhinos, and tapirs, and also included bizarre, large-bodied, browsing groups, the chalicotheres and especially at this time the brontotheres. Forest adapted lemuriform primates were another very diverse and abundant group. On the other hand, the rodents and artiodactyls, two orders destined eventually to become amongst the most diverse of all placental groups, remained relatively modest during the

Eocene. This was also the time of appearance of two radically new kinds of placental body plan. Bats are recorded from the Early Eocene and rapidly gained a worldwide distribution including, uniquely for placentals, dispersal to Australia. Amphibious members of both cetaceans and sirenians appeared in the Early Eocene, and by the Middle to Late Eocene permanently aquatic versions had evolved.

Far less is known about Eocene mammals in the southern continents, although from what the fossil record does reveal, it was quite as rich as in Laurasia. At present, Early and Middle Eocene mammal fossils of Africa include isolated teeth of creodonts, 'condylarths', and basal primates, which presumably entered the continent from Eurasia by an unknown route at an unknown time, and the undoubtedly indigenous afrotherian order Proboscidea. It is not until the Late Eocene and the fossil fauna of the Jebel Qatrini Formation of Fayum in Egypt that a clear African picture emerges. The herbivore fauna was dominated by the afrotherian orders Proboscidea and Hyracoidea, and the embrithopod *Arsinoitherium*, a fauna no doubt culminating from the radiation in isolation of these orders throughout the Eocene and possibly earlier. The few non-afrotherian placentals included the continued presence of creodonts and primates, amongst the latter of which were early anthropoids inexplicably contemporaneous with those of Asia. Marine mammals found here, Cetacea, and Sirenia, are palaeobiogeographically easier to account for.

In South America, the xenungulates and meridiungulate groups radiated throughout the Eocene. One of the more unexpected features of meridiungulate evolution is the presence of high crowned, hypsodont teeth in Mid-Eocene notoungulates (MacFadden 1997). These kinds of teeth are associated with grazing, suggesting that grasslands had started to spread in South America 10–15 Ma earlier than in the northern continents. The carnivorous mammals of South America were represented not by placentals, but by the borhyaenid marsupials.

Antarctica was still attached to South America at this time, and to judge from the presence of meridiungulates and xenarthrans from the mid-Eocene of Seymour Island, off the Antarctic Peninsula, the two landmasses had a contiguous placental fauna, differing only at a low taxonomic level (Reguero *et al.* 2002). However, the point has been made elsewhere that absolutely nothing at all is known about any possible Eocene mammals in eastern Antarctica, the side of the Transantarctic Range closest to Australia, and the virtual absence of placental mammals in Australia itself, apart from bats.

The Eocene–Oligocene: transition and *La Grande Coupure*

By the middle of the Eocene, 50 Ma, placental mammal diversity was at one of its two peaks. Shortly thereafter, however, a general, worldwide cooling period commenced which, with occasional interruptions and brief reversals, lasted for the rest of the Cenozoic to the present day, and which marked the start of the replacement of the archaic placental fauna by the essentially modern one (Janis 1993; Prothero 1994). The fall in mean temperature was not regular, but came as three pulses. The first cooling period occurred about 50 Ma, when average terrestrial temperatures fell by 7–11 °C. This was followed by a rebound almost back to what it had been, before a second drop around 40 Ma. Again there was a brief rebound, the Late Eocene warming phase, although the temperature never regained its former value. Finally, soon after the Eocene–Oligocene boundary, 33 Ma, the largest fall of the sequence occurred, during which average terrestrial temperatures fell by 8–12 °C within the space of a million years. This last episode coincided with the final loss of contact between South America and Antarctica, which affected patterns of oceanic currents, including the establishment of the circum-polar current that isolated Antarctica from the warming influence of the oceans, and was at least partly the cause of the onset of the present-day Antarctic glaciation. The principal cause of the cooling was probably a reduction in the level of CO_2 in the atmosphere, reducing the greenhouse effect that had characterised the first half of the Eocene (Pearson and Palmer 2000). The cause of the CO_2 reduction was complex and involved the balance between input from volcanic out-gassing on the one hand, and sequestration and release rates of organic carbon on the other (DeConto and Pollard 2003).

In North America, conditions had become drier and more seasonal as well as generally cooler, as indicated by the profound effect on the vegetation (Wolfe 1978, 1992). The moist tropical and semi-tropical forests became much more restricted, and in the higher latitudes they were replaced by deciduous, temperate forests. The associated dramatic changes in the mammalian fauna occurred in a stepwise fashion during the Late Eocene and earliest Oligocene (Janis 1993; Prothero 1994). At the ordinal level, it was the archaic taxa that had been in existence since the Palaeocene that were most affected. The last of the plesiadapiforms, 'condylarths', dinoceratans, tillodonts, and taeniodonts disappeared, and other groups such as haplorhine primates survived but with drastically reduced diversity. Along with these, many of the Eocene groups of the modern orders also went extinct or severely declined, such as the brontotheriid and ceratomorph perissodactyls, and the protoceratid and oromerycid artiodactyls. However, the fauna was enriched over the same period of time by the appearance of many new families of modern kinds of mammals, presumed to have entered from Asia. There were rodents such as beavers, squirrels, and cricetids; rabbits; modern-type carnivores including bears, dogs, and nimravids; and amongst the large herbivores, members of the pig, camel, and deer families of artiodactyls, and the rhinocerotid perissodactyls. The outcome by the close of the Oligocene was a fauna of distinctly modern aspect.

The Late Eocene–Oligocene history of mammals in Europe was different to that of North America and was marked by an even more profound faunal interchange, referred to by Hans Stehlin as *La Grande Coupure*—the great cut (Prothero 1994; Agustí and Antón 2002). With the high sea-level characteristic of the Middle and Late Eocene, Western Europe had become an archipelago in a warm, tropical sea. It had become isolated from North America by the complete opening of the Atlantic Ocean, and from Asia by the Turgai Straits that connected the Arctic Ocean west of Siberia to the Tethys Sea. An endemic fauna had evolved, including families of rodents, perissodactyls, artiodactyls, and haplorhine primates unique, or almost unique to Europe. At first, the effect of the Late Eocene temperature fall seems to have been less extreme because of the ameliorating effect of the surrounding sea. Tropical rain forest with its rich fauna of arboreal and browsing mammals persisted. However, by the Early Oligocene, 30 Ma, the fall in sea-level had caused the coalescence of the western European islands, and even more significantly had caused the drying up of the Turgai Straits, thereby establishing a land connection with Asia. The eventual effect was one of the largest turnovers of mammal faunas of all. Around 60% of the Late Eocene species went extinct, including tropical tree dwellers like the haplorhine primates, some of the uniquely European rodents such as the forest browsing theridomyids, and the browsing palaeothere perissodactyls. The archaic insectivorous leptictidans disappeared, as did most of the hyaenodontid creodonts. In their place came a great wave of immigrants from Asia, most of them directly, others possibly via North America. Notable amongst the new arrivals were families of Carnivora such as bears, weasels, cats, and the cat-like nimravids; the perissodactyl families of rhinos, indricotheriids, and chalicotheres; and the artiodactyl anthracotheriids and ruminants. New families of Asian rodents including squirrels and cricetids amongst several others, and the lagomorph pikas arrived.

Clearly a key to understanding the events of the Eocene–Oligocene transition lies in Asia, but the picture here is nowhere near as well-known as in North America and Europe. Certainly there was a faunal change from archaic to modern groups, much as in North America (Prothero 1994; Lucas 2001), and by the Early Oligocene, the fauna was dominated by modern eulipotyphlan, rodent, lagomorph, perissodactyl, artiodactyl, and carnivoran families as in the rest of the northern hemisphere. Relative dating is difficult, however, and while there is little doubt that these new taxa occurred in Asia before Europe, it is less certain that this is the case with respect to North America.

The Late Eocene and earliest Oligocene of Africa is known adequately only from the Fayum deposits of Egypt (Rasmussen *et al.* 1992). This part of Africa was in the tropical belt along the southern shore of the Tethys Sea, and appears to have been relatively immune from the effects of the cooling episodes of the time elsewhere. A fauna unique to Africa, whose presence is hinted at by the few earlier Eocene fossils of North Africa, persisted (Maglio

1978). It was dominated by the hyracoids, proboscideans, and embrithopods that were part of the afrotherian radiation, and African taxa of hyaenodontid creodonts, anthracotheriid artiodactyls, rodents, and primates, which had entered the continent earlier, by an unknown route from the north. This fauna remained unchanged until at least the Oligocene–Miocene boundary, to judge from the Chilga fauna of Ethiopia (Kappelman *et al.* 2003).

The Eocene–Oligocene transition is represented in South America by the Tinguiririan beds of Chile (Wyss *et al.* 1994; Flynn and Wyss 1998), in which a number of archaic groups disappeared, and more modern versions of the South American indigenous meridiungulates appeared. It was around this time that the New World rodents arrived, probably by rafting from Africa across what was still a relatively narrow Atlantic Ocean (Wyss *et al.* 1993). The platyrrhine anthropoids are not recorded until the Late Oligocene when they too arrived, probably also from Africa.

Finally, and needless to say, the flourishing Eocene flora and fauna of Antarctica disappeared permanently and without trace with the formation of the ice-cap.

The Miocene: second flourishing and second decline

The Early Miocene was a time of temporarily increased global temperature and relative dryness. The climate change was accompanied by the expansion once again of areas of tropical and subtropical forest, while sea level falls and tectonic events resulted in the opening up of several land connections. The consequence was a period during which several major groups of placentals spread and diversified (Potts and Behrensmeyer 1992). The most important dispersal event was the immigration into the hitherto largely isolated Africa of many groups of Eurasian mammals which had never occurred there before, but which were to be the basis of the development of the rich modern African fauna. According to Maglio (1978), 29 new families of mammals entered Africa at this time, compared to a mere 14 families remaining from the Oligocene. Amongst the most prominent arrivals were rhinocerotid and chalicothere Perissodactyla, and giraffid, bovid, suid, and tragulid families of Artiodactyla. These various ungulates were accompanied by the main families of the Carnivora, felids, canids, viverrids, and later mustelids. Several families of Eurasian rodents and Eulipotyphla were introduced. A number of the African mammals made the reverse migration, northwards from Africa, and of these the most prominent were proboscideans, with representatives of the gomphotheres, mastodonts, and deinotheres arriving in Eurasia in what has been termed the 'proboscidean event' of about 20 Ma (Agustí 1999; Rögl 1999). They were accompanied by African lineages of anthracotheriid artiodactyls, and members of the last group of surviving hyaenodontid creodonts, which still flourished in Africa. By mid-Miocene, other groups of African origin had also spread into Eurasia, including the pliopithecid anthropoids, dryopithecid apes, and hyracoids.

Relatively free dispersal across the Bering Straits between Asia and North America was also possible during the Early and Middle Miocene, allowing the introduction of, notably, proboscideans, felid carnivores, and antilocaprid deer into North America.

The warming phase reached its peak in mid-Miocene, about 15 Ma, which coincides with the second peak of Cenozoic mammal diversity. As at the time of the Eocene peak, the rise in temperature is believed to have resulted from an increase in CO_2 levels (Pearson and Palmer 2000). The associated rise in plant productivity would have created the opportunity for diversification of herbivorous mammals and their predators. In North America the diversity of ungulates, dominated by tayassuid, camelid and antilocaprid artiodactyls, equids, and proboscideans increased from about 30 genera at the commencement of the Miocene to a peak of about 65 at the mid-Miocene temperature maximum. Most of the increase was in mammals with cheek teeth that were intermediate between the low crowned version of pure browsers and the high crowned, hypsodont version found in grazers (Janis *et al.* 2000, 2002). In Africa, the Miocene was marked by the radiation of the newly introduced bovids, suids, and giraffids amongst the artiodactyls, and rhinocerotid perissodactyls.

From this time on, temperatures started to fall, and dry, more seasonal conditions began to prevail.

During the second half of the Miocene, the tropical and subtropical forests were increasingly displaced by the spread of more open woodland, and eventually grassland savannas. The mammal diversity fell, and there was an especially large drop in browsing ungulates, and forest adapted forms. Close to the end of the Miocene, about 7 Ma, there was a dramatic acceleration of this trend, and it was associated with a floral revolution. The photosynthetic pathway of most modern grasses is C_4 metabolism, which is more efficient under conditions of low CO_2 levels coupled with high temperatures. The proportion of C_4 plants in a fossil animal's diet can be estimated from the $^{13}C/^{12}C$ ratio preserved in its bones, and Cerling *et al.* (1997, 1998) revealed that there was a worldwide increase in the utilisation of C_4 plants by mammals about 7 Ma. It coincides with a worldwide faunal change in which hypsodont-toothed mammals became far and away the dominant ungulates: equids and proboscideans in North America; antelopes, hippos, and giraffids in Africa; hypsodont deer, hippos, and giraffids in Eurasia; several kinds of notoungulates and rodents in South America.

This inferred spread of temperate grasslands was part of the final important episode of the Miocene, known as the Messinian Crisis. After a sequence of relatively minor glaciations and intervening warmer times, the end of the Miocene was marked by a much more substantial arctic glaciation. As well as the direct effect on mammalian faunas, the associated fall in sea level combined with the continuing northward movement of Africa to cause the closing off of the Mediterranean Sea from the Atlantic Ocean. Evaporation followed creating a huge, Dead Sea-like hypersaline lake surrounded by highly arid, desert conditions. It created a land bridge between North Africa and the Iberian Peninsula, across which passed a range of taxa. They included the hippos, camels, cercopithecoid monkeys, and gerbils entering into Europe (Agustí 1999).

The Plio-Pleistocene: exchanges and extinctions

For the first 2 million years of the Pliocene, after the Messinian Crisis, the world underwent another brief warm phase in which tropical and subtropical forests expanded once more and mammal diversity was high. It lasted from 5 to 3 Ma and was followed by the extended cool period marked by the cycle of Plio-Pleistocene ice ages and intervening interglacials that continues at the present time.

Amongst many minor extinctions, radiations, and dispersal events recorded in the Plio-Pleistocene, there was one major episode, referred to as the Great American Biotic Interchange, which concerns the exchange of faunas following the connection of South America and North America by the Isthmus of Panama (Simpson 1980; Stehli and Webb 1985). Prior to this event, the isolated South American mammalian fauna was completely different from anywhere else. All the carnivores, large and small, were marsupials. The large grazing and browsing herbivores were the archaic placental meridiungulates, of which only the litopterns and notoungulates still existed at the time of the interchange. The Xenarthra occupied the specialist roles of ant-eating on the part of armadillos and anteaters, and leaf-browsing by the glyptodonts, ground sloths, and tree sloths. Apart from the inevitable bats, two other groups of placentals had arrived in South America from elsewhere, radiated, and became a fully integrated part of the indigenous fauna. As described earlier, hystricognath rodents are first recorded in the Early Oligocene about 31 Ma, and platyrrhine primates 25 Ma in the Late Oligocene. Both are believed to have come from West Africa by means of rafting across the then much narrower Atlantic Ocean.

Even before the formation of the Isthmus, a handful of mammals had dispersed between North and South America. Two kinds of ground sloths appear in the fossil record of Florida about 8 Ma, and at about the same time a procyonid raccoon occurred in Argentina. They are presumed to have made the journey by island-hopping across the Caribbean Islands. The next record of an exchange is the occurrence of cricetid rodents in South America around 3 Ma, which was only shortly before the formation of the land bridge. The bridge was finally completed 2.5 Ma as a consequence of tectonic changes, and the first actual wave of immigrants crossed in either direction. Going northwards were more xenarthrans, consisting of additional species of ground sloths, armadillos, and the glyptodont *Glyptotherium*. They were accompanied by capybaran rodents and also one of the

flightless, predatory phorusrhacid birds. A second wave shortly afterwards, about 1.9 Ma, saw a second contingent of xenarthrans including a giant anteater *Myrmecophaga*, a rhinoceros-like notoungulate *Mixotoxodon* and the first didelphid marsupial enter southern USA and Mexico. There are also four genera of New World monkeys living in the forests of Central America today, although no fossil record of exactly when they arrived there.

The taxonomic traffic over this period in the opposite, southerly direction was considerably more extensive. At the time of completion of the landbridge, 2.5 Ma, there is evidence for a mustelid skunk, a tayassuid peccary, and the horse genus *Hippidion* in South America and these were shortly afterwards joined by a procession of dogs, cats, bears, more horses, gomphothere elephants, tapirs, camels, deer, shrews, and several rodent groups.

The subsequent history of the respective immigrants was just as asymmetrical as the initial invasions. In South America, the northern immigrants had a higher speciation rate and a lower extinction rate than the native taxa, leading eventually to the present day situation where 44% of the families and 54% of the genera of mammals are of North American origin, either directly or by evolution within South America after the interchange (Marshall 1988). In fact, only two of the immigrant families have become extinct, the horses and the gomphothere elephants. In contrast, many of the indigenous South American groups have disappeared, including the large carnivorous borhyaenid marsupials, and both the meridiungulate groups Litopterna and Notoungulata. The invasion by southern mammals into North America had vastly less ultimate consequence. The ground sloths, glyptodonts, and toxodonts disappeared. Of the rest of the immigrants, only three genera of anteaters, two of tree sloths, and two of armadillos still occur, in the company of the four monkeys mentioned, a few rodents, and about half a dozen genera of didelphid marsupials. Even then, they are largely restricted to the tropical forests of Central America, only a few extending even as far as southern Mexico and Florida. The porcupine *Erethizon dorsatum*, and the Virginia opossum *Didelphis virginiana* alone have spread deep into North America, and the latter was largely due to human intervention.

There have been a number of attempts to explain the details of the Great American Biotic Interchange, traditionally by invoking competition: supposedly inferior South American mammals succumbing to the competitive onslaught of superior North American forms (Simpson 1980). However, there are serious anomalies to such a simple explanation. For one thing, the extinctions of the South American taxa do not always coincide with the entry of their supposed North American competitors. The borhyaenids, which occupied the large carnivore niches, had disappeared prior to the Pliocene, with the single exception of the marsupial sabre-tooths *Thylacosmilus*, leaving the phorusrhacid birds as the most abundant top carnivores. The meridiungulates, too, had already undergone a severe decline in diversity before the interchange. Of the two orders remaining by this time, litopterns were represented by *Macrauchenia* and a few related genera, and Notoungulata by the one family Toxodontidae. In this case, both groups survived until well after the interchange and are recorded in the Late Pleistocene. Furthermore, a close relative of *Toxodon* was one of the forms that entered North America during the exchange. A second complication to a simple competitive explanation is that the interchange was associated with considerable environmental changes.

All more recent authors, such as Webb (1985*a*, 1991), Marshall (1988), and Vrba (1992, 1993) have appreciated that the actual pattern of immigrations and extinctions must have resulted from a complex interplay of environmental changes, ecological opportunism, and biotic interactions, and have pointed to the effect that the alternation between interglacial and glacial episodes might have had. Webb (1991) proposed a two-phase model (Fig. 7.27). The first phase occurred when the land bridge formed, 2.5 Ma. It was during an interglacial period of warm, humid conditions, and continuous dense rain-forest covered much of South America, the Isthmus, and southernmost North America, much as is the case today. Mammals that were adapted to rain-forest conditions could readily disperse from the much larger area of South American rain-forest into the far smaller area of rain-forest that occupied the low latitudes of North America, where most of the survivors are still to be found.

The second phase of the model commenced with the onset of an arctic glaciation 2.4 Ma. The tropical regions became cooler and drier, and the forests were largely transformed into savannah and light, dry woodland. Conditions now favoured expansion southwards of mammals from the very large area of subtropical and temperate North America that were already adapted to this kind of habitat. Once into South America by this route, they could extend their range to higher latitudes as the next warmer and wetter, interglacial phase returned,

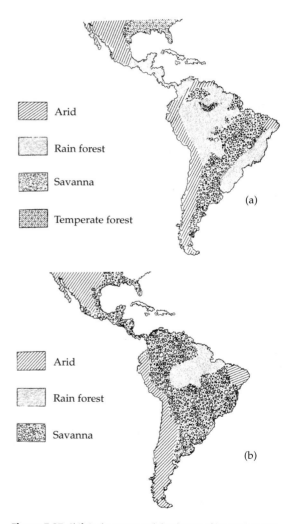

Arid

Rain forest

Savanna

Temperate forest

(a)

Arid

Rain forest

Savanna

(b)

Figure 7.27 Shift in the extent and distribution of biomes between interglacial and glacial phases across Central and South America. (a) Interglacial phase. (b) Glacial phase (Webb 1991).

and thus remain in their preferred habitat. However, no such option of extending their geographical range was available to the rain-forest fauna of Central America, which merely experienced a decline in the area of their available habitat.

The Great American Biotic Interchange was unique and dramatic because it brought together two biotas that had been virtually completely isolated from one another for the whole of the previous 60–70 Ma. There were nevertheless significant biotic changes in other parts of the world. The severe glaciation 2.4 Ma, shortly after the establishment of the Panamanian land bridge, was the start of the particular cycle of ice ages and interglacial periods that is still the dominant feature of the Earth's climate. In Europe, approximately 17 such ice ages are recorded in the 1.7 Ma of the Pleistocene. Typically, an ice age is associated with a reduction in the snowline by about 1000 m and a fall in average global temperature of about 6 °C. Sea-levels fall by between 100 and 150 m as water is locked up in the ice caps. The effects on the environment are worldwide, but differ from place to place. In the higher latitudes of the Holarctic continents, North America and Eurasia, there is an oscillation between temperate forest and steppe. In lower latitudes warm, moist canopied forest alternates with drier, more open woodland and grassland conditions, and it was indeed during the Pleistocene that the fully modern version of savannah, prairie, steppe, and pampas developed. Superimposed on these general changes were particular events affecting the environment. The uplifting of the Himalayas and associated formation of the Tibetan Plateau had a profound effect both directly by the increase in the area of habitable upland, and indirectly by modifying the pattern of the Monsoon winds affecting the Asian continent. Further north, the Bering Strait dried up from time to time, allowing free dispersal of tundra and cool forest mammals, with the result that the Holarctic mammalian fauna became relatively homogenous.

The outcome of this complex oscillation of environmental factors on the mammalian fauna was a series of waves of interchanges and extinctions. An enormous amount of detailed information is available from the Pleistocene palaeoenvironmental and fossil record (e.g. Potts and Behrensmeyer 1992;

Graham 1997; Agustí and Wenderli 1995; Agustí and Antón 2002), and there is space here for no more than a single taste of it. Of the 130 genera of Early Pleistocene mammals in North America, Webb (1985b) found that 40 of them were new. Of these, 20 were immigrants, composed of 8 from South America, and 12 from Asia via the Bering land bridge; the latter notably included mammoths and musk oxen. The other new genera, many of them cricetid rodents, had presumably evolved *in situ*. In similar vein, Maglio (1978) quotes figures for the Early Pleistocene of Africa. Fifty-three new genera appeared, of which about half were immigrants from the north, including members of such disparate taxa as bovids, primates, bats, pigs, and rodents. Simultaneously, there were 33 generic extinctions.

The end-Pleistocene: megafaunal extinction

The end of the Pleistocene was marked by the abrupt extinction of many mammals, with a strong bias against species of larger body size, the mammalian megafauna (Martin and Klein 1984; McPhee 1999). The timing and severity of this extinction varies from continent to continent (Fig. 7.28), and its unusual features led to a controversy concerning the cause that has continued for the last century and a half (Grayson 1984). Was it due to the direct effect of human over-hunting (Martin and Steadman 1999), or was it due to the kind of environmental change that caused all the mass extinction events prior to this one (Webb 1985b)? Leaving aside the case of Africa and southern Asia, all mammals of body size greater than 1000 kg disappeared and about 80% of those weighing 100–1000 kg. Africa is the most glaring exception, with only about 10% loss of its mammalian megafauna; elephants, hippos, and rhinos continue to represent mammals of over 1 tonne body weight. Were this continent to be considered in isolation, no particularly significant increase in extinction would even be noted at the end of the Pleistocene. Survival of mammals over 1 tonne in weight also occurred in southern Asia, but the fossil record is presently too poor to be able to quantify the extinction pattern.

The timing of the extinction was also variable. In North America it is well dated to within a narrow

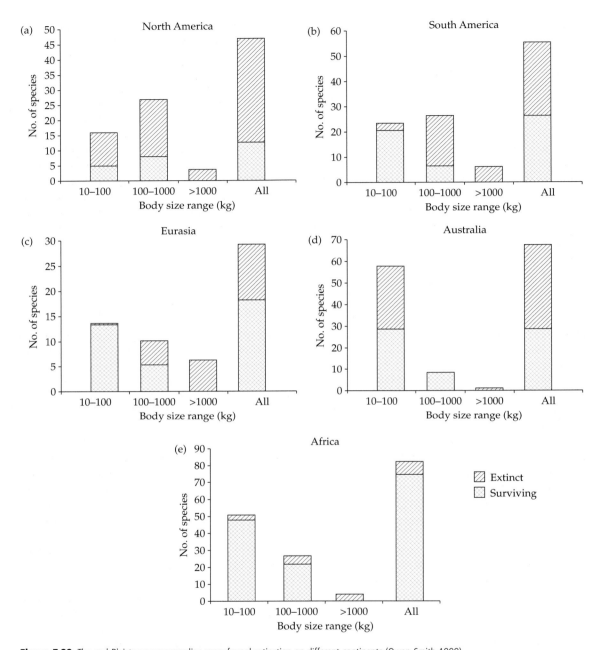

Figure 7.28 The end-Pleistocene mammalian megafaunal extinction on different continents (Owen-Smith 1999).

window of time between 11,500 and 10,500 years ago. The dating for South America is less precise, but probably slightly later (Less and Fariña 1996). In contrast, the northern Eurasian extinction was spread over a far longer period of some tens of

thousands of years (Stuart 1999), beginning about 50,000 years ago and not being completed until about 10,000 years ago. The Australian megafaunal extinction was completed much earlier, between 51,000 and 38,000 years ago, although there is little agreement about exactly when within that time frame (Brook and Bowman 2002), nor whether it occurred rapidly or spread over as much as 10,000 years.

The arguments in favour of human overkill as the cause of the megafaunal extinctions are twofold (Martin 1984; Alroy 1999*a*). The first is the claim that the timing of the extinction in each region coincides with the timing of the arrival of human populations. This applies particularly well to North America, where the Clovis people evidently crossed the Bering land-bridge from Asia, migrated down an ice-free corridor along the western side of Alaska and Canada, and entered the southern part of the continent around 12–10,000 years ago. Details of human entry into South America are less clear, but probably occurred as a continuation of the southerly migration through North America. In northern Eurasia the evidence is more equivocal. Neanderthals persisted there until about 30,000 years ago, at which time they were replaced by modern humans (Stuart 1999). This was as much as 20,000 years before many of the megafaunal extinctions actually occurred. The time of arrival of humans into Australia is disputed and estimates vary from as long as 70,000 to as recently as 40,000 years ago (Brook and Bowman 2002). Thus, while the possibility of a close overlap between human arrival and the Australian extinction certainly exists (Miller *et al.* 1999), it cannot be claimed to be substantiated by the evidence. More limited support for the overkill hypothesis is also provided by the much more accurately measured correlation between human appearance and mammal extinctions on various islands, notably Madagascar where most of the extinction of giant lemurs, hippopotami, the mysterious *Plesiorycteropus*, and also the elephant bird *Aepyornis* occurred soon after the arrival of humans about 1,500 years ago (Dewar 1984; Burney 1999). The reason offered by overkill enthusiasts for the low level of extinction of large mammals in Africa is attributed paradoxically to the fact that humans had been there for so long that they and the mammals had co-evolved behavioural strategies, allowing them to coexist.

The second kind of argument for the overkill hypothesis is the nature of the extinction itself. The end-Pleistocene was a time of net increase in habit-

able area, as the ice retreated, and so a diversification rather than an extinction phase would be expected if climate change alone were affecting the fauna. Furthermore, none of the numerous earlier episodes of mammalian extinction had ever had such a marked discriminatory effect against large body size. Human hunting activity, however, would be expected to focus more on larger mammals because of the greater ease of finding and killing them, and because of the greater resource each one represents. The large mammals that did survive also supports the hypothesis. They occupy habitats that are particularly hostile to humans, such as the musk ox and caribou of the tundra, and mountain dwelling bovids, (Alroy 1999*a*), or else inaccessible such as the tropical forests of southern Asia or arboreal and nocturnal species (Johnson 2002).

The arguments in favour of climatic change as the cause of the extinction are first that a climatic change of the same order of magnitude as those that had caused earlier severe extinctions did indeed occur (Webb 1985*a*), and second that small populations of humans possessing very limited hunting technologies could not plausibly have had such a large effect. A number of authors have proposed more specific theories of how the climate change might have resulted in the end-Pleistocene extinction, although the picture is so complex that it is most unlikely a single, climatic factor could ever be identified as the cause (Graham 1997). Guthrie (1984) argued that the end-Pleistocene saw a continuation of the increasing seasonality that had been developing. The non-growing cold season in the temperate latitudes, and the non-growing dry season in the tropics, both increased in length. This, he proposed, caused reduced plant diversity and increased plant zonation, resulting in the change from highly mosaic habitats to much more homogeneous ones dominated by large, continuous areas of, respectively, grassland, broad-leafed forest, and coniferous forest, as is seen today. Large herbivores would be the most directly affected by such a shift, followed inevitably by their large carnivore predators. Graham and Lundelius (1984) pointed to evidence of the breaking up of co-adapted biotas at this time, which they claim would have led to a variety of possible kinds of environmental stresses on individual species, such as the need for a herbivore to shift from its preferred food plants to others for which it is less well adapted. Guthrie (2003) observed a decrease in mean body size of horses in North America 12,000 years ago, which suggests a climatic shift had occurred.

It is hard to believe that either human activity or environmental change alone could be the whole explanation, and Owen-Smith (1999), for example, explicitly combined the two. He proposed that early humans hunted the very large herbivores at a greater rate than their slow reproductive rates could replace. Removal of these key species caused a shift in the habitat from mixed open grassland and light forest to more dense, closed-canopied forest, so in their turn the large numbers of open area grazers and low browsers were affected. Others have proposed that the extinction was indeed caused by humans, but indirectly by their effect on the environment from the use of fire to clear areas *eg*. Miller *et al*. 1999.

The most recent approach to the question is the construction of computer models to test hypotheses. Mossiman and Martin (1975) created an early model, which pointed towards an overkill explanation, but whose assumptions were probably unrealistically simple (Whittington and Dyke 1984). Using what he took to be more appropriate and detailed assumptions about population sizes and reproductive rates of herbivores, human population sizes, and hunting efficiency, Alroy (2001) also found that most of the observed pattern of large mammal extinctions in North America could be simulated without the need to assume any environmental changes at all, which he took as powerful corroboration of the overkill

hypothesis. However, like all such models this one is very sensitive to some of the values attributed to the parameters (Brook and Bowman 2002), and suffers from the inevitable over-simplification compared to the real thing.

Others have attempted to test the hypotheses by making explicit predictions. Beck (1996) considered the Blitzkrieg version of the overkill hypothesis for North America. If the humans had entered the southern part of the continent from the vicinity of present day Edmonton, and then migrated like a 'bow-wave' from northwest to southeast, exterminating species of large mammals on the way, there should be a significantly higher number of last fossil occurrences of species in the southeast than in the northwest segment of USA. In fact, not only did he not find this to be so, but to a small extent the very reverse is true. However, it is another measure of the very intractability of the problem that this result can quite readily be explained other than by rejecting the hypothesis. Perhaps the movement of the human population was not a simple bow-wave, but included advances along some lines such as the Pacific coast and reflexions back northwards along other lines. Perhaps there are reasons for a systematically biased fossil record of last occurrences, which are not necessarily the same as sites of final extinction.

Apart from the science of this issue being difficult, there are also contemporary socio-political overtones. Was the end-Pleistocene megafaunal extinction the last of the many acts of biotic reorganisation by unbridled Nature, or the first of the many acts of global devastation by unconstrained Humanity?

References

Abdala, F. Redescription of *Platycraniellus elegans* from the Early Triassic of the Karoo Basin of South Africa, with some considerations about skull morphology in non-mammalian cynodonts. Unpublished manuscript.

Abdala, F. 1998. An approach to the phylogeny of gomphodont cynodonts based on dental characters. *J. Afr. Earth Sci.* **27**(1A): 1–2.

Abdala, F. and Giannini, N.P. 2000. Gomphodont cynodonts from the Chañares Formation: the analysis of an ontogenetic sequence. *J. Vert. Paleontol.* **20**: 501–506.

Abdala, F. and Giannini, N.P. 2002. Chiniquodont cynodonts: systematics and morphometric considerations. *Palaeontology* **45**: 1151–1170.

Abdala, F. and Ribeiro, A.M. 2003. A new traversodontid cynodont from the Santa Maria Formation (Ladinian-Carnian) of southern Brazil, with a phylogenetic analysis of Gondwanan traversodontids. *Zool. J. Linn. Soc.* **139**: 529–545.

Agustí, J. 1999. A critical re-evaluation of the Miocene mammal units in Western Europe: dispersal events and problems of correlation. In J. Agustí, L. Rook, and P. Andrews (eds) *The evolution of Neogene terrestrial ecosystems in Europe*. Cambridge University Press: Cambridge, pp. 84–112.

Agustí, J. and Antón, M. 2002. *Mammoths, sabertooths, and hominids: 65 million years of mammalian evolution in Europe*. Columbia University Press: New York.

Agustí, J. and Wenderlin, L. (eds) 1995. *Influence of climate on faunal evolution in the Quaternary*. Krakow Institute of Systematics and Evolution of Mammals: Krakow, *Acta Zool. Cracow.* **38**.

Ahlberg, P.E. 1995. *Elginerpeton pancheni* and the earliest tetrapod clade. *Nature* **373**: 420–425.

Ahlberg, P.E. 1998. Postcranial stem reptile remains from the Devonian of Scat Craig, Morayshire, Scotland. *Zool. J. Linn. Soc.* **122**: 99–141.

Ahlberg, P.E. and Milner, A.R. 1994. The origin and diversification of tetrapods. *Nature* **368**: 507–514.

Allin, E.F. 1975. Evolution of the mammalian middle ear. *J.Morphol.* **147**: 403–438.

Allin, E.F. and Hopson, J.A. 1992. Evolution of the auditory system in Synapsida ('mammal-like reptiles' and primitive mammals) as seen in the fossil record. In D.B. Webster, R.R. Fay and A.N. Popper (eds) *The evolutionary biology of hearing*. Springer-Verlag: New York, pp. 587–614.

Allman, J. 1990. *Neuroscience* **2**: 257.

Allman, J. 1999. *Evolving brains*. Scientific American Library: New York.

Alroy, J. 1999a. Putting North America's end-Pleistocene megafaunal extinction in context: large-scale analyses of spatial patterns, extinctions rates, and size distributions. In R.D.E. McPhee (ed) *Extinctions in near time: causes, contexts, and consequences*. Kluwer Academic: New York, pp. 105–143.

Alroy, J. 1999b. The fossil record of North American mammals: evidence for a Paleocene evolutionary radiation. *Syst. Biol.* **48**: 107–118.

Alroy, J. 2001. A multiple overkill simulation of the end-Pleistocene megafaunal mass extinction. *Science* **292**: 1893–1896.

Amalitzky, V.P. 1922. Diagnoses of new forms of vertebrates and plants from the Upper Permian of North Dvinia. *Izv. Ross. Akad. Nauk Ser. 6.* **25**: 1–12.

Amrine-Madsen, H., Scally, M., Westerman, M., Stanhope, M.J., Krajewski, C., and Springer, M.S. 2003. Nuclear gene sequences provide evidence for the monophyly of australidelphian marsupials. *Mol. Phylogenet. Evol.* **28**: 186–196.

Anderson, J.M. and Cruickshank, A.R.I. 1978. The biostratigraphy of the Permian and the Triassic. Part 5. A review of the classification and distribution of Permo-Triassic tetrapods. *Palaeontol. Afr.* **21**: 15–44.

Aplin, K.P. 1987. Basicranial anatomy of the Early Miocene diprotodontian *Wynardia bassiana* (Marsupialia: Wynyardiidae) and its implications for wynyardiid phylogeny and classification. In M. Archer (ed) *Possums and opossums: studies in evolution*. Surrey Beatty and Sons: Sydney, pp. 369–391.

Aplin, K.P. and Archer, M. 1987. Recent advances in marsupial systematics with a new syncretic classification. In M. Archer (ed) *Possums and opossums: studies in evolution*. Surrey Beatty and Sons: Chipping Norton, Australia, pp. xv–lxxii.

Archer, M. 1982. Review of the dasyurid (Marsupialia) fossil record, integration of data bearing on phylogenetic interpretation, and suprageneric classification. In M. Archer (ed) *Carnivorous marsupials*. Royal Zoological Society of New South Wales: Sydney, pp. 397–443.

Archer, M., Flannery, T.F., Ritchie, A., and Molnar, R. 1985. First Mesozoic mammal from Australia—an early Cretaceous monotreme. *Nature* **318**: 363–366.

Archer, M., Tedford, R.H., and Rich, T.H. 1987. The Pilkipildridae, a new family and four new species of petauroid possums (Marsupialia: Phalangerida) from the Australian Miocene. In M. Archer (ed) *Possums and opossums: studies in evolution*. Surrey Beatty and Sons: Chipping Norton, Australia, pp. 607–627.

Archer, M., Hand, S., and Godthelp, H. 1988. A new order of Tertiary zalambdodont marsupials. *Science* **239**: 1528–1531.

Archer, M., Hand, S.J., and Godthelp, H. 1991. *Australia's lost world: prehistoric animals of Riversleigh*. Indiana University Press: Bloomington.

Archer, M., Jenkins, F.A. Jr, Hand, S.J., Murray, P., and Godthelp, H. 1992. Description of the skull and non-vestigeal dentition of a Miocene platypus (*Obdurodon dicksoni* n.sp.) from Riversleigh, Australia and the problem of monotreme origins. In A.Augee (ed). *Platypus and echidnas*. Royal Zoological Society of Sydney: Sydney, pp. 15–27.

Archer, M., Godthelp, H., and Hand, S. 1993. Early Eocene marsupial from Australia. In F. Schrenk and K. Ernst (eds) *Kaupia: Darmstädter Beiträge zur Naturgesschichte Monument Grube Messel*. Hessisches Landesmuseum Darmstadt: Darmstadt, pp. 193–200.

Archer, M., Arena, D.A., Bassarova, M., Black, K., Brammall, J.R., Cooke, B.N., Creaser, P.H., Crosby, K., Gillespie, K.A., Godthelp, H., Gott, M., Hand, S.J., Kear B.P., Krikmann, A., Mackness, B.S., Muirhead, J., Musser, A.M., Myers, T.J., Pledge, N.S., Wang, Y., and Wroe, S. 1999. The evolutionary history and diversity of Australian mammals. *Aust. Mamm.* **21**: 1–45.

Archibald, J.D. 1991. Survivorship patterns of non-marine vertebrates across the Cretaceous–Tertiary boundary in the western U.S. *Fifth Symposium on Mesozoic Terrestrial Ecosystems and Biota. Extended Abstracts*. Contributions from the Paleontological Museum, University of Oslo, No. 364, pp. 1–2.

Archibald, J.D. 1996. *Dinosaur extinction and the end of an era: what the fossils say*. Columbia University Press: New York.

Archibald, J.D. 1998. Archaic ungulates ("Condylarthra"). In C.M. Janis, K.M. Scott, and L.L. Jacobs (eds) *Evolution of Tertiary mammals of North America, Vol. 1: Terrestrial carnivores, ungulates, and ungulate-like mammals*. Cambridge, University Press: Cambridge, pp. 292–331.

Archibald, J.D. 1999. Divergence times of eutherian mammals. *Science* **285**: 2031a.

Archibald, J.D. 2003. Timing and biogeography of the eutherian radiation: fossils and molecules compared. *Mol. Phylogenet Evol.* **28**: 350–359.

Archibald, J.D. and Averianov, A.O. 2003. The Late Cretaceous placental mammal. *Kulbeckia. J. Vert. Paleontol.* 23: 404–419.

Archibald, J.D. and Bryant, L.J. 1990. Differential Cretaceous/Tertiary extinctions of nonmarine vertebrates: evidence from northeastern Montana. *Geol. Soc. Am. Spec. Pap.* **247**: 549–562.

Archibald, J.D. and Deutschman, D.H. 2001. Quantitative analysis of the timing of the origin and diversification of the extant placental orders. *J. Mamm. Evol.* **8**: 107–124.

Archibald, J.D. and Lofgren, D.L. 1990. Mammalian zonation near the Cretaceous/Tertiary boundary. In T.M. Bown and K.D. Rose (eds) *Dawn of the age of mammals in the northern part of the Rocky Mountain interior, North America. Geol. Soc. Spec. Pap.* **243**: 31–50.

Archibald, J.D., Averianov, A.O., and Ekdale, E.G. 2001. Late Cretaceous relatives of rabbits, rodents, and other extant eutherian mammals. *Nature* **414**: 62–65.

Arnason, U., Adegoke, J.A., Bodin, K., Born, E.W., Esa, Y.B., Gullberg, A., Nilsson, M., Short, R.V., Xu, X., and Janke, A. 2002. Mammalian mitogenomic relationships and the root of the eutherian tree. *Proc. Natl. Acad. Sci. USA.* **99**: 8151–8156.

Asher, R.J., Novacek, M.J., and Geisler, J.H. 2003. Relationships of endemic African mammals and their fossil relatives based on morphological and molecular evidence. *J. Mamm. Evol.* **10**: 131–194.

Askin, R.A., 1990. Campanian to Paleocene spore and pollen assemblages of Seymour Island, Antarctica. *Rev. Paleobot. Palynol.* **65**: 105–113.

Averianov, A. and Kielan-Jaworowska, Z. 1999. Marsupials from the Late Cretaceous of Uzbekistan. *Acta Palaeontol. Pol.* **44**: 71–81.

Averianov, A.O. and Skutschas, P.P. 2000. A new genus of Eutherian mammal from the Early Cretaceous of Transbaikalia, Russia. *Acta Palaeontol. Pol.* **46**: 431–436.

Averianov, A.O., Archibald, J.D., and Martin, T. 2003. Placental nature of the alleged marsupial from the Cretaceous of Madagascar. *Acta Palaeontol. Pol.* **48**: 149–151.

Ax, P. 1987. *The phylogenetic system: the systematization of organisms on the basis of their phylogenesis.* Wiley and Sons: Chichester.

Babot, M.J., Powell, J.E., and Muizon, C. de 2002. *Callistoe vincei,* a new Proborhyaenidae (Borhyaenidae, Metatheria, Mammalia) from the Early Eocene of Argentina. *Geobios* **35**: 615–629.

Bajpai, S. and Gingerich, P.D. 1998. A new Eocene archaeocete (Mammalia, Cetacea) from India and the time of origin of whales. *Proc. Natl. Acad. Sci.* **95**: 15464–15468.

Bakker, R.T. 1968. The superiority of dinosaurs. *Discovery* **3**: 11–22.

Bakker, R.T. 1971. Dinosaur physiology and the origin of mammals. *Evolution* **25**: 636–658.

Bakker, R.T. 1974. Experimental and fossil evidence for the evolution of tetrapod bioenergetics. In D. Gates and R. Schmerl (eds) *Perspectives in biophysical ecology.* Springer-Verlag: New York, pp. 365–399.

Bakker, R.T. 1975. Dinosaur renaissance. *Sci. Am.* **232**: 58–78.

Bakker, R.T. 1980. Dinosaur heresy–dinosaur renaissance. In R.D.K. Thomas and E.C. Olson (eds) *A cold look at the warm-blooded dinosaurs.* Westview Press: Boulder, Colorado, pp. 351–462.

Barghusen, H.R. 1973. The adductor jaw musculature of *Dimetrodon* (Reptilia, Pelycosauria). *J. Paleontol.* **47**: 823–834.

Barghusen, H.R. 1975. A review of fighting adaptations in dinocephalians (Reptilia, Therapsida). *Paleobiology* **1**: 295–311.

Battail, B. 1982. Essai de phylogénie des cynodontes (Reptilia, Therapsida). *Geobios, Lyon Mem. Spec.* **6**: 157–167.

Battail, B. 1991. Les cynodontes (Reptilia, Therapsida): une phylogénie. *Bull. Mus. Natl. Hist. Nat., Paris* 4th Ser. **C13** (1–2): 17–105.

Battail, B. and Surkov, M.V. 2000. Mammal-like reptiles from Russia. In M.J. Benton, M.A. Shishkin, D.M. Unwin, and E.N. Kurkin (eds) *The Age of dinosaurs in Russia and Mongolia.* Cambridge University Press: Cambridge, pp. 86–119.

Beard, K.C. 1993. Phylogenetic systematics of the Primatomorpha, wi th special reference to Dermoptera. In F.S. Szalay, M.J. Novacek, and M.C. McKenna (eds) *Mammal phylogeny: placentals.* Springer-Verlag: New York, pp. 129–150.

Beard, K.C. 1998. East of Eden: Asia as an important center of taxonomic origination in mammalian evolution. In K.C. Beard and M.R. Dawson (eds) *Dawn of the age of mammals in Asia. Bull. Carnegie Mus. Nat. Hist.* **34**: 5–39.

Beard, K.C. 2002*a*. Basal anthropoids. In W. Hartwig (ed) *The primate fossil record.* Cambridge University Press: Cambridge, pp. 133–150.

Beard, K.C. 2002*b*. East of Eden at the Paleocene/Eocene boundary. *Science* 295: 2028–2029.

Beard, K.C., Qi, T., Dawson, M.R., Wang, B., and Li, C. 1994. A diverse new primate fauna from Middle Eocene fissure-fillings in southeastern China. *Nature* **368**: 604–609.

Beard, K.C., Tong, Y., Dawson, M.R., Wang, J., and Huang, X. 1996. Earliest complete dentition of an anthropoid primate from the late middle Eocene of Shanxi Province, China. *Nature* **272**: 82–85.

Beck, M.W. 1996. On discerning the cause of Late Pleistocene megafaunal extinctions. *Paleobiology* **22**: 91–103.

Begun, D.R. 2002. The Pliopithecoidea. In W. Hartwig (ed) *The primate fossil record.* Cambridge University Press: Cambridge, pp. 221–240.

Behrensmeyer, A.K., Damuth, J.D., DiMichele, W.A., Potts, R., Sues, H.-D., and Wing, S.L. (eds) 1992. *Terrestrial ecosystems through time: evolutionary paleoecology of terrestrial plants and animals.* Chicago University Press: Chicago.

Bennett, A.F. 1991. The evolution of activity capacity. *J. Exp. Biol.* **160**: 1–23.

Bennett, A.F. and Ruben, J. 1979. Endothermy and activity in vertebrates. *Science* **206**: 649–654.

Bennett, A.F. and Ruben, J. 1986. The metabolic and thermoregulatory status of therapsids. In N.Hotton III, P.D. McLean, J.J. Roth, and E.C. Roth (eds) *The ecology and biology of mammal-like reptiles.* Smithsonian Institution Press: Washington, pp. 207–218.

Bennett, A.F., Hicks, J.W., and Cullum, A.J. 2000. An experimental test of the thermoregulatory hypothesis for the evolution of endothermy. *Evolution* **54**: 1768–1773.

Bennett, S.C. 1996. Aerodynamics and thermoregulatory function of the dorsal sail of *Edaphosaurus. Paleobiology* **22**: 496–505.

Benton, M.J. 1986. The Late Triassic tetrapod extinction events. In K. Padian (ed) *The beginning of the age of dinosaurs: faunal changes across the Triassic–Jurassic boundary.* Cambridge University Press: Cambridge, pp. 303–320.

Benton, M.J. 1994. Late Triassic to Middle Jurassic extinctions among continental tetrapods: testing the pattern. In N.C. Fraser and H.-D. Sues (eds) *In the shadow of the dinosaurs: early Mesozoic tetrapods.* Cambridge University Press: Cambridge, pp. 366–397.

Benton, M.J. 1999. Early origins of modern birds and mammals: molecules vs. morphology. *Bioessays* **21**: 1043–1051.

Benton, M.J. 2000. Stems, nodes, crown clades, and rank-free lists: is Linnaeus dead? *Biol. Rev.* **75**: 633–648.

Benton, M.J. and Twitchett, R.J. 2003. How to kill (almost) all life: the end-Permian extinction event. *TREE* **18**: 358–365.

Berman, D.S. 2000. Origin and early evolution of the amniote occiput. *J. Paleontol.* **74**: 938–956.

Berman, D.S., Reisz, R.R., Bolt, J.R., and Scott, D. 1995. The cranial anatomy and relationships of the synapsid *Varanosaurus* (Eupelycosauria: Ophiacodontidae) from the Early Permian of Texas and Oklahoma. *Ann. Carnegie Mus.* **64**: 99–133.

Berman, D.S., Sumida, S.S., and Lombard, R.E. 1997. Biogeography of primitive amniotes. In S.S. Sumida and K.L. M. Martin (eds) *Amniote origins: completing the transition to land*. Academic Press: London, pp. 85–139.

Berta, A. 2002. Pinniped evolution. In W.F. Perrin, B. Würsig, and J.G.M. Thewissen (eds) *Encyclopedia of marine mammals*. Academic Press: San Diego, pp. 921–928.

Berta, A. and Ray, C.E. 1990. Skeletal morphology and locomotor capabilities of the archaic pinniped *Enaliarctos mealsi*. *J. Vert. Paleontol.* **10**: 141–157.

Berta, A. and Sumich, J.L. 1994. *Marine mammals: evolutionary biology*. Academic Press: San Diego.

Black, C.C. 1978. Anthracotheriidae. In V.J. Maglio and H.B.S. Cooke (eds) *Evolution of African mammals*. Harvard University Press: Cambridge, MA, pp. 423–434.

Blob, R. 2001. Evolution of hindlimb posture in nonmammalian therapsids: biomechanical tests of paleontogical hypotheses. *Paleobiology* **27**: 14–38.

Bloch, J.I. and Boyer, D.M. 2003*a*. Grasping primate origins. *Science* **298**: 1606–1610.

Bloch, J.I. and Boyer, D.M. 2003*b*. Response to comments on 'Grasping primate origins'. *Science* **300**: 741c.

Bloch, J.I. and Silcox, M.T. 2001. New basicrania of Paleocene–Eocene *Ignacius*: re-evaluation of the plesiadapiform–dermopteran link. *Am. J. Anthropol.* **116**: 184–198.

Bonaparte, J.F. 1963. Descripción del esqueleto postcraneano de *Exaeretodon* (Cynodontia-Traversodontidae). *Acta Geol,. Lilloana* **4**: 5–51.

Bonaparte, J.F. 1970. Annotated list of the South American Triassic tetrapods. *Proceedings of the Second Gondwana Symposium*. International Union of Geological Sciences Commission on Stratigraphy. C.S.I.R. Pretoria, pp. 665–682.

Bonaparte, J.F. 1980. El primer ictidosaurio (Reptilia-Therapsida) de América del Sur, *Chaliminia musteloides*, del Triásico Superior de La Rioja, República Argentina. *Actas, II Congreso Argentino de Paleontología y Biostratigrafía y Congreso Latinoamericano de Paleontología* (*Asociación Paleontologíca Argentina, Buenos Aires*), **1**: 123–133.

Bonaparte, J.F. 1990. New Late Cretaceous mammals from the Los Alamitos Formation, northern Patagonia. *Natl Geogr. Res.* **6**: 63–94.

Bonaparte, J.F. 1992. Una nueva especie de Triconodonta (Mammalia) de la formación Los Alamitos, Provincia de Río Negro y comentarios sobre su fauna de mamíferos. *Ameghiniana* **29**: 99–110.

Bonaparte, J.F. 1994. Approach to the significance of the Late Cretaceous mammals of South America. *Berlin. Geowiss. Abt.* **E13**: 31–44.

Bonaparte, J.F. and Barberena, M.C. 1975. A possible mammalian ancestor from the Middle Triassic of Brazil (Therapsida–Cynodontia). *J. Paleontol.* **49**: 931–936.

Bonaparte, J.F. and Barberena, M.C. 2001. On two advanced carnivorous cynodonts from the Late Triassic of Southern Brazil. *Bull. Mus. Comp. Zool. Harv.* **156**: 59–80.

Bonaparte, J.F. and Crompton, A.W. 1994. A juvenile probainognathid cynodont skull from the Ischigualasto Formation and the origin of mammals. *Rev. Mus. Arg. Cien. Nat. Bernardino Rivadavia Inst. Nac. Invest. Cienc. Nat.* **1**: 1–12.

Bonaparte, J.F., Ferigolo, J., and Ribeiro, A.M. 2001. A primitive Late Triassic 'ictidosaur' from Rio Grande do Sul, Brazil. *Palaeontology* **44**: 623–635.

Bonaparte, J.F., Martinelli, A.G., Schultz, C.L., and Rubert, R. 2003. The sister group of mammals: small cynodonts from the Late Triassic of southern Brazil. *Rev. Bras. Paleontol.* **5**: 5–27.

Boonstra, L.D. 1965. The girdles and limbs of the Gorgonopsia of the *Tapinocephalus* zone. *Ann. S. Afr. Mus.* **48**: 237–249.

Boonstra, L.D. 1972. Discard the names Theriodontia and Anomodontia: a new classification of the Therapsida. *Ann. S. Afr. Mus.* **59**: 315–338.

Boulter, M.C., Spicer, R.A., and Thomas, B.A. 1988. Patterns of plant extinction from some palaeobotanical evidence. In G.P. Larwood (ed) *Extinction and survival in the fossil record. Syst. Assoc. Spec. Pap.* **34**: 1–36.

Bowen, G.J., Clyde, W.C., Koch, P.L., Ting, S., Alroy, J., Tsubamoto, T., and Wang, Y. 2002. Mammalian dispersal at the Paleocene/Eocene boundary. *Science* **295**: 2062–2065.

Bown, T.M. and Fleagle, J.G. 1993. Systematics, biostratigraphy, and dental evolution of the Palaeothentidae, later Oligocene to early-middle Miocene (Deseadan–Santacrucian) caenolestid marsupials of South America. *Paleontol. Soc. Mem.* **29**: 1–76.

Boyden, A. and Gemeroy, D. 1950. The relative position of Cetacea among the orders of Mammalia as indicated by precipitin tests. *Zoologica* **35**: 145–151.

Bramble, D. M. 1978. Origin of the mammalian feeding complex: models and mechanisms. *Paleobiology* **4**: 271–301.

Briggs, D.E.G. and Crowther, P.R. 2001. *Palaeobiology II.* Blackwell Science: Oxford.

Brink, A.S. 1956*a*. On *Aneugomphius ictidoceps* Broom and Robinson. *Palaeontol. Afr.* **4**: 97–115.

Brink, A.S. 1956*b*. Speculations on some advanced mammalian characteristics in the higher mammal-like reptiles. *Palaeontol. Afr.* **4**: 77–96.

Brink, A.S. 1960. On some small therocephalians. *Palaeontol. Afr.* **7**: 155–182.

Brink, A.S. 1963*a*. On *Bauria cynops* Broom. *Palaeontol. Afr.* **8**: 39–56.

Brink, A.S. 1963*b*. Two cynodonts from the Ntawere Formation in the Luangwa Valley of Northern Rhodesia. *Palaeontol. Afr.* **8**: 77–96.

Brink, A.S. 1965. A new ictidosuchid (Scaloposauria) from the *Lystrosaurus*-zone. *Palaeontol. Afr.* **9**: 129–138.

Brinkman, D. 1981. The structure and relationships of the dromasaurs (Reptilia: Therapsida). *Breviora* **465**: 1–34.

Broili, F. and Schröder, J. 1934. Zur Osteologie des Kopfes von *Cynognthus*. *Sitz. Bayer. Akad. Wiss. for 1934*: 163–177.

Bromham, L., Phillips, M.J., and Penny, D. 1999. Growing up with dinosaurs: molecular dates and the mammalian radiation. *TREE* **14**: 113–118.

Brook, B.W. and Bowman, M.J.S. 2002. Explaining the Pleistocene megafaunal extinctions: models, chronologies and assumptions. *Proc. Natl. Acad. Sci. USA.* **99**: 14624–14627.

Broom, R. 1910. A comparison of the Permian reptiles of North America with those of South Africa. *Bull. Am. Mus. Nat. Hist.* **28**: 197–234.

Broom, R. 1932. *The mammal-like reptiles of South Africa.* H.F. & G. Witherby: London.

Broom, R. 1948. A contribution to our knowledge of the vertebrates of the Karroo Beds of South Africa. *Trans. R. Soc. Edinb.* **61**: 577–629.

Buckland, W. 1824. Notice on *Megalosaurus*. *Trans. Geol. Soc. Lond.* **2**: 390–396.

Burney, D.A. 1999. Rates, patterns, and processes of landscape transformation and extinction in Madagascar. In R.D.E. McPhee (ed) *Extinctions in near time: causes, contexts, and consequences.* Kluwer Academic: New York, pp. 145–164.

Butler, P.M. 1984. Macroscelidea, Insectivora and Chiroptera from the Miocene of East Africa. *Palaeovertebrata* **14**: 117–198.

Butler, P.M. 1988. Phylogeny of the insectivores. In M.J. Benton (ed) *The phylogeny and classification of the tetrapods*, Vol. 2. Oxford University Press: Oxford. pp. 117–141.

Butler, P.M. 1990. Early trends in the evolution of tribosphenic molars. *Biol. Rev.* **65**: 529–552.

Butler, P.M. 1995. Fossil Macroscelidea. *Mamm. Rev.* **25**: 3–14.

Butler, P.M. 2000. Review of the early allotherian mammals. *Acta Palaeontol. Pol.* **45**: 317–342.

Butler, P.M. and Kielan-Jaworowska, Z. 1973. Is *Deltatheridium* a marsupial? *Nature.* **245**: 105–106.

Butler, P.M. and MacIntyre, G.T. 1994. Review of the British Haramiyidae (?Mammalia, Allotheria), their molar occlusion and relationships. *Phil. Trans. R. Soc.* **B345**: 433–458.

Calder, W.A. III 1984. *Size, function, and life history.* Harvard University Press: Cambridge, MA.

Cao, Y., Fujiwara, M., Nikaido, M., Okada, N., and Hasegawa, M. 2000. Interordinal relationships and timescale of eutherian evolution as inferred from mitochondrial genome data. *Gene* **259**: 149–158.

Carrier, D.R. 1987. The evolution of locomotor stamina in tetrapods: circumventing a mechanical constraint. *Paleobiology* **13**: 326–341.

Carroll, R.L. 1964. The earliest reptiles. *Zool. J. Linn. Soc. Lond.* **45**: 61–83.

Carroll, R.L. 1969. Problems of the origin of reptiles. *Biol. Rev.* **44**: 393–432.

Carroll, R.L. 1970*a*. Quantitative aspects of the amphibian–reptile transition. *Formaet Functio* **3**: 165–178.

Carroll, R.L. 1970*b*. The ancestry of reptiles. *Phil. Trans. R. Soc.* **B257**: 267–308.

Carroll, R.L. 1988. *Vertbrate paleontology and evolution.* Freeman: New York.

Case, E.C. 1897. Foramina perforating the cranial region of a Permian reptile and on a cast of its brain cavity. *Am. J. Sci. 4th Ser.* **3**: 321–326.

Cerling, T.E., Ehleringer, J.R., and Harris, J.M. 1998. Carbon dioxide starvation, the development of C_4 ecosystems, and mammalian evolution. *Phil. Trans. R. Soc.* **B353**: 159–171.

Cerling, T.E., Harris, J.M., MacFadden, B.J., Leakey, M.G., Quade, J., Eisenmann, V., and Ehleringer, J.R. 1997. Global vegetation change through the Miocene/Pliocene boundary. *Nature* **389**: 153–158.

Charig, A.J. 1984. Competition between therapsids and archosaurs during the Triassic period: a review and synthesis of current theories. *Symp. Zool. Soc. Lond.* **52**: 597–628.

Chiappe, L.M. and Witmer, L.M. (eds) 2002. *Mesozoic birds: above the heads of dinosaurs.* University of California Press: Berkeley.

Chinsamy, A. 1997. Assessing the biology of fossil vertebrates through bone histology. *Palaeontol. Afr.* **33**: 29–35.

Chow, M. and Rich, T.H.V. 1982. *Shuotherium dongi*, n.gen. and sp., a therian with pseudo-tribosphenic molars from the Jurassic of Sichuan, China. *Aust. Mamm.* **5**: 127–142.

Chow, M. and Wang, B. 1979. Relationships between pantodonts and tillodonts and classification of the order Pantodonta. *Vert. PalAs.* **17**: 37–48.

Chudinov, P.K. 1960. Upper Permian therapsids from the Esheevo locality. *Paleontol. Zh.* **85**: 81–94.

Chudinov, P.K. 1983. Early therapsids. *Trudy Paleont. Inst. Akad. Nauk SSSR (Moscow)* **202**: 1–227.

Cifelli, R.L. 1993a. Early Cretaceous mammals from North America and the evolution of marsupial dental characters. *Proc. Nat. Acad. Sci. USA.* **90**: 9413–9416.

Cifelli, R.L. 1993b. The phylogeny of the native South American ungulates. In F.S. Szalay, M.J. Novacek, and M.C. McKenna (eds) *Mammal phylogeny: placentals.* Springer-Verlag: New York, pp. 195–216.

Cifelli, R.L. 1993c. Theria of metatherian-eutherian grade and the origin of marsupials. In F.S. Szaley, M.C. McKenna, and M.J. Novacek (eds) *Mammalian phylogeny: Mesozoic differentiation, multituberculates, monotremes, early therians, and marsupials.* Springer-Verlag: New York, pp. 205–215.

Cifelli, R.L. 1999. Tribosphenic mammal from the North American Early Cretaceous. *Nature* **401**: 363–366.

Cifelli, R.L. and Davis, B.M. 2003. Marsupial origins. *Science* **302**: 1899–1900.

Cifelli, R.L. and Eaton, J.G. 1987. Marsupials from the earliest Late Cretaceous of western U.S. *Nature* **325**: 520–522.

Cifelli, R. L. and Muizon, C. de 1997. Dentition and jaw of *Kokopellia juddi*, a primitive marsupial or near marsupial from the medial Cretaceous of Utah. *J. Mamm. Evol.* **4**: 241–258.

Cifelli, R.L. and Schaff, C.R. 1998. Arctostylopida. In C.M. Janis, K.M. Scott, and L.L. Jacobs (eds) *Evolution of Tertiary mammals of North America, Vol. 1: Terrestrial carnivores, ungulates, and ungulate-like mammals.* Cambridge University Press: Cambridge, pp. 332–336.

Cifelli, R.L., Schaff, C.R., and McKenna, M.C. 1989. The relationships of the Arctostylopidae (Mammalia): new data and interpretation. *Bull. Mus. Comp. Zool. Harv.* **152**: 1–44.

Clack, J.A. 2002. *Gaining ground: the origin and evolution of tetrapods.* Indiana University Press: Bloomington.

Clack, J.A. and Carroll, R.L. 2000. Early Carboniferous tetrapods. In H. Heatwole and R.L. Carroll (eds) *Amphibian biology, Vol 4. Palaeontology: the evolutionary history of amphibians.* Surrey Beatty & Sons: Chipping Norton, Australia, pp. 1030–1043.

Clack, J.A., Ahlberg, P.E., Finney, S.M., Alonso, P.D., Robinson, J., and Ketcham, R.A. 2003. A uniquely specialized ear in a very early tetrapod. *Nature* **425**: 65–69.

Clark, J.M. and Hopson, J.A. 1985. Distinctive mammal-like reptile from Mexico and its bearing on the phylogeny of Tritylodontidae. *Nature* **315**: 398–400.

Clemens, W.A. 1966. Fossil mammals of the type Lance Formation, Wyoming. Part II. Marsupialia. *Univ. Calif. Public Geol. Sci.* **62**: vi, 122.

Clemens, W.A. 1979. Marsupialia. In J.A. Lillegraven, Z. Kielan-Jaworowska, and W.A. Clemens (eds) *Mesozoic mammals: the first two-thirds of mammalian history.* California University Press: Berkeley, pp. 192–200.

Clemens, W.A. 1986. On Triassic and Jurassic mammals. In K. Padian (ed) *The beginning of the age of dinosaurs.* Cambridge University Press: Cambridge, pp. 237–246.

Clemens, W.A. and Kielan-Jaworowska, Z. 1979. Multituberculata. In J.A. Lillegraven, Z. Kielan-Jaworowska, and W.A. Clemens (eds) *Mesozoic mammals: the first two-thirds of mammalian history.* University of California Press: Berkeley, pp. 99–149.

Clemens, W.A. and Lillegraven, J.A. 1986. New Late Cretaceous, North American mammals that fit neither the marsupial nor eutherian moulds. *Contrib. Geol. Univ. Wyom. Spec. Pap.* **3**: 55–85.

Clemens, W.A. and Mills, J.R.E. 1971. Review of *Peramus tenuirostris* Owen (Eupantotheria, Mammalia). *Bull. Br. Mus. Nat. Hist.(Geol.)* **20**: 87–113.

Clementz, M.T., Hoppe, K.A. and Koch, P.L. 2003. A palaeoecological paradox: the habit and dietary preferences of the extinct tethythere *Desmostylius*, inferred from stable isotope analysis. *Paleobiology* **29**: 506–519.

Cluver, M.A. 1975. A new dicynodont reptile from the *Tapinocephalus* Zone (Karoo System, Beaufort Series) of South Africa, with evidence of the jaw adductor musculature. *Ann. S. Afr. Mus.* **67**: 7–23.

Cluver, M.A. 1978. The skeleton of the mammal-like reptile *Cistecephalus* with evidence for a fossorial mode of life. *Ann. S. Afr. Mus.* **76**: 213–246.

Cluver, M.A. and King, G.M. 1983. A reassessment of the relationships of Permian Dicynodontia (Reptilia: Therapsida) and a new classification of dicynodonts. *Ann. S. Afr Mus.* **91**: 195–273.

Coates, M.I. 1996. The Devonian tetrapod *Acanthostega gunnari* Jarvik: postcranial anatomy, basal tetrapod interrelationships and patterns of skeletal evolution. *Trans. R. Soc. Edinb. Earth Sci.* **83**: 363–421.

Coates, M.I. and Clack, J.A. 1990. Polydactyly in the earliets known tetrapod limbs. *Nature* **347**: 66–69.

Coates, M.I. and Clack, J.A. 1991. Fish-like gills and breathing in the earliest known tetrapod. *Nature* **352**: 234–236.

Coates, M.I. and Clack, J.A. 1995. Romer's gap: tetrapod origin and terrestriality. *Bull. Mus. Natl. Hist. Nat. Paris 4th Ser.* **17**: 373–388.

Colbert, E.H. 1948. The mammal-like reptile *Lycaenops*. *Bull. Am. Mus. Nat. Hist.* **89**: 357–404.

Colbert, E.H. 1982. Triassic vertebrates in the Transantarctic Mountains. In M.D. Turner and J.F. Splettstoesser (eds) *Geology of the Central Transantarctic Mountains.* American Geophysical Union: Washington, D. C, pp. 11–35.

Colbert, E.H. 2000. Asiatic dinosaur rush. In M.J. Benton, M.A. Shiskin, D.M. Unwin, and E.N. Kurochkin (eds) *The age of dinosaurs in Russia and Asia.* Cambridge University Press: Cambridge, pp. 211–234.

Cooke, H.B.S. 1978. Africa: the physical setting. In V.J.Maglio and H.B.S. Cooke (eds) *Evolution of African mammals.* Harvard University Press: Cambridge, MA, pp. 17–45.

Coombs, M.C. 1989. Interrelationships and diversity in the Chalicotheriidae. In D.R. Protherol and R.M. Schoch (eds) *The evolution of perissodactyls.* Oxford university Press: New York. pp. 438–457.

Coombs, M.C. 1998. Chalicotherioidea. In C.M. Janis, K.M. Scott and L.L. Jacobs (eds) *Evolution of Tertiary mammals of North America. Volume 1: terrestrial carnivores, ungulates, and ungulate-like mammals.* Cambridge University Press: Cambridge. pp. 560–568.

Coppens, I. and Beden, M. 1978. Moeritherioidea. In V.J. Maglio and H.B.S. Cooke (eds) *Evolution of African mammals.* Harvard University Press: Cambridge, MA, pp. 333–335.

Corneli, P.S. 2003. Complete mitochondrial genomes and eutherian evolution. *J. Mamm. Evol.* **9**: 281–305.

Cosgriff, J.W., Hammer, W.R., and Ryan, W.J. 1982. The Pangaean reptile *Lystrosaurus mccaigi* in the Lower Triassic of Antarctica. *J. Paleontol.* **56**: 371–385.

Court, N. 1990. Periotic anatomy of *Arsinoitherium* (Mammalia, Embrithopoda) and its phylogenetic implications. *J. Vert. Paleontol.* **10**: 170–182.

Court, N. 1992. A unique form of dental bilophodonty and a functional interpretation of peculiarities in the masticatory system of *Arsinoitherium* (Mammalia: Embrithopoda). *Hist. Biol.* **6**: 91–111.

Court, N. 1995. A new species of *Numidotherium* (Mammalia: Proboscidea) from the Eocene of Libya and the early phylogeny of the Proboscidea. *J. Vert. Paleontol.* **15**: 650–671.

Covert, H.H. 2002. The earliest fossil primates and the evolution of prosimians: introduction. In W. Hartwig (ed) *The primate fossil record.* Cambridge University Press: Cambridge, pp. 13–20.

Cowles, R.B. 1958. Possible origin of dermal temperature regulation. *Evolution* **12**: 347–357.

Cox, C.B. 1959. On the anatomy of a new dicynodont genus with evidence of the position of the tympanum. *Proc. Zool. Soc. Lond.* **132**: 321–367.

Cox, C.B. 1965. New Triassic dicynodonts from South America, their origins and relationships. *Phil. Trans. R. Soc. B* **248**: 457–516.

Cox, C.B. 1972. A new digging dicynodont from the Upper Permian of Tanzania. In K.A. Joysey and T.S. Kemp (eds) *Studies in vertebrate evolution.* Oliver and Boyd: Edinburgh, pp. 173–190.

Cox, C.B. 1991. The Pangaea dicynodont *Rechnisaurus* and the comparative biostratigraphy of Triassic dicynodont faunas. *Palaeonology* **34**: 767–748.

Cox, C.B. 1998. The jaw function and adaptive radiation of the dicynodont mammal-like reptiles of the Karoo basin of South Africa. *Zool. J. Linn. Soc.* **122**: 349–384.

Crochet, J.-Y. and Sigé, B. 1996. Un marsupial ancien (transition Crétacé-Tertiare) á denture évoluée en Amérique du Sud (Chulpas, Formation Umayo, Pérou). *Neues Jr. Geol.Paläontol.*: 622–634.

Crompton, A.W. 1955. On some Triassic cynodonts from Tanganyika. *Proc. Zool. Soc. Lond.* **125**: 617–669.

Crompton, A.W. 1958. The cranial morphology of a new genus and species of ictidosaurian. *Proc. Zool. Soc. Lond.* **130**: 183–216.

Crompton, A.W. 1963. On the lower jaw of *Diarthrognathus* and the origin of the mammalian lower jaw. *Proc. Zool. Soc. Lond.* **140**: 697–753.

Crompton, A.W. 1971. The origin of the tribosphenic molar. In D.M. Kermack and K.A. Kermack (eds) *Early mammals. Zool. J. Linn. Soc.* **50** (Suppl. 1): 65–87.

Crompton, A.W. 1972*a*. Postcranial occlusion in cynodonts and tritylodonts. *Bull. Br. Mus Nat. Hist. (Geol.)* **21**: 29–71.

Crompton, A.W. 1972*b*. The evolution of the jaw articulation of cynodonts. In K.A. Joysey and T.S. Kemp (eds) *Studies in vertebrate evolution.* Oliver and Boyd: Edinburgh, pp. 231–253.

Crompton, A.W. 1974. The dentitions and relationships of the southern African mammals *Erythrotherium parringtoni* and *Megazostrodon rudnerae. Bull. Br. Mus. Nat. Hist. (Geol.)* **24**: 399–437.

Crompton, A.W. and Ellenberger, F. 1957. On a new cynodont from the Molteno Beds and the origin of tritylodontids. *Ann. S. Afr. Mus.* **44**: 1–14.

Crompton, A.W. and Hylander, W.L. 1986. Changes in mandibular function following the acquisition of a dentary-squamosal jaw articulation. In N. Hotton III, P.D. MacLean, J.J. Roth, and E.C. Roth (eds) *The ecology and biology of mammal-like reptiles.* Smithsonian Institution Press: Washington, pp. 263–282.

Crompton, A.W. and Jenkins, F.A. Jr. 1967. American Jurassic symmetrodonts and Rhaetic "Pantotheres". *Science* **155**: 1006–1009.

Crompton, A.W. and Jenkins, F.A. Jr. 1968. Molar occlusion in late Triassic mammals. *Biol. Rev.* **43**: 427–458.

Crompton, A.W. and Jenkins, F.A. Jr. 1973. Mammals from reptiles: a review of mammalian origins. *Annu. Rev. Earth Planet. Sci.* **1**: 131–155.

Crompton, A.W. and Jenkins, F.A. Jr. 1978. Mesozoic mammals. In V.J. Maglio and H.B.S. Cooke (eds) *Evolution of African mammals.* Harvard University Press: Cambridge, MA, pp. 46–55.

Crompton, A.W. and Jenkins, F.A. Jr. 1979. Triconodonta. In J.A. Lillegraven, Z. Kielan-Jaworowska, and W.A. Clemens (eds) *Mesozoic mammals: the first two-thirds of mammalian history.* California University Press: Berkeley. pp. 74–90.

Crompton, A.W. and Kielan-Jaworowska, Z. 1978. Molar structure and occlusion in Cretaceous therian mammals. In P.M. Butler and K.A. Joysey (eds) *Studies in the development, function and evolution of teeth.* Academic Press: London. pp. 249–287.

Crompton, A.W. and Luo, Z. 1993. Relationships of the Liassic mammals *Sinoconodon, Morganucodon oehleri,* and *Dinnetherium.* In F.S. Szalay, M.C. McKenna, and M.J. Novacek (eds) *Mammalian phylogeny: Mesozoic differentiation, multituberculates, monotremes, early therians, and marsupials.* Springer-Verlag, New York, pp. 30–43.

Crompton, A.W. and Sun, A.-L. 1985. Cranial structure and relationships of the Liassic mammal *Sinoconodon.* *Zool. J. Linn. Soc.* **85**: 99–119.

Crompton, A.W., Taylor, C.R., and Jagger, J.A. 1978. Evolution of homeothermy in mammals. *Nature* **272**: 333–336.

Cruickshank, A.R.I. 1978. Feeding adaptations in Triassic dicynodonts. *Palaeontol. Afr.* **21**: 121–132.

Cuneo, N.R. 1996. Permian phylogeography in Gandwana. *Palaeogeog. Palaeoclimatol. Palaeoecol.* **125**: 75–104.

Currie, P.J. 1979. The osteology of haplodontine sphenacodonts (Reptilia: Pelycosauria). *Palaeontogr. Abt. A.* **163**: 130–168.

Dagosto, M. 2002. The origin and diversification of anthropoid primates. In W. Hartwig (ed) *The primate fossil record.* Cambridge University Press: Cambridge, pp. 125–132.

Dagosto, M., Gebo, D.L., Beard, K.C., and Qi, T. 1996. New primate postcranial remains from the middle Eocene Shanghuang fissures, southeastern China. *Am. J. Phys. Anthropol.* **99**: 92–93.

Damiani, R., Modesto, S., Yates, A. and Neveling, J. 2003. Earliest evidence of cynodont burrowing. *Proc. R. Soc. Lond.* **B270**: 1747–1751.

Dashzeveg, D. and Kielan-Jaworowska, Z. 1984. The lower jaw of an aegialodontid mammal from the Early Cretaceous of Mongolia. *Zool. J. Linn. Soc.* **82**: 217–227.

De Klerk, W.J. 2003. A dicynodont trackway from the *Cistecephalus* Assemblage Zone in the Karoo, east of Graaf-Reinet, South Africa. *Palaeontol. Afr.* **38**: 73–91.

DeConto, R.M. and Pollard, D. 2003. Rapid Cenozoic glaciation of Antarctica induced by declining atmospheric CO_2. *Nature* **421**: 245–249.

Delsuc, F., Catzeflis, M., Stanhope, M.J., and Douzery, J.P. 2001. The evolution of armadillos, anteaters and sloths depicted by nuclear and mitochondrial phylogenies: implications for the staus of the enigmatic fossil *Eutamandua. Proc. R. Soc. Lond.* **B268**: 1605–1615.

Desmond, A. 1985. Interpreting the origin of mammals: new approaches to the history of palaeontology. *Zool. J. Linn. Soc.* **82**: 7–16.

Dewar, R.E. 1984. Extinctions in Madagascar: the loss of the subfossil fauna. In P.S. Martin and R.G. Klein (eds) *Quaternary extinctions: a prehistoric revolution.* Arizona University Press: Tucson, pp. 574–593.

Dilkes, D.W. and Reisz, R.R. 1996. First record of a basal synapsid ('mammal-like reptile') in Gondwana. *Proc. R. Soc. Lond.* **B263**: 1165–1170.

Domning, D. P. 2001. The earliest known fully quadrupedal sirenian. *Nature* **413**: 625–627.

Domning, D.P. 2002*a*. Desmostylia. In W.F. Perrin, B. Würsig, and J.G.M. Thewissen (eds) *Encyclopedia of marine mammals.* Academic Press: San Diego, pp. 19–32.

Domning, D.P. 2002*b*. Sirenian evolution In W.F. Perrin, B. Würsig, and J.G.M. Thewissen (eds) *Encyclopedia of marine mammals.* Academic Press: San Diego.

Domning, D.P., Ray, C.E., and McKenna, M.C. 1986. Two new Oligocene desmostylians and a discussion of tethytherian systematics. *Smiths. Contrib. Paleobiol.* **52**: 1–69.

Ducrocq, S. 1999. *Siamopithecus eocaenus,* a late Eocene anthropoid primate from Thailand: its contribution to the evolution of anthropoids in Southeast Asia. *J. Hum. Evol.* **36**: 613–635.

Ducrocq, S. 2001. Palaeogene anthropoid primates from Africa and Asia: new phylogenetical evidence. *C.R. Acad. Sci. Paris, Sci. Terre Planet.* **332**: 351–356.

Ducrocq, S., Buffetaut, E., Buffetaut-Tong, H., Jaeger, J.-J., Jongkanjanasoontorn, Y., and Suteethorn, V. 1992. First fossil flying lemur: a dermopteran from the late Eocene of Thailand. *Palaeontology* **35**: 373–380.

Dumont, E.R., Strait, S.G., and Friscia, A.R. 2000. Abderitid marsupials from the Miocene of Patagonia: an assessment of form, function, and evolution. *J. Paleontol.* **74**: 1161–1172.

Durand, J.F. 1991. A revised description of the skull of *Moschorhinus* (Therapsida, Therocephalia). *Ann. S. Afr. Mus.* **99**: 381–413.

Duvall, D. 1986. A new question of pheromones: aspects of possible chemical signaling and reception in the mammal-like reptiles. In N. Hotton III, P.D. MacLean, J.J. Roth, and E.C. Roth (eds) *The ecology and biology of mammal-like reptiles*. Smithsonian Institution Press: Washington, pp. 219–238.

Efremov, I.A. 1940. *Ulemosaurus svijagensis* Riab., ein Dinocephale aus den Ablagerungen des Perm der USSR. *Nova Acta Leopold.* **9**: 12–205.

Eisenberg, J.E. 1981. *The mammalian radiations: an analysis of trends in evolution, adaptation, and behaviour*. The Athlone Press: London.

Eisenberg, J.E. 1990. The behavioural/ecological significance of body size in the Mammalia. In J. Damuth and B.J. McFadden (eds) *Body size in mammalian paleobiology: estimation and biological implications*. Cambridge University Press: Cambridge, pp. 25–37.

Eizirik, E., Murphy, W.J., and O'Brien, S.J. 2001. Molecular dating and biogeography of the early placental mammal radiation. *J. Hered.* **92**: 212–219.

Elinson, R.P. 1989. Egg evolution. In D.B. Wake and G. Roth (eds) *Complex organismal functions: integration and evolution in vertebrates*. John Wiley & Sons: Chichester, pp. 251–262.

Elliot, D.H., Askin, R.A., Kyte, F.T., and Zinsmeister, W.J. 1994. Iridium and dinocysts at the Cretaceous–Tertiary boundary on Seymour Island, Antarctica: implications for the K–T event. *Geology* **22**: 675–678.

Erwin, D.H., Bowring, S.A., and Yugan, J. 2002. End-Permian mass extinctions: a review. *Geol. Soc. Am. Spec. Pap.* **356**: 363–383.

Farmer, C.G. 2000. Parental care: the key to understanding endothermy and other convergent features in birds and mammals. *Am. Nat.* **155**: 326–334.

Fischer, M.S. 1989. Hyracoids, the sister-group of perissodactyls. In D.R. Prothero and R.M. Schoch (eds) *The evolution of perissodactyls*. Oxford University Press: Oxford, pp. 37–56.

Fischer, M.S. and Tassy, P. 1993. The interrelation between Proboscidea, Sirenia, Hyracoidea, and Mesaxonia: the morphological evidence. In F.S. Szaley, M.J. Novacek, and M.C. McKenna (eds) *Mammal phylogeny: placentals*. Springer-Verlag: New York, pp. 217–234.

Flynn, J.J. 1998. Early Cenozoic Carnivora ('Miacoidea'). In C.M. Janis, K.M. Scott, and L.L. Jacobs (eds) *Evolution of Tertiary mammals of North America, Vol. 1: Terrestrial carnivores, ungulates, and ungulate-like mammals*. Cambridge University Press: Cambridge, pp. 110–123.

Flynn, J.J. and Wyss, A.R. 1998. Recent advances in South American mammalian paleontology. *TREE* **13**: 449–454.

Flynn, J.J. and Wyss, A.R. 1999. New marsupials from the Eocene-Oligocene transition of the Andean Main Range, Chile. *J. Vert. Paleontol.* **19**: 533–549.

Flynn, J.J., Jacobs, L.L., and Cheema, I.U. 1986. Baluchimyinae, a new ctenodactyloid rodent subfamily from the Miocene of Baluchistan. *Am. Mus. Novit.* **2841**: 1–58.

Flynn, J.J., Neff, N.A., and Tedford, R.H. 1988. Phylogeny of the Carnivora. In M.R. Benton (ed) *The phylogeny and classification of the tetrapods, Vol. 2: Mammals*. Oxford University Press: Oxford, *Systematics Association Special Volume* **35B**, pp. 73–115.

Flynn, J.J., Wyss, A.R., Charrier, R. and Swisher, C.C. 1995. An early Miocene anthropoid skull from the Chilean Andes. *Nature* **373**: 603–607.

Flynn, J.J., Parrish, J.M., Rakotosamimanana, B., Simpson, W.F., and Wyss, A.R. 1999. A Middle Jurassic mammal from Madagascar. *Nature* **401**: 57–60.

Foote, M., Hunter, J.P., Janis, M., and Sepkoski, J.J. Jr 1999. Evolutionary and preservational constraints on origins of biologic groups: divergence times of eutherian mammals. *Science* **283**: 1310–1314.

Fostowicz, F.L. and Kielan-Jaworowska, Z. 2002. Lower incisor in zalambdolestid mammals (Eutheria) and its phylogenetic implications. *Acta Palaeontol. Pol.* **47**: 177–180.

Fourie, H. 2001. Morphology and function of the postcrania of selected genera of Therocephalia (Amniota: Therapsida). PhD Thesis. University of the Witwatersrand, Johannesburg.

Fox, R. and Youzwyshyn, G.P. 1994. New primitive carnivorans (Mammalia) from the Paleocene of Western Canada, and their bearing on the relationships of the order. *J. Vert. Paleontol.* **14**: 382–404.

Fox, R.C. 1974. *Deltatheroides*-like mammal from the Upper Cretaceous of North America. *Nature* **249**: 392.

Fox, R.C. 1984. *Paranyctoides maleficius* (new species), an early eutherian mammal from the Cretaceous of Alberta. *Spec. Publ. Carnegie Mus. Nat. Hist.* **9**: 9–20.

Frazzetta, T.H. 1969. Adaptive problems and possibilities in the temporal fenestration of tetrapod skulls. *J. Morphol.* **125**: 145–158.

Froehlich, D.J. 2002. Quo vadis eohippus? The systematics and taxonomy of the early Eocene equids (Perissodactyla). *Zool. J. Linn. Soc.* **134**: 141–256.

Frolich, L.M. 1997. The role of the skin in the origin of amniotes: permeability barrier, protective covering and mechanical support. In S.S. Sumida and K.L.M. Martin (eds) *Amniote origins: completing the transition to land*. Academic Press: London, pp. 327–352.

Gambaryan, P.P. and Kielan-Jaworowska, Z. 1995. The masticatory musculature of Asian taeniolabidoid multituberculate mammals. *Acta Palaeontol. Pol.* **40**: 45–108.

Gambaryan, P.P. and Kielan-Jaworowska, Z. 1997. Sprawling versus parasagittal stance in multituberculate mammals. *Acta Palaeontol. Pol.* **42**: 13–44.

Gardiner, B. G. 1982. Tetrapod classification. *Zool. J. Linn. Soc.* **74**: 207–232.

Gatesy, J. and O'Leary, M.A. 2001. Deciphering whale origins with molecules and fossils. *TREE* **16**: 562–570.

Gaudin, T.J. and Branham, D.G. 1998. The phylogeny of the Myrmecophagidae (Mammalia, Xenarthra, Vermilingua) and the relationships of *Eurotamandua* to the Vermilingua. *J. Mamm. Evol.* **5**: 237–265.

Gauthier, J.A., Kluge, G., and Rowe, T. 1988. The early evolution of the Amniota. In M.J. Benton (ed) *The phylogeny and classification of the tetrapods*, Vol. 1. Oxford University Press: Oxford, *Systematics Association Special Volume* **35A**, pp. 103–156.

Gayet, M., Rage, J.-C., Sempere, T., and Gagnier, P.-Y. 1992. Moldalités des échanges de vertebras continentaux entre l'Amerique du Nord et l'Amerique du Sud au Cretacé supérieur et au Paléocène. *Bull. Soc. Geol. Fr.* **163**: 781–791.

Gayet, M., Marshall, L.G., Sempere, T., Meunier, F.J., Cappetta, H., and Rage, J.-C. 2001. Middle Maastrichtian vertebrates (fishes, amphibians, dinosaurs and other reptiles, mammals) from Pajcha Pata (Bolivia). Biostratigraphic, palaeoecologic and palaeobiogeographic implications. *Palaeogeogr. Palaeoclimatol. Palaeoecol.* **169**: 39–68.

Gebo, D.L. 2002. Adapiformes: phylogeny and adaptation. In W. Hartwig (ed) *The primate fossil record*. Cambridge University Press: Cambridge. pp. 21–43.

Gentry, A.W. and Hooker, J.J. 1988. The phylogeny of the Artiodactyla. In M.J. Benton (ed) *The phylogeny and classification of the tetrapods*, Vol. 2: *Mammals*. Oxford University Press: Oxford, pp. 235–272.

Gheerbrant, E. 1992. The Paleocene mammals from the Ouarzazate Basin (Morocco): I. General introduction and Palaeoryctidae. *Palaeontogr. Abt. A. Palaeozool. Stratig.* **224**: 67–132.

Gheerbrant, E. 1995. The Paleocene mammals from the Ouarzazate Basin (Morocco): III. Adapisoriculidae and other mammals (Carnivora, ?Creodonta, Condylarthra, ?Ungulata and incertae sedis. *Palaeontogr. Abt. A Palaeozool. Stratigr.* **237**: 39–132.

Gheerbrant, E. and Astibia, H. 1994. Un nouveau mammifère du Maastrichian de Laño (Pays Basque espagnol). *C. R. Acad. Sci. Paris, Ser.* II **318**: 1125–1131.

Gheerbrant, E. and Hartenberger, J.-L. 1999. *Palaont. Zeit.* Nouveau mammifere insectivore (?Lipotyphla, ?Erinaceomorpha,) de l'Eocene inférieur de Chambi (Tunisie). **73**: 143–156.

Gheerbrant, E., Sudre, J., and Capetta, H. 1996. A Palaeocene proboscidean from Morocco. *Nature* **383**: 68–70.

Gheerbrant, E., Sudre, J., Capetta, H. and Bignot, G. 1998. *Phosphatherium escuillei*, from the Thanetian of the Ouled Abdoun Basin (Morocco), oldest known proboscidean (Mammalia) from Africa. *Geobios* **31**: 247–269.

Gheerbrant, E., Sudre, J., Iarochene, M., and Moumni, A. 2001. First ascertained African 'condylarth' mammals (Primitive ungulates: cf. Bulbulodentata) and cf. Phenacodonta) from the earliest Ypresian of the Ouled Abdoun Basin, Morocco. *J. Vert. Paleontol.* **21**: 107–118.

Gheerbrant, E., Sudre, J., Cappetta, H., Iarochène, M., Amaghzaz, M., and Bouya, B. 2002. A new large mammal from the Ypresian of Morocco: evidence of surprising diversity of early proboscideans. *Acta Palaeontol. Pol.* **47**: 493–506.

Gill, P.G. 1974. Resorption of premolars in the early mammal *Kuehneotherium praecursoris*. *Archs Oral Biol.* **19**: 327–328.

Gillette, D.D. and Ray, C.E. 1981. Glyptodonts of North America. *Smiths. Contrib. Paleobiol.* **40**: 1–255.

Gingerich, P.D., Smith, B.H., and Simons, E.L. 1990. Hind limbs of Eocene *Basilosaurus*: evidence of feet in whales. *Science* **249**: 154–157.

Gingerich, P.D., Haq, M. ul, Zalmout, I.S., Khan, I.H., and Malkani, M.S. 2001. Origin of whales from early artiodactyls: hands and feet of Eocene Protocetidae from Pakistan. *Science* **293**: 2239–2242.

Godinot, M. and Mahboubi, M. 1992. Earliest known simian primate found in Algeria. *Nature* **357**: 324–326.

Godthelp, H., Archer, M., Cifelli, R., Hand, S. J., and Gilkeson, C.F. 1992. Earliest known Australian Tertiary mammal fauna. *Nature* **356**: 514–516.

Godthelp, H., Wroe, S., and Archer, M. 1999. A new marsupial from the early Eocene Tingamarra Local Fauna of Murgon, southeastern Queensland: a prototypical Australian marsupial? *J. Mamm. Evol.* **6**: 289–313.

Goin, F.J. and Carlini, A.A. 1995. An early Tertiary microbiotheriid marsupial from Antarctica. *J. Vert. Paleontol.* **15**: 205–207.

Goin, F.J. and Candela, A.M. 1996. A new Early Eocene polydolopimorphian (Mammalia, Marsupialia) from Patagonia. *J. Vert. Paleont.* **16**: 292–296.

Goin, F.J., Case, J.A., Woodburne, M.O., Vizcaíno, S.F., and Reguero, M.A. 1999. New discoveries of 'opossum-like' marsupials from Antarctica (Seymour Island, Medial Eocene). *J. Mamm. Evol.* **6**: 335–365.

Golovneva, L.B. 1994. The flora of the Maastrichtian-Danian deposits of the Koryak Upland, northeast Russia. *Cret. Res.* **15**: 89–100.

Golubev, V.K. 2000. The faunal assemblages of Permian terrestrial vertebrates from Eastern Europe. *Paleontol. J.* **34** (Suppl. 2): S111–S224.

Gow, C.E. 1978. The advent of herbivory in certain reptilian lineages during the Triassic. *Palaeontol. Afr.* **21**: 133–141.

Gow, C.E. 1980. The dentitions of the Tritheledontidae (Therapsida: Cynodontia). *Proc. R. Soc. Lond.* **B208**: 461–481.

Gow, C.E. 2001. A partial skeleton of the tritheledontid *Pachygenelus* (Therapsida: Cynodontia). *Palaeontol. Afr.* 93–97.

Graham, R.W. 1997. The Pleistocene terrestrial mammal fauna of North America. In C.M. Janis, K.M. Scott, and L.L. Jacobs (eds) *Evolution of Tertiary mammals of North America. Vol. 1: Terrestrial carnivores, ungulates, and ungulate-like mammals.* Cambridge University Press: Cambridge, pp. 66–71.

Graham, R.W. and Lundelius, E.L., Jr 1984. Coevolutionary disequilibrium and Pleistocene extinctions. In P.S. Martin and R.G. Klein (eds) *Quaternary extinctions: a prehistoric revolution.* Arizona University Press: Tucson, pp. 223–249.

Graur, D. and Higgins, D. 1994. Molecular evidence for the inclusion of cetaceans within the order Artiodactyla. *Mol. Biol. Evol.* **11**: 357–364.

Graybeal, A., Rosowski, J.J., Ketton, D.R., and Crompton, A.W. 1989. Inner-ear structure in *Morganucodon*, an early Jurassic mammal. *Zool. J. Linn. Soc.* **96**: 107–117.

Grayson, D.K. 1984. Nineteenth-century explanations of Pleistocene extinctions: a review and analysis. In P.S. Martin and R.G. Klein (eds) *Quarternary extinctions: a prehistoric revolution.* Arizona University Press: Tucson. pp. 5–39.

Gregory, W.K. 1910. The orders of mammals. *Bull. Am. Mus. Nat. Hist.* **27**: 1–524.

Gregory, W.K. 1947. The monotremes and the palimpsest theory. *Bull. Am. Mus. Nat. Hist.* **88**: 1–52.

Gregory, W.K. and Simpson, G.G. 1926. Cretaceous mammal skulls from Mongolia. *Am. Mus. Novit.* **225**: 1–20.

Grine, F.E. 1997. Dinocephalians are not anomodonts. *J. Vert. Paleont.* **17**: 177–188.

Gunnell, G.F. 1998. Creodonta. In C.M. Janis, K.M. Scott, and L.L. Jacobs (eds) *Evolution of Tertiary mammals of North America, Vol. 1: Terrestrial carnivores, ungulates, and ungulate-like mammals.* Cambridge University Press: Cambridge, pp. 91–109.

Gunnell, G.F. and Rose, K.D. 2002. Tarsiiformes: evolutionary history and adaptation In W. Hartwig (ed) *The primate fossil record.* Cambridge University Press: Cambridge, pp. 45–82.

Guthrie, R.D. 1984. Mosaics, allelochemics, and nutrients: an ecological theory of Late Pleistocene megafaunal extinctions. In P.S. Martin and R.G. Klein (eds) *Quaternary extinctions: a prehistoric revolution.* Arizona University Press: Tucson, pp. 256–298.

Guthrie, R.D. 2003. Rapid body size decline in Alaskan Pleistocene horses before extinction. *Nature* **426**: 169–171.

Hahn, G. 1973. Neue Zähne von Haramiyiden aus der Deutschen Ober-Trias und ihre beziehungen zu den Multituberculaten. *Palaeontogr. Abt. A* **142**: 1–15.

Hahn, G. and Hahn, R., 1983. Multituberculata. In F.Westphal (ed) *Fossilium Catalogus 1: Animalia, part 127.* Kugler Publications: Amsterdam. pp. 409.

Hahn, G., Hahn, R., and Godefroit, P. 1994. Zur Stellung der Dromatheriidae (Ober-Trias) zwischen den Cynodontia und den Mammalia. *Geol. Palaeontol.* **28**: 141–159.

Hahn, G., Sigogneua-Russell, D. and Wouters, G. 1989. New data on Therioteinidae – their relations with Paulchoffatiidae and Haramiyidae. *Geol. Paleont.* **23**: 205–215.

Hall, L.S. 1987. Syndactyly in marsupials—problems and prophesies. In M. Archer (ed) *Possums and opossums: studies in evolution.* Surrey Beatty and Sons: Chipping Norton, Australia, pp. 245–255.

Hallam, A. 1994. An outline of Phanerozoic biogeography. Oxford University Press: Oxford.

Hallam, A. and Wignall, P.B. 1997. *Mass extinctions and their aftermath.* Oxford University Press: Oxford.

Hammer, W.R. 1995. New therapsids from the Upper Fremouw Formation (Triassic) of Antarctica. *J. Vert. Paleontol.* **15**: 105–112.

Hammer, W.R. and Cosgriff, J.W. 1981. *Myosaurus gracilis*, an anomodont reptile from the Lower Triassic of Antarctica and South Africa. *J. Paleontol.* **55**: 410–424.

Hand, S., Novacek, M., Godthelp, H., and Archer, M. 1994. First Eocene bat from Australia. *J. Vert. Paleontol.* **14**: 375–381.

Harrison, T. 2002. Late Oligocene to middle Miocene catarrhines from Afro-Arabia. In W. Hartwig (ed) *The primate fossil record.* Cambridge University Press: Cambridge, pp. 311–338.

Haughton, S. and Brink, A.S. 1954. A bibliographic list of the Reptilia from the Karroo beds of South Africa. *Palaeontol. Afr.* **2**: 1–187.

Hayes, J.P. and Garland, T. Jr 1995. The evolution of endothermy: testing the aerobic capacity model. *Evolution* **49**: 836–847.

Heatwole, H.E. and Carroll, R.L. (eds) 2000. *Amphibian biology, Vol. 4. Palaeontology: the evolutionary history of amphibians.* Surrey Beatty & Sons: Chipping Norton, Australia.

Hedges, S.B. and Kumar, S. 1999. Divergence times of eutherian mammals. *Science* **285**: 2031a.

Heinrich, W.-D. 1999. First haramiyid (Mammalia, Allotheria) from the Mesozoic of Gondwana. *Mitt. Mus. Naturkunde Berlin. Geowiss. Reiche* **2**: 159–170.

Heissig, K. 1989. The Rhinocerotidae. In D.R. Prothero and R.M. Schoch (eds) *The evolution of perissodactyls.* Oxford University Press: Oxford, pp. 399–417.

Henkel, S. and Krusat, B. 1980. Die Fossil—Lagerstätte in der Kohlengrube Guimarota (Portugal) und der erste Fund eines Docodontiden-Skelettes: *Berlin. Wiss. Abt. A.* **20**: 209–216.

Hennig, W. 1966. *Phylogenetic systematics.* University of Illinois Press: Urbana.

Hillenius, W.J. 1992. The evolution of nasal turbinates and mammalian endothermy. *Paleobiology* **18**: 17–29.

Hillenius, W.J. 1994. Turbinates in therapsids: evidence for Late Permian origins of mammalian endothermy. *Evolution* **48**: 207–229.

Hillenius, W.J. 2000. Septomaxilla of nonmammalian synapsids: soft tissue correlates and a new functional interpretation. *J. Morphol.* **245**: 29–50.

Holbrook, L.T. 2001. Comparative osteology of early Tertiary tapiromorphs (Mammalia, Perissodactyla). *Zool. J. Linn. Soc.* **132**: 1–54.

Honey, J.G., Harrison, J.A., Prothero, D.R., and Stevens, M.S. 1998. Camelidae. In C.M. Janis, K.M. Scott, and L.L. Jacobs (eds) *Evolution of Tertiary mammals of North America, Vol. 1: Terrestrial carnivores, ungulates, and ungulatelike mammals.* Cambridge University Press: Cambridge, pp. 439–462.

Hooker, J.J. 1994. The beginning of the equoid radiation. *Zool. J. Linn. Soc.* **112**: 29–63.

Hopson, J.A. 1966. The origin of the mammalian middle ear. *Am. Zool.* **6**: 437–450.

Hopson, J.A. 1967. Comments on the competitive inferiority of the multituberculates. *Syst. Zool.* **16**: 352–355.

Hopson, J.A. 1971. Postcanine replacement in the gomphodont cynodont *Diademodon.* In D.M. Kermack and K.A. Kermack (eds) *Early mammals, Suppl. 1. Zool. J. Linn. Soc.* **50**: 1–22.

Hopson, J.A. 1973. Endothermy, small size, and the origin of mammalian reproduction. *Am. Nat.* **107**: 446–452.

Hopson, J.A. 1975. Patterns of evolution in the manus and pes of non-mammalian therapsids. *J. Vert. Paleontol.* **15**: 615–639.

Hopson, J.A. 1979. Paleoneurology. In C. Gans, R.G. Northcutt and P. Ulinski (eds) *Biology of the Reptilia, Vol 9.* Academic Press: London, pp. 39–146.

Hopson, J.A. 1991. Systematics of the nonmammalian Synapsida and implications for patterns of evolution in synapsids. In H.-P. Schultze and L. Trueb (eds) *The origin of higher groups of tetrapods: controversy and consensus.* Cornell University Press: Ithaca. pp. 635–693.

Hopson, J.A. 2001. Origin of mammals. In D.E.G. Briggs and P.R. Crowther (eds) *Palaeobiology II.* Blackwell Science: Oxford. pp. 88–94.

Hopson, J.A. and Barghusen, H.R. 1986. An analysis of therapsid relationships. In N.Hotton III, P.D.McLean, J.J.Roth, and E.C. Roth (eds). *The ecology and biology of mammal-like reptiles.* Smithsonian Institution Press: Washington, pp. 83–106.

Hopson, J.A. and Crompton, A.W. 1969. Origin of mammals. *Evol. Biol.* **3**: 15–72.

Hopson, J.A. and Kitching, J.W. 1972. A revised classification of cynodonts (Reptilia: Therapsida). *Palaeontol. Afr.* **14**: 71–85.

Hopson, J.A. and Kitching, J.W. 2001. A probainognathian cynodont from South Africa and the phylogeny of nonmammalian cynodonts. *Bull. Mus. Comp. Zool. Harv.* **156**: 5–35.

Hopson, J.A. and Rougier, G.W. 1993. Braincase structure in the oldest known skull of a therian mammal: implications for mammalian systematics and cranial evolution. *Am. J. Sci.* **293**: 268–299.

Hopson, J.A., Kielan-Jaworowska, Z., and Allin, E.F. 1989. The cryptic jugal of multituberculates. *J. Vert. Paleontol.* **20**: 201–209.

Horovitz, I. and Sánchez-Villagra, M.R. 2003. A morphological analysis of marsupial mammal higher-level phylogenetic relationships. *Cladistics* **19**: 181–212.

Hotton, N. III 1974. A new dicynodont from the *Cynognathus* zone deposits of South Africa. *Ann. S. Afr. Mus.* **64**: 157–166.

Hotton, N. III 1986. Dicynodonts and their role as primary consumers. In N. Hotton III, J.J. Roth, and E.C.Roth (eds) *The ecology and biology of mammal-like reptiles.* Smithsonian Institute Press: Washington, pp. 71–82.

Hu, Y.-M., Wang, Y.-Q., Luo, Z.-H., and Li, C.-K. 1997. A new symmetrodont mammal from China and its implications for mammalian evolution. *Nature* **390**: 137–142.

Huchon, D., Madsen, O., Sibbald, J.B., Ament, K., Stanhope, M.J., Catzeflis, F., Jong, W.W. de, and Douzery, E.J.P. 2002. Rodent phylogeny and a timescale for the evolution of Glires: evidence from an extensive taxon sampling using three nuclear genes. *Mol. Biol. Evol.* **19**: 1053–1065.

Hunt, R.M., Jr 1998. Ursidae. In C.M. Janis, K.M. Scott, and L.L. Jacobs (eds) *Evolution of Tertiary mammals of North America, Vol. 1: terrestrial carnivores, ungulates, and ungulate-like mammals.* Cambridge University Press: Cambridge, pp. 174–195.

Hurlbert, S.H. and Archibald, J.D. 1995. No statistical support for sudden (or gradual) extinction of dinosaurs. *Geology* **23**: 881–884.

Hurum, J.H. 1994. The snout and orbit of Mongolian multituberculates studied by serial sections. *Acta Palaeontol. Pol.* **39**: 181–221.

Hurum, J.H., Presley, R., and Kielan-Jaworowska, Z. 1996. The middle ear in multituberculate mammals. *Acta Palaeontol. Pol.* **41**: 253–275.

Ivakhnenko, M.F. 1990. The late Palaeozoic faunal assemblage of tetrapods from the deposits of the Mezen' River. *Paleontol. J.* **23**: 81–90.

Ivakhnenko, M.F. 1991. Elements of the Early Permian tetrapod faunal assemblages of Eastern Europe. *Paleontol. J.* **24**: 104–112.

Ivakhnenko, M.F. 1994. A new Late Permian dromasaurian (Anomodontia) from Eastern Europe. *Paleontol. J.* **28**: 96–103.

Ivakhnenko, M.F. 1995. New primitive therapsids from the Permian of Eastern Europe. *Paleontol. J.* **30**: 337–343.

Ivakhnenko, M.F. 1996. Primitive anomodonts, venyukoviids, from the Late Permian of Eastern Europe. *Paleontol. J.* **30**: 575–582.

Ivakhnenko, M.F. 1999. Biarmosuches from the Ocher Faunal Assemblage of Eastern Europe. *Paleontol. J.* **33**: 289–296.

Ivakhnenko, M.F. 2000a. Estemmenosuches and primitive theriodonts from the Late Permian. *Paleontol. J.* **34**: 189–197.

Ivakhnenko, M.F. 2000b. The Nikkasauridae-problematic primitive therapsids from the Late Permian of the Mezen localities. *Paleontol. J.* **34** (Suppl. 2): S179–S186.

Ivakhnenko, M.F. 2003. Eotherapsids from the East European Placket (Late Permian). *Paleont. J.* **37**, Suppl. 4: S339–S465.

Ivakhnenko, M.F., Golubev, V.K., and Gubin, Y. M. et al. 1997. Permian and Triassic tetrapods of Eastern Europe. *Trudy Paleontol. Inst. Ross. Akad. Nauk* **268**: 1–216.

Jaeger, J.-J. 1988. Rodent phylogeny: new data and old problems. In M.J. Benton (ed) *The phylogeny and classification of tetrapods, Vol. 2; Mammals. Systematics Association Special Volume* **35B**, pp. 177–199.

Jaeger, J.J., Courtillot, V., and Tapponier, P. 1989. Paleontological view of the ages of the Deccan Traps, and the Cretaceous/Tertiary boundary, and the India–Asia collision. *Geology* **17**: 316–319.

Janis, C.M. 1989. A climatic explanation for patterns of evolutionary diversity in ungulate mammals. *Paleontology* **32**: 463–481.

Janis, C.M. 1993. Tertiary mammal evolution in the context of changing climates, vegetation, and tectonic events. *Annu. Rev. Ecol. Syst.* **24**: 467–500.

Janis, C.M. and Farmer, C. 1999. Proposed habits of early tetrapods: gills, kidneys and the water–land transition. *Zool. J. Linn. Soc.* **126**: 117–126.

Janis, C.M. and Manning, E. 1998. Antilocapridae. In C.M. Janis, K.M. Scott, and L.L. Jacobs (eds) *Evolution of Tertiary mammals of North America, Vol. 1: Terrestrial carnivores, ungulates, and ungulatelike mammals.* Cambridge University Press: Cambridge, pp. 491–508.

Janis, C.M. and Scott, K.M. 1988. The phylogeny of the Ruminantia (Artiodactyla, Mammalia). In M.J. Benton (ed) *The phylogeny and classification of the tetrapods, Vol. 2: Mammals.* Oxford University Press: Oxford, pp. 273–282.

Janis, C.M., Archibald, J.D., Cifelli, R.L., Lucas, S.G., Schaff, C.R., Schoch, R.M., and Williamson, T.E. 1998a. Archaic ungulates and ungulate-like mammals. In C.M. Janis, K.M. Scott, and L.L. Jacobs (eds) *Evolution of Tertiary mammals of North America, Vol. 1: Terrestrial carnivores, ungulates, and ungulate-like mammals.* Cambridge University Press: Cambridge, pp. 247–259.

Janis, C.M., Baskin, J.A., Berta, A., Flynn, J.J., Gunnell, G.F., Hunt, R.M. Jr, Martin, L.D., and Munthe, K. 1998b. Carnivorous mammals. In C.M. Janis, K.M. Scott, and L.L. Jacobs (eds) *Evolution of Tertiary mammals of North America, Vol. 1: Terrestrial carnivores, ungulates, and ungulate-like mammals.* Cambridge University Press: Cambridge, pp. 73–90.

Janis, C.M., Effinger, J.A., Harrison, J.A., Honey, J.G., Kron, G., Lander, B., Manning, E., Prothero, D.R., Stevens, M.S., Stucky, R.K., Webb, S.D., and Wright, D.B. 1998c. Artiodactyla. In C.M. Janis, K.M. Scott, and L.L. Jacobs (eds) *Evolution of Tertiary mammals of North America, Vol. 1: Terrestrial carnivores, ungulates, and ungulate-like mammals.* Cambridge University Press: Cambridge, pp. 337–357.

Janis, C.M., Damuth, J., and Theodor, J.M. 2000. Miocene ungulates and terrestrial primary productivity: where have all the browsers gone? *Proc. Natl. Acad. Sci. USA.* **97**: 7899–7904.

Janis, C.M., Damuth, J., and Theodor, J.M. 2002. The origins and evolution of the North American grassland biome: the story from the hoofed mammals. *Palaeogeogr. Palaeoclimat. Palaeoecol.* **177**: 183–198.

Jarvik, E. 1980. *Basic structure and evolution of vertebrates, Vol. 1.* Academic Press: London.

Jarvik, E. 1996. The Devonian tetrapod *Ichthyostega. Fossil. Strat.* **40**: 1–213.

Jefferies, R.P.S. 1979. The origin of chordates—a methodological essay. In M.R. House (ed) *The origin of major invertebrate groups.* Academic Press: London., pp. 443–477.

Jenkins, F.A. Jr. 1969. Occlusion in *Docodon* (Mammalia, Docodonta). *Postilla* No. 139, 1–24.

Jenkins, F.A. Jr 1970. The Chañares (Argentina) Triassic reptile fauna VII. The postcranial skeleton of the traversodontid *Massetognathus pascuali* (Therapsida, Cynodontia). *Breviora* **352**: 1–28.

Jenkins, F.A. Jr 1971a. Limb posture and locomotion in the Virginia opossum (*Didelphis virginiana*) and in other non-cursorial mammals. *J. Zool. Lond.* **165**: 303–315.

Jenkins, F.A. Jr 1971*b*. The postcranial skeleton of African cynodonts. *Bull. Peabody Mus. Nat. Hist.* **36**: 1–216.

Jenkins, F.A. Jr 1974. The movement of the shoulder in claviculate and aclaviculate mammals. *J. Morphol.* **144**: 71–84.

Jenkins, F.A. Jr and Bramble, D.M. 1989. Structural and functional integration across the reptile-mammal boundary. In D.B. Wake and G. Roth (eds) *Complex organismal functions: integration and evolution in vertebrates.* John Wiley & Sons: Chichester, pp. 133–146.

Jenkins, F.A. Jr and Crompton, A.W. 1979. Triconodonta. In J.A. Lillegraven, Z. Kielan-Jaworowska and W.A. Clemens (eds) *Mesozoic mammals: the first two-thirds of mammalian history.* University of California Press: Berkeley, pp. 74–90.

Jenkins, F.A. Jr and Krause, D.W. 1983. Adaptations for climbing in North American multituberculates (Mammalia). *Science* **220**: 715.

Jenkins, F.A. Jr and Parrington, F.R. 1976. The postcranial skeleton of the Triassic mammals *Eozostrodon, Megazostrodon* and *Erythrotherium. Phil. Trans. R. Soc.* **B273**: 387–431.

Jenkins, F.A. Jr and Schaff, C.R. 1988. The Early Cretaceous mammal *Gobiconodon* (Mammalia, Triconodonta) from the Cloverly Formation in Montana. *J. Vert. Paleontol.* **6**: 1–24.

Jenkins, F.A. Jr, Crompton, A.W., and Downs, W. 1983. Mesozoic mammals from Arizona: new evidence on mammalian evolution. *Science* **222**: 1233–1235.

Jenkins, F.A. Jr, Gatesey, S.M., Shubin, N.H., and Amaral, W.W. 1997. Haramiyids and Triassic mammal evolution. *Nature* **385**: 715–718.

Jepson, G.L. 1966. Early Eocene bat from Wyoming. *Science* **154**: 1333–1339.

Jerison, H. J. 1973. *Evolution of the brain and intelligence.* Academic Press: London.

Ji, Q., Luo, X.-Z., and Ji, S. 1999. A Chinese triconodont mammal and mosaic evolution of the mammalian skeleton. *Nature* **398**: 326–330.

Ji, Q., Luo, Z.-X., Wible, J.R., Zhang, J.-P., and Georgi, J.A. 2002. The earliest known eutherian mammal. *Nature* **416**: 816–822.

Johnson, C.N. 2002. Determinants of loss of mammal species during the Late Quaternary 'megafauna' extinctions: life history and ecology, but not body size. *Proc. R. Soc. Lond.* **B269**: 2221–2227.

Johnson, K.R. 1993. High latitude deciduous forests and the Cretaceous–Tertiary boundary in New Zealand. *Geol. Soc. Am. Abt Program.* **25**: 295.

Johnson, K.R. and Hickey, L.J. 1990. Megafloral change across the Cretaceous/Tertiary boundary in the northern Great Plains and Rocky mountains, U.S.A. *Geol. Soc. Am. Spec. Pap.* **247**: 433–444.

Jong, W.W. de, Leunissen, J.A.M., and Wistow, G.J. 1993. Eye lens crystallins and the phylogeny of placental orders: evidence for a macroscelid–paenungulate clade. In F.S. Szalay, M.J. Novacek, and M.C. McKenna, (eds) *Mammal phylogeny: placentals.* Springer-Verlag: New York, pp. 5–12.

Kappelman, J., Rasmussen, D.T., Sanders, W.J., Fesaha, M., Bown, T., Copeland, P., Crabaugh, J., Fleagle, J., Glanz, M., Gordon, A., Jacobs, B., Maga, M., Muldoon, K., Pan, A., Pyne, L., Richmond, B., Ryan, T., Seiffert, E.R., Sen, S., Todd, L., Wiemann, M.C., and Winkler, A. 2003. Oligocene mammals from Ethiopia and faunal exchange between Afro-Arabia and Eurasia. *Nature* **426**: 549–552.

Kauffman, E.G. and Hart, M.B. 1996. Cretaceous bio-events. In O.H. Walliser (ed) *Global events and event stratigraphy in the Phanerozoic.* Springer-Verlag: Berlin, pp. 285–312.

Kay, R.F., Thorington, R.W., and Houde, P. 1990. Eocene plesiadapiform shows affinities with flying lemurs not primates. *Nature* **345**: 342–344.

Kay, R.F., Ross, C., and Williams, B.A. 1997. Anthropoid origins. *Science* **275**: 797–804.

Kemp, T.S. 1969*a*. The atlas-axis complex of the mammal-like reptiles. *J. Zool. Lond.* **159**: 223–248.

Kemp, T.S. 1969*b*. On the functional morphology of the gorgonopsid skull. *Phil. Trans. R. Soc.* **B256**: 1–83.

Kemp, T.S. 1972*a*. The jaw articulation and musculature of the whaitsiid Therocephalia. In K.A. Joysey and T.S. Kemp (eds) *Studies in vertebrate evolution.* Oliver & Boyd: Edinburgh, pp. 213–230.

Kemp, T.S. 1972*b*. Whaitsiid Therocephalia and the origin of cynodonts. *Phil. Trans. R. Soc.* **B264**: 1–54.

Kemp, T.S. 1978. Stance and gait in the hindlimb of a therocephalian mammal-like reptile. *J. Zool. Lond.* **186**: 143–161.

Kemp, T.S. 1979. The primitive cynodont *Procynosuchus*: functional anatomy of the skull and relationships. *Phil. Trans. R. Soc.* **B285**: 73–122.

Kemp, T.S. 1980*a*. Aspects of the structure and functional anatomy of the Middle Triassic cynodont *Luangwa. J. Zool. Lond.* **191**: 193–239.

Kemp, T.S. 1980*b*. Origin of the mammal-like reptiles. *Nature* **283**: 378–380.

Kemp, T.S. 1980*c*. The primitive cynodont *Procynosuchus*: structure, function and evolution of the postcranial skeleton. *Phil. Trans. R. Soc.* **288**: 217–258.

Kemp, T.S. 1982. *Mammal-like reptiles and the origin of mammals.* Academic Press: London.

Kemp, T.S. 1983. The relationships of mammals. *Zool. J. Linn. Soc.* **77**: 353–384.

Kemp, T.S. 1986. The skeleton of a baurioid therocephalian therapsid from the Lower Triassic (*Lystrosaurus* zone) of South Africa. *J. Vert. Paleontol.* **6**: 215–232.

Kemp, T.S. 1988*a*. A note on the Mesozoic mammals. In M.J. Benton (ed) *The phylogeny and classification of the tetrapods, Vol. 2: Mammals.* Oxford University Press: Oxford, pp. 23–29.

Kemp, T.S. 1988*b*. Haemothermia or Archosauria? The interrelationships of mammals, birds and crocodiles. *Zool. J. Linn. Soc.* **92**: 67–104.

Kemp, T.S. 1988*c*. Interrelationships of the Synapsida. In M.J. Benton (ed) *The phylgeny and classification of the tetrapods, Vol.2: Mammals.* Oxford University Press: Oxford, 23–29.

Kemp, T.S. 1999. *Fossils and evolution.* Oxford University Press: Oxford.

Kent, D.V., Cramer, B.S., Lanci, L., Wang, D., Wright, J.D., and Van der Voo, R. 2003. A case for a comet impact trigger for the Paleocene/Eocene thermal maximum and carbon isotope excursion. *Earth Planet. Lett.* **211**: 13–26.

Kermack, K.A. and Kielan-Jaworowska, Z. 1971. Therian and non-therian mammals. In D.M. Kermack and K.A.Kermack (eds) *Early mammals. Zool. J. Linn. Soc. Lond.* **50**(Suppl. 1) 103–115.

Kermack, K.A., Lees, P.M., and Mussett, F. 1965. *Aegialodon dawsoni,* a new trituberculosectorial tooth from the Lower Wealden. *Proc. R. Soc. Lond.* **B162**: 535–554.

Kermack, D.M., Kermack, K.A., and Mussett, F. 1968. The Welsh pantothere *Kuehneotherium praecursoris. J. Linn. Soc. Zool.* **53**: 87–175.

Kermack, K.A., Mussett, F., and Rigney, H.W. 1973. The lower jaw of *Morganucodon. Zool. J. Linn. Soc.* **53**: 86–175.

Kermack, K.A., Musett, F., and Rigney, H.W. 1981. The skull of *Morganucodon. Zool. J. Linn. Soc.* **71**: 1–158.

Kermack, K. A., Kermack, D.M., Lees, P. M., and Mills, J.R.E. 1998. New multituberculate-like teeth from the Middle Jurassic of England. *Acta Paleontol. Pol.* **43**: 581–606.

Keyser, A.W. and Cruickshank, A.R.I. 1979. The origins and classification of Triassic dicynodonts. *Trans. Geol. Soc. S. Afr.* **82**: 81–108.

Kielan-Jaworowska, Z. 1975. Evolution of the therian mammals in the Late Cretaceous of Asia. Part I. Deltatheriidae. *Palaeontol. Pol.* **33**: 103–132.

Kielan-Jaworowska, Z. 1977. Evolution of therian mammals in the Late Cretaceous of Asia. Part II. Postcranial skeleton in *Kennalestes* and *Asioryctes.* Results of the Polish-Mongolian palaeontological expeditions, Part VII. *Palaontol. Pol.* **37**: 55–83.

Kielan-Jaworowska, Z. 1978. Evolution of the therian mammals in the late Cretaceous of Asia. Part III. Postcranial skeleton in Zalambdalestidae. *Palaeontol. Pol.* **38**: 5–41.

Kielan-Jaworowska, Z. 1979. Pelvic structure and nature of reproduction in Multituberculata. *Nature* **277**: 402–403.

Kielan-Jaworowska, Z. 1981. Evolution of the therian mammals in the Late Cretaceous of Asia. Part IV. Skull structure in *Kennalestes* and *Asioryctes. Palaeontol. Pol.* **42**: 25–78.

Kielan-Jaworowska, Z. 1983. Multituberculate endocranial casts. *Palaeovertebrata* **13**: 1–12.

Kielan-Jaworowska, Z. 1984*a*. Evolution of the therian mammals in the Late Cretaceous of Asia. Part VII. Synopsis. *Palaeontol. Pol.* **46**: 173–183.

Kielan-Jaworowska, Z. 1984*b*. Evolution of the therian mammals in the Late Cretaceous of Asia. Part VI. Endocranial casts of eutherian mammals. *Palaeontol. Pol.* **46**: 157–171.

Kielan-Jaworowska, Z. 1986. Brain evolution in Mesozoic mammals. *Contrib.Geol.Univ.Wyom. Spec. Pap.* **3**: 21–34.

Kielan-Jaworowska, Z. 1989. Postcranial skeleton of a Cretaceous mammal. *Acta Palaeontol. Pol.* **34**: 75–85.

Kielan-Jaworowska, Z. 1997. Characters of multituberculates neglected in phylogenetic analyses of early mammals. *Lethaia* **29**: 249–266.

Kielan-Jaworowska, Z. and Bonaparte, J.F. 1996. Partial dentary of a multituberculate mammal from the Late Cretaceous of Argentina and its taxonomic implications. *Museo Argentino de Ciencias Naturales 'Bernardino Rivadavia' e Instituto Naciolal de Investigacion de las Ciencias Naturales, Nueva Serie* **145**: 1–9.

Kielan-Jaworowska, Z. and Cifelli, R. L. 2001. Primitive boreosphenidan mammal (?Deltatheroidea) from the Early Cretaceous of Oklahoma. *Acta. Palaeontol. Pol.* **46**: 377–391.

Kielan-Jaworowska, Z. and Dashzeveg, D. 1989. Eutherian mammals from the Early Cretaceous of Mongolia. *Zool. Script.* **18**: 347–355.

Kielan-Jaworowska, Z. and Dashzeveg, D. 1998. New Early Cretaceous amphilestid ('triconodont') mammals from Mongolia. *Acta Palaeontol. Pol.* **43**: 413–438.

Kielan-Jaworowska, Z. and Gambaryan, P.D. 1994. Postcranial anatomy and habits of Asian multituberculate mammals. *Fossil. Strat.* **36**: 1–92.

Kielan-Jaworowska, Z. and Hurum, J.H. 1997. Djadochtatheria—a new suborder of multituberculate mammals. *Acta Palaeontol. Pol.* **42**: 201–242.

Kielan-Jaworowska, Z. and Hurum, J.H. 2001. Phylogeny and systematics of multituberculate mammals. *Palaeontology* **44**: 389–429.

Kielan-Jaworowska, Z. and Nessov, L.A. 1990. On the metatherian nature of the Deltatheroidea, a sister group of the Marsupialia. *Lethaia* **23**: 1–10.

Kielan-Jaworowska, Z. and Nessov, L.A. 1992. Multituberculate mammals from the Cretaceous of Uzbekistan. *Acta Palaeont. Pol.* **37**: 1–17.

Kielan-Jaworowska, Z. and Qi, T. 1990. Fossorial adaptations of a taeniolabidoid multituberculate from Eocene of China. *Vert. PalAs.* **28**: 81–94.

Kielan-Jaworowska, Z., Bown, T.M., and Lillegraven, J.A. 1979a. Eutheria. In J.A.Lillegraven, Z. Kielan-Jaworowska and W.A. Clemens (eds) *Mesozoic mammals: the first two-thirds of mammalian history.* California University Press: Berkeley, pp. 221–276.

Kielan-Jaworowska, Z., Eaton, J.G., and Bown, T.M. 1979b. Theria of the metatherian–eutherian grade. In J.A. Lillegraven, Z. Kielan-Jaworowska and W.A. Clemens (eds) *Mesozoic mammals: the first two-thirds of mammalian history.* University of California Press: Berkeley, pp. 182–191.

Kielan-Jaworowska, Z., Presley, R., and Poplin, C. 1986. The cranial vascular system in taeniolabidoid multituberculate mammals. *Phil. Trans.R. Soc.* **B313**: 525–602.

Kielan-Jaworowska, Z., Crompton, A.W., and Jenkins, F.A. Jr 1987. The origin of egg-laying mammals. *Nature* **326**: 871–873.

Kielan-Jaworowska, Z., Cifelli, R.L., and Luo, Z.-X. 1998. Alleged Cretaceous placental from down under. *Lethaia* **31**: 267–268.

Kielan-Jaworowska, Z., Novacek, M.J., Trofimov, B.A., and Dashzeveg, D. 2000. Mammals from the Mesozoic of Mongolia. In M.J. Benton, M.A. Shishkin, D.M. Unwin, and E.N. Kurochkin (eds) *The age of dinosaurs in Russia and Mongolia.* Cambridge University Press: Cambridge, pp. 573–626.

Kielan-Jaworowska, Z., Cifelli, R.L., and Luo, Z.-X. 2002. Dentition and relationships of the Jurassic mammal *Shuotherium. Acta Palaeontol. Pol.* **47**: 479–486.

Kielan-Jaworowska, Z., Cifelli, R.L., and Luo, Z.-X. 2004. *Mammals from the age of dinosaurs: origins, evolution, and structure.* Columbia University Press: New York.

Killian, K.J., Buckley, T.R., Stewart, N., Munday, B.L., and Jirtle, R.L. 2001. Marsupials and eutherians reunited:- genetic evidence for the Theria hypothesis of mammalian evolution. *Mamm Genome* **12**: 513–517.

King, G.M. 1981a. The functional anatomy of a Permian dicynodont. *Phil. Trans. R. Soc.* **B291**: 243–322.

King, G.M. 1981b. The postcranial skeleton of *Robertia broomiana*, an early dicynodont (Reptilia: Therapsida) from the South African Karroo. *Ann. S. Afr. Mus.* **84**: 203–231.

King, G.M. 1983. First mammal-like reptile from Australia. *Nature* **306**: 209.

King, G.M. 1985. The postcranial skeleton of *Kingoria nowacki* (von Huene) (Therapsida: Dicynodontia). *Zool. J. Linn. Soc.* **84**: 263–289.

King, G.M. 1988. *Anomodontia. Encyclopedia of Paleoherpetology, Part 17C.* Gustav Fischer Verlag: Stuttgart.

King, G.M. 1990. *The dicynodonts: a study in palaeobiology.* Chapman and Hall: London.

King, G.M. 1991. The aquatic *Lystrosaurus*: a palaeontological myth. *Hist. Biol.* **4**: 285–321.

King, G.M. 1994. The early anomodont *Venjukovia* and the evolution of the anomodont skull. *Zool. J. Linn. Soc.* **232**: 651–673.

King, G.M. and Cluver, M.A. 1991. The aquatic *Lystrosaurus*: an alternative lifestyle. *Hist. Biol.* **4**: 323–341.

King, G.M. and Jenkins, I. 1997. The dicynodont *Lystrosaurus* from the Upper Permian of Zambia: evolutionary and stratigraphical implications. *Palaeontology* **40**: 149–156.

King, G.M. and Rubidge, B.S. 1993. A taxonomic revision of small dicynodonts with postcanine teeth. *Zool. J. Linn. Soc.* **107**: 131–154.

King, G.M., Oelofson, B.W., and Rubidge, B.S. 1989. The evolution of the dicynodont feeding system. *Zool. J. Linn. Soc.* **96**: 185–211.

Kirk, E.C., Cartmill, M. Kay, R.F., and Lemelin, P. 2003. Comment on 'Grasping primate origins'. *Science* **300**: 741b.

Kirsch, J.A. 1977. The comparative serology of Marsupialia and a classification of marsupials. *Aust. J. Zool.* **38**: (Suppl.) 1–152.

Kirsch, J.A., Dickerman, A.W., Reig, O.A., and Springer, M.S. 1991. DNA hybridization evidence for the Australasian affinity of the American marsupial *Dromiciops australis. Proc. Natl. Acad. Sci. USA.* **88**: 10465–10469.

Kitching, I.J., Forey, P.L., Humphries, C.J., and Williams, D.M. 1998. *Cladistics: the theory and practice of parsimony analysis.* Oxford University Press: Oxford.

Koenigswald, W. von 1980. Das skelett eines Pantolestide (Proteutheria, Mamm.) aus dem mittleren Eozän von Messel bei Darmstadt. *Paläontol. J.* **54**: 267–287.

Koenigswald, W. von and Storch, G. 1992. The marsupials: inconspicuous opossums. In S. Schaal and W. Zeigler (eds) *Messel: an insight into the history of life and of the Earth.* Clarendon Press: Oxford, pp. 159–78

Koenigswald, W. von, Rensberger, J.M., and Pfretzschner, H.U. 1987. Change in the tooth enamel of early Paleocene mammals allowed increased diet diversity. *Nature* **328**: 150–152.

Koenigswald, W. von, Storch, G., and Richter, G. 1992. Primitive insectivores, extraordinary hedgehogs, and long fingers. In S. Schaal and W. Ziegler (eds) *Messel: an insight into the history of life and of the Earth.* Oxford University Press: Oxford, pp. 159–178.

Koteja, P. 2000. Energy assimilation, parental care and the evolution of endothermy. *Proc. R. Soc. Lond.* **B267**: 479–484.

Krajewski, C., Wroe, S., and Westerman, M. 2000. Molecular evidence for the pattern and timing of cladogenesis in dasyurid marsupials. *Zool. J. Linn. Soc.* **130**: 375–404.

Krause, D.W. 1982. Jaw movement, dental function, and diet in the Paleocene multituberculate *Ptilodus Paleobiol.* **8**: 265–313.

Krause, D.W. 1986. Competitive exclusion and taxonomic displacement in the fossil record: the case of rodents and multituberculates in North America. In K.M. Flanagan and J.A. Lillegraven (eds) *Vertebrate phylogeny and philosophy. University of Wyoming: Laramie. Contrib. Geol. Spec. Pap.* **3**: pp. 95–117.

Krause, D.W. 2001. Fossil molar from a Madagascan marsupial. *Nature* **412**: 497–498.

Krause, D.W. and Jenkins, F.A. Jr 1983. The postcranial skeleton of North American multituberculates. *Bull. Mus. Comp. Zool. Harvd.* **150**: 199–246.

Krause, D.W. and Kielan-Jaworowska, Z. 1993. The endocranial cast and encephalization quotient of *Ptilodus* (Multituberculata, Mammalia). *Palaeovertebrata* **22**: 99–112.

Krause, D.W. and Maas, M.C. 1990. The biogeographic origins of the Late Paleocene-Early Eocene mammalian immigrants to the Western Interior of North America. *Geol. Soc. Am. Spec. Pap.* **243**: 71–105.

Krause, D.W., Kielan-Jaworowska, Z., and Bonaparte, J.F. 1992. *Ferugliotherium* Bonaparte, the first known multituberculate from South Africa. *J. Vert. Paleont.* **12**: 351–376.

Krause, D.W., Prasad, G.V.R., Koenigswald, W. von, Sahni, A., and Grine, F.E. 1997. Cosmopolitanism among Gondwana Late Cretaceous mammals. *Nature* **390**: 504–507.

Krause, D.W., Gottfried, M.D., O'Connor, P.M., and Roberts, E.M. 2003. A Cretaceous mammal from Tanzania. *Acta Palaeontol.Pol.* **48**: 321–330.

Krebs, B. 1991. Das Skelett von *Henkelotherium guimarotae* gen et sp. nov. (Eupantotheria, Mammalia) aus dem Oberen Jura von Portugal. *Berlin. Geowiss. Abt.* **A133**: 1–110.

Krishtalka, L., Emry, J.E., and Sutton, J.F. 1982. Oligocene multituberculates (Mammalia: Allotheria): youngest known record. *J. Paleontol.* **56**: 791–794.

Kron, D.G. 1979. Docodonta. In J.A. Lillegraven, Z. Kielan-Jaworowska and W.A.Clemens (eds) *Mesozoic mammals: the first two-thirds of mammalian history.* University of California Press: Berkeley, pp. 91–98.

Kron, D.G. and Manning, E. 1998. Anthracotheriidae. In C.M. Janis, K.M. Scott, and L.L. Jacobs (eds) *Evolution of Tertiary mammals of North America, Vol. 1: Terrestrial carnivores, ungulates, and ungulatelike mammals.* Cambridge University Press: Cambridge, pp. 381–388.

Krusat, G. 1980. Contribução para o conhecimento da fauna do Kimeridgiano da lignite Guimarota (Leira,

Portugal). VI Parte. *Haldanodon exspectatus* Küand Krusat 1972 (Mammalia, Docodonta). *Memórias dos Serviços Geológicos de Portugal,* 27: 1–79.

Krusat, 1980. The Kimmeridgian fauna of the Guimaroto lignite mine, Leiria, Portugal. Part IV, *Haldanodon expectatus* (Mammalia, Docodonta). *Memórias dos Serviçs Geológicos de Portugal,* **27**: 1–79.

Krusat, G. 1991. Functional morphology of *Haldanodon expectatus* (Mammalia, Docodonta) from the Upper Jurassic of Portugal. In Z. Kielan-Jarworowska, N. Heintz, and H.A. Nakram (eds) *Fifth symposium on Mesozoic terrestrial ecosystems and biota. Contrib. Paleontol. Mus. Oslo* **363**: 37–38.

Kühne, W.G. 1949. On a triconodont tooth of a new pattern from a fissure-filling in South Glamorgan. *Proc. Zool. Soc. Lond.* **119**: 345–350.

Kühne, W.G. 1950. A symmetrodont tooth from the Rhaeto-Lias. *Nature* **166**: 696–697.

Kühne, W.G. 1956. *The Liassic therapsid* Oligokyphus. British Museum (Natural History): London.

Kumar, S. and Hedges, B. 1998. A molecular timescale for vertebrate evolution. *Nature* **392**: 917–920.

Lander, B. 1998. In C.M. Janis, K.M. Scott, and L.L. Jacobs (eds) *Evolution of Tertiary mammals of North America, Vol. 1: Terrestrial carnivores, ungulates, and ungulatelike mammals.* Cambridge University Press: Cambridge, pp. 402–425.

Langer, M. C. 2000. The first record of dinocephalians in South America: Late Permian (Rio do Rasto Formation) of the Parana basin, Brazil. *Neues Jahrbuch Geol. Palaeont., Abh* **215** (1): 69–95.

Langston, W. Jr 1965. *Oedaleops campi* (Reptilia: Pelycosauria), a new genus and species from the Lower Permian of New Mexico, and the family Eothyrididae. *Texas Mem. Mus. Bull.* **9**: 1–48.

Langston, W. Jr and Reiss, R.R. 1981. *Aerosaurus wellesi,* new species, a varanopseid mammal-like reptile from the Lower Permian of New Mexico. *J Paleont.* **1**: 73–96.

Latimer, E.M., Gow, C.E., and Rubidge, B.S. 1995. Dentition and feeding niche of *Endothiodon* (Synapsida: Anomodontia). *Palaeontol. Afr.* **32**: 75–82.

Lauder, G.V. and Gillis, G.B. 1997. Origin of the amniote feeding mechanism. In S.S. Sumida and K.L.M. Martin (eds) *Amniote origins: completing the transition to land.* Academic Press: London, pp. 169–206.

Laurin, M. 1993. Anatomy and relationships of *Haptodus garnettensis,* a Pennsylvanian synapsid. *J. Vert. Paleontol.* **13**: 200–229.

Laurin, M. 1998. New data on the cranial anatomy of *Lycaenops* (Synapsida, Gorgonopsidae), and reflections on the possible presence of streptostyly in gorgonopsians. *J. Vert. Paleontol.* **18**: 765–776.

Laurin, M. and Reisz, R.R. 1995. A reevaluation of early amniote phylogeny. *Zool. J. Linn. Soc.* **113**: 165–223.

Laurin, M. and Reisz, R.R. 1996. The osteology and relationships of *Tetraceratops insignis*, the oldest known therapsid. *J. Vert. Paleontol.* **16**: 95–102.

Laurin, M. and Reisz, R.R. 1997. A new perspective on tetrapod phylogeny. In S.S. Sumida and K.L.M. Martin (eds) *Amniote origins: completing the transition to land.* Academic Press: London, pp. 9–59.

Leakey, M.G., Ungar, P.S., and Walker, A.C. 1995. A new genus of large primate from the Late Oligocene of Lothidok, Turkana District, Kenya. *J. Hum. Evol.* **28**: 519–531.

Lebedev, O.A. and Coates, M.I. 1995. The postcranial skeleton of the Devonian tetrapod *Tulerpeton curtum* Lebedev. *Zool. J. Linn. Soc.* **114**: 307–348.

Lessa, E.P. and Fariña, R.A. 1996. Reassessment of extinction patterns among the Late Pleistocene mammals of South America. *Palaeontology* **39**: 651–662.

Lewis, G.E. 1986. *Nearctylodon broomi*, the first Nearctic tritylodont. In N. Hotton III, P.D. MacLean, J.J. Roth, and E.C. Roth (eds) *The ecology and biology of mammal-like reptiles.* Smithsonian Institution Press: Washington, DC. pp. 295–303.

Li, C-K. and Ting, S-Y. 1993. New cranial and postcranial evidence for the affinities of the eurymylids (Rodentia) and mimotonids (Lagomorpha). In F.S. Szalay, M.J. Novacek, and M.C. McKenna (eds) *Mammal phylogeny: placentals.* Springer-Verlag: New York, pp. 151–158.

Li, J. and Cheng, Z. 1995. A new Late Permian vertebrate fauna from Dashankou, Gansu with comments on Permian and Triassic vertebrate assemblage zones of China. In A.L. Sun and Y.Q. Wang (eds) *Short Papers of the Sixth Symposium on Mesozoic Terrestrial Ecosystems and Biota.* Ocean Press: Beijing, pp. 33–37.

Li, J., Rubidge, B.S., and Cheng, Z. 1996. A primitive anteosaurid dinocephalian from China-implications for the distribution of earliest therapsid faunas. *S. Afr. J. Sci.* **92**: 252–253.

Lillegraven, J.A. 1969. Latest Cretaceous mammals of upper part of Edmonton Formation of Alberta, Canada, and review of marsupial-placental dichotomy in mammalian evolution. *Univ. Kansas. Paleontol. Contrib. Art.* **50**: 1–122.

Lillegraven, J.A. 1979a. Introduction. In J.A. Lillegraven, Z. Kielan-Jaworowska, and W.A. Clemens (eds) *Mesozoic mammals: the first two-thirds of mammalian history.* California University Press: Berkeley, pp. 1–6.

Lillegraven, J.A. 1979b. Reproduction in Mesozoic mammals. In J.A. Lillegraven, Z. Kielan-Jaworowska, and W.A. Clemens (eds) *Mesozoic mammals: the first two-thirds of mammalian history.* California University Press: Berkeley, pp. 26–276.

Lillegraven, J.A. and Krusat, G. 1991. Cranio-mandibular anatomy of *Haldanodon expectatus* (Docodonta: Mammalia) from the Late Jurassic of Portugal and its implications to the evolution of mammalian characters. *Contrib. Geol. Univ. Wyoming* **28**: 39–138.

Lillegraven, J.A., Kielan-Jaworowska, Z., and Clemens, W.A. (eds) 1979. *Mesozoic mammals: the first two-thirds of mammalian history.* California University Press: Berkeley.

Lillegraven J.A., Thompson, S.D., McNab, B.K., and Patton, J.L. 1987. The origin of eutherian mammals. *Biol. J. Linn. Soc.* **32**: 281–336.

Linnaeus, C. 1758. *Systema naturae.* 10th edition. Laurentii Salvii: Stockholm.

Lin, Y.-H., McLenachan, P.A., Gore, A.R., Phillips, M.J., Ota, R., Hendy, M.D., and Penny, D. 2002. Four new mitochondrial genomes and the increased stability of evolutionary trees of mammals from improved taxon sampling. *Mol. Biol. Evol.* **19**: 2060–2070.

Long, J., Archer, M., Flannery, T., and Hand, S. 2002. *Prehistoric mammals of Australia and New Guinea.* University of New South Wales Press: Sydney.

Lucas, S.G. 1993. Pantodonts, tillodonts, uintotheres, and pyrotheres are not ungulates. In F.S. Szalay, M.J. Novacek, and M.C. McKenna (eds) *Mammal phylogeny: placentals.* Springer-Verlag: New York, pp. 182–194.

Lucas, S.G. 1998. Pantodonta. In C.M. Janis, K.M. Scott, and L.L. Jacobs (eds) *Evolution of Tertiary mammals of North America, Vol. 1: Terrestrial carnivores, ungulates, and ungulate-like mammals.* Cambridge University Press: Cambridge, pp. 274–283.

Lucas, S.G. 2001. *Chinese fossil vertebrates.* Columbia University Press: New York.

Lucas, S.G. and Hunt, A.P. 1994. The chronology and paleobiology of mammalian origins. In N.C. Fraser and H.-D. Sues (eds) *In the shadow of the dinosaurs: early Mesozoic tetrapods.* Cambridge University Press: Cambridge, pp. 335–351.

Lucas, S.G. and Luo, Z.-X. 1993. *Adelobasileus* from the Upper Triassic of West Texas: the oldest mammal. *J. Vert. Paleontol.* **13**: 309–334.

Lucas, S.G. and Schoch, R.M. 1998a. Tillodontia. In C.M. Janis, K.M. Scott, and L.L. Jacobs (eds) *Evolution of Tertiary mammals of North America, Vol. 1: Terrestrial carnivores, ungulates, and ungulate-like mammals.* Cambridge University Press: Cambridge, pp. 268–273.

Lucas, S.G. and Schoch, R.M. 1998b. Dinocerata. In C.M. Janis, K.M. Scott, and L.L. Jacob (eds) *Evolution of Tertiary mammals of North America, Vol. 1: Terrestrial carnivores, ungulates, and ungulate-like mammals.* Cambridge University Press: Cambridge. pp. 284–291.

Lucas, S.G., Schoch, R.M. and Williamson, T.E. 1998. Taeniodonta. In C.M. Janis, K.M. Scott and L.L. Jacobs (eds) *Evolution of Tertiary mammals of North America. Volume 1: terrestrial carnivores, ungulates, and ungulate-like mammals.* Cambridge University Press: Cambridge. pp. 260–267.

Luckett, W.P. 1980. The suggested evolutionary relationships and classification of tree shrews. In W.P. Luckett (ed) *Comparative biology and evolutionary relationships of tree shrews.* Plenum: New York, pp. 3–31.

Luckett, W.P. 1993. An ontogenetic assessment of dental homologies in therian mammals. In F.S. Szalay, M.C. Novacek, and M.C. McKenna (eds) *Mammal phylogeny: Mesozoic differentiation, multituberculates, monotremes, early therians, and marsupials.* Springer-Verlag: New York, pp. 182–204.

Luo, Z.-X., 1994. Sister-group relationships of mammals and transformations of diagnostic mammalian characters. In N.C. Fraser and H.-D. Sues (eds) *In the shadow of the dinosaurs: early Mesozoic tetrapods.* Cambridge University Press: Cambridge, pp. 98–128.

Luo, Z.-X., 1999. Palaeobiology: a refugium for relicts. *Nature* **400**: 23–25.

Luo, Z.-X., 2001. The inner ear and its bony housing in tritylodontids and implicatiuons for evolution of the mammalian middle ear. *Bull. Mus. Comp. Zool. Harvd.* **156**: 81–97.

Luo, Z.-X., Cifelli, R.L., and Kielan-Jaworowska, Z. 2001a. Dual origin of tribosphenic mammals. *Nature* **409**: 53–57.

Luo, Z.-X., and Crompton, A.W. 1994. Transformation of the quadrate (incus) through the transition from non-mammalian cynodonts to mammals. *J. Vert. Paleontol.* **14**: 341–374.

Luo, Z.-X., Crompton, A.W., and Lucas, S.G. 1995. Evolutionary origins of the mammalian promontorium and cochlea. *J. Vert. Paleontol.* **15**: 113–121.

Luo, Z.-X., Crompton, A.W., and Sun, A.-L. 2001b. A new mammal from the Early Jurassic and evolution of mammalian characteristics. *Science* **292**: 1535–1540.

Luo, Z.-X., Ji, Q., Wible, J.R., and Yuan, C.-X. 2003. An Early Cretaceous tribosphenic mammal and metatherian evolution. *Science* **302**: 1934–1940.

Luo, Z.-X., Kielan-Jaworowska, Z., and Cifelli, R. L. 2002. In quest for a phylogeny of Mesozoic mammals. *Acta. Palaeontol. Pol.* **47**: 1–78.

Lydekker, R. 1887. Catalogue of the fossil Mammalia in the British Museum (Natural History) Cromwell Road, S.W. Part 5. Containing the Group Tillodontia, and the Orders Sirenia, Cetacea, Edentata, Marsupialia, Monotremata, and supplement. British Museum (Natural History): London.

Maas, M.C., Thewissen, J.G.M., and Kappelman, J. 1998. *Hypsamasia seni* (Mammalia: Embrithopoda) and other mammals from the Eocene Kartal Formation of Turkey.

In K.C. Beard and M.R. Dawson (eds) *Dawn of the age of mammals in Asia. Bull. Carnegie Mus. Nat. Hist.* **34**: 286–297.

Macdonald, D. (ed) 2001. *The new encyclopedia of mammals.* Oxford University Press: Oxford.

MacFadden, B.J. 1992. *Fossil horses: systematics, paleobiology, and evolution of the family Equidae.* Cambridge University Press: Cambridge.

MacFadden, B.J. 1997. Origin and evolution of the grazing guild in New World terrestrial mammals. *TREE* **12**: 182–187.

MacFadden, B.J. 1998. Equidae. In C.M. Janis, K.M. Scott, and L.L. Jacobs (eds) *Evolution of Tertiary mammals of North America, Vol. 1: Terrestrial carnivores, ungulates, and ungulate-like mammals.* Cambridge University Press: Cambridge, pp. 537–559.

MacFadden, B.J. and Frailey, C.D. 1984. *Pyrotherium*, a large enigmatic ungulate (Mammalia, *incertae sedis*) from the Desidean (Oligocene) of Salla, Bolivia. *Palaeontology* **27**: 867–874.

MacLeod, K.G., Smith, R.M.H., Koch, P.L., and Ward, P.D. 2000. Timing of mammal-like reptile extinctions across the Permian–Triassic boundary in South Africa. *Geology* **28**: 227–230.

MacPhee, R.D.E. 1994. Morphology, adaptations, and relationships of *Plesiorycteropus*, and a diagnosis of a new order of eutherian mammals. *Bull. Am. Mus. Nat. Hist.* **220**: 1–214.

MacPhee, R.D.E. (ed) 1999. *Extinctions in near time: causes, contexts, and consequences.* Kluwer Academic, New York.

MacPhee, R.D.E. and Novacek, M.J. 1993. Definition and relationships of Lipotyphla. In F.S. Szalay, M.J. Novacek, and M.C. McKenna (eds) *Mammal phylogeny: placentals.* Springer-Verlag: New York, pp. 13–31.

MacPhee, R.D.E., Cartmill, M., and Rose, K.D. 1989. Craniodental morphology and relationships of the supposed Eocene dermopteran *Plagiomene* (Mammalia). *J. Vert. Paleontol.* **9**: 329–349.

Mader, B.J. 1998. Brontotheriidae. In C.M. Janis, K.M. Scott, and L.L. Jacobs (eds) *Evolution of Tertiary mammals of North America, Vol. 1: Terrestrial carnivores, ungulates, and ungulatelike mammals.* Cambridge University Press: Cambridge, pp. 525–536.

Madsen, O., Scally, M., Douady, C.J., Kao, D.J., DeBry, R.W., Adkins, R., Amrine, H.M., Stanhope, M.J., de Jong, W.W., and Springer, M.S. 2001. Parallel adaptive radiations in two major clades of placental mammals. *Nature* **409**: 610–614.

Maglio, V.J. 1978. Patterns of faunal evolution. In V.J. Maglio and H.B.S. Cooke (eds) *Evolution of African mammals.* Harvard University Press: Cambridge, MA, pp. 603–619.

Mahboubi, M., Ameur, R., Crochet, J.Y., and Jaeger, J.J. 1984. Earliest known proboscidean from Early Eocene of north-west Africa. *Nature* **308**: 543–544.

Mahboubi, M., Ameur, R., Crochet, J.Y., and Jaeger, J.J. 1986. A new Eocene mammal locality in northwestern Africa. *Palaeontogr. Abt.* **A.192**: 15–49.

Maier, W., Van den Heever, J., and Durand, F. 1995. New therapsid specimens and the origin of the secondary hard and soft palate of mammals. *J. Zool. Syst. Evol. Res.* **34**: 9–19.

Maio, D. 1993. Cranial morphology and multituberculate relationships. In F.S. Szalay, M.J. Novacek and M.C. McKenna (eds) *Mammal phylogeny: Mesozoic differentiation, multituberculates, monotremes, early therians, and marsupials.* Springer-Verlag: New York. pp. 63–74.

Maisch, M.W. 2002. A new basal lystrosaurid dicynodont from the Upper Permian of South Africa. *Palaeontology* **45**: 343–359.

Marcus and Berger 1984. The significance of radiocarbon dates for Rancho La Brea. In P.S. Martin and R.G. Klein (eds) *Quaternary extinctions: a prehistoric revolution.* Arizona University Press: Tucson, pp. 159–188.

Marrivaux, L., Welcomme, J.-L., Antoine, P.-O., Métais, G., Baloch, I.M., Benammi, M., Chaimanee, Y., Ducrocq, S., and Jaeger, J.J. 2001. A fossil lemur from the Oligocene of Pakistan. *Science* **294**: 587–591.

Marshall, L.G. 1978. Evolution of the Borhyaenidae, extinct South American predaceous marsupials. *Univ. Calif. Public Geol. Sci.* **117**: 1–89.

Marshall, L.G. 1980. *Marsupial palaeobiogeography.* In L.L. Jacobs (ed) *Aspects of vertebrate history: essays in honour of Edwin Harris Colbert.* Museum of Northern Arizona Press. pp. 345–386.

Marshall, L.G. 1987. Systematics of Itaborian (Middle Paleocene) age 'Opossum-like' marsupials from the limestone quarry at São José de Itaboraí, Brazil. In M. Archer (ed) *Possums and opossums: studies in evolution.* Surrey Beatty and Sons: Chipping Norton, Australia, pp. 91–160.

Marshall, L.G. 1988. Land mammals and the Great American Interchange. *Am. Sci.* **76**: 380–388.

Marshall, L.G. and Cifelli, R.L.1990. Analysis of changing diversity patterns in Cenozoic land mammal age faunas, South America. *Palaeovertebrata* **19**: 169–210.

Marshall, L.G. and Kielan-Jaworowska, Z. 1992. Relationships of the dog-like marsupials, deltatheroidans and early tribosphenic mammals. *Lethaia* **25**: 361–374.

Marshall, L. G. and Muizon, C. de 1992. Atlas photographique (MEB) des Metatheria et de quelques Eutheria du Paléocène inférieur de la formation Santa Lucia à Tiupampa (Bolivie). *Bull. Mus. Natl. Hist. Nat. Paris 4ᵉ sér.,* **14**, section C, no. 1: 63–91.

Marshall, L.G., Muizon, C. de, and Sigé, B. 1983. Late Cretaceous mammals (Marsupialia) from Bolivia. *Geobios* **16**: 739–745.

Marshall, L.G., Case, J.A., and Woodburne, M.O. 1990. Phylogenetic relationships of the families of marsupials. In H.H.Genoways (ed) *Current mammalogy, Vol 2.* Plenum: New York, pp. 433–505.

Marshall, L.G., Muizon, C. de, and Sigogneau-Russell, D. 1995. *Pucadelphys andinus* (Marsupialia, Mammalia) from the early Paleocene of Bolivia. *Mem. Mus. Natl. Hist. Nat.* **165**: 1–164.

Marshall, L.G., Sempere, T., and Butler, R.F. 1997. Chronostratigraphy of the mammal-bearing Paleocene of South America. *J. S. Am. Earth Sci.* **10**: 49–70.

Martin, L.D. 1998*a*. Felidae. In C.M. Janis, K.M. Scott, and L.L. Jacobs (eds) *Evolution of Tertiary mammals of North America, Vol. 1: Terrestrial carnivores, ungulates, and ungulatelike mammals.* Cambridge University Press: Cambridge, pp. 236–242.

Martin, L.D. 1998*b*. Nimravidae. In C.M. Janis, K.M. Scott, and L.L. Jacobs (eds) *Evolution of Tertiary mammals of North America, Vol. 1: Terrestrial carnivores, ungulates, and ungulatelike mammals.* Cambridge University Press: Cambridge, pp. 228–235.

Martin, M.S. and Klein, P.S. 1984. *Quaternary extinctions: a prehistoric revolution.* Arizona University Press: Tucson.

Martin, P.S. 1984. Prehistoric overkill: the global model. In P.S. Martin and R.G. Klein (eds) *Quaternary extinctions: a prehistoric revolution.* Arizona University Press: Tucson, pp. 354–403.

Martin, P.S. and Steadman, D.W. 1999. Prehistoric extinctions on islands and continents. In R.D.E. McPhee (ed) *Extinctions in near time: causes, contexts, and consequences.* Kluwer Academic: New York, pp. 17–55.

Martin, R.D. 1993. Primate origins: plugging the gaps. *Nature* **363**: 223–234.

Martin, T. 2002. New stem-lineage representative of Zatheria (Mammalia) from the Late Jurassic of Portugal. *J. Vert. Paleontol.* **22**: 332–348.

Martinez, R.N. and Forster, C.A. 1996. The skull of *Probelesodon sanjuanensis,* sp.nov., from the Late Triassic Ischigualasto Formation of Argentina. *J. Vert. Palaeontol.* **16**: 285–291.

Martinez, R.N., May, C.L., and Forster, C.A. 1996. A new carnivorous cynodont from the Ischigualasto Formation (Late Triassic, Argentina), with comments on eucynodont phylogeny. *J. Vert. Paleontol.* **16**: 271–284.

McKenna, M.C. 1975. Towards a phylogenetic classification of the Mammalia. In W.P. Luckett and F.S. Szalay (eds) *Phylogeny of the Primates.* Plenum: New York. pp. 21–46.

McKenna, M.C. and Bell, S.K. 1997. *Classification of mammals above the species level.* Columbia University Press: New York.

McKenna, M.C. and Manning, E. 1977. Affinities and paleobiogeographic significance of the Mongolian Palaeogene genus *Phenacolophus. Geobios, Mem. Spec.* **1**: 61–85.

McKenna, M.C. and Meng, J. 2001. A primitive relative of rodents from the Chinese Paleocene. *J. Vert. Paleontol.* **21**: 565–572.

McKenna, M.C., Chow, M., Suyin, T., and Zhexi, L. 1989. *Radinskya yupingae*, a perissodactyl-like mammal from the late Paleocene of southern China. In D.R. Prothero and R.M. Schoch (eds) *The evolution of perissodactyls.* Oxford University Press: Oxford, pp. 13–23.

McKenna, M.C., Xue, X., and Zhou, M. 1984. *Prosarcodon lonanensis*, a new Palaeocene palaeoryctid insectivore from Asia. *Am. Mus. Novit.* **2780**: 1–17.

McNab, B. K. 1978. The evolution of homeothermy in the phylogeny of mammals. *Am. Nat.* **112**: 1–21.

Mendrez, C. H. 1972. On the skull of *Regisaurus jacobi*, a new genus and species of Bauriamorpha Watson and Romer 1956 (= Scaloposauria Boonstra 1953), from the *Lystrosaurus*-zone of South Africa. In K.A. Joysey and T.S. Kemp (eds) *Studies in vertebrate evolution.* Oliver and Boyd: Edinburgh, pp. 191–212.

Mendrez, Ch.H. 1975. Principales variations du palais chez les thérocéphales Sud-Africains (Pristerosauria et Scaloposauria) au cours du Permien Supérieur et du Trias Inférieur. *Colloque International C.N.R.S. No. 218 (Paris).* Problèmes actuels de paléontologie-évolution des vertébrés. pp. 379–408.

Mendrez-Carroll, CH.H. 1979. Nouvelle édude du crâne du type de *Scaloposaurus constrictus* Owen, 1876, de la zone à *Cistecephalus* (Permien supérieure) d'Afrique australe. *Bull. Mus. Natl. Hist. Nat., Paris, 4ᵉ sér.,1, Section C, No. 3.* 155–201.

Meng, J. and Miao, D. 1992. The breast-shoulder apparatus of *Lambdopsalis bulla* (Multituberculata) and its systematic and functional implications. *J. Vert. Paleontol.* **12**: 43A.

Meng, J. and Wyss, A.R. 1995. Monotreme affinities and low frequency hearing suggested by multituberculate ear. *Nature* **377**: 141–144.

Meng, J. and Wyss, A.R. 1997. Multituberculate and other mammal hair recovered from Palaeogene excreta. *Nature* **385**: 712–714.

Meng, J. and Wyss, A.R. 2001. The morphology of *Tribosphenomys* (Rodentiformes, Mammalia): phylogenetic implications for basal Glires. *J. Mamm. Evol.* **8**: 1–71.

Meng, J., Dawson, M., and Zhai, R.-J. 1994. Primitive fossil rodent from Inner Mongolia and its implications for mammalian phylogeny. *Nature* **370**: 134–136.

Miao, D. 1993. Cranial morphology and multituberculate relationships. In F.S. Szalay, M.C. McKenna, and M.J. Novacek (eds) *Mammalian phylogeny: Mesozoic differentiation, multituberculates, monotremes, early therians, and marsupials.* Springer-Verlag: New York, pp. 30–43.

Miao, D. and Lillegraven, J.A. 1986. Discovery of three ear ossicles in a multituberculate mammal. *Natl. Geogr. Res.* **2**: 500–507.

Miller, G.H., Magee, J.W., Johnson, B.J., Fogel, M.L., Spooner, N.A., McCulloch, M.T. and Ayliffe, L.K. 1999. Pleistocene extinction of *Genyornis newtoni*: human impact on Australian megafauna. *Science* **283**: 205–208.

Mills, J.R.E. 1971. The dentition of *Morganucodon*. In D.M. Kermack and K.A. Kermack (eds) *Early mammals. Zool. J. Linn. Soc. Lond.* **50**(Suppl. 1): 29–63.

Milner, A.R. 1987. The Westphalian tetrapod fauna: some aspects of its geography and ecology. *J. Geol. Soc. Lond.* **144**: 495–506.

Milner, A.R. 1993. Biogeography of Palaeozoic tetrapods. In J.A. Long (ed) *Palaeozoic vertebrate biostratigraphy.* Belhaven: London, pp. 324–353.

Modesto, S.P., Rubidge, B., and Welman, J. 1999. The most basal anomodont therapsid and the primacy of Gondwana in the evolution of anomodonts. *Proc. R. Soc. Lond.* **B266**: 331–337.

Modesto, S.P., Sidor, C.A., Rubidge, B.S., and Welman, J. 2001. A second varanopseid skull from the Upper Permian of South Africa: implications for Late Permian 'pelycosaur' evolution. *Lethaia* **34**: 249–259.

Modesto, S.P. 1994. The Lower Permian synapsid *Glaucosaurus* from Texas. *Paleontology* **37**: 51–60.

Modesto, S.P. 1995. The skull of the herbivorous synapsid *Edaphosaurus boanerges* from the Lower Permian of Texas. *Paleontology* **38**: 213–239.

Modesto, S.P. and Reisz, R.R. 1990. A new skeleton of *Ianthasaurus hardestii*, a primitive edaphosaur (Synapsida: Pelycosauria) from the Upper Pennsylvanian of Kansas. *Can. J. Earth Sci.* **27**: 834–844.

Modesto, S.P. and Rybczynski, N. 2000. The amniote faunas of the Russian Permian: implications for Late Permian terrestrial vertebrate biogeography. *The age of dinosaurs in Russia and Mongolia.* Cambridge University Press: Cambridge, pp. 17–34.

Mosimann, J.E. and Martin, P.S. 1975. Simulating overkill by paleoindians. *Am. Sci.* **63**: 304–313.

Muirhead, J. 1997. Two new early Miocene thylacines from Riversleigh, northwestern Queensland. *Mem. Queensld. Mus.* **41**: 367–377.

Muirhead, J. 2000. Yaraloidea (Marsupialia, Peramelemorphia), a new superfamily of marsupial and a description and analysis of the cranium of the Miocene *Yarala burchfieldi. J. Paleontol.* **74**: 512–523.

Muirhead, J. and Filan, S.L. 1995. *Yarala burchfieldi*, a plesiomorphic bandicoot (Marsupialia, Peramelemorphia) from Olig-Miocene deposits of Riversleigh, northerwestern Queensland. *J. Paleont.* **69**: 127–134.

Muirhead, J. and Wroe, S. 1998. A new genus and species, *Badjcinus turnbulli* (Thylacinidae, Marsupialia), from the late Oligocene of Riversleigh, northern Australia, and an investigation of thylacinid phylogeny. *J. Vert. Paleontol.* **18**: 612–626.

Muizon, C. de 1991. La fauna de mammiferos de Tiupampa (Paleoceno Inferior, Formacion Santa Lucia), Bolivia. In R. Suarez-Soruco (ed) *Fosiles y facies de Bolivia, Vol. 1: vertebrados.* Santa Cruz, Bolivia. *Revi. Técni. YPFB.* **12**: 575–624.

Muizon, C. de 1998. *Mayulestes ferox*, a borhyaenid (Metatheria, Mammalia) from the early Palaeocene of Bolivia: phylogenetic and paleobiologic implications. *Geodiversitas* **20**: 19–142.

Muizon, C. de and Cifelli, R.L. 2000. The 'condylarths' (archaic Ungulata, Mammalia) from the early Palaeocene of Tiupampa (Bolivia): implications on the origin of the South American ungulates. *Geodiversitas* **22**: 47–150.

Muizon, C. de and Cifelli, R.L. 2001. A new basal 'didelphoid' (Marsupialia, Mammalia) from the Early Paleocene of Tiupampa (Bolivia). *J. Vert. Paleontol.* **21**: 87–97.

Muizon, C. de and Marshall, L.G. 1987. Le plus ancien Pantodonte (Mammalia) du Crétacé supérieur de Bolivie. *C. R. Hebd. Scé. Acad. Sci. Paris.* **304**: 205–208.

Muizon, C. de and Marshall, L.G. 1992. *Alcidedorbignya inopinata* (Mammalia: Pantodonta) from the Early Paleocene of Bolivia: phylogenetic and paleobiogeographic implications. *J. Paleontol.* **66**: 499–520.

Muizon, C. de, Cifelli, R.L., and Paz, C. 1997. The origin of the dog-like borhyaenid marsupials of South America. *Nature* **389**: 486–489.

Munthe, K. 1998. Canidae. In C.M. Janis, K.M. Scott and L.L. Jacobs (eds) *Evolution of Tertiary mammals of North America, Vol. 1: Terrestrial carnivores, ungulates, and ungulate-like mammals.* Cambridge University Press: Cambridge, pp. 124–143.

Murphy, W.J., Eizirik, E., Johnson, W.E., Zhang, Y.P., Ryder, O.A., and O'Brien, S.J. 2001a. Molecular phylogenetics and the origins of placental mammals. *Nature* **409**: 614–618.

Murphy, W.J., Eizirik, E., O'Brien, S.J., Madsen, O., Scally, M., Douady, C.J., Teeling, E., Ryder, O.,

Stanhope, M.J., Jong, W.W. de, and Springer, M.S. 2001b. Resolution of the early placental mammal radiation using Bayesian phylogenetics. *Science* **294**: 2348–2351.

Murray, P. 1991. The Pleistocene megafauna of Australia. In Vickers-Rich, P., Monaghan, J.M., Baird, R.F. and Rich, T.H. (eds) *Vertebrate palaeontology of Australia.* Pioneer Design Studio and Monash University: Melbourne. pp. 1071–1164.

Murray, P., Wells, R. and Plane, M. 1987. The cranium of the Miocene thylacoleonid *Wakaleo vanderleuri:* click go the shears – a fresh bite at thylacoleonid systematics. In M. Archer (ed) *Possums and opossums: studies in evolution.* Surrey Beatty and Sons: Chipping Norton, Australia. pp. 433–466.

Musser, A.M. and Archer, M. 1998. New information about the skull and dentary of the Miocene platypus *Obdurodon dicksoni*, and a discussion of ornithorhynchid relationships. *Phil. Trans. R. Soc.* **B353**: 1063–1078.

Naylor, G.J.P. and Adams, D.C. 2001. Are the fossil data really at odds with the molecular data? Morphological evidence for Cetartiodactyla re-examined. *Syst. Biol.* **50**: 444–453.

Nedin, C. 1991. The dietary niche of the extinct Australian marsupial lion *Thylacoleo carnifex* Owen. *Lethaia* **24**: 115–118.

Nessov, L.A., Archibald, J.D., and Kielan-Jaworowska, Z. 1998. Ungulate-like mammals from the Late Cretaceous of Uzbekistan and a phylogenetic analysis of Ungulatomorpha. In K.C. Beard and M.R. Dawson (eds) *Dawn of the age of mammals in Asia. Bull. Carnegie Mus. Nat. Hist.* **34**: 40–88.

Nicholls, J.D. and Fleming, R.F. 1990. Plant microfossil record of the terminal Cretaceous event in the western United States and Canada. *Geol. Soc. Am. Spec. Pap.* **247**: 445–455.

Novacek, M.J. 1980. Cranio-skeletal features in tupaiids and selected eutherians as phylogenetic evidence. In W.P. Luckett (ed) *Comparative biology and evolutionary relationships of tree shrews. Advances in Primatology*, Vol. 4. Plenum: New York, pp. 35–93.

Novacek, M.J. 1986a. The skull of leptictid insectivorans and the higher level classification of eutherian mammals. *Bull. Am. Mus. Nat. Hist.* **183**: 1–111.

Novacek, M.J. 1986b. The primitive eutherian dental formula. *J. Vert. Paleontol.* **6**: 191–196.

Novacek, M.J., Rougier, G.W., Wible, J.R., McKenna, M.C., Dashzeveg, D., and Horovitz, I. 1997. Epipubic bones in eutherian mammals from the Late Cretaceous of Mongolia. *Nature* **389**: 483–486.

Novacek, M.J., Wyss, A.R., and McKenna, M.C. 1988. The major groups of eutherian mammals. In M.J. Benton (ed) *The phylogeny and classification of the tetrapods, Vol. 2: Mammals*. Oxford University Press: Oxford, pp. 31–71.

O'Leary, M. A. 1999. Parsimony analysis of total evidence from extinct and extant taxa and the cetacean-artiodactyl question (Mammalia, Ungulata). *Cladistics* **15**: 315–330.

Olsen, P.E. and Sues, H.-D. 1986. Correlation of continental Late Triassic and Early Jurassic sediments, and the patterns of the Triassic–Jurassic tetrapod transition. In K. Padian (ed) *The beginning of the age of dinosaurs*. Cambridge University Press: Cambridge, pp. 321–351.

Olson, E.C. 1962. Late Permian terrestrial vertebrates, U.S.A. and U.S.S.R. *Trans. Am. Phil. Soc.* **52**: 3–224.

Olson, E.C. 1968. The family Caseidae. *Field. Geol.* **17**: 225–349.

Olson, E.C. 1974. On the source of therapsids. *Ann. S. Afr. Mus.* **64**: 27–46.

Olson, E.C. 1975. Permo-Carboniferous paleoecology and morphotypic series. *Am. Zool.* **15**: 371–389.

Olson, E.C. 1986. Relationships and ecology of the early therapsids and their predecessors. In N. Hotton III, P.D. McLean, J.J. Roth and E.C. Roth (eds) *The ecology and biology of mammal-like reptiles*. Smithsonian Institution Press: Washington, pp. 47–60.

Olson, E.C. and Vaughn, P.P. 1970. The changes of terrestrial vertebrates and climates during the Permian of North America. *Forma Funct.* **3**: 113–138.

Owen, R. 1871. *Monograph of the fossil Mammalia of the Mesozoic formations*. Palaeontological Society: London.

Owen-Smith, N. 1999. The interactions of humans, megaherbivores, and habitats in the late Pleistocene extinction event. In R.D.E. MacPhee (ed) *Extinctions in near time: causes, contexts, and consequences*. Kluwer Academic: New York, pp. 57–69.

Packard, M.J. and Seymour, R.S. 1997. Evolution of the amniote egg. In S.S. Sumida and K.L.M. Martin (eds) *Amniote origins: completing the transition to land*. Academic Press: London, pp. 265–290.

Paddle, R. 2000. *The last Tasmanian tiger*. Cambridge University Press: Cambridge.

Palma, R.E. and Spotorno, A.E. 1999. Molecular systematics of marsupials based on the rRNA 12S mitochondrial gene: the phylogeny of Didelphimorphia and of the living fossil microbiotheriid *Dromicops gliroides* Thomas. *Mol. Phylogenet. Evol.* **13**: 525–535.

Panchen, A.L. and Smithson, T.R. 1988. The relationships of the earliest tetrapods. In M.J. Benton (ed) *The phylogeny and classification of the tetrapods, Vol. 1: amphibians, reptiles, birds.*, pp. 1–58. Oxford University Press: Oxford, Systematics Association Special Volume **35A**.

Parrish, J.M., Parrish, J.T., and Zeigler, A.M. 1986. Permian-Triassic paleogeography and paleoclimatology and implications for therapsid distribution. In N. Hotton III, P.D. MacLean, J.J. Roth, and E.C. Roth (eds) *The ecology and biology of mammal-like reptiles*. Smithsonian Institution: Washington, pp. 109–131.

Parker, P. 1977. An ecological comparison of marsupial and placental patterns of reproduction. In B. Stonehouse and D. Gilmore (eds) *The biology of marsupials*. University Park Press: Baltimore, pp. 273–286.

Parrington, F.R. 1934. On the cynodont genus *Galesaurus*, with a note on the functional significance of the changes in the evolution of the theriodont skull. *Ann. Mag. Nat. Hist., ser.10*, **13**: 38–67.

Parrington, F.R. 1947. On a collection of Rhaetic mammalian teeth. *Proc. Zool. Soc. Lond.* **116**: 707–728.

Parrington, F.R. 1971. On the Upper Triassic mammals. *Phil. Trans. R. Soc.* **B261**: 231–272.

Pascual, R. and Carlini, A.A. 1987. A new superfamily in the extensive radiation of South American Paleogene marsupials. *Fieldiana. Zool.* **39**: 99–110.

Pascual, R. and Ortiz, J.E. 1990. Evolving climates and mammal faunas in Cenozoic South America. *J. Hum. Evol.* **19**: 23–60.

Pascual, R., Archer, M., Ortiz-Jaureguizar, E.O., Prado, J.L., Godthelp, H., and Hand, S.J. 1992. First discovery of monotremes in South America. *Nature* **365**: 704–705.

Pascual, R., Goin, F.J., and Carlini, A.A. 1994. New data on the Groeberiidae: unique Late Eocene-Early Oligocene South American Marsupials. *J. Vert. Paleontol.* **14**: 247–259.

Pascual, R., Goin, F.J., Krause, D.W., Ortiz-Jaureguizar, E., and Carlini, A. A. 1999. The first gnathic remains of *Sudamerica*: implications for gondwanathere relationships. *J. Vert. Paleontol.* **19**: 373–382.

Pascual, R., Goin, F.J., Ardolino, A., and Puerta, P.F. 2000. A highly derived docodont from the Patagonian Late Cretaceous: evolutionary implications for Gondwanan mammals. *Geodiversitas* **22**: 395–414.

Pascual, R., Goin, F.J., Balarino, L., and Sauthier, D.E.U. 2002. New data on the Paleocene monotreme *Monotrematum sudamericanum*, and the convergent evolution of triangulate molars. *Acta Palaeontol. Pol.* **47**: 487–492.

Paton, R. L. 1974. Lower Permian pelycosaurs from the English midlands. *Palaeontology* **17**: 541–552.

Paton, R.L., Smithson, T.R., and Clack, J.A. 1999. An amniote-like skeleton from the Early Carboniferous of Scotland. *Nature* **398**: 508–513.

Patterson, B. 1956. Early Cretaceous mammals and the evolution of mammalian molar teeth. *Fieldiana. Geol.* **13**: 1–105.

Pearson, P.N. and Palmer, M.R. 2000. Atmospheric carbon dioxide concentrations over the past 60 million years. *Nature* **406**: 695–699.

Penny, D., Hasegawa, M., Waddell, P.J., and Hendy, M.D. 1999. Mammalian evolution: timing and implications from using logdeterminant transform for proteins of differing amino acid composition. *Syst. Biol.* **48**: 76–93.

Pettigrew, J.P., Jamieson, B.G.M., Robson, S.K., Hall, L.S., McNally, K.I., and Cooper, H.M. 1989. Phylogenetic relations between microbats, megabats and primates. *Phil. Trans. R. Soc.* **B325**: 489–554.

Phillips, M.J., Lin, Y.H., Harrison, G.L., and Penny, D. 2001. Mitochondrial genomes of a bandicoot and a brushtail possum confirm the monophyly of australidelphian marsupials. *Proc. R. Soc. Lond.* **B268**: 1533–1538.

Pledge, N. S. 1987. *Muramura williamsi*, a new genus and species of ?wynardiid (Marsupialia: Vombatoidea) from the Middle Miocene Etadunna formation of South Australia. In M.Archer (ed) *Possums and opossums: studies in evolution.* Surrey Beatty and Sons: Chipping Norton, Australia, pp. 393–400.

Pond, C. M. 1984. Physiological and ecological importance of energy storage in the evolution of lactation: evidence for a common pattern of anatomical organization of adipose tissue in mammals. *Symp. Zool. Soc. Lond.* **51**: 1–32.

Potts, R. and Behrensmeyer, A.K. 1992. Late Cenozoic terrestrial ecosystems. In A.K. Behrensmeyer, J.D. Damuth, W.A. DiMichele, R. Potts, H.-D. Sues, and S.L. Wing (eds) *Terrestrial ecosystems through time.* Chicago University Press: Chicago, pp. 419–541.

Pough, F.H., Janis, C.M. and Heiser, J.B. 2001. *Vertebrate life*, 6th edition. Prentice Hall.

Prasad, G.V.R., Jaeger, J.J., Sahni, A., Gheerbrant, E., and Khajuria, C.K. 1994. Eutherian mammals from the Upper Cretaceous (Maastrichtian) Intertrappean Beds of Naskal, Andhra Pradesh, India. *J. Vert. Paleontol.* **14**: 260–277.

Prothero, D.R. 1993. Ungulate phylogeny: molecular vs. morphological evidence. In F.S. Szalay, M.J. Novacek, and M.C. McKenna (eds) *Mammal phylogeny: placentals.* Springer-Verlag: New York, pp. 173–181.

Prothero, D.R. 1994. *The Eocene–Oligocene transition: Paradise lost.* Columbia University Press: New York.

Prothero, D.R. 1997. The chronological, climatic, and paleogeographic background to North American mammal evolution. In C.M. Janis, K.M. Scott, and L.L. Jacobs (eds) *Evolution of Tertiary mammals of North America. Vol. 1: Terrestrial carnivores, ungulates, and ungulate-like mammals.* Cambridge University Press: Cambridge. pp. 9–36.

Prothero, D.R. and Schoch, R.M. 1989. Origin and evolution of the Perissodactyla: summary and synthesis. In D.R. Prothero and R.M. Schoch (eds) *The evolution of perissodactyls.* Oxford University Press: Oxford, pp. 504–529.

Prothero, D.R., Manning, E.M., and Fischer, M. 1988. The phylogeny of the ungulates. In M.R. Benton (ed) *The phylogeny and classification of the tetrapods, Vol. 2:* Oxford University Press: Oxford, *Mammals. Systematics Association Special Volume* **35B**, pp. 201–234.

Pumo, D.E., Finamore, P.S., Franek, W.R., Phillips, C.J., Tarzami, S., and Balzarano, D. 1998. Complete mitochondrial genome of a Neotropical fruit bat. *J. Mol. Evol.* **15**: 709–717.

Quiroga, J.C. 1980. The brain of the mammal-like reptile *Probainognathus jenseni* (Therapsida, Cynodontia). A correlative paleoneurological approach to the neocortex at the reptile-mammal transition. *J. Hirnforsch.* **21**: 299–326.

Quiroga, J.C. 1984. The endocranial cast of the mammal-like reptile *Therioherpeton cargnini* (Therapsida, Cynodontia) from the Middle Triassic of Brazil. *J. Hirnforsch.* **25**: 285–290.

Radinsky, L.B. and Ting, S. 1984. The skull of *Ernanodon*, an unusual fossil mammal. *J. Mamm.* **65**: 155–158.

Rage, J.-C. 1988. Gondwana, Tethys, and terrestrial vertebrates during the Mesozoic and Cainozoic. In M.G. Audley-Charles (ed) *Gondwana and Tethys.* Oxford University Press: Oxford, pp. 255–273.

Rana, R.S. and Wilson, G.P. 2003. New Late Cretaceous mammals from the Intertrappean beds of Ranhapur, India. *Acta Palaeontol. Pol.* **48**: 331–348.

Rasmussen, D.T. 1989. The evolution of the Hyracoidea: a review of the fossil evidence. In D.R. Prothero and R.M. Schoch (eds) *The evolution of perissodactyls.* Oxford University Press: Oxford, pp. 57–78.

Rasmussen, D.T. 2002. Early catarrhines of the African Eocene and Oligocene. In W. Hartwig (ed) *The primate fossil record.* Cambridge University Press: Cambridge, pp. 203–220.

Rasmussen, D.T. and Simons, E.L. 2000. Ecomorphological diversity among Paleogene hyracoids (Mammalia): a new cursorial browser from the Fayum, Egypt. *J. Vert. Palaeontol.* **20**: 167–176.

Rasmussen, D.T., Bown, T.M., and Simons, E.L. 1992. The Eocene–Oligocene transition in continental Africa. In D.R. Prothero and W.A. Berggren (eds) *Eocene–Oligocene climatic and biotic evolution.* Princeton University Press: Princeton, pp. 548–566.

Rauhut, O.W.M., Martin, T., Ortiz-Jaureguizar, E., and Puerta, P. 2002. A Jurassic mammal from South America. *Nature* **416**: 165–168.

Rauscher, B. 1987. *Priscileo pitikantensis*, a new genus and species of thylacoleonid marsupial (Marsupialia: Thylacoleonidae) from the Miocene Etadunna Formation, South Australia. In M. Archer (ed) *Possums and opossums: studies in evolution.* Surrey Beatty and Sons: Chipping Norton, Australia, pp. 423–432.

Ray, S. 2000.Endothiodont dicynodonts from the Late Permian Kundaram Formation, India. *Palaeontology* **43**: 375–404.

Ray, S. and Chinsamy, A. 2003. Functional aspects of the postcranial anatomy of the Permian dicynodont *Diictodon* and their ecological implications. *Palaeontology* **46**: 151–183.

Reguero, M.A., Marenssi, S.A., and Santillana, S.N. 2002. Antarctic Peninsula and South America (Patagonia) Paleogene terrestrial faunas and environments: biogeographic relationships. *Palaeogeogr. Palaeoclimatol. Palaeoecol.* **179**: 189–210.

Reichert, C. 1837. Über die Visceralbogen der Wirbeltiere in Allgemeinen und deren Metamorphosen bei den Vögeln und Säugetiereen. *Arch. Anat. Physiol. Wissent. Med.* **1837**: 120–220.

Reig, O.A., Kirsch, J.A.W., and Marshall, L.G. 1987. Systematic relationships of the living and Neocenozoic American `opossum-like' marsupials (Suborder Didelphimorphia), with comments on the classification of these and of the Cretaceous and Paleogene New World and European metatherians. In M.Archer (ed) *Possums and opossums: studies in evolution*. Surrey Beatty and Sons: Chipping Norton, Australia, pp. 1–89.

Reisz, R.R. 1972. Pelycosaurian reptiles from the Middle Pennsylvanian of North America. *Bull. Mus. Comp. Zool. Harv.* **144**: 27–62.

Reisz, R.R. 1986. Pelycosauria. *Handbook of paleoherpetology*, Part 17A. Gustav Fischer Verlag: Stuttgart.

Reisz, R.R. and Berman, D.S. 1986. *Ianthasaurus hardestii* n. sp., a primitive edaphosaur (Reptilia: Pelycosauria) from the Upper Pennsylvanian Rock Lake Shale near Garnett, Kansas. *Can. J. Earth Sci.* **23**: 77–91.

Reisz, R.R. and Berman, D.S. 2001. The skull of *Mesenosaurus romeri*, a small varanopseid (Synapsida: Eupelycosauria) from the Upper Permian of the Mezen River Basin, Northern Russia. *Ann. Carnegie Mus.* **70**: 113–132.

Reisz, R.R. and Dilkes, D.W. 1992. The taxonomic position of *Anningia megalops*, a small amniote from the Permian of South Africa. *Can. J. Earth Sci.* **29**: 1605–1608.

Reisz, R.R. and Sues, H.-D. 2000. Herbivory in late Paleozoic and Triassic terrestrial vertebrates. In H.-D. Sues (ed) *Evolution of herbivory in terrestrial vertebrates: perspectives from the fossil record*. Cambridge University Press: Cambridge, pp. 9–41.

Reisz, R.R., Berman, D.S., and Scott, D. 1992. The cranial anatomy and relationships of *Secodontosaurus*, an unusual mammal-like reptile (Synapsida: Sphenacodontidae) from the early Permian of Texas. *Zool. J. Linn. Soc.* **104**: 127–184.

Reisz, R.R., Dilkes, D.W., and Berman, D.S. 1998. Anatomy and relationships of *Elliotsmithia longiceps* Broom, a small

synapsid (Eupelycosauria: Varanopseidae) from the Late Permian of South Africa. *J. Vert. Paleontol.* **18**: 602–611.

Renaut, A.J. and Hancox, P.J. 2001. Cranial description and taxonomic re-evaluation of *Kannemeyeria argentinensis* (Therapsida: Dicynodontia). *Palaeontol. Afr.* **37**: 81–91.

Renfree, M.B. 1993. Ontogeny, genetic control, and phylogeny of female reproduction in monotreme and therian mammals. In F.S. Szalay, M.J. Novacek, and M.C. McKenna (eds) *Mammal phylogeny: Mesozoic differentiation, multituberculates, monotremes, early therians, and marsupials*. Springer-Verlag: New York, pp. 4–20.

Rich, T.H. 1991. The history of mammals in *Terra Australis*. In P. Vickers-Rich, J.M. Monaghan, R.F. Baird, and T.H. Rich (eds) *Vertebrate palaeontology of Australia*. Monash University: Melbourne, pp. 893–1070.

Rich, T.H., Vickers-Rich, P., and Flannery, T.F. 1999. Divergence times of eutherian mammals. *Science* **285**: 2031a.

Rich, T.H., Flannery, T.F., Trusler, P., Kool, L., Klaveren, N.A., and Vickers-Rich, P. 2002. Evidence that monotremes and ausktribosphenids are not sister groups. *J. Vert. Paleontol.* **22**: 466–479.

Rich, T.H., Flannery, T.F., Trusler, P., Kool, L., van Klaveren, N., and Vickers-Rich, P. 2001a. A second placental mammal from the Early Cretaceous Flat Rocks site, Victoria, Australia. *Rec. Queen Vict. Mus.* **110**: 1–9.

Rich, T.H., Vickers-Rich, P., Constantine, A., Flannery, T.F., Kool, L., and van Klaveren, N. 1997. A tribosphenic mammal from the Mesozoic of Australia. *Science* **278**: 1438–1442.

Rich, T.H., Vickers-Rich, P., Trusler. P., Flannery, T.F., Cifelli, R., Constantine, A., Kool, L., and Klaveren, N. van 2001b. Monotreme nature of the Australian Early Cretaceous mammal *Teinolophos*. *Acta Palaeontol. Pol.* **46**: 113–118.

Rögl, F. 1999. Mediterranean and paratethys palaeogeography during the Oligocene and Miocene. In J. Agustí, L. Rook, and P. Andrews (eds) *The evolution of Neogene terrestrial ecosystems in Europe*. Cambridge University Press: Cambridge, pp. 8–22.

Romer, A.S. 1922. The locomotor apparatus of certain primitive and mammal-like reptiles. *Bull. Am. Mus. Nat. Hist.* **46**: 517–606.

Romer, A.S. 1966. *Vertebrate paleontology, 3rd edn*. Chicago University Press: Chicago.

Romer, A.S. 1967. The Chañares (Argentina) Triassic reptiles *Massetognathus pascuali* and *M. terrugii*. *Breviora* No. **264**: 1–25.

Romer, A.S. 1969. The Chañares (Argentina) Triassic reptile fauna V. A new chiniquodont cynodont *Probelesodon lewisi*-cynodont ancestry. *Breviora* **333**: 1–24.

Romer, A.S. 1970. The Chañares (Argentina) Triassic reptilian fauna VI. A chiniquodontid cynodont with an incipient squamosal-dentary jaw articulation. *Breviora* **344**: 1–18.

Romer, A.S. and Lewis, A.D. 1973. The Chañares (Argentina) Triassic reptile fauna XIX. Postcranial material of the cynodonts *Probelesodon* and *Probainognathus*. *Breviora* **407**: 1–26.

Romer, A.S. and Price, L.W. 1940. Review of the Pelycosauria. *Spec. Pap. Geol. Soc. Am.* **28**: 1–538.

Rose, K.D. 1982. Skeleton of *Diacodexis*, oldest known artiodactyl. *Science* **216**: 621–623.

Rose, K.D. 1987. Climbing adaptations in the Early Eocene mammal *Chriacus* and the origin of Artiodactyla. *Science* **236**: 314–316.

Rose, K. D. 1999. *Eurotamandua* and Palaeodonta: convergent or related? *Paleont. Zeits.* **73**: 395–401.

Rose, K.D. 2001. The ancestry of whales. *Science* **293**: 2216–2217.

Rose, K.D. and Emry, R.J. 1993. Relationships of Xenarthra, Pholidota, and fossil "edentates": the morphological evidence. In F.S. Szalay, M.J. Novacek, and M.C. McKenna (eds). *Mammal phylogeny: placentals.* Springer-Verlag: New York, pp. 81–102.

Rose, K.D. and Lucas, S.G. 2000. An early Paleocene palaeanodont (Mammalia, ?Pholidota) from New Mexico, and the origin of Palaeanodonta. *J. Vert. Paleontol.* **20**: 139–156.

Rose, K.D. and Simons, E.L. 1977. Dental function in the Plagiomenidae: origin and relationships of the mammalian order Dermoptera. *Univ. Mich. Contr. Mus. Paleont.* **24**: 221–236.

Ross, C., Williams, B., and Kay, R.F. 1998. Phylogenetic analysis of anthropoid relationships. *J. Hum. Evol.* **35**: 221–306.

Rougier, G.W. 1993. *Vincelestes neuquenianus* Bonaparte (Mammalia, Theria) un primitivo mamifero del Cretacico Inferior de la Cuenca Neuquuina. *Ph.D. Thesis, Universidad Nacional de Buenos Aires: Buenos Aires,* 720 pp.

Rougier, G.W., Wible, J.R., and Hopson, J.A. 1996. Basicranial anatomy of *Priacodon fruitaensis* (Triconodontidae, Mammalia) from the Late Jurassic of Colorado, and a reappraisal of mammaliaform interrelationships. *Am. Mus. Novit.* **3183**: 1–38.

Rougier, G.W., Wible, J.R., and Novacek, M.J. 1998. Implications of *Deltatheridium* specimens for early marsupial history. *Nature* **396**: 459–463.

Rougier, G.W., Novacek, M.J., and Dashzeveg, D. 1997. A new multituberculate from the Late Cretaceous locality Ukhaa Tolgod, Mongolia: considerations on multituberculate interrelationships. *Am. Mus. Novit.* **3191**: 1–26.

Rowe, T. 1986. Osteological diagnosis of Mammalia, L. 1758, and its relationships to extinct Synapsida. *Ph.D. Thesis, University of California, Berkeley.* 446 pp.

Rowe, T. 1988. Definition, diagnosis, and the origin of Mammalia. *J. Vert. Paleontol.* **8**: 241–264.

Rowe, T. 1993. Phylogenetic systematics and the early history of mammals. In F.S. Szalay, M.J. Novacek, and M.C. McKenna (eds) *Mammal phylogeny: placentals.* Springer-Verlag: New York. pp. 129–145.

Rowe, T. 1996. Coevolution of the mammalian middle ear and neocortex. *Science* **273**: 651–654.

Ruben, J. 1995. The evolution of endothermy in mammals and birds: from physiology to fossils. *Annu. Rev. Physiol.* **57**: 69–95.

Rubidge, B.S. 1984. The cranial morphology and palaeoenvironment of *Eodicynodon* Barry (Therapsida: Dicynodontia). *Navors. Nas. Mus. Bloemfontein* **4**: 325–404.

Rubidge, B.S. 1990*a*. Redescription of the cranial morphology of *Eodicynodon oosthuizeni* (Therapsida: Dicynodontia). *Navors. Nas. Mus. Bloemfont.* **7**: 1–25.

Rubidge, B.S. 1990*b*. A new vertebrate biozone at the base of the Beaufort Group, Karoo Sequence (South Africa). *Palaeontol. Afr.* **27**: 17–20.

Rubidge, B. S. 1991. A new primitive dinocephalian mammal-like reptile from the Permian of South Africa. *Palaeontology* **34**: 547–559.

Rubidge, B.S. 1993. New South African fossil links with the earliest mammal-like reptile (therapsid) faunas from Russia. *S. Afr. J. Sci.* **89**: 460–461.

Rubidge, B.S. 1994. *Australosyodon*, the first primitive anteosaurid dinocephalian from the Upper Permian of Gondwana. *Palaeontology* **37**: 579–594.

Rubidge, B.S. 1995. Biostratigraphy of the *Eodicynodon* Assemblage Zone. In B.S. Rubidge (ed) *Biostratigraphy of the Beaufort Group (Karoo Supergroup). South African Committee for Stratigraphy, Biostratigraphic Series No. 1.* Council for Geosciences: Pretoria, pp. 3–7.

Rubidge, B.S. and Hopson, J.A. 1990. A new anomodont therapsid from South Africa and its bearing on the ancestry of Dicynodontia. *Suid-Afr. Tydskrif Wetensk.* **86**: 43–45.

Rubidge, B.S. and Hopson, J.A. 1996. A primitive anomodont therapsid from the base of the Beaufort Group (Upper Permian) of South Africa. *Zool. J. Linn. Soc.* **117**: 115–139.

Rubidge, B.S. and Sidor, C.A. 2001. Evolutionary patterns among Permo-Triassic therapsids. *Annu. Rev. Ecol. Syst.* **32**: 449–480.

Rubidge, B.S. and Sidor, C.A. 2002. On the cranial morphology of *Burnetia* and *Proburnetia* (Therapsida: Burnetiidae). *J. Vert. Paleontol.* **22**: 257–267.

Rubidge, B.S. and Van den Heever, J. A. 1997. Morphology and systematic position of the dinocephalian *Styracocephalus platyrhynchus. Lethaia* **30**: 157–168.

Rubidge, B.S., Kitching, J.W., and Van den Heever, J.A. 1983. First record of a therocephalian (Therapsida: Pristerognathidae) from the Ecca of South Africa. *Navors. Nas. Mus. Bloemfont.* **4**: 229–235.

Rubidge, B.S., King, G.M. and Hancox, P.J. 1994. The post-cranial skeleton of the earliest synapsid *Eodicynodon* from the Upper Permian of South Africa. *Palaeontology* **37**: 397–408.

Russell, D.E., Louis, P., and Savage, D.E. 1973. Chiroptera and Dermoptera of the French Early Eocene. *Univ. Calif. Public. Geol. Sci.* **95**: 1–57.

Rybczynski, N. 2000. Cranial anatomy and phylogenetic position of *Suminia getmanovi*, a basal anomodont (Amniota: Therapsida) from the Late Permian of eastern Europe. *Zool. J. Linn. Soc.* **130**: 329–373.

Rybczynski, N. and Reisz, R.R. 2001. Earliest evidence for efficient oral processing in a terrestrial herbivore. *Nature* **411**: 684–687.

Salemi, M. and Vandamme, A.-M. (eds) 2003. *The phylogenetic handbook.* Cambridge University Press: Cambridge.

Sánchez-Villagra, M.R. 2001. The phylogenetic relation-ships of argyrolagid marsupials. *Zool. J. Linn. Soc.* **131**: 481–496.

Sánchez-Villagra, M.R. and Kay, R.F. 1997. A skull of *Proargyrolagus*, the oldest argyrolagid (Late Oligocene Salla Beds, Bolivia), with brief comments concerning its paleobiology. *J. Vert. Paleontol.* **17**: 717–724.

Sánchez-Villagra, M.R., Kay, R.F., and Anaya-Daza, F. 2000. Cranial anatomy and palaeobiology of the Miocene marsupial *Hondalagua altiplanensis* and a phylogeny of argyrolagids. *Palaeontology* **43**: 287–301.

Sánchez-Villagra, M.R., Aguilera, O., and Horovitz, I. 2003. The anatomy of the world's largest extinct rodent. *Science* **301**: 1708–1710.

Sarich, V.M. and Wilson, A.C. 1967. Immunological time scale for hominid evolution. *Science* **158**: 1200–1203.

Savage, R.J.G. and Long, M.R. 1986. *Mammal evolution: an illustrated guide.* British Museum (Natural History): London.

Savage, R. J.G., Domning, D.P., and Thewissen, J.G.M. 1994. Fossil Sirenia of the West Atlantic and Caribbean region. V. The most primitive known sienian, *Prorastomus sirenoides. J. Vert. Paleontol.* **14**: 427–449.

Scally, M., Madsen, O., Douady, C.J., Jong, W.W. de, Stanhope, M.J., and Springer, M.S. 2001. Molecular evidence for the major clades of placental mammals. *J. Mamm. Evol.* **8**: 239–277.

Schaal, S. and Zeigler, W. (eds) (1992). *Messel: an insight into the history of life and the Earth.* Oxford University Press: Oxford.

Schoch, R.M. 1986. Systematics, functional morphology and macroevolution of the extinct mammalian order Taeniodonta. *Bull. Peabody Mus. Nat. Hist.* **42**: 1–307.

Schoch, R.M. 1989. A review of the tapiroids. In D.R. Prothero and R.M. Schoch (eds) *The evolution of peris-sodactyls.* Oxford University Press: Oxford, pp. 298–321.

Schoch, R.M. and Lucas, S.G. 1982. The distribution and paleobiogeography of the Tillodonta (Mammalia: Eutheria). *Geol. Soc. Am. Progr. and Abt.* 14: 349.

Schoch, R.M. and Lucas, S.G. 1985. The phylogeny and classification of the Dinocerata (Mammalia, Eutheria). *Bull. Geol. Inst. Univ. Uppsala* **11**: 31–58.

Schutt, W.A. and Simmons, N.B. 1998. Morphology and homology of the chiropteran calcar, with comments on the relationships of *Archaeopteropus. J. Mamm. Evol.* **5**: 1–32.

Scotese, C.R. and McKerrow, W.S. 1990. Revised world maps and introduction. In W.S. McKerrow and C.R. Scotese (eds) *Palaeozoic palaeogeography and biogeography. Geol. Soc. Mem.* **12**: 1–21.

Seiffert, E.R. and Simons, E.L. 2000. *Widanelfarasia*, a diminutive placental from the late Eocene of Egypt. *Proc. Nat. Acad. Sci. USA.* **97**: 2646–2651.

Seiffert, E.R., Simons, E.L., and Attia, Y. 2003. Fossil evid-ence for an ancient divergence of lorises and galagos. *Nature* **422**: 421–424.

Sereno, P. and McKenna, M.C. 1995. Cretaceous multi-tuberculate skeleton and the early evolution of the mammalian shoulder girdle. *Nature* **377**: 144–147.

Sheehan, P.M., Fastovsky, D.E., Hoffman, R.G., Berghaus, C.B., and Ganriel, D.L. 1991. Sudden extinction of the dinosaurs: latest Cretaceous, Upper Plains, U.S.A. *Science* **254**: 835–839.

Shoshani, J. 1986. Mammalian phylogeny: comparison of morphological and molecular results. *Mol. Biol. Evol.* **3**: 222–242.

Shoshani, J., West, R.M., Court, N., Savage, R.J.G., and Harris, J.M. 1996. The earliest proboscideans: general plan, taxon-omy, and palaeoecology. In J. Shoshani and P. Tassy (eds) *The Proboscidea: evolution and palaeoecology of elephants and their relatives.* Oxford University Press: Oxford, pp. 57–75.

Shubin, N.H., Crompton, A.W., Sues, H.-D., and Olsen, P.E. 1991. New fossil evidence on the sister-group of mammals and early Mesozoic faunal distributions. *Science* **251**: 1063–1065.

Sidor, C.A. and Hopson, J.A. 1998. Ghost lineages and "mammalness": assessing the temporal pattern of char-acter acquisition in the Synapsida. *Paleontology* **24**: 254–273.

Sidor, C.A. and Welman, J. 2003. A second specimen of *Lemurosaurus pricei* (Therapsida: Burnetiamorpha). *J. Vert. Paleontol.* **23**: 631–642.

Sigé, B. 1972. La faunule de mammifères du Crétacé supérieur de Laguna Umayo (Andes péruviennes). *Bull. Mus. Natl. Hist. Nat. 99: Sci. Terre.* **19**: 375–409.

Sigogneau, D. 1970. Révision systématique des gorgonopsiens sud-africaines. *Cahiers Paleont.* Centre National de la Recherche Scientifique: Paris.

Sigogneau, D. and Chudinov, 1972. Reflections on some Russian eotheriodonts. *Palaeovertebrata* **5**: 79–109.

Sigogneau-Russell, D. 1983. Nouveau taxons de mammifères rhétiens. *Acta. Palaeontol. Pol.* **28**: 233–249.

Sigogneau-Russell, D. 1989a. Haramiyidae (Mammalia, Allotheria) en provenance du Trias supérieur de Lorraine (France). *Palaeontogr. Abt.* **A206**: 137–198.

Sigogneau-Russell, D. 1989b. *Theriodonta I. Phthinosuchia, Biarmosuchia, Eotitanosuchia, Gorgonopsia. Encyclopedia of paleoherpetology,* Part 17B/I. Gustav Fischer Verlag: Stuttgart.

Sigogneau-Russell, D. 1991. First evidence of Multituberculata (Mammalia) in the Mesozoic of Africa. *Neues Jahrbuch Palaontol. Monatsh.* 119–125.

Sigogneau-Russell, D. 1995. Further data and reflexions on the tribosphenid mammals (Tribotheria) from the Early Cretaceous of Morocco. *Bull. Mus. Natl. Hist. Nat.,* Paris, 4th Ser., 16, 1994 (1995), section C, **2–4**: 291–312.

Sigogneau-Russell, D. 1998. Discovery of a Late Jurassic Chinese mammal in the upper Bathonian of England. *C. R. Acad. Sci. Paris* **327**: 571–576.

Sigogneau-Russell, D. and Hahn, R. 1995. Reassessment of the Late Triassic symmetrodont mammal *Woutersia. Acta Palaeontol. Pol.* **40**: 245–260.

Sigogneau-Russell, D. and Russell, D.E. 1974. Étude du premier caséidé (Reptilia, Pelycosauria) d'Europe occidentale. *Bull. Mus. Natl. Hist. Nat.3ʳᵈ Ser. Sci. Terre,* **230**: 145–216.

Sigogneau-Russell, D., Frank, R.M., and Hemmerle, J. 1986. A new family of mammals from the lower part of the French Rhaetic. In K. Padian (ed) *The beginning of the age of dinosaurs: faunal change across the Triassic–Jurassic boundary.* Cambridge University Press: Cambridge, pp. 99–108.

Sigogneau-Russell, D., Hooker, J.J., and Ensom, P.C. 2001. The oldest tribosphenic mammal from Laurasia (Purbeck Limestone Group, Berriasian, Cretaceous, U.K.) and its bearing on the 'dual origin' of Tribospshenida. *C. R. Acad. Sci. Paris, Earth Planet. Sci.* **333**: 141–147.

Simmons, N.B. 1993. Phylogeny of Multituberculata. In F.S. Szalay, M.C. McKenna, and M.J.Novacek (eds) *Mammalian phylogeny: Mesozoic differentiation, multituberculates, monotremes, early therians, and marsupials.* Springer-Verlag: New York, pp. 146–163.

Simmons, N.B. 2000. Bat phylogeny: an evolutionary context for comparative studies. In R.A. Adams and S.C. Pedersen (eds) *Ontogeny, functional ecology, and evolution of bats.* Cambridge University Press: Cambridge., pp. 9–58.

Simmons, N.B. and Geisler, J.H. 1998. Phylogenetic relationships of *Icaronycteris, Archaeonycteris, Hassianycteris,* and *Palaeochiropteryx* to extant bat lineages, with comments on the evolution of echolocation and foraging strategies in Microchiroptera. *Bull. Am. Mus. Nat. Hist.* **235**: 1–182.

Simms, M.J., Ruffell, A.H., and Johnson, L.A. 1994. Biotic and climatic changes in the Carnian (Triassic) of Europe and adjacent areas. In N.C. Fraser and H.-D. Sues (eds) *In the shadow of the dinosaurs: early Mesozoic tetrapods.* Cambridge University Press: Cambridge, pp. 352–365.

Simons, E.L. 1992. The fossil history of primates. In S. Jones, R. Martin, and D. Pilbeam (eds) *The Cambridge encyclopedia of human evolution.* Cambridge University Press: Cambridge, pp. 199–208.

Simons, E.L. 1995. Crania of *Apidium*: primitive anthropoidean (Primates, Parapithecidae) from the Egyptian Oligocene. *Am. Mus. Novit.* **3124**: 1–10.

Simons, E.L. 1997. Preliminary description of the cranium of *Proteopithecus sylviae,* an Egyptian late Eocene anthropoidean primate. *Proc. Natl. Acad. Sci. USA.* **94**: 14970–14975.

Simons, E.L. and Rasmussen, D.T. 1996. Skull of *Catopithecus browni,* an early Tertiary catarrhine *Amer. J. Phys. Anthropol.* **100**: 261–292.

Simons, E.L., Holroyd, P.A., and Bown, T.M. 1991. Early Tertiary elephant shrews from Egypt and the origin of the Macroscelidea. *Proc. Nat. Acad. Sci. USA.* 88: 9734–9737.

Simpson, G.G. 1927. Mesozoic Mammalia. IX. The brain of Jurassic mammals. *Am. J. Sci.* **214**: 259–268.

Simpson, G.G. 1928. *A catalogue of the Mesozoic Mammalia in the Geological Department of the British Museum.* Oxford University Press: Oxford.

Simpson, G.G. 1929. American Mesozoic Mammalia. *Mem. Peabody Mus.* **3**: 1–235.

Simpson, G.G. 1937. Skull structure of the Multituberculata. *Bull. Am. Mus. Nat. Hist.* **73**: 727–763.

Simpson, G.G. 1944. *Tempo and mode in evolution.* Columbia University Press: New York.

Simpson, G.G. 1945. The principles of classification and a classification of mammals. *Bull. Am. Mus. Nat. Hist.* **85**: 1–350.

Simpson, G.G. 1953. *The major features of evolution.* Columbia University Press: New York.

Simpson, G.G. 1960. Diagnoses of the classes Reptilia and Mammalia. *Evolution* **14**: 388–392.

Simpson, G.G. 1970. Additions to our knowledge of the Argyrolagidae (Mammalia, Marsupialia) from the Late Cenozoic of Argentina. *Breviora* **361**: 1–9.

Simpson, G.G. 1970. The Argyrolagidae, extinct southern American marsupials. *Bull. Mus. Comp. Zool. Harvd.* **139**: 1–86.

Simpson, G.G. 1980. *Splendid isolation: the curious history of South American mammals.* Yale University Press: New Haven.

Slaughter, B.H. 1971. Mid-Cretaceous (Albian) therians of the Butler Farm local fauna, Texas. In D.M. Kermack and K.A. Kermack (eds) *Early mammals. Zool. J. Linn. Soc.* **50** (Suppl. 1), 31–144.

Smith, A.G., Smith, D.G., and Funnell, B.M. 1994. *Atlas of Mesozoic and Cenozoic coastlines.* Cambridge University Press: Cambridge.

Smith, R.M.H. 1987. Helical burrow casts of therapsid origin from the Beaufort group (Permian) of South Africa. *Palaeogeogr. Palaeoclimatol. Palaeoecol.* **60**: 155–170.

Smith, R.M.H. 1993. Sedimentology and ichnology of palaeosurfaces in the Beaufort Group (Late Permian), Karoo Sequence, South Africa. *Palaios* **8**: 339–357.

Smith, R.M.H. 1995. Changing fluvial environments across the Permian-Triassic boundary in the Karoo Basin, South Africa and possible causes of tetrapod extinctions. *Palaeogeogr. Palaeoclimatol. Palaeoecol.* **117**: 81–104.

Smith, R.M.H. and Ward, P.D. 2001. Pattern of vertebrate extinctions across an event bed at the Permian–Triassic boundary in the Karoo basin of South Africa. *Geology* **29**: 1147–1150.

Smithson, T.R., Carroll, R.L., Panchen, A.L., and Andrews, S.M. 1994. *Westlothiana lizziae* from the Visean of East Kirton, West Lothian, Scotland, and the amniote stem. *Trans.R. Soc. Edinb. Earth Sci.* **84**: 417–431.

Springer, M.S., Kirsch, J.A., and Case, J.A. 1997a. The chronicle of marsupial evolution. In T.J. Givnish and K.J. Sytsma (eds) *Molecular evolution and adaptive radiation.* Cambridge University Press: Cambridge, pp. 129–161.

Springer, M.S., Cleven, G.C., Madsen, O., Jong, W.W. de, Waddell, V.G., Amrine, H.M., and Stanhope, M.J. 1997b. Endemic African mammals shake the phylogenetic tree. *Nature* **388**: 61–64.

Springer, M.S., Westerman, M., Kavanagh, J.R., Burk, A., Woodburne, M.O., Kao, D.J., and Krajewski, C. 1998. The origin of the Australasian marsupial fauna and the phylogenetic affinities of the enigmatic monito del monte and marsupial mole. *Proc. R. Soc. Lond.* **B265**: 2381–2386.

Springer, M.S., Murphy, W.J., Eizirik, E., and O'Brien, S.J. 2003. Placental mammal diversification and the Cretaceous-Tertiary boundary. *Proc. Natl. Acad. Sci.* **100**: 1056–1061.

Stanhope, M., Waddell, V., Madsen, O., de Jong, W., Hedges, S., Cleve, G., Kao, D., and Springer, M. 1998.

Molecular evidence for multiple origins of Insectivora and a new order of endemic African insectivore mammals. *Proc. Natl. Acad. Sci. USA.* **95**: 9967–9972.

Stehli, F.G. and Webb, S.D. (eds) 1985. *The Great American Interchange.* Plenum: New York.

Stewart, J.R. 1997. Morphology and evolution of the egg of oviparous amniotes. In S.S. Sumida and K.L.M. Martin (eds) *Amniote origins: completing the transition to land.* Academic Press: London, pp. 291–326.

Storch, G. 1981. *Eurotamandua joresi,* ein Myrmecophagide aus dem Eozän der 'Grube Messel' bei Darmstadt (Mammalia: Xenarthra). *Senckenbergiana Lethaea* **61**: 247–289.

Storch, G. and Richter, G. 1992a. The ant-eater *Eurotamandua*: a South American in Europe. In S. Schaal and W. Ziegler (eds) *Messel: an insight into the history of life and of the Earth.* Oxford University Press: Oxford, pp. 209–216.

Storch, G. and Richter, G. 1992b. Pangolins: almost unchanged for 50 million years. In S. Schaal and W. Ziegler (eds) *Messel: an insight into the history of life and of the Earth.* Oxford University Press: Oxford.

Stuart, A.J. 1999. Late Pleistocene megafaunal extinctions: a European perspective. In R.D.E. MacPhee (ed) *Extinctions in near time: causes, contexts, and consequences.* Kluwer Academic: New York, pp. 257–269.

Stucky, R.K. 1998. Eocene bunodont and selenodont Artiodactyla ('dichobunids'). In C.M. Janis, K.M. Scott, and L.L. Jacobs (eds) *Evolution of Tertiary mammals of North America, Vol. 1: Terrestrial carnivores, ungulates, and ungulate-like mammals.* Cambridge University Press: Cambridge, pp. 358–374.

Sues, H.-D. 1983. *Advanced mammal-like reptiles from the early Jurassic of Arizona.* Ph.D. thesis, Harvard University.

Sues, H.-D. 1985. The relationships of the Tritylodontidae (Synapsida). *Zool. J. Linn. Soc.* **85**: 205–217.

Sues, H.-D. 1986a. Locomotion and body form in early therapsids (Dinocephalia, Gorgonopsia, and Therocephalia). In N. Hotton III, P.D. MacLean, J.J. Roth, and E.C. Roth (eds) *The ecology and biology of mammal-like reptiles.* Smithsonian Institution Press: Washington, pp. 61–70.

Sues, H.-D. 1986b. Relationships and biostratigraphic significance of the Tritylodontidae (Synapsida) from the Kayenta Formation of northeastern Arizona. In K. Padian (ed) *The beginning of the age of dinosaurs: faunal change across the Triassic–Jurassic boundary.* Cambridge University Press: Cambridge, pp. 279–284.

Sues, H.-D. 2001. On *Microconodon,* a Late Triassic cynodont from the Newark Supergroup of eastern North America. *Bull. Mus. Comp. Zool. Harvd.* **156**: 37–48.

Sues, H.-D. and Boy, J.A. 1988. A procynosuchid cynodont from central Europe. *Nature* **331**: 523–524.

Sues, H.D. and Munk, W. 1996. A remarkable assemblage of terrestrial tetrapods from the Zechstein (Upper Permian, Tatarian) near Korbach (northwestern Hesse). *Palaeontol. Zeitsch* **70**: 213–223.

Sues, H.-D, Olsen, P.E., and Carter, J.G. 1999. A Late Triassic traversodont cynodont from the Newark Supergroup of North Carolina. *J. Vert. Paleontol.* **19**: 351–354.

Sumida, S.S. 1997. Locomotor features of taxa spanning the origin of amniotes. In S.S. Sumida and K.L.M. Martin (eds) *Amniote origins: completing the transition to land.* Academic Press: London, pp. 353–398.

Sun, A. 1991. A review of Chinese therocephalian reptiles. *Vert. PalAs.* **29**: 85–94.

Sweet, A.R., Braman, D., and Lerbekmo, J.F. 1993. Northern mid-continental Maastrichtian and Paleocene extinction events. *Geol. Assoc. Can. Progr. Abt.* p. 103.

Swisher, C.C. III and Prothero, D.R. 1990. Single-Crystal ^{40}Ar/^{39}Ar dating of the Eocene–Oligocene transition in North America. *Science* **249**: 760–762.

Szalay, F.S. 1969. Origin and evolution of function of the mesonychid condylarth feeding mechanism. *Evolution* **23**: 703–720.

Szalay, F.S. 1982. A new appraisal of marsupial phylogeny and classification. In M. Archer (ed) *Carnivorous marsupials.* Royal Zoological Society of New South Wales: Sydney, pp. 621–640.

Szalay, F.S. 1994. *Evolutionary history of the marsupials and an analysis of osteological characters.* Cambridge University Press: Cambridge.

Szalay, F.S. and Delson, E. 1979. *Evolutionary history of the primates.* Academic Press: New York.

Szalay, F.S. and Lucas, S.G. 1993. Cranioskeletal morphology of archontans, and diagnoses of Chiroptera, Volitantia, and Archonta. In R.D.E. MacPhee (ed) *Primates and their relatives in phylogenetic perspective.* Plenum: New York, pp. 187–226.

Szalay, F.S. and McKenna, M.C. 1971. Beginning of the age of mammals in Asia: the Late Paleocene Gashato fauna, Mongolia. *Bull. Am. Mus. Nat. Hist.* **144**: 269–318.

Szalay, F.S. and Sargis, E.J. 2001. Model-based analysis of postcranial osteology of marsupials from the Palaeocene of Itaboraí (Brazil) and the phylogenetics and biogeography of Metatheria. *Geodiversitas* **23**: 139–302.

Szalay, F.S. and Trofimov, B.A. 1996. The Mongolian Late Cretaceous *Asiatherium*, and the early phylogeny and paleobiogeography of Metatheria. *J. Vert. Paleontol.* **16**: 474–509.

Tabuce, R., Coiffait, B., Coiffait, P.E., Mahboubi, M., and Jaeger, J.J. 2001. A new genus of Macroscelidea (Mammalia) from the Eocene of Algeria: a possible origin for elephant shrews. *J. Vert. Paleontol.* **21**: 535–546.

Takei, M., Anaya, F., Shigehara, N., and Setoguchi, T. 2000. New fossil material of the earliest New World monkey, *Branisella boliviana*, and the problem of platyrrhine origins. *Am. J. Phys. Anthropol.* **111**: 263–282.

Tassy, P. 1996. Who is who among the Proboscidea? In J. Shoshani and P. Tassy (eds) *The Proboscidea: evolution and palaeoecology of elephants and their relatives.* Oxford University Press: Oxford, pp. 39–48.

Tassy, P. and Shoshani, J. 1988. The Tethytheria: elephants and their relatives. In M.J. Benton (ed) *The phylogeny and classification of the tetrapods*, Vol. 2: *Mammals*. Oxford University Press: Oxford, pp. 283–316.

Tatarinov, L.P. 1968. Morphology and systematics of the North Dvina cynodonts (Reptilia, Therapsida: Upper Permian). *Postilla* **126**: 1–15.

Tatarinov, L.P. 1974. Theriodonts of the USSR. *Tr. Paleontol. Inst. Akad. Nauk SSSR (Moscow)* **143**: 1–240.

Tatarinov, L.P. and Matchenko, E.N. 1999. A find of an aberrant tritylodont (Reptilia, Cynodontia) in the Lower Cretaceous of the Kemerovo region. *Palaeontol. J.* **33**: 422–428.

Taveré, S., Marshall, C.R., Will, C.O., Soligo, C., and Martin, R.D. 2002. Using the fossil record to estimate the age of the last common ancestor of extant primates. *Nature* **416**: 726–729.

Tchudinov, P.K. 1960. Upper Permian therapsids of the Ezhovo locality. *Paleontol. J.* **2**: 85–98.

Tchudinov, P.K. 1983. Early therapsids. *Tr.Paleontol. Inst. Akad. Nauk SSSR (Moscow)*. **202**: 1–230.

Tedford, R.H. and Woodburne, M.O. 1987. The Iliariidae, a new family of vombatiform marsupials from Miocene strata of South Australia and an evaluation of the homology of molar cusps in the Diprotodontia. In M. Archer (ed) *Possums and opossums: studies in evolution.* Surrey Beatty and Sons: Chipping Norton, Australia, pp. 401–418.

Teeling, E.C., Scally, M., Kao, D.J., Romagnoli, M.L., Springer, M.S., and Stanhope, M.J. 2000. Molecular evidence regarding the origin of echolocation and flight in bats. *Nature* **403**: 188–192.

Teeling, E.C., Madsen, O., Van Den Bussche, R.A., Jong, W. de, Stanhope, M.J. and Springer, M.S. 2002. Microbat paraphyly and the convergent evolution of a key innovation in the Old World rhinolophoid microbats. *Proc. Natl. Acad. Sci. USA.* **99**: *1431–1436.*

Temple-Smith, P. 1987. Sperm structure and marsupial phylogeny. In M. Archer (ed) *Possums and opossums: studies in evolution.* Surrey Beatty and Sons: Chipping Norton, Australia, pp. 171–193.

Thewissen, J.G.M. 1990. Evolution of Paleocene and Eocene Phenacodontidae (Mammalia, Condylarthra). *Univ. Mich. Pap. Paleontol.* **29**: 1–107.

Thewissen, J.G.M. 1998. Cetacean origins: evolutionary turmoil during the invasion of the oceans. In J.G.M. Thewissen (ed) *The emergence of whales: evolutionary patterns in the origin of Cetacea.* Plenum: New York, pp. 451–464.

Thewissen, J.G.M. and Domning, D.P. 1992. The role of phenacodontids in the origin of the modern orders of ungulate mammals. *J. Vert. Paleontol.* **12**: 494–504.

Thewissen, J.G.M. and Gingerich, P.D. 1989. Skull and endocranial cast of *Eoryctes melanus*, a new palaeoryctid (Mammalia: Insectivora) from the early Eocene of western North America. *J. Vert. Paleontol.* **9**: 459–470.

Thewissen, J.G.M. and Simons, E.L. 2001. Skull of *Megalohyrax eocaenus* (Hyracoidea, Mammalia) from the Oligocene of Egypt. *J. Vert. Paleontol.* **21**: 98–106.

Thewissen, J.G.M. and Williams, E.M. 2002. The early radiations of Cetacea (Mammalia): evolutionary pattern and developmental correlations. *Annu. Rev. Ecol. Syst.* **33**: 73–90.

Thewissen, J.G.M., Hussain, S.T., and Arif, M. 1994. Fossil evidence for the origin of aquatic locomotion in archaeocete whales. *Science* **263**: 210–212.

Thewissen, J.G.M., Madar, S.I., and Hussain, S.T. 1996. *Ambulocoetus natans*, an Eocene cetacean (Mammalia) from Pakistan. *Courier Forsch. Inst. Senckenberg* **191**: 1–86.

Thewissen, J.G.M., Williams, E.M., and Hussain, S.T. 2001. Skeletons of terrestrial cetaceans and the relationship of whales to artiodactyls. *Nature* **413**: 277–281.

Thomason, J.J. and Russell, A.P. 1986. Mechanical factors in the evolution of the mammalian secondary palate: a theoretical analysis. *J. Morphol.* **189**: 199–213.

Thomson, K.S. 1993. The origin of tetrapods. *Am. J. Sci.* **293A**: 33–62.

Thulborn, R.A. 1983. A mammal-like reptile from Austalia. *Nature* **303**: 330–331.

Thulborn, T. and Turner, S. 2003. The last dicynodont: an Australian Cretaceous relic. *Proc. R. Soc. Lond.* **B270**: 985–993.

Ting, S. 1998. Paleocene and early Eocene land mammal ages of Asia. In K.C. Beard and M.R. Dawson (eds) *Dawn of the age of mammals in Asia. Bull. Carnegie Mus. Nat. Hist.* **34**: 124–147.

Trofimov, B.A. and Szalay, F.S. 1994. New Cretaceous marsupials from Mongolia and the early radiation of Metatheria. *Proc. Nat. Acad. Sci. USA.* **91**: 12569–12573.

Turner, A. and Antón, M. 1997. *The big cats and their fossil relatives.* Columbia University Press: New York.

Tyndale-Biscoe, C.H., and Renfree, M.B. 1987. *Reproductive physiology of marsupials.* Cambridge University Press: Cambridge.

Uhen, M.D. 1998. Middle to Late Eocene basilosaurines and dorudontines. In J.G.M. Thewissen (ed) *The emergence of whales: evolutionary patterns in the origin of Cetacea.* Plenum: New York, pp. 29–61.

V'iuschkov, B.P. 1969. New dicynodonts from the Triassic of southern Cisuralia. *Palaeontol. J.* **3**: 237–242.

Van den Heever, J.A. 1994. The cranial anatomy of the early Therocephalia (Amniota: Therapsida). *Ann. Univ. Stellenbosch* **1**: 1–59.

Van Valen, L. 1971. Towards the origin of artiodactyls. *Evolution* **25**: 523–529.

Van Valen, L. and Sloan, R.E. 1966. The extinction of the multituberculates. *Syst. Zool.* **15**: 261–278.

Vicaíno, S.F. and Scillato Yané, G.J. 1995. An Eocene tardigrade (Mammalia, Xenarthra) from Seymour Island, West Antarctica. *Antarct. Sci.* **7**: 407–408.

Vickers-Rich, P., Monaghan, J.M., Baird, R.F. and Rich, T.H. (eds) *Vertebrate palaeontology of Australia.* Pioneer Design Studio and Monash University: Melbourne.

Visscher, H., Brinkhuis, H., Dilcher, D.L., Elsik, W.C., Eshet, Y., Looy, C.V., Rampino, M.R., and Traverse, A. 1996. The terminal Paleozoic fungal event: evidence of terrestrial ecosystem destabilization and collapse. *Proc. Natl. Acad. Sci. USA.* **93**: 2155–2158.

Vrba, E.S. 1992. Mammals as a key to evolutionary theory. *J. Mamm.* **73**: 1–28.

Vrba, E.S. 1993. Mammal evolution in the African Neogene and a new look at the Great American Interchange. In P. Goldblatt (ed) *Biological relationships between Africa and South America.* Yale University Press: New Haven.

Waddell, P.J., Okada, N., and Hasegawa, M. 1999. Towards resolving the interordinal relationships of placental mammals. *Syst. Biol.* **48**: 1–5.

Wake, D.B. and Roth, G. 1986. Introduction. In D.B. Wake and G. Roth (eds) *Complex organismal functions: integration and evolution in vertebrates.* John Wiley & Sons: Chichester, pp. 1–5.

Waldman, M. and Savage, R.J.G. 1972. The first Jurassic mammals from Scotland. *J. Geol. Soc. Lond.* **128**: 119–125.

Wang, Y., Hu, Y., Chow, M., and Li, C. 1998. Chinese Paleocene mammals and their correlation. In K.C. Beard and M.R. Dawson (eds) *Dawn of the age of mammals in Asia. Bull. Carnegie Mus. Nat. Hist.* **34**: 89–123.

Wang, Y.-Q., Clemens, W.A., Hu, Y.-M., and Li, C.-K. 1998. A probable pseudotribosphenic upper molar from the Late Jurassic of China and the early radiation of the Holotheria. *J. Vert. Paleontol.* **18**: 777–787.

Wang, Y.-Q., Hu, Y.-M., Meng, J., and Li, C.-K. 2001. An ossified Meckel's cartilage in two Cretaceous mammals and the origin of the mammalian middle ear. *Science* **294**: 357–361.

Watson, D.M.S. 1931. On the skeleton of a bauriamorph reptile. *Proc. Zool. Soc. Lond.* 163–1205.

Watson, D.M.S. and Romer, A.S. 1956. A classification of therapsid reptiles. *Bull. Mus. Comp. Zool. Harvd.* **111**: 37–89.

Webb, S. D. 1985*a*. Late Cenozoic mammal dispersals between the Americas. In F.G. Stehli and S.D. Webb (eds) *The Great American Interchange*. Plenum: New York, pp. 357–386.

Webb, S.D. 1985*b*. Main pathways of mammalian diversification in North America. In F.G. Stehli and S.D. Webb (eds) *The Great American Interchange*. Plenum: New York, pp. 201–217.

Webb, S.D. 1991. Ecogeography and the Great American Interchange. *Paleobiology* **17**: 266–280.

Wells, N.A. and Gingerich, P.D. 1983. Review of Eocene Anthracobunidae (Mammalia, Proboscidea with a new genus and species, *Jozaria palustris*, from the Kuldana Formation of Kohat (Pakistan). *Univ. Michigan Contr. Mus. Paleont.* **26**: 117–139.

Whittington, S.L. and Dyke, B. 1984. Simulating overkill: experiments with the Mosimann and Martin model. In P.S. Marten and R.G. Klein (eds) *Quaternary extinctions: a prehistoric revolution*. University of Arizona Press: Tucson.

Wible, J.R. 1991. Origin of Mammalia: the craniodental evidence reexamined. *J. Vert. Paleontol.* **11**: 1–28.

Wible, J.R. 1993. Cranial circulation and relationships of the colugo. *Am. Mus. Novit.* **3072**: 1–27.

Wible, J.R. and Hopson, J.A. 1993. Basicranial evidence for early mammal phylogeny. In F.S. Szalay, M.J. Novacek, and M.C. McKenna (eds) *Mammal phylogeny: Mesozoic differentiation, multituberculates, monotremes, early therians, and marsupials*. Springer-Verlag: New York, pp. 45–62.

Wignall, P.B. 2001. Large igneous provinces and mass extinctions. *Earth Sci. Rev.* **53**: 1–33.

Williams, E.M. 1998. Synopsis of the earliest cetaceans: Pakicetidae, Ambulocetidae, Remingtonocetidae, and Protocetidae. In J.G.M. Thewissen (ed) *The emergence of whales: evolutionary patterns in the origin of Cetacea*. Plenum: New York, pp. 1–28.

Wing, S.L. and Sues, H.-D. 1992. Mesozoic and early Cenozoic terrestrial ecosystems. In A.K. Behrensmeyer, J.D. Damuth, W.A. DiMichele, R.Potts, H.-D. Sues, and S.L. Wing (eds) *Terrestrial ecosystems through time: evolutionary paleoecology of terrestrial plants and animals*. Chicago University Press: Chicago, pp. 327–416.

Wing, S.L., Alroy, J., and Hickey, L.J. 1995. Plant and animal diversity in the Paleocene to Early Eocene of the Bighorn basin. *Palaegeogr. Palaeoclimatol. Palaeoecol.* **115**: 117–155.

Wolfe, J.A. 1978. A paleobotanical interpretation of Tertiary climates in the Northern Hemisphere. *Am. Sci.* **66**: 694–703.

Wolfe, J.A. 1992. Climatic, floristic, and vegetational changes near the Eocene-Oligocene boundary in North America. In D.R. Prothero and W.A. Berggren (eds) *Eocene–Oligocene climatic and biotic evolution*. Princeton University Press: Princeton, pp. 421–436.

Woodburne, M.O. and Case, J.A. 1996. Dispersal, vicariance, and the Late Cretaceous to Early Tertiary land mammal biogeography from South America to Australia. *J. Mamm. Evol.* **3**: 121–161.

Woodburne, M.O. and Zinsmeister, W.J. 1982. Fossil land mammals from Antarctica. *Science* **218**: 284–286.

Woodburne, M.O. and Zinsmeister, W.J. 1984. The first land mammal from Antarctica and its biogeographic implications. *J. Paleontol.* **58**: 913–948.

Woodburne, M.O., Rich, T.H., and Springer, M.S. 2003. The evolution of tribosphamy and the antiquity of mammalian clades. *Mol. Phylogenet. Evol.* **28**: 360–385.

Woodburne, R.H., Pledge, N.S., and Archer, M. 1987. The Miralinidae, a new family and two new species of phalangeroid marsupials from Miocene strata of South Australia. In M. Archer (ed) *Possums and opossums: studies in evolution*. Surrey Beatty and Sons: Chipping Norton, Australia, pp. 581–602.

Woodburne, R.O. 1987. The Ektopodontidae, an unusual family of Neogene phalangeroid marsupials. In M. Archer (ed) *Possums and opossums: studies in evolution*. Surrey Beatty and Sons: Chipping Norton, Australia, pp. 603–606.

Wright, D.B. 1998. Tayassuidae. In C.M. Janis, K.M. Scott, and L.L. Jacobs (eds) *Evolution of Tertiary mammals of North America, Vol. 1: Terrestrial carnivores, ungulates, and ungulatelike mammals*. Cambridge University Press: Cambridge, pp. 389–401.

Wroe, S. 1996. *Muribacinus gadyuli* (Thylacinidae, Marsupialia), a very plesiomorphous thylacinid from the Miocene of Riversleigh, northwestern Queensland, and the problem of paraphyly for the Dasyuridae. *J. Paleontol.* **70**: 1032–1044.

Wroe, S. 1999. The geologically oldest dasyurid, from the Miocene of Riversleigh, north-west Queensland. *Palaeontology* **42**: 501–527.

Wroe, S., Crowther, M., Dortch, J. and Chong, J. 2003. The size of the largest marsupial and why it matters. *Proc. R. Soc. Lond. (Suppl)* **B271**: S34–S36.

Wyss, A.R. and Flynn, J.J. 1993. A phylogenetic analysis and definition of the Carnivora. In Szalay, F.S.,

Novacek, M.J., and McKenna, M.C. (eds) *Mammal phylogeny: placentals.* Springer-Verlag: New York, pp. 32–52.

Wyss, A.R., Flynn, J.J., Norfell, M.A., Swisher, C.C. III, Charrier, R., Novacek, M.J., and McKenna, M.C. 1993. South America's earliest rodent and recognition of a new interval of mammalian evolution. *Nature* **365**: 434–437.

Wyss, A.R., Flynn, J.J., Norrell, M.A., Swisher, C.C. III, Novacek, M.J., McKenna, M.C., and Charrier, R. 1994. Paleogene mammals from the Andes of central Chile: a preliminary taxonomic, biostratigraphic, geochronologic assessment. *Am. Mus. Novit.* **3098**: 1–31.

Zeigler, A.M., Hulver, M.L., and Rowley, D.B. 1997. Permian world topography and climate. In I.P. Martini (ed) *Late glacial and postglacial environmental changes: Quaternary, Carboniferous-Permian.* Oxford University Press: Oxford. pp. 111–146.

Zeigler, R. 1999. Order Marsupialia: *Amphiperatherium,* the last European opossum. In G.E. Rösssner and K. Heissig (eds) *The Miocene land mammals of Europe.* Pfeil: Munich. pp. 49–52.

Zhang, F.-K., Crompton, A.W., Luo, Z.-X., and Schaff, C.R. 1998. Pattern of dental replacement of *Sinoconodon* and its implications for evolution of mammals. *Vert, PalAs.* **36**: 197–217.

Zhou, Z., Barrett, P.M., and Hilton, J. 2003. An exceptionally preserved Lower Cretaceous ecosystem. *Nature* **421**: 807–814.

Index

Note: *Page numbers in italics denote illustration*

Abderites 205
Abderitidae 206
Abdounodus 236
Acanthostega 14, *15*
Acrobatidae 216
Adapiformes 269, 271
Adelobasileus 138–40, 184
Aegialodon 167, 169
Aegialodontidae 167, 169
Aegyptopithecus 272
Aepyornis 289
aerobic capacity hypothesis 125–6
Aerosaurus 21, 22
Afrosoricida 257, 259
Afrotheria 224, 226, 238, 251–9, 275, 277, 279
   cladogram *252*
Allotheria 181
Allqokirus 203
Alphadon 196, *197*
   marshi 189
Ambondro 178, 180
   dentition *179*
Ambulocetus 265, *266*
Ameridelphia 194, 198–208
Amniotes 3, 16, 17
   evolution 18–9
   feeding mechanism 90
   skull 18
   stem-group 16, 18
Amniotic egg 19
Amphibians 19
Amphilestes 150, 152
Amphilestidae 152
Amphitherium 167, 178
   dentition *168*
Anagale 232
Anagalida 231
Ancestral amniote grade
   feeding mechanism 90
   locomotion 101
Andinodelphys 199, 203, 220
Andrewsarchus 235, 236

Ankotarinja 211, 212
Anningia 22
Anomaluridae 274
Anomocephalus 40, 41, 79, 80
Anomodontia 28, 35, 39–43
   cranial characteristics 39
Antarctodolops 204, 206
Anteosaurus 37
   skull *35, 36*
Anthracotheriids 264
Anthracobunids 253
Anthropoidea 271–2
Antilocapra 264
Antilocapridae 264
Apatemyida 232
Apidium 271
Aquiladelphis 198
Archaeocetes 265
Archaeolambda 240
Archaeonycteris 268
Archaeopteropus 268
Archeothyris 3, 16, *17*, 18, 22–3
Archonta 222
Arctocyon 234, *235*
Arctocyonidae 234, 236
Arctostylopida 242
Argyrolagoidea 206
Argyrolagus 207
Arsinoitherium 254, 257
Artiocetus 265, *266*
Artiodactyls 262–4, 265, 267
   phylogeny *263*
Arctognathus 53
Asfaltomylos *179*, 180
Asiatherium 171, *172*
Asioryctes 171, 227, 228
Asioryctida 228–9
Astrapotheria 245
Astrapotherium 243, 244, 245
Atokatheridium 174
Aulacephalodon 46, 49
Ausktribosphenos 171, 178, 180, 277
   dentition *179*

Australidelphia 194, 208–16
Australosphenida 169, 173, 178–80
   dentition *179*
Australosyodon 37
Austrotriconodon *150*, 152
Avenantia 38

Badjcinus 211, 212
Baluchitherium 262
Barbourofelis 258, 259
Barinya 211, 212
Barunlestes 171, 229
   skull 227
Barytherium 256
Basal Metabolic Rate (BMR) 122
Basiliosaurus 265, *266*
Batodon 268
   tenuis 189
Bauria 58–9
Baurioidea 57, 59
Bemalambda 238, 240
Biarmosuchia 28, 31–33
Biarmosuchian grade
   feeding mechanism 94
Biarmosuchus 31, *32*, 79, 80
Biseridens 33
Bishops 180
Bistius 173, *174*
Bobbschaefferia 209
Bocatherium 70, *71*
Bondesius 164, 166
Boreosphenida 161, 169, 178
Boreoeutheria 226, 277
Borhyaenoidea
   See Sparassodontia
Branisella 271
Brasilitherium 73, 75, 76
Brasilodon 73, 75
Brithopia 36, 37
Brithopian Dinocephalia 28
Brontops 261
Brontotheres 260

Bugtilemur 271
Bulganbaatar 157, 159
Bunodontia 264
Burnetia 33, 38
Burnetiidae 33
Burramyidae 216
Buxolestes 231

Caenolestes 205
Caenolestidae 192, 206
Callistoe 202, 203
Camelidae 264
Caniformes 259–60
Cantius 269
Capybara 274
Carnivora 247–50, 259–60
Carodnia 243, 244, 247
Caroloameghina 204, 205
Carolopaulacoutia 205, 206
Carpolestes 233, 234, 269
Casea broilii 21, 22
   rutena 21, 22
Caseasauria 20, 21, 22
Caseidae 20, 21, 22
Casineria 16
Catarrhini 272
Catopithecus 272
Cervidae 264
Cetacea 264–5
Cetartiodactyla 262
Chalicothere 262
Chalicotherium 261, 262
Chambilestes 259
Chambius 257
Chilicebus 271
Chiniquodon 66, 70
   feeding mechanism 98
Chiniquodontidae 70
Chironectes 192
Chiroptera 268
Chriacus 235, 236
Chrysochlorida 259
Chrysochloris 224
Chulpasia 199
Chulsanbaatar 159, 160
Cimolestes 189, 228, 229
Cimolodonta 155
Cistecephalus 46, 48
Cistecephaloides 48
Cladosictis 202, 204
Cladotheria 161
Clelandina 54
Cocomys 273
Competitive exclusion hypothesis
   184–5
Condylarthra 234–8, 257
   carnivorous 236

herbivorous 236–8
correlated progression 133–4
Coryphodon 240
Cotylorhynchus hancocki 22
Creodonta 247–50
Cricodon 67
Crusafontia 165
Ctenacodon 155
Ctenadactyla 274
Cynocephalus 272
Cynognathidae 65, 67, 75–6
Cynognathus 65, 67
Cynodonts 60–75
   characteristics 60
   cladogram 77
   interrelationships 75–80

Dasyuromorphia 192, 212
Dasyurus 212
Daouitherium 253, 255, 256
Deinotherium 255
Deltatheridium 173, 174
Deltatheroida 171–3
Deltatheroides 173, 174
Dermoptera 272
Dermotherium 272
Desmostylia 257
Diacodexis 262, 263, 267
Diadectomorphs 18
Diademodon 65, 67, 68
   tooth replacement 121
Diademodontoidea 65, 67–9
Diarthrognathus 72, 73, 74
Dichobunidae 262
Dicroidium 86
Dicynodon 49, 50
   trigonocephalus 45
Dicynodonts 43–52
   evolutionary radiation 45–7
   phylogeny 46
Didelphidae 196, 198
Didelphimorphia 191–2, 193, 199–201
   dentition 200
   skull 200
Didelphis 159, 160
   virginiana 183, 191–2, 285
Didelphodon 197, 198
Didymoconidae 230
Diictodon 48
Diictodontoidea 48
Dimetrodon 24–5, 26
   brain 118
   feeding mechanism 92
   locomotion 101, 103–5
Dinanomodon 49
Dinnetherium 146–7
Dinocephalia 30, 33–9

characteristics 34
cladogram 36
Dinocerata 240, 242
Diprotodon optatum 213, 214
Diprotodontia 193, 213
Diprotodontidae 213, 214
Djadochtatherioids 156–7
Djarthia 208, 209, 212, 219
Djilgaringa 214
Docodon 147, 148
Docodonta 147–9
Domnina 267, 268
Dromasauroidea 43
Dromiciops 194, 195, 196, 209,
   210, 219
   gliroides 192
Dryolestida 166–7
Dvinia 61, 62

East of Eden hypothesis 279–80
Echidna 1
Ecteninion 66, 70
Ectoconus 236, 237, 238
Edaphosauria 24
Edaphosauridae 24
Edaphosaurus 23, 24, 81
Edentata 222
Ektopodon 214
Ektopodontidae 216
Elephantosaurus 51
Elephantulus rufescens 224
Eleutherodon 139, 141
Elginerpeton 14
Elliotsmithia 21, 22
Elomeryx 263
Embrithopoda 257
Emydops 48
Enaliarctos 258, 260
endothermy 121–3, 124
   mammals 128–9
   synapsids 126–8
Endothiodontoidea 47
Ennatosaurus 22
Eodicynodon 43, 44, 45, 46, 79
Eomaia 169, 170, 183, 226, 274
Eomanis 267
Eorcytes 231
Eosimias 271
Eotalpa 267, 268
Eothyrididae 20, 21
Eothyris 17, 20, 90
Eotitanops 260
Eotitanosuchus 31, 33
Eozostrodon 142
Epicynodontia 62–4
Epidolops 204, 205
Epitheria 222

Equoidea 262
Equus 262
Erethizon 285
Erinaceidae 268
Ernanodon 247, *248*
Erythrotherium 145
Escavadon 247, *248*
Estemmenosuchia 38–9
Estemmenosuchus 29, 30, 37
Esthonyx 240
Euarchontoglires 226, 268–74
Euchambersia 57, *58*
Eucynodont grade
  feeding mechanism 98–100
  locomotion 110–2
Eucynodontia 64–75
Eulipotyphla 268
Eupantotheria 161, 166–7, 178
Eupelycosauria 22–6
  feeding mechanism 92
Eurotamandua 251
Eurymylids 273
Eutherapsida 79
Eutheria
  See Placentalia
Eutheriodontia 60, 79
Eutherocephalia 57
Eutriconodonta 150–2, 178
Exaeretodon 69

Falpetrus 173, *174*
Feliformes 259
Ferugliotherium 152
Ferungulata 222
Folivores 267
fossils
  dating 7, 9
  morphological characters 9

Galechirus 41, 43
Galeops 41, 43
Galepus 43
Galesauridae 63–4
Galesaurus 63, 86
Garden of Eden hypothesis 276–7,
  278
Gashatostylops 241
Glasbiidae 198
Glasbius 197, 204, 217
Glaucosaurus 23, 24
Glires 222, 229, 273–4
Global Stratotype Section and Point
  (GSSP) 6
Glossopteris 85
Glyptodonta 250–1
Glyptotherium 251
Gobiconodon 150, 151, 152

ostromi 183
Gobiotherodon 163
Golden spike 6
Gomphodont cynodonts 76
Gomophothere Land Bridge 256
Gomphotherium 255, 256
Gorgonpsidae 52
Gorgonopsia 28, 52–4
  brain *119*
Great American Biotic Interchange
  286
Groeberia 206–8, *207*
Groeberida 206
Grube Messel Eocene Locality 4
Gypsonictops 227, 228, 229
  illuminatus 189

Hadrocodium 149, 183
Haemothermia 1
Haldanodon 147, *148*, 149
Hapalodectes 236
Haplorhini 271
Haptodus 25–6
Haramiya 140
Haramiyida 140
Harmiyavia 139
Harpagolestes 235, 236
Henkelotherium 165, 166–7, 183
Heomys 231, 232, 273
Herodotius 257
Hesperocyon 258, 260
Himalayacetus 265
Hippidion 285
Hippopotamidae 264
Hipposaurus 33
Holoclemensia 173, *174*
Holotheria 161
Homo sapiens 272
Homogalax 260, 261
Hoplophoneus 258, 259
Hunter-Schreger bands 186
Hyaenodontidae 248, 250
Hylonomus 16, 18
Hyopsodontids 236
Hyopsodus 236, 237
Hypsilolambda 238
Hyrachyus 262
Hyracoidea 252, 253
Hyracotherium 260, 260
Hystricognatha 273–4

Ianthasaurus 23, 24
Icaronycteris 267, 268
Ichthyostega 14, 15
Ictidosuchops 55
Ictidorhinus 32, 33

Ilariidae 215
Indricotherium 262
inertial homeothermy 123
Inostrancevia 52, *54*
Insectivora 225
Ischigualastia 50, 52

Jeholodens 150, 151–2
Jonkeria 35, *36*, 37

K/T fern-spike 188
K/T mass extinction 186–9
Kamoyapithecus 272
Kannemeyeria 50, 51
Kannemeyeriids 86
  Triassic radiation 51–2
Karanisia 271
Kawingasaurus 48
Kayentatherium 71, 72
Keeura 212
Kennalestes 171, 227, 229
  brain *119*
Kermackia 173, *174*
Khasia 199, 209, *210*, 219
Kielantherium 168, 169
Klohnia 207, 208
Kwazulusaurus 49, 50
Kingoria 46, 47–8
Kingorioidea 47–8
Kokopellia 171, *172*, 196, 198
Kombuisia 48, 86
Kryptobaatar 160–1
Kuehneotherium 161, 162, *163*, 166,
  175, 181
Kulbeckia 229
Kühne, Walter 147

La Grande Coupure 282–3
Lagomorpha 274
Lambdopsalis 158, 159, 160
Laurasiatheria 226, 259–68
Lemurosaurus 32, 33
Leptictida 229
Leptictidium 230, *231*
Leptictis 230
Leonardus 165
Leontocephalus 53
Limnoscelis 17
Litopterna 245
Lofochaius 240
Long fuse hypothesis 277–9
Luangwa 68, 69, 96
Lumkuia 66, 70
Lycaenops 54
Lycosuchidae 57
Lycosuchus 55, 57
Lystrosaurus 46, 49, 50–1, 86

Macrauchenia 244, 245
Macropodidae 216
Macroscelidea 257
mammal-like reptiles
　See Synapsida
Mammalia
　See Mammals
Mammaliaformes 2, 138
Mammaliamorpha 2
Mammalian grade
　feeding mechanism 100
　locomotion 112–3
Mammalian megafaunal extinction
　287–90
Mammals 14
　bird relationship 1
　brain 118–20
　characteristics 1, 88
　classification *12–13*
　definition 1–3
　ears 113–6
　　evolution *115*
　evolution 3–4, 134–6
　growth and development 120–1
　habitat 130–3
　macroevolutionary issues 4–5
　Mesozoic
　　See Mesozoic mammals
　olfaction 116–8
　phylogeny 11, 75, *223*
　Sister group
　temperature regulation 132–3
Marambiotherium 210
Marsasia 171, 198
Marsupials 1, 171, 173, *194*
　cladogram *195*
　classification *13*, 191–3
　Cretaceous radiation 196–8
　dentition 190, *191*
　evolution 216–21
　　Antarctica 218
　　Australia 218–221
　　South America 217–8
　interrelationships 193–6
Molecular phylogeny 195–196
　reproduction 190
Massetognathus 68–9
　locomotion 110
Mayulestes 202, 203
Megachiroptera 268
Megalohyrax 253, *254*
Megalonyx 251
Megatheriids 251
Megatherium 251
Megazostrodon 130, 145
Meiostylodon 240
Meniscotherium 237, 238

Meridiungulata 242, *243*
Mesenosaurus 22
Mesonychids 224, 236
Mesonyx 235, 236
Mesotherium 246
Mesozoic mammals 3, *12*
　diet 183, 184, 185–6
　evolution 181–3
　interrelationships 180–1
　nocturnality 183–4
　phylogeny *180*
Metacheiromys 247, *248*
Metatheria
　　See Marsupials
Miacidae 250
Miacis 249
Miacoidea 250
Microbiotheria 192, 209–10
Microbiotherium 209–10
Microsorex hoyi 183
Microurania 29, 30
Mimotona 231, *232*, 274
Minchenella 253
miniaturisation
　mammals 135–6
　thermoregulation hypothesis
　　123–4
Mioclaenidae 236, *237*
Mioclaenus 243
Miocochilius 246
Miralina 214
Mixodonta 231
Mixotoxodon 245, 285
Moeritherium 253, 255, 256, 257
Molecular taxonomy 9, 11
Monotremata 1, 173–8
　Mesozoic mammals
　　relationship 175
　Theria relationship 176
Monotrematum 176, *177*
Montanalestes 169, *170*, 228
Morganucodon 70, 74, 76, 77, 78, 138,
　　142–3, 145, 146, 147, 167
　brain 120
　feeding mechanism 100
　growth and development 121
　habitat 130
　Monotremes relationship
　　175–6
　teeth *148*
Morganucodon oehleri 143
Morganucodonta 142–7
　locomotion 146
Moromorpha 262
Moropus 262
Moschops 34, 38
Moschowhaitsia 57

Moschorhinus 57, *58*, 86
Miralinidae 216
Multituberculates 152–61
　brain 160
　dentition 154
　evolution 154–5
　extinction 161
　jaw musculature *156*
　locomotion 157–9
　skull *153*, 154
Muramura 213, *214*
Murtoilestes 169, 226, 228
Mylodontids 251
Myosaurus 45
Myrmecobiidae 212
Myrmecophaga 285
Myrmecobius fasciatus 192

Nemegtbaatar 157
　brain 160
　locomotion *158*, 159
　skull *153*
Neotherapsida 79
Niaftasuchus 29, 31
Nikkasauridae 31
Nikkasaurus 29, 31
Nimravidae 259
nocturnalisation
　thermoregulation hypothesis
　　124–5
North American Land Mammal
　　Ages (NALMA) 7
Notharctus 269
Notioprogonia 245
Notoryctemorphia 192–3, 213
Notoryctes 194, 213
Notostylops 246
Notoungulata 245
Numbats
　See Myrmecobiidae
Numidotherium 255, 256

Obdurodon 176, *177*
Ocher/Isheevo fauna 38
Oedaleops 20, *21*, 22
Oligokyphus 71, *72*, 76
Oliveria 55, 57
Omomyoidea 269, 271
Onychodectes 238, *239*
Ophiacodon 23–4
Ophiacodontidae 22–4
Oreodontidae 264
Ornithorhynchus 174, 176
Orycteropus afer 257
Otsheria 39, 42
Otsheria netzvetajevi 41, 42
Oudenodon 46, 49

Overkill hypothesis 289, 290
Oxyaena 248, *249*
Oxyaenidae 248, 250

Pachygenelus 72, *73*, 74, 75
 feeding mechanism 100
Pachyrukhos 246
Paenungulata 222, 252–3
Pakicetus 265, *266*
Palaeanodonta 247
Palaeoamasia 254
Palaeomastodon 256
Palaeoryctida 229–30
Palaeotheriidae 260
Palaeotherium 262
Paleoparadoxia 254, 257
Paleothyris 16
Palorchestes 213
Palorchestidae 213
Pantodonta 238, *239*, 240
Pantolestida 231
Pappictidops 247, 250
Pappotherium 173, *174*
Parabradyosaurus 30
Paraceratherium 262
Parageogale 259
Paramys 273
parental provision hypothesis 126
Pariadens 196
Parictis 260
Pascualgnathus 69
Patagonia 208
Patene 203
Patriofelis 249
Patriomanis 268
Patronomodon 40, *41*, 79
Paucituberculata 192, 204–8
Paucitubulidentata 194
Paulacoutoia 243
Paulchoffatia 153, *154*
Paurodontidae 166
Pediomyidae 198
Pediomys 197, 198
Pelycosaurs 19–26, 27
 feeding mechanism 90, 92
 forelimb musculature *102*
 hindlimb musculature *105*
 jaw musculature *91*
 palaeoecology 80–3
 phylogeny *20*
Peradectes 199, 201
Peramelemorphia 193, 212
Peramus 161, 178
 dentition 167, *168*
Periptychids 236
Perissodactyla 260–2
 cladogram *261*

Perrodelphys 206
Petauridae 216
Pezosiren 254, 256
Phalanger 206
Phalangeridae 216
Phanerozoic
 stratigraphy chart *10*
Phascolotherium 152
Phascolarctidae 216
Phenacodontidae 238, 260
Phenacodus 237, 238
Phenacolophids 260
Pholidota 222, 267–8
Phorusrhacus 218
Phosphatherium 253, *255*
Phthinosuchus 29, 30
Phyllophaga 251
Physiological constraints
 hypothesis 185–6
Pilgrimella 252
Pilkipildridae 216
Pinnipedia 260
Placentals 1, 169–71, 173
 Cretaceous radiation 274–9
 Divergent dates 275
 Eocene radiation 280–3
 fossil record
  incompleteness of 275
 interrelationships 222–226
 Miocene radiation 283–4
 Molecular phylogeny 224–226
 Palaeocene radiation 279–80
 Pleistocene radiation 284–90
 reproduction 190
Placerias 52
Plagiaulacida 154
Plagiaulacoidea 154
Plagiaulax 153
Plagiomene 272
Platycraniellus 63, 64
Platyrrini 271
Plesiadapiformes 232–4, 269
Plesiadapis 233, 234
Plesiorycteropus 257, 289
Polydolopoids 204, 206
Polydolops 204, 206
Preadaptation
 definition 16
Primates 269–72
 phylogeny *270*
Priscileo 213
Pristerodon 49
Pristinictis 249, 250
Pristerodontoidea 48–51
Pristerognathidae 56, 57
Pristerognathus 55
Pristinictis 247

Proailurus 258, 259
Proargyrolagus 207
Probainognathia 66, 69–70
Probainognathus 70, 78
 brain *119*
 skull *66*
Probelesodon 70
Proborhyaena 204
Proboscidea 252, 253, *255-6*
Proburnetia 32, 33
Proconsul 272
Procynosuchia 60–2
Procynosuchian grade
 feeding mechanism 95, 97
Procynosuchus 60–2, 64, 75
 brain *119*
 feeding mechanism 95, 97
 jaw musculature *96*
 locomotion 110
Prodesmatochoerus 263
Prodinoceros 240, *241*
Prokennalestes 169, *170*, 228
Prolicaphrium 243
Proliopithecoidea 272
Prorastomus 256
Prosimians 269, 271
Proteopithecus 271
Prostegotherium 250
Prothylacinus 203, 204
Protictis 249
Protoclepsydrops 16
Protorothyridids 16, 18
Protorothyris 17
Protoungulatum 230, 234, *235*
Prozostrodon 74
Ptilodontoidea 154, 155
Ptilodus
 brain 160
 dentition 155–6
 locomotion 158
 skull *153*
Pucadelphys 199, *200*, 201, 203
Pucanodus 242
Purgatorius 189, 232–3
Pyrotheria 242, 245
Pyrotherium 243, *244*

Radinskya 260, *261*
Radiocarbon dating 7
Radiometric dating 7, 9
Ragnarok 236
Ravenictis 247, 250
Regisaurid 56
Regisaurus 57
Reigitherium 147
Reiszia 29
Repenomamus 152

Rhinocerotoids 262
Rhopaladonta 30
Riebeekosaurus 38
Riograndia 73, 74
Roberthoffsteteria
    nationalgeographica 204
Robertia 44, 45, 48
Rodentia 273–4
Rodhocetus 265, *266*
Rubidgeinae 52
Ruminantia 264
Rusconodon 69

Saharagalago 271
Sarcophilus 212
Scaloposaurs 57
Scalaposaurus 58
Scandentia 272–3
Scylacosauridae 57
Secodontosaurus 25, 26
Selenodontia 264
Seymourids 18
Shansiodon 51
Shuotherium 166, 178
Siamopithecus 271
Sillustania 199
Sinclairella 232
Sinoconodon 3, 141–2, 145, 180, 183
    growth and development 121
    skull *141*
    tooth replacement *121*
Sinodelphys 171, *172*, 216, 274
Sinognathus 64
Sinomylus 273
Sinopa 249, 250
Sirenia 251, 256
Smilodon 259
Solendon 268
Solenodontidae 268
Soricidae 268
Spalacotherium 163–4
Sparassocynidae 201
Sparassocynus 200, 201
Sparassodontia 201–4
Sphenacodon 25, 26
Sphenacodontia 24–6
    feeding mechanism 92–3
    locomotion 101
Sphenacodontidae *20*, 26
    Therapsids
        common characteristics 27
Sphenacodontine grade
    feeding mechanism 92–3
Stagodontidae 198
Stahleckeria 50, 51
Stahleckeriines 51–2

Steropodon 176, *177*, 178, 180
Strepsirhini 269, 271
Struthiocephalus 36
Stylinodon 238, *239*
Styracocephalus 36, 38
Sudamerica 152, *153*, 154
Suidae 264
Suiformes 264
Suminia 41, 42
Symmetrodonta 162–4, 166
Synapsida 9, 16, 19, 88
    characteristics 19
    cladogram *89*
    classification 11, *12*
    definition 1
    palaeoecology 80–3
Syodon 35
Szalinia 199

Tachyglossus 174
Taeniodonta 238, *239*
Taeniolabidoidea 154
Talpidae 268
Tapinocephalidae 38
Tapinocephalus 38
Tapiroids 262
Tarsipedidae 216
Tarsius 271
Tayassuidae 264
Teilhardina 269
Teinolophos 176, *177*, 180
Tenrecida 259
    thermoregulation 124
Tethytheria 222
Tetraceratops 27–8, *30*, 83
Tetragonius 50, 51
Tetrapods 14–6, 18
    evolution 14–16
    terrestrial adaptation 15, 19
Therapsida 26–75
    cladogram *78*
    extinction 84–7
    interrelationships 78–80
    jaw musculature *93*
    limb function *107*
    palaeoecology 83–4
    Sphenacodontidae
        similar characteristics 27
Theria 2, 162
Theriodontia 79
Theriognathus 57, *58*
Therioherpeton 73, 74–5, 76
Theriomorpha 2
Theriiformes 2
Theroteinus 139, 141
Therocephalia 30, 56–9

    characteristics 56
    jaw mechanism 59
    locomotion 59
Therocephalian grade
    feeding mechanism 94–5
Thoatherium 244
Thomasia 139, 140
Thrinaxodon 63, 64, 65, 70,
        75, 86
    feeding mechanism *96*, 97–8
    locomotion 110
Thylacinidae 212
Thylacinus 212
    cynocephalus 192
Thylacoleo 213
    carnifex 215
Thylacoleonidae 213
Thylacosmilus 203, 204, 286
Thylacotinga 208–9
Thyrohyrax 253
Tillodonta 240
Timescale
    chronometric 7–8
    chronostatic 6–7
    geological 6, *8*
Tingamarra 208–9
Tinodon 163, *166*
Titanoides 239
Titanophoneus 34, 36, 37, 79
Titanosuchia 36, 37–8
Toxodon 245, *246*, 286
Traversodontidae 65, 67
Traversodontoides 59
Trechnotheria 161
Tribosphenic tooth 3, 173, 178
Tribosphenida 161, 167, 169–73
    minor groups 173
    phylogeny *175*
Tribosphenomys 273
Tribotherium 169, *174*, 228
Triconodon 150, 151, 152
    brain *119*, 120
Triconodonta 150–2
Triisodontids 236
Trinity Theria 173
Trirachodon 67
Tritheledonta 66, 72, 77
Tritheledontian grade
    feeding mechanism 100
    locomotion 112
Tritylodontid grade
    feeding mechanism 100
    locomotion 112
Tritylodontidae 66, 70, 76–7
Trogosus 240, *241*
Tubulidentata 222, 257

Tulerpeton 14, 18
Tylopoda 264

Ukhaatherium 229
Uintatherium 240, *241*
Ulemica 41, 42, 79
  invisa 40, *41*
Ulemosaurus 35, *36*, 38

Varanodon 22, 26
Varanopseidae 22, 90
Varanosaurus 22, 24
Venyukovia 39, 42
  invisa 40
  prima 40, 42
Vermilingua 251

Victoriapithecus 272
Victorlemoinea 245
Vincelestes 167, *168*
Vinceria 51
Vombatidae 216

Wakaleo 213, *215*
Watongia 30
Westlothiana *17*, 18, 90
Whaitsiidae 60
whales
  hippopotamus relationship 224,
    264, 267
Widanelfasia 257, 259
Woutersia *164*, 166
Wynyardiidae 213

Xenarthra 222, 225–6, 250–1
Xenungulata 242, 247

Yalkaparidon 210, *211*, 212
Yalkaparidontia 210–12
Yarala *211*, 212

Zaglossus 174
Zalambdalestes 171
Zalambdalestidae 229
Zambiasaurus 51
Zatheria 161
Zhangheotherium *164*, 166
Zhelestidae 227, 230
Zygolophodon 256